Springer Optimization and Its Applications

VOLUME 62

Aims and Scope
Optimization has been expanding in all directions at an astonishing rate during the last few decades. New algorithmic and theoretical techniques have been developed, the diffusion into other disciplines has proceeded at a rapid pace, and our knowledge of all aspects of the field has grown even more profound. At the same time, one of the most striking trends in optimization is the constantly increasing emphasis on the interdisciplinary nature of the field. Optimization has been a basic tool in all areas of applied mathematics, engineering, medicine, economics, and other sciences.

The series *Springer Optimization and Its Applications* publishes undergraduate and graduate textbooks, monographs and state-of-the-art expository work that focus on algorithms for solving optimization problems and also study applications involving such problems. Some of the topics covered include nonlinear optimization (convex and nonconvex), network flow problems, stochastic optimization, optimal control, discrete optimization, multi-objective programming, description of software packages, approximation techniques and heuristic approaches.

For further volumes:
http://www.springer.com/series/7393

Ding-Zhu Du • Ker-I Ko • Xiaodong Hu

Design and Analysis of Approximation Algorithms

Springer

Ding-Zhu Du
Department of Computer Science
University of Texas at Dallas
Richardson, TX 75080
USA
dzdu@utdallas.edu

Ker-I Ko
Department of Computer Science
State University of New York at Stony Brook
Stony Brook, NY 11794
USA
keriko@cs.sunysb.edu

Xiaodong Hu
Institute of Applied Mathematics
Academy of Mathematics and Systems Science
Chinese Academy of Sciences
Beijing 100190
China
xdhu@amss.ac.cn

ISSN 1931-6828
ISBN 978-1-4899-9844-6 ISBN 978-1-4614-1701-9 (eBook)
DOI 10.1007/978-1-4614-1701-9
Springer New York Dordrecht Heidelberg London

Printed on acid-free paper

Springer is part of Springer Science+Business Media (www.springer.com)

Preface

An approximation algorithm is an efficient algorithm that produces solutions to an optimization problem that are guaranteed to be within a fixed ratio of the optimal solution. Instead of spending an exponential amount of time finding the optimal solution, an approximation algorithm settles for near-optimal solutions within polynomial time in the input size. Approximation algorithms have been studied since the mid-1960s. Their importance was, however, not fully understood until the discovery of the **NP**-completeness theory. Many well-known optimization problems have been proved, under reasonable assumptions in this theory, to be intractable, in the sense that optimal solutions to these problems are not computable within polynomial time. As a consequence, near-optimal approximation algorithms are the best one can expect when trying to solve these problems.

In the past decade, the area of approximation algorithms has experienced an explosive rate of growth. This growth rate is partly due to the development of related research areas, such as data mining, communication networks, bioinformatics, and computational game theory. These newly established research areas generate a large number of new, intractable optimization problems, most of which have direct applications to real-world problems, and so efficient approximate solutions to them are actively sought after.

In addition to the external, practical need for efficient approximation algorithms, there is also an intrinsic, theoretical motive behind the research of approximation algorithms. In the design of an exact-solution algorithm, the main, and often only, measure of the algorithm's performance is its running time. This fixed measure often limits our choice of techniques in the algorithm's design. For an approximation algorithm, however, there is an equally important second measure, that is, the performance ratio of the algorithm, which measures how close the approximation algorithm's output is to the optimal solution. This measure adds a new dimension to the design and analysis of approximation algorithms. Namely, we can now study the tradeoff between the running time and the performance ratio of approximation algorithms, and apply different design techniques to achieve different tradeoffs between these two measures. In addition, new theoretical issues about the approximation to an optimization problem need to be addressed: What is the performance ratio of an approximation algorithm for this problem based on certain types of design strategy? What is the best performance ratio of any polynomial-time approximation algorithm for this problem? Does the problem have a polynomial-time approximation scheme or a fully polynomial-time approximation scheme? These questions are not only of significance in practice for the design of approximation algorithms; they are also of great theoretical interest, with intriguing connections to the **NP**-completeness theory.

Motivated by these theoretical questions and the great number of newly discovered optimization problems, people have developed many new design techniques for approximation algorithms, including the greedy strategy, the restriction method, the relaxation method, partition, local search, power graphs, and linear and semidefinite programming. A comprehensive survey of all these methods and results in a single book is not possible. We instead provide in this book an intensive study of the main methods, with abundant applications following our discussion of each method. Indeed, this book is organized according to design methods instead of application problems. Thus, one can study approximation algorithms of the same nature together, and learn about the design techniques in a more unified way. To this end, the book is arranged in the following way: First, in Chapter 1, we give a brief introduction to the concept of **NP**-completeness and approximation algorithms. In Chapter 2, we give an in-depth analysis of the greedy strategy, including greedy algorithms with submodular potential functions and those with nonsubmodular potential functions. In Chapters 3, 4, and 5, we cover various restriction methods, including partition and Guillotine cut methods, with applications to many geometric problems. In the next four chapters, we study the relaxation methods. In addition to a general discussion of the relaxation method in Chapter 6, we devote three chapters to approximation algorithms based on linear and semidefinite programming, including the primal-dual schema and its equivalence with the local ratio method. Finally, in Chapter 10, we present various inapproximability results based on recent work in the **NP**-completeness theory. A number of examples and exercises are provided for each design technique. They are drawn from diverse areas of research, including communication network design, optical networks, wireless ad hoc networks, sensor networks, bioinformatics, social networks, industrial engineering, and information management systems.

This book has grown out of lecture notes used by the authors at the University of Minnesota, University of Texas at Dallas, Tsinghua University, Graduate School of Chinese Academy of Sciences, Xi'an Jiaotong University, Zhejiang University, East China Normal University, Dalian University of Technology, Xinjiang University, Nankai University, Lanzhou Jiaotong University, Xidian University, and Harbin Institute of Technology. In a typical one-semester class for first-year graduate students, one may cover the first two chapters, one or two chapters on the restriction method, two or three chapters on the relaxation method, and Chapter 10. With more advanced students, one may also teach a seminar course focusing on one of the greedy, restriction, or relaxation methods, based on the corresponding chapters of this book and supplementary material from recent research papers. For instance, a seminar on combinatorial optimization emphasizing approximations based on linear and semidefinite programming can be organized using Chapters 7, 8, and 9.

This book has benefited much from the help of our friends, colleagues, and students. We are indebted to Peng-Jun Wan, Weili Wu, Xiuzhen Cheng, Jie Wang, Yinfeng Xu, Zhao Zhang, Deying Li, Hejiao Huang, Hong Zhu, Guochuan Zhang, Wei Wang, Shugang Gao, Xiaofeng Gao, Feng Zou, Ling Ding, Xianyue Li, My T. Thai, Donghyun Kim, J. K. Willson, and Roozbeh Ebrahimi Soorchaei, who made much-valued suggestions and corrections to the earlier drafts of the book. We are also grateful to Professors Frances Yao, Richard Karp, Ronald Graham, and Fan Chung for their encouragement. Special thanks are due to Professor Andrew Yao and the Institute for Theoretical Computer Science, Tsinghua University, for the generous support and stimulating environment they provided for the first two authors during their numerous visits to Tsinghua University.

Dallas, Texas Ding-Zhu Du
Stony Brook, New York Ker-I Ko
Beijing, China Xiaodong Hu
August 2011

Contents

1
Introduction

It is the mark of an educated mind to rest satisfied with the degree of precision which the nature of the subject admits and not to seek exactness where only an approximation is possible.

— Aristotle

A man only becomes wise when he begins to calculate the approximate depth of his ignorance.

— Gian Carlo Menotti

When exact solutions are hard to compute, approximation algorithms can help. In this chapter, we introduce the basic notions of approximation algorithms. We study a simple optimization problem to demonstrate the tradeoff between the time complexity and performance ratio of its approximation algorithms. We also present a brief introduction to the general theory of computational complexity and show how to apply this theory to classify optimization problems according to their approximability.

1.1 Open Sesame

As legend has it, Ali Baba pronounced the magic words "open sesame" and found himself inside the secret cave of the Forty Thieves, with all their precious treasures laid before him. After the initial excitement subsided, Ali Baba quickly realized that he had a difficult optimization problem to solve: He had only brought a single knapsack with him. Which items in the cave should he put in the knapsack in order to maximize the total value of his find?

In modern terminology, what Ali Baba faced is a *resource management problem*. In this problem, one is given a fixed amount S of resources (the total volume of the knapsack) and a set of n tasks (the collection of treasures in the cave). Completing each task requires a certain amount of resources and gains a certain amount of profit. The problem is to maximize the total profit, subject to the condition that the total resources used do not exceed S. Formally, we can describe Ali Baba's problem as follows:

> Given n items $I_1, I_2, \ldots, I_n$, a volume s_i and a value c_i for each item I_i, $1 \le i \le n$, and an integer S, find a subset A of items that maximizes the total value $\sum_{I_i \in A} c_i$, subject to the condition that the total volume $\sum_{I_i \in A} s_i$ does not exceed S.

We can introduce, for each $1 \le i \le n$, a 0–1 variable x_i to represent item I_i in the following sense:

$$x_i = \begin{cases} 1, & \text{if } I_i \in A, \\ 0, & \text{if } I_i \notin A. \end{cases}$$

Then, Ali Baba's problem can be reformulated as a 0–1 integer programming problem:

> KNAPSACK: Given $2n + 1$ positive integers S, $s_1, s_2, \ldots, s_n$ and $c_1, c_2, \ldots, c_n$,
>
> $$\begin{aligned} \text{maximize} \quad & c(\boldsymbol{x}) = c_1x_1 + c_2x_2 + \cdots + c_nx_n, \\ \text{subject to} \quad & s_1x_1 + s_2x_2 + \cdots + s_nx_n \le S, \\ & x_1, x_2, \ldots, x_n \in \{0, 1\}. \end{aligned}$$

Notation. (1) In this book, we will use the following notation about an optimization problem Π: On an input instance I of Π, we write *Opt*(I) to denote the optimal solution of the instance I, and *opt*(I) to denote the optimum value of the objective function on input I. When there is no confusion, we write *Opt* and *opt* for *Opt*(I) and *opt*(I), respectively. In addition, for convenience, we often write, for an objective function $f(\boldsymbol{x})$, f^* to denote the optimum value of the function f, and $\boldsymbol{x}^*$ to denote the value of $\boldsymbol{x}$ that achieves the optimum value f^*. For instance, for the problem KNAPSACK above, we write *opt* or c^* to denote the maximum value of $c(\boldsymbol{x})$ under the given constraints, and *Opt* or $\boldsymbol{x}^*$ to denote the value of $(x_1, x_2, \ldots, x_n)$ that makes $\sum_{i=1}^{n} c_ix_i = c^*$.

(2) For the sets of numbers, we write $\mathbb{N}$ to denote the set of natural numbers (i.e., the set of nonnegative integers), $\mathbb{Z}$ the set of integers, $\mathbb{Z}^+$ the set of positive integers, $\mathbb{R}$ the set of real numbers, and $\mathbb{R}^+$ the set of positive integers.

Following the above convention, let *opt* denote the optimum value of the objective function $c(\boldsymbol{x})$. Without loss of generality, we may assume that $s_k \le S$ for all

$k = 1, \ldots, n$. In fact, if $s_k > S$, then we must have $x_k = 0$, and so we need not consider the kth item at all. This assumption implies that $opt \geq \max_{1 \leq k \leq n} c_k$.

There are many different approaches to attacking the KNAPSACK problem. First, let us use the dynamic programming technique to find the exact solutions for KNAPSACK.

To simplify the description of the algorithm, we first define some notations. For any subset $I \subseteq \{1, \ldots, n\}$, let S_I denote the sum $\sum_{k \in I} s_k$. For each pair (i, j), with $1 \leq i \leq n, 0 \leq j \leq \sum_{i=1}^{n} c_i$, if there exists a set $I \subseteq \{1, 2, \ldots, n\}$ such that $\sum_{k \in I} c_k = j$ and $S_I \leq S$, then let $a(i, j)$ denote such a set I with the minimum S_I. If such an index subset I does not exist, then we say that $a(i, j)$ is undefined, and write $a(i, j) = nil$.

Using the above notation, it is clear that $opt = \max\{j \mid a(n, j) \neq nil\}$. Therefore, it suffices to compute all values of $a(i, j)$. The following algorithm is based on this idea.[1]

Algorithm 1.A (*Exact Algorithm for* KNAPSACK)

Input: Positive integers $S, s_1, s_2, \ldots, s_n, c_1, c_2, \ldots, c_n$.

(1) Let $c_{sum} \leftarrow \sum_{i=1}^{n} c_i$.

(2) **For** $j \leftarrow 0$ **to** c_{sum} **do**
  if $j = 0$ **then** $a(1, j) \leftarrow \emptyset$
    else if $j = c_1$ **then** $a(1, j) \leftarrow \{1\}$ **else** $a(1, j) \leftarrow nil$.

(3) **For** $i \leftarrow 2$ **to** n **do**
  for $j \leftarrow 0$ **to** c_{sum} **do**
   if $[a(i-1, j-c_i) \neq nil]$ and $[S_{a(i-1,j-c_i)} \leq S - s_i]$
    and $[a(i-1, j) \neq nil \Rightarrow S_{a(i-1,j)} > S_{a(i-1,j-c_i)} + s_i]$
     then $a(i, j) \leftarrow a(i-1, j-c_i) \cup \{i\}$
     else $a(i, j) \leftarrow a(i-1, j)$.

(4) Output $c^* \leftarrow \max\{j \mid a(n, j) \neq nil\}$. ■

It is not hard to verify that this algorithm always finds the optimal solutions to KNAPSACK (see Exercise 1.1).

Next, we consider the time complexity of Algorithm 1.A. Since Ali Baba had to load the treasures and leave the cave before the Forty Thieves came back, he needed an efficient algorithm. It is easy to see that, for any $I \subseteq \{1, \ldots, n\}$, it takes time $O(n \log S)$ to compute S_I.[2] Thus, Algorithm 1.A runs in time $O(n^3 M \log(MS))$ where $M = \max\{c_k \mid 1 \leq k \leq n\}$ (note that $c_{sum} = O(nM)$). We note that

[1] We use the standard pseudocodes to describe an algorithm; see, e.g., Cormen et al. [2001].

[2] In the rest of the book, we write $\log k$ to denote $\log_2 k$.

the input size of the problem is $n \log M + \log S$ (assuming that the input integers are written in the binary form). Therefore, Algorithm 1.A is *not* a polynomial-time algorithm. It is actually a *pseudo-polynomial-time* algorithm, in the sense that it runs in time polynomial in the maximum input value but not necessarily polynomial in the input size. Since the input value could be very large, a pseudo polynomial-time algorithm is usually not considered as an efficient algorithm. To be sure, if Ali Baba tried to run this algorithm, then the Forty Thieves would definitely have come back before he got the solution—even if he could calculate as fast as a modern digital computer.

As a compromise, Ali Baba might find a fast approximation algorithm more useful. For instance, the following is such an approximation algorithm, which uses a simple greedy strategy that selects the *heaviest* item (i.e., the item with the greatest *density* c_i/s_i) first.

Algorithm 1.B (*Greedy Algorithm for* KNAPSACK)

Input: Positive integers $S, s_1, s_2, \ldots, s_n, c_1, c_2, \ldots, c_n$.

(1) Sort all items in the nonincreasing order of c_i/s_i. Without loss of generality, assume that $c_1/s_1 \geq c_2/s_2 \geq \cdots \geq c_n/s_n$.

(2) **If** $\sum_{i=1}^{n} s_i \leq S$ **then** output $c_G \leftarrow \sum_{i=1}^{n} c_i$

$$\textbf{else}\ \ k \leftarrow \max\left\{ j \;\middle|\; \sum_{i=1}^{j} s_i \leq S < \sum_{i=1}^{j+1} s_i \right\};$$

$$\text{output } c_G \leftarrow \max\left\{ c_{k+1}, \sum_{i=1}^{k} c_i \right\}. \qquad \blacksquare$$

It is clear that this greedy algorithm runs in time $O(n \log(nMS))$ and hence is very efficient. The following theorem shows that it produces an approximate solution not very far from the optimum.

Theorem 1.1 *Let opt be the optimal solution of the problem* KNAPSACK *and* c_G *the approximate solution obtained by* Algorithm 1.B. *Then opt* $\leq 2c_G$ *(and we say that the* performance ratio *of* Algorithm 1.B *is bounded by the constant* 2*).*

Proof. For convenience, write c^* for *opt*. If $\sum_{i=1}^{n} s_i \leq S$, then $c_G = c^*$. Thus, we may assume $\sum_{i=1}^{n} s_i > S$. Let k be the integer found by Algorithm 1.B in step (2). We claim that

$$\sum_{i=1}^{k} c_i \leq c^* < \sum_{i=1}^{k+1} c_i. \tag{1.1}$$

The first half of the above inequality holds trivially. For the second half, we note that, in step (1), we sorted the items according to their density, c_i/s_i. Therefore, if we are allowed to cut each item into smaller pieces, then the most efficient way of using the knapsack is to load the first k items, plus a portion of the $(k+1)$st item that fills the knapsack, because replacing any portion of these items by other items decreases the total density of the knapsack. This shows that the maximum total value c^* we can get is less than $\sum_{i=1}^{k+1} c_i$.

We can also view the above argument in terms of linear programming. That is, if we replace the constraints $x_i \in \{0,1\}$ by $0 \le x_i \le 1$, then we obtain a linear program which has the maximum objective function value $\hat{c} \ge c^*$. It is easy to check that the following assignment is an optimal solution to this linear program[3]:

$$x_j = \begin{cases} 1, & \text{for } j = 1, 2, \ldots, k, \\ (S - \sum_{i=1}^{k} s_i)/s_{k+1}, & \text{for } j = k+1, \\ 0, & \text{for } j = k+2, \ldots, n. \end{cases}$$

Therefore,

$$c^* \le \hat{c} = \sum_{i=1}^{k} c_i + \frac{c_{k+1}}{s_{k+1}}\left(S - \sum_{i=1}^{k} s_i\right) < \sum_{i=1}^{k} c_i + \frac{c_{k+1}}{s_{k+1}} s_{k+1} = \sum_{i=1}^{k+1} c_i.$$

Finally, it is obvious that, from (1.1), we have

$$c_G = \max\left\{c_{k+1}, \sum_{i=1}^{k} c_i\right\} \ge \frac{1}{2}\sum_{i=1}^{k+1} c_i > \frac{c^*}{2}. \qquad \square$$

The above two algorithms demonstrate an interesting tradeoff between the running time and the accuracy of an algorithm: If we sacrifice a little in the accuracy of the solution, we may get a much more efficient algorithm. Indeed, we can further explore this idea of tradeoff and show a spectrum of approximation algorithms with different running time and accuracy.

First, we show how to generalize the above greedy algorithm to get better approximate solutions—with worse, but still polynomial, running time. The idea is as follows: We divide all items into two groups: those with values $c_i \le a$ and those with $c_i > a$, where a is a fixed parameter. Note that in any feasible solution $I \subseteq \{1, 2, \ldots, n\}$, there can be at most $opt/a \le 2c_G/a$ items that have values c_i greater than a. So we can perform an exhaustive search over all index subsets $I \subseteq \{1, 2, \ldots, n\}$ of size at most $2c_G/a$ from the second group as follows: For each subset I, use the greedy strategy on the first group to get a solution of the total volume no greater than $S - S_I$, and combine it with I to get an approximate solution. From Theorem 1.1, we know that our error is bounded by the value of a single item of the first group, which is at most a. In addition, we note that there are at most $n^{2c_G/a}$ index subsets of the second group to be searched through, and so the running time is still a polynomial function in the input size.

In the following, we write $|A|$ to denote the size of a finite set A.

Algorithm 1.C (*Generalized Greedy Algorithm for* KNAPSACK)

Input: Positive integers $S, s_1, s_2, \ldots, s_n, c_1, c_2, \ldots, c_n$, and a constant $0 < \varepsilon < 1$.

[3] See Chapter 7 for a more complete treatment of linear programming.

(1) Run Algorithm 1.B on the input to get value c_G.

(2) Let $a \leftarrow \varepsilon c_G$.

(3) Let $I_a \leftarrow \{i \mid 1 \le i \le n, c_i \le a\}$. (Without loss of generality, assume that $I_a = \{1, \ldots, m\}$, where $m \le n$.)

(4) Sort the items in I_a in the nonincreasing order of c_i/s_i. Without loss of generality, assume that $c_1/s_1 \ge c_2/s_2 \ge \cdots \ge c_m/s_m$.

(5) **For** each $I \subseteq \{m+1, m+2, \ldots, n\}$ with $|I| \le 2/\varepsilon$ **do**

$$\textbf{if } \sum_{i \in I} s_i > S \textbf{ then } c(I) \leftarrow 0$$

$$\textbf{else if } \sum_{i=1}^{m} s_i \le S - \sum_{i \in I} s_i$$

$$\textbf{then } c(I) \leftarrow \sum_{i=1}^{m} c_i + \sum_{i \in I} c_i$$

$$\textbf{else } \; k \leftarrow \max\left\{ j \;\middle|\; \sum_{i=1}^{j} s_i \le S - \sum_{i \in I} s_i < \sum_{i=1}^{j+1} s_i \right\};$$

$$c(I) \leftarrow \sum_{i=1}^{k} c_i + \sum_{i \in I} c_i.$$

(6) Output $c_{GG} \leftarrow \max\{c(I) \mid I \subseteq \{m+1, m+2, \ldots, n\}, |I| \le 2/\varepsilon\}$. ■

Theorem 1.2 *Let opt be the optimal solution to* KNAPSACK *and* c_{GG} *the approximation obtained by* Algorithm 1.C. *Then* $opt \le (1+\varepsilon)c_{GG}$. *Moreover,* Algorithm 1.C *runs in time* $O(n^{1+2/\varepsilon}\log(nMS))$.

Proof. For convenience, write $c^* = opt$ and let $I^* = Opt$ be the optimal index set; that is, $\sum_{i \in I^*} c_i = c^*$ and $\sum_{i \in I^*} s_i \le S$. Define $\overline{I} = \{i \in I^* \mid c_i > a\}$. We have already shown that $|\overline{I}| \le c^*/a \le 2c_G/a = 2/\varepsilon$. Therefore, in step (5) of Algorithm 1.C, the index set I will eventually be set to $\overline{I}$. Then, the greedy strategy, as shown in the proof of Theorem 1.1, will find $c(\overline{I})$ with the property

$$c(\overline{I}) \le c^* \le c(\overline{I}) + a.$$

Since c_{GG} is the maximum $c(I)$, we get

$$c(\overline{I}) \le c_{GG} \le c^* \le c(\overline{I}) + a \le c_{GG} + a.$$

Let I_G denote the set obtained by Algorithm 1.B on the input. Let $\overline{I}_G = \{i \in I_G \mid c_i > a\}$. Then $|\overline{I}_G| \le c_G/a = 1/\varepsilon$. So, we will process set $\overline{I}_G$ in step (5) and get $c(\overline{I}_G) = c_G$. It means $c_{GG} \ge c_G$, and so

$$c^* \le c_{GG} + a = c_{GG} + \varepsilon c_G \le (1+\varepsilon)c_{GG}.$$

Note that there are at most $n^{2/\varepsilon}$ index sets I of size $|I| \le 2/\varepsilon$. Therefore, the running time of Algorithm 1.C is $O(n^{1+2/\varepsilon}\log(nMS))$. □

By Theorem 1.2, for any fixed $\varepsilon > 0$, Algorithm 1.C runs in time $O(n^{1+2/\varepsilon}\log(nMS))$ and hence is a polynomial-time algorithm. As ε decreases to zero, however, the running time increases exponentially with respect to $1/\varepsilon$. Can we slow down the speed of increase of the running time with respect to $1/\varepsilon$? The answer is yes. The following is such an approximation algorithm:

Algorithm 1.D (*Polynomial Tradeoff Approximation for* KNAPSACK)
Input: Positive integers $S, s_1, s_2, \ldots, s_n, c_1, c_2, \ldots, c_n$, and an integer $h > 0$.

(1) **For** $k \leftarrow 1$ **to** n **do**

$$c'_k \leftarrow \left\lfloor \frac{c_k n(h+1)}{M} \right\rfloor, \text{ where } M = \max_{1\le i\le n} c_i.$$

(2) Run Algorithm 1.A on the following instance of KNAPSACK:

$$\begin{aligned} &\text{maximize} && c'_1x_1 + c'_2x_2 + \cdots + c'_nx_n \\ &\text{subject to} && s_1x_1 + s_2x_2 + \cdots + s_nx_n \le S, \\ & && x_1, x_2, \ldots, x_n \in \{0,1\}. \end{aligned} \tag{1.2}$$

Let $(x_1^*, \ldots, x_n^*)$ be the optimal solution found by Algorithm 1.A (i.e., the index set corresponding to the optimum value $opt' = (c')^*$ of (1.2)).

(3) Output $c_{PT} \leftarrow c_1x_1^* + \cdots + c_nx_n^*$. ■

Theorem 1.3 *The solution obtained by* Algorithm 1.D *satisfies the relationship*

$$\frac{opt}{c_{PT}} \le 1 + \frac{1}{h},$$

where opt is the optimal solution to the input instance.

Proof. For convenience, let $c^* = opt$ and $I^* = Opt$ be the optimal index set of the input instance; that is, $c^* = \sum_{k\in I^*} c_k$. Also, let J^* be the index set found in step (2); that is, $J^* = \{k \mid 1 \le k \le n, x_k^* = 1\}$. Then, we have

$$\begin{aligned} c_{PT} &= \sum_{k\in J^*} c_k = \sum_{k\in J^*} \frac{c_k n(h+1)}{M} \cdot \frac{M}{n(h+1)} \\ &\ge \sum_{k\in J^*} \left\lfloor \frac{c_k n(h+1)}{M} \right\rfloor \cdot \frac{M}{n(h+1)} \\ &= \frac{M}{n(h+1)} \sum_{k\in J^*} c'_k \ge \frac{M}{n(h+1)} \sum_{k\in I^*} c'_k \\ &\ge \frac{M}{n(h+1)} \sum_{k\in I^*} \left(\frac{c_k n(h+1)}{M} - 1 \right) \\ &\ge c^* - \frac{M}{h+1} \ge c^* \left(1 - \frac{1}{h+1} \right). \end{aligned}$$

In the above, the second inequality holds because J^* is the optimal solution to the modified instance of KNAPSACK; and the last inequality holds because $M = \max_{1\le i\le n}\{c_i\} \le c^*$. Thus,

$$\frac{c^*}{c_{PT}} \le \frac{1}{1-1/(h+1)} = 1 + \frac{1}{h}. \qquad \square$$

We note that in step (2), the running time for Algorithm 1.A on the modified instance is $O(n^3 M' \log(M'S))$, where $M' = \max\{c'_k \mid 1 \le k \le n\} \le n(h+1)$. Therefore, the total running time of Algorithm 1.D is $O(n^4 h \log(nhS))$, which is a polynomial function with respect to n, $\log S$, and $h = 1/\varepsilon$. Thus, the tradeoff between running time and approximation ratio of Algorithm 1.D is better than that of the generalized greedy algorithm.

From the above analysis, we learned that if we turn our attention from the optimal solutions to the approximate solutions, then we may find many new ideas and techniques to attack the problem. Indeed, the design and analysis of approximation algorithms are very different from that of exact (or, optimal) algorithms. It is a cave with a mother lode of hidden treasures. Let us say "Open Sesame" and find out what they are.

1.2 Design Techniques for Approximation Algorithms

What makes the design and analysis of approximation algorithms so different from that of algorithms that search for exact solutions?[4]

First, they study different types of problems. Algorithms that look for exact solutions work only for tractable problems, but approximation algorithms apply mainly to intractable problems. By tractable problems, we mean, in general, problems that can be solved exactly in polynomial time in the input size. While tractable problems, such as the minimum spanning-tree problem, the shortest-path problem, and maximum matching are the main focus of most textbooks for algorithms, most intractable problems are not discussed in these books. On the other hand, a great number of problems we encounter in the research literature, such as the traveling salesman problem, scheduling, and integer programming, are intractable. That is, no polynomial-time exact algorithms have been found for them so far. In addition, through the study of computational complexity theory, most of these problems have proven unlikely to have polynomial-time exact algorithms at all. Therefore, approximation algorithms seem to be the only resort.

Second, and more importantly, they emphasize different aspects of the performance of the algorithms. For algorithms that look for exact solutions, the most important issue is the efficiency, or the running time, of the algorithms. Data structures and design techniques are introduced mainly to improve the running time. For approximation algorithms, the running time is, of course, still an important issue. It,

[4] We call such algorithms *exact algorithms*.

however, has to be considered together with the performance ratio (the estimate of how close the approximate solutions are to the optimal solutions) of the algorithms. As we have seen in the study of the KNAPSACK problem, the tradeoff between the running time and performance ratio is a critical issue in the analysis of approximation algorithms. Many design techniques for approximation algorithms aim to improve the performance ratio with the minimum extra running time.

To illustrate this point, let us take a closer look at approximation algorithms. First, we observe that, in general, an optimization problem may be formulated in the following form:

$$\begin{array}{ll} \text{minimize (or, maximize)} & f(x_1, x_2, \ldots, x_n) \\ \text{subject to} & (x_1, x_2, \ldots, x_n) \in \Omega, \end{array} \tag{1.3}$$

where f is a real-valued function and Ω a subset of $\mathbb{R}^n$. We call the function f the *objective function* and set Ω the *feasible domain* (or, the *feasible region*) of the problem.

The design of approximation algorithms for such a problem can roughly be divided into two steps. In the first step, we convert the underlying intractable problem into a tractable variation by perturbing the input values, the objective function, or the feasible domain of the original problem. In the second step, we design an efficient exact algorithm for the tractable variation and, if necessary, convert its solution back to an approximate solution for the original problem. For instance, in Algorithm 1.D, we first perturb the inputs c_i into smaller c_i', and thus converted the original KNAPSACK problem into a tractable version of KNAPSACK in which the maximum parameter c_i' is no greater than $n(h+1)$. Then, in the second step, we use the technique of dynamic programming to solve the tractable version in polynomial time, and use the optimal solution $(x_1^*, x_2^*, \ldots, x_n^*)$ with the tractable version of KNAPSACK as an approximate solution to the original instance of KNAPSACK.

It is thus clear that in order to design good approximation algorithms, we must know how to perturb the original intractable problem to a tractable variation such that the solution to the tractable problem is closely related to that of the original problem. A number of techniques for such perturbation have been developed. The perturbation may act on the objective functions, as in the greedy strategy and the local search method. It may involve changes to the feasible domain, as in the techniques of restriction and relaxation. It may sometimes also perform some operations on the inputs, as in the technique of power graphs. These techniques are very different from the techniques for the design of efficient exact algorithms, such as divide and conquer, dynamic programming, and linear programming. The study of these design techniques forms an important part of the theory of approximation algorithms. Indeed, this book is organized according to the classification of these design techniques. In the following, we give a brief overview of these techniques and the organization of the book (see Figure 1.1).

In Chapter 2, we present a theory of *greedy strategies*, in which we demonstrate how to use the notions of independent systems and submodular potential functions

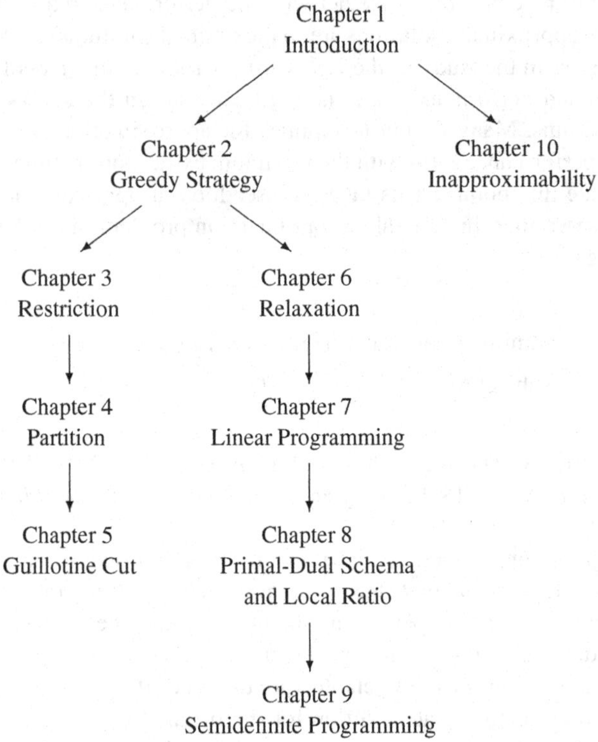

Figure 1.1: Relationships among chapters.

to analyze the performance of greedy algorithms. Due to space limits, we will omit the related but more involved method of *local search*.

The technique of *restriction* is studied in Chapters 3–5. The basic idea of restriction is very simple: If we narrow down the feasible domain, the solutions may become easier to find. There are many different ways to restrict the feasible domains, depending on the nature of the problems. We present some simple applications in Chapter 3. Two of the most important techniques of restriction, *partition* and *Guillotine cut*, are then studied in detail in Chapters 4 and 5, respectively.

In Chapters 6–9, we study the technique of *relaxation*. In contrast to restriction, the technique of relaxation is to enlarge the feasible domain to include solutions which are considered infeasible in the original problem so that different design techniques can be applied. A common implementation of the relaxation technique is as follows: First, we formulate the problem into an integer programming problem (i.e., a problem in the form of (1.3) with $\Omega \subseteq \mathbb{Z}^n$). Then, we relax this integer program into a linear program by removing the integral constraints on the variables. After we solve this relaxed linear program, we round the real-valued solution into integers and use them as the approximate solution to the original problem. *Linear programming*,

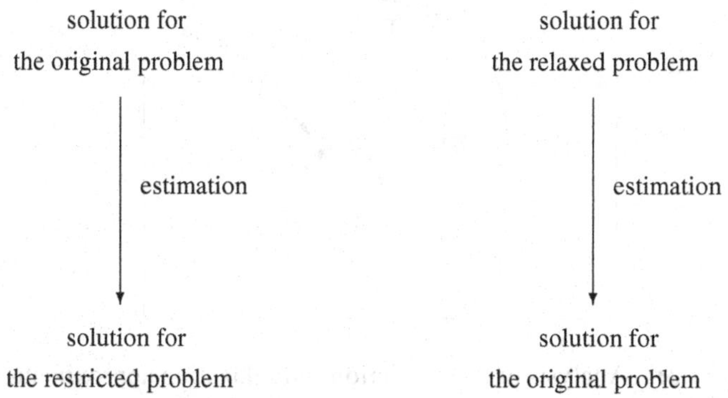

Figure 1.2: Analysis of approximation algorithms based on restriction and relaxation.

the *primal-dual method*, and the *local ratio method* are the main techniques in this approach. We study these techniques in Chapters 7 and 8. In addition to the linear programming technique, it has recently been found that semidefinite programming can also be applied in such a relaxation approach. We present the theory of semidefinite programming and its application to approximation algorithms in Chapter 9.

We remark that an important step in the analysis of approximation algorithms is the estimation of the errors created by the perturbation of the feasible domain. For the algorithms based on the restriction and relaxation techniques, this error estimation often uses similar methods. To analyze an algorithm designed with the restriction technique, one usually takes an optimal solution for the original problem and modifies it to meet the restriction, and then estimates the errors that occurred in the modification. For the algorithms designed with the relaxation technique, the key part of the analysis is about *rounding* the solution, or estimating the errors that occurred in the transformation from the solution for the relaxed problem to the solution for the original problem. Therefore, in both cases, a key step in the analysis is the estimation of the change of solutions from those in a larger (or, relaxed) domain to those in a smaller (or, restricted) domain (see Figure 1.2).

To explain this observation more clearly, let us consider a minimization problem $\min_{\boldsymbol{x}\in\Omega} f(\boldsymbol{x})$ as defined in (1.3), where $\boldsymbol{x}$ denotes a vector $(x_1, x_2, \ldots, x_n)$ in $\mathbb{R}^n$. Assume that $\boldsymbol{x}^* \in \Omega$ satisfies $f(\boldsymbol{x}^*) = \min_{\boldsymbol{x}\in\Omega} f(\boldsymbol{x})$. Suppose we restrict the feasible domain to a subregion Γ of Ω and find an optimal solution $\boldsymbol{y}^*$ for the restricted problem; that is, $f(\boldsymbol{y}^*) = \min_{\boldsymbol{x}\in\Gamma} f(\boldsymbol{x})$. Then, we may analyze the performance of $\boldsymbol{y}^*$ as an approximate solution to the original problem in the following way (see Figure 1.3):

(1) Consider a minimum solution $\boldsymbol{x}^*$ of $\min_{\boldsymbol{x}\in\Omega} f(\boldsymbol{x})$.

(2) Modify $\boldsymbol{x}^*$ to obtain a feasible solution $\boldsymbol{y}$ of $\min_{\boldsymbol{x}\in\Gamma} f(\boldsymbol{x})$.

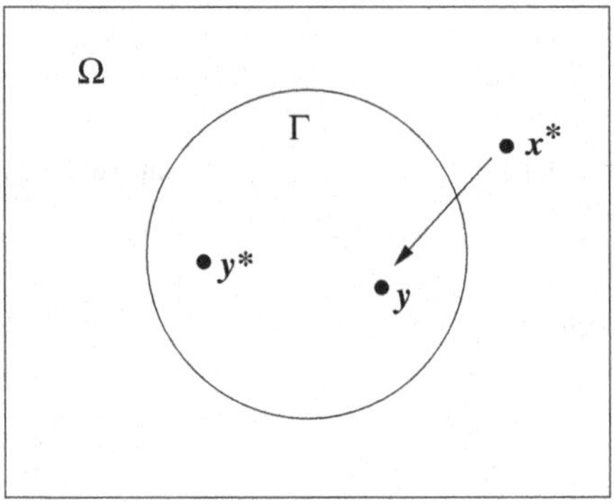

Figure 1.3: Analysis of the restriction and relaxation approximations.

(3) Estimate the value of $f(\boldsymbol{y})/f(\boldsymbol{x}^*)$, and use it as an upper bound for the performance ratio for the approximate solution $\boldsymbol{y}^*$, since $\boldsymbol{y} \in \Gamma$ implies

$$\frac{f(\boldsymbol{y}^*)}{f(\boldsymbol{x}^*)} \leq \frac{f(\boldsymbol{y})}{f(\boldsymbol{x}^*)}.$$

Similarly, consider the problem $\min_{\boldsymbol{x}\in\Gamma} f(\boldsymbol{x})$. Suppose we relax the feasible region Γ to a bigger region Ω, and find the optimal solution $\boldsymbol{x}^*$ for the relaxed problem; that is, $f(\boldsymbol{x}^*) = \min_{\boldsymbol{x}\in\Omega} f(\boldsymbol{x})$. Then, we can round $\boldsymbol{x}^*$ into a solution $\boldsymbol{y} \in \Gamma$ and use it as an approximate solution to the original problem. The analysis of this relaxation algorithm can now be done as follows:

- Estimate the value $f(\boldsymbol{y})/f(\boldsymbol{x}^*)$, and use it as an upper bound for the performance ratio for the approximate solution $\boldsymbol{y}$, since, for any optimal solution $\boldsymbol{y}^*$ for the original problem, we have $f(\boldsymbol{x}^*) \leq f(\boldsymbol{y}^*)$, and hence

$$\frac{f(\boldsymbol{y})}{f(\boldsymbol{y}^*)} \leq \frac{f(\boldsymbol{y})}{f(\boldsymbol{x}^*)}.$$

Thus, in both cases, the analysis of the performance of the approximate solution is reduced to the estimation of the ratio $f(\boldsymbol{y})/f(\boldsymbol{x}^*)$.

Notice, however, a critical difference in the above analyses. In the case of the restriction algorithms, the change from $\boldsymbol{x}^*$ to $\boldsymbol{y}$ is part of the analysis of the algorithm, and we are not concerned with the time complexity of this change. On the other hand, in the case of the relaxation algorithms, this change is a step in the approximation algorithm, and has to be done in polynomial time. As a consequence, while the method of rounding for the analysis of the relaxation algorithms may, in general, be applied to the analysis of the restriction algorithms, the converse may not be true; that is, the analysis techniques developed for the restriction algorithms are not necessarily extendable to the analysis of the relaxation algorithms.

1.3 Heuristics Versus Approximation

In the literature, the word "heuristics" often appears in the study of intractable problems and is sometimes used interchangeably with the word "approximation." In this book, however, we will use it in a different context and distinguish it from approximation algorithms. The first difference between heuristics and approximation is that approximation algorithms usually have guaranteed (worst-case) performance ratios, while heuristic algorithms may not have such guarantees. In other words, approximations are usually justified with theoretical analysis, while heuristics often appeal to empirical data.

The second difference is that approximation usually applies to optimization problems, while heuristics may also apply to *decision problems*. Let us look at an example. First, we define some terminologies about Boolean formulas. A *Boolean formula* is a formula formed by operations $\vee$ (OR), $\wedge$ (AND), and $\neg$ (NOT) over Boolean constants 0 (FALSE) and 1 (TRUE) and Boolean variables. For convenience, we also use $+$ for OR and $\cdot$ for AND, and write $\bar{x}$ to denote $\neg x$. An *assignment* to a Boolean formula ϕ is a function mapping each Boolean variable in ϕ to a Boolean constant 0 or 1. A *truth assignment* is an assignment that makes the resulting formula TRUE. We say a Boolean formula is *satisfiable* if it has a truth assignment. For instance, the Boolean formula

$$(v_1\bar{v}_2 + \bar{v}_1 v_3 \bar{v}_4 + v_2\bar{v}_3)(\bar{v}_1\bar{v}_3 + \bar{v}_2\bar{v}_4)$$

over the variables $v_1, \ldots, v_4$ is satisfiable, since the assignment $\tau(v_1) = \tau(v_3) = 1$ and $\tau(v_2) = \tau(v_4) = 0$ is a truth assignment for it.

Now, consider the following problem.

> SATISFIABILITY (SAT): Given a Boolean formula, determine whether it is satisfiable.

This is not an optimization problem. Therefore, it does not make much sense to try to develop an approximation algorithm for this problem, though there are a number of heuristics, such as the resolution method, developed for this problem. Such heuristics may work efficiently for a large subset of the input instances, but they do not guarantee to solve all instances in polynomial time.

Although approximations and heuristics are different concepts, their ideas and techniques can often be borrowed from each other. Theoretical analysis of approximation algorithms could provide interesting ideas for heuristic algorithms. In addition, for some decision problem, we may first convert it into an equivalent optimization problem, and then adapt the approximation algorithms for the optimization problem to heuristic algorithms for the original decision problem. For instance, we may use the approximation algorithms for integer programming to develop a heuristic algorithm for SAT as follows.

We first convert the problem SAT into an optimization problem. Let $v_1, v_2, \ldots, v_n$ be Boolean variables and $\boldsymbol{v}$ the vector $(v_1, v_2, \ldots, v_n)$ in $\{0,1\}^n$. Let $y_1, y_2, \ldots, y_n$ be real variables and $\boldsymbol{y}$ the vector $(y_1, y_2, \ldots, y_n)$ in $\mathbb{R}^n$. For each Boolean function $f(\boldsymbol{v})$, we define a real function $F_f(\boldsymbol{y})$ recursively as follows:

(1) Initially, if $f(\boldsymbol{v}) = v_i$, then set $F_f(\boldsymbol{y}) \leftarrow y_i$; if $f(\boldsymbol{v}) = 0$, then set $F_f(\boldsymbol{y}) \leftarrow 0$; and if $f(\boldsymbol{v}) = 1$, then set $F_f(\boldsymbol{y}) \leftarrow 1$.

(2) Inductively, if $f(\boldsymbol{v}) = g(\boldsymbol{v}) \vee h(\boldsymbol{v})$, then set $F_f(\boldsymbol{y}) \leftarrow F_g(\boldsymbol{y}) + F_h(\boldsymbol{y}) - F_g(\boldsymbol{y}) \cdot F_h(\boldsymbol{y})$; if $f(\boldsymbol{v}) = g(\boldsymbol{v}) \wedge h(\boldsymbol{v})$, then set $F_f(\boldsymbol{y}) \leftarrow F_g(\boldsymbol{y}) \cdot F_h(\boldsymbol{y})$; and if $f(\boldsymbol{v}) = \neg g(\boldsymbol{v})$, then set $F_f(\boldsymbol{y}) \leftarrow 1 - F_g(\boldsymbol{y})$.

The above construction converts the decision problem SAT into an equivalent optimization problem, in the sense that a Boolean formula $f(\boldsymbol{v})$ is satisfiable if and only if the following 0–1 integer program has a positive maximum objective function value:

$$\begin{array}{ll} \text{maximize} & F_f(\boldsymbol{y}) \\ \text{subject to} & \boldsymbol{y} \in \{0,1\}^n. \end{array}$$

Although this new problem is still intractable, it is nevertheless an optimization problem, and the approximation techniques for 0–1 integer programming are applicable. These approximation algorithms could then be studied and developed into a heuristic for the decision version of SAT.

Historically, heuristic algorithms have appeared much earlier than approximation algorithms. The first documented approximation algorithm was discovered by Graham [1966] for a scheduling problem, while heuristic algorithms probably existed, at least in the informal form, as early as the concept of algorithms was developed. The existence of the rich families of heuristics and their wide applications encourage us to develop them into new approximation algorithms. For instance, an important idea for many heuristics is to link the discrete space of a combinatorial optimization problem to the continuous space of a nonlinear optimization problem through geometric, analytic, or algebraic techniques, and then to apply the nonlinear optimization algorithms to the combinatorial optimization problems. Researchers have found that this approach often leads to very fast and effective heuristics for combinatorial optimization problems of a large scale. However, most of these heuristics, with a few exceptions such as the interior point method for linear programming, though working well in practice, do not have a solid theoretical foundation. Theoretical analyses for these algorithms could provide new, surprising approximation algorithms.

1.4 Notions in Computational Complexity

Roughly speaking, the main reason for studying approximation algorithms is to find efficient, but not necessarily optimal, solutions to intractable problems. We have informally defined an intractable problem to be a problem which does not have a polynomial-time algorithm. From the theoretical standpoint, there are, in this informal definition, several important issues that have not been clearly addressed. For instance, why do we identify polynomial-time computability with tractability? Does polynomial-time computability depend on the computational model that we use to implement the algorithm? How do we determine, in general, whether a problem has a polynomial-time algorithm? These fundamental issues have been carefully examined in the theory of computational complexity. We present, in this and the next sections, a brief summary of this theory. The interested reader is referred to Du and Ko [2000] for more details.

The time complexity of an algorithm refers to the running time of the algorithm as a function of the input size. As a convention, in the worst-case analysis, we take the maximum running time over all inputs of the same size n as the time complexity of the algorithm on size n. In order to estimate the running time of an algorithm, we must specify the computational model in which the algorithm is implemented. Several standard computational models have been carefully studied. Here, we consider only two simple models: the pseudocode and the Turing machine.

We have already used pseudocodes to express algorithms in Section 1.1. Pseudocodes are an informal high-level programming language, similar to standard programming languages such as `Pascal`, `C`, and `Java`, without complicated language constructs such as advanced data structures and parameter-passing schemes in procedure calls. It is an abstract programming language in the sense that each variable in a procedure represents a memory location that holds an integer or a real number, without a size limit. We assume the reader is familiar with such high-level programming languages and understands the basic syntax and semantics of pseudocodes. The reader who is not familiar with pseudocodes is referred to any standard algorithm textbook.

When an algorithm is expressed in the form of a program in pseudocode, it is natural to use the number of statements or the number of arithmetic and comparison operations as the basic measure for the time complexity of the algorithm. This time complexity measure is simple to estimate but does not reflect the exact complexity of the algorithm. For instance, consider the following simple procedure that computes the function $f(a, m) = a^m$, where a and m are two positive integers:

$b \leftarrow 1$;
For $k \leftarrow 1$ **to** m **do** $b \leftarrow b \cdot a$;
Output b.

It is not hard to see that, on any input (a, m), the number of operations to be executed in the above algorithm is $O(m)$, independent of the size n of the other input number a. However, a detailed analysis shows that the size of b increases from 1 bit to about nm bits in the computation of the algorithm, and yet we counted only one unit of time for the multiplication of b and a, no matter how large b is. This does not seem to reflect the real complexity of the algorithm. A more accurate estimate of the time complexity should take into account the size of the operands of the arithmetic operations. For instance, the *logarithmic cost* measure counts $O(\log n)$ units of time for each arithmetic or comparison operation that is executed on operands whose values are at most n. Thus, the time complexity of the above algorithm for a^m, under the logarithmic cost measure, would be $O(m^2 \log a)$.

We note that even using the logarithmic cost measure does not give the time complexity of the algorithm completely correctly. Indeed, the logarithmic cost measure is based on the assumption that arithmetic or comparison operations on operands of n bits can be executed in $O(n)$ units of time (in other words, these operations can be implemented in linear time). This assumption is plausible for simple operations, but not for more complicated operations such as multiplication and division. Indeed, no linear-time multiplication algorithm is known. The best algorithm known today for multiplying two n-bit integers requires $\Omega(n \log n)$ units of time. Therefore, the logarithmic cost measure tends to underestimate the complexity of an algorithm with heavy multiplications.

To more accurately reflect the exact complexity of an algorithm, we usually use a primitive computational model, called the *Turing machine*. We refer the reader to textbooks of theory of computation, for instance, Du and Ko [2000], for the definition of a Turing machine. Here, it suffices to summarize that (1) all input, output, and temporary data of the computation of a Turing machine are stored on a finite number of tapes, with one single character stored in one cell of the tape, and (2) each instruction of the Turing machine works on one cell of the tape, either changing the character stored in the cell or moving its tape head to one of its neighboring cells. That is, the complexity measure of the Turing machine is a *bit-operation* measure, which most closely represents our intuitive notion of time complexity measure.

The instructions of Turing machines are very simple and so it makes the analysis of the computation of a Turing machine easier. In particular, it allows us to prove lower bounds of a problem, which is difficult to do for more complicated computational models. However, one might suspect whether we can implement sophisticated algorithms with, for instance, advanced data structures and complicated recursive calls in such a simplistic machine and, even if so, whether the implementation is as efficient as more general models. It turns out that Turing machines, though primitive, can simulate all known computational models efficiently in the following sense: For any algorithm that can be implemented in the model in question with time complexity $t(n)$, there is a Turing machine implementing this algorithm in time $p(t(n))$, where p is a polynomial function depending on the model but independent of the algorithms. In fact, a widely accepted hypothesis, called the *extended Church–Turing thesis*, states that a Turing machine can simulate any reasonable deterministic computational model within polynomial time. In other words, polynomial-time computability is a notion that is independent of the computational models used to implement the algorithms.

Based on the extended Church–Turing thesis, we now formally identify the class of tractable problems with the following complexity class:

P: the class of all decision problems that are solvable in polynomial time by a deterministic Turing machine.

In other words, we say a problem is *tractable* if there is a Turing machine M that solves the problem in polynomial time in the input size (i.e., M runs in time $O(n^k)$, where n is the input size and k is a constant). We note that the composition of two polynomial functions is still a polynomial function. Thus, the combination of two polynomial-time algorithms is still a polynomial-time algorithm. This reflects the intuition that the combination of two tractable algorithms should be considered tractable.

Now, let us go back to our choice of using pseudocodes to describe algorithms. From the above discussion, we may assume (and, in fact, prove) that the logarithmic cost measure of a pseudocode procedure and the bit-operation complexity of an equivalent Turing machine program are within a polynomial factor. Therefore, in order to demonstrate that a problem is tractable, we can simply present the algorithm in a pseudocode procedure and perform a simple time analysis of the procedure. On the other hand, to show that a problem is intractable, we usually use Turing machines as the computational model.

1.5 NP-Complete Problems

In the study of computational complexity, an optimization problem is usually formulated into an equivalent decision problem, whose answer is either YES or NO. For instance, we can formulate the problem KNAPSACK into the following decision problem:

> KNAPSACK$_D$: Given $2n+2$ integers: $S, K, s_1, s_2, \ldots, s_n, c_1, c_2, \ldots,$ c_n, determine whether there is a sequence $(x_1, x_2, \ldots, x_n) \in \{0,1\}^n$ such that $\sum_{i=1}^n s_i x_i \le S$ and $\sum_{i=1}^n c_i x_i \ge K$.

It is not hard to see that KNAPSACK and KNAPSACK$_D$ are equivalent, in the sense that they are either both tractable or both intractable.

Proposition 1.4 *The optimization problem* KNAPSACK *is polynomial-time solvable if and only if the decision problem* KNAPSACK$_D$ *is polynomial-time solvable.*

Proof. Suppose the optimization problem KNAPSACK is polynomial-time solvable. Then, we can solve the decision problem KNAPSACK$_D$ by finding the optimal solution *opt* of the corresponding KNAPSACK instance and then answering YES if and only if $opt \ge K$.

Conversely, suppose KNAPSACK$_D$ is solvable in polynomial time by a Turing machine M. Assume that M runs in time $O(N^k)$, where N is the input size and k is a constant. Now, on input $I = (S, s_1, \ldots, s_n, c_1, \ldots, c_n)$ to the problem KNAPSACK, we can binary search for the maximum K such that M answers YES on input $(S, K, s_1, \ldots, s_n, c_1, \ldots, c_n)$. This maximum value K is exactly the optimal solution *opt* for input I of the problem KNAPSACK. Note that K satisfies $K \le M_2 = \sum_{i=1}^n c_i$. Thus, the above binary search needs to simulate M for at most $\lfloor \log M_2 + 1 \rfloor = O(N)$ times, where N is the size of input I. So, we can solve KNAPSACK in time $O(N^{k+1})$. □

From the discussion of the last section, in order to prove a problem intractable, we need to show that (the decision version of) the problem is not in **P**. Unfortunately, for a great number of optimization problems, there is strong evidence, both empirical and mathematical, suggesting that they are likely intractable, but no one is able to find a formal proof that they are not in **P**. Most of these problems, however, share a common property called **NP**-*completeness*. That is, they can be solved by *nondeterministic* algorithms in polynomial time and, furthermore, if any of these problems is proved to be not in **P**, then all of these problems are not in **P**.

A nondeterministic algorithm is an algorithm that can make *nondeterministic moves*. In a nondeterministic move, the algorithm can assign a value of either 0 or 1 to a variable nondeterministically, so that the computation of the algorithm after this step branches into two separate *computation paths*, each using a different value for the variable. Suppose a nondeterministic algorithm executes nondeterministic moves k times. Then it may generate 2^k different deterministic computation paths, some of which may output YES and some of which may output NO. We say the nondeterministic algorithm *accepts* the input (i.e., answers YES) if at least one of the computation paths outputs YES; and the nondeterministic algorithm *rejects* the input if all computation paths output NO. (Thus, the actions of accepting and rejecting an input by a nondeterministic algorithm A are not symmetric: If we change each answer YES of a computation path to answer NO, and each NO to YES, the collective solution of A does not necessarily change from accepting to rejecting.) On each input x accepted by a nondeterministic algorithm A, the running time of A on x is the length of the shortest computation path on x that outputs YES. The time complexity of algorithm A is defined as the function

$$t_A(n) = \text{the maximum running time on any } x \text{ of length } n \text{ that is accepted by the algorithm } A.$$

For instance, the following is a nondeterministic algorithm for KNAPSACK (more precisely, for the decision problem KNAPSACK$_D$):

Algorithm 1.E (*Nondeterministic Algorithm for* KNAPSACK$_D$)

Input: Positive integers $S, s_1, s_2, \ldots, s_n, c_1, c_2, \ldots, c_n$, and an integer $K > 0$.

(1) **For** $i \leftarrow 1$ **to** n **do**
  nondeterministically select a value 0 or 1 for x_i.

(2) **If** $\sum_{i=1}^{n} x_i s_i \leq S$ and $\sum_{i=1}^{n} x_i c_i \geq K$ **then** output YES
  else output NO. ■

It is clear that the above algorithm works correctly. Indeed, it contains 2^n different computation paths, each corresponding to one choice of $(x_1, x_2, \ldots, x_n) \in \{0, 1\}^n$. If one choice of $(x_1, x_2, \ldots, x_n)$ satisfies the condition of step (2), then the algorithm accepts the input instance; otherwise, it rejects. In addition, we note that in this algorithm, all computation paths have the same running time, $O(n)$. Thus, this is a linear-time nondeterministic algorithm.

The nondeterministic Turing machine is the formalism of nondeterministic algorithms. Corresponding to the deterministic complexity class **P** is the following nondeterministic complexity class:

NP: the class of all decision problems that are computable by a nondeterministic Turing machine in polynomial time.

We note that in a single path of a polynomial-time nondeterministic algorithm, there can be at most a polynomial number of nondeterministic moves. It is not hard to see that we can always move the nondeterministic moves to the beginning of the algorithm without changing its behavior. Thus, all polynomial-time nondeterministic algorithms M_N have the following common form:

Assume that the input x has n bits.

(1) Nondeterministically select a string $y = y_1 y_2 \cdots y_{p(n)} \in \{0,1\}^*$, where p is a polynomial function.

(2) Run a polynomial-time deterministic algorithm M_D on input (x, y).

Suppose M_D answers YES on input (x, y); then we say y is a *witness* of the instance x. Thus, a problem Π is in **NP** if there is a two-step algorithm for Π in which the first step nondeterministically selects a potential witness y of polynomial size, and the second step deterministically verifies that y is indeed a witness. We call such an algorithm a *guess-and-verify* algorithm.

As another example, let us show that the problem SAT is in **NP**.

Algorithm 1.F (*Nondeterministic Algorithm for* SAT)

Input: A Boolean formula ϕ over Boolean variables $v_1, v_2, \ldots, v_n$.

(1) Guess n Boolean values $b_1, b_2, \ldots, b_n$.

(2) Verify (deterministically) that the formula ϕ is TRUE under the assignment $\tau(v_i) = b_i$, for $i = 1, \ldots, n$. If so, output YES; otherwise, output NO. ■

The correctness of the above algorithm is obvious. To show that SAT is in **NP**, we only need to check that the verification of whether a Boolean formula containing no variables is TRUE can be done in deterministic polynomial time.

We have seen that problems in **NP**, such as KNAPSACK and SAT, have simple polynomial-time nondeterministic algorithms. However, we do not know of any physical devices to implement the nondeterministic moves in the algorithms. So, what is the exact relationship between **P** and **NP**? This is one of the most important open questions in computational complexity theory. On the one hand, we do not know how to find efficient deterministic algorithms to simulate a nondeterministic algorithm. A straightforward simulation by the deterministic algorithm that runs the verification step over all possible guesses would take an exponential amount of time. On the other hand, though many people believe that there is no polynomial-time deterministic algorithm for every problem in **NP**, no one has yet found a formal proof for that.

Without a proof for **P** $\neq$ **NP**, how do we demonstrate that a problem in **NP** is likely to be intractable? The notion of **NP**-*completeness* comes to help.

For convenience, we write in the following $x \in A$ to denote that the answer to the input x for the decision problem A is YES (that is, we identify the decision problem with the set of all input instances which have the answer YES). We say a decision problem A is *polynomial-time reducible* to a decision problem B, denoted by $A \leq_m^P B$, if there is a polynomial-time computable function f from instances of A to instances of B (called the *reduction function* from A to B) such that $x \in A$ if and only if $f(x) \in B$. Intuitively, a reduction function f reduces the membership problem of whether $x \in A$ to the membership problem of whether $f(x) \in B$. Thus, if there is a polynomial-time algorithm to solve problem B, we can combine the function f with this algorithm to solve problem A.

Proposition 1.5 *(a) If $A \leq_m^P B$ and $B \in$ **P**, then $A \in$ **P**.*
(b) If $A \leq_m^P B$ and $B \leq_m^P C$, then $A \leq_m^P C$.

The above two properties justify the use of the notation $\leq_m^P$ between decision problems: It is a partial ordering for the *hardness* of the problems (modulo polynomial-time computability).

We can now define the term **NP**-completeness: We say a decision problem A is **NP**-*hard* if, for any $B \in$ **NP**, $B \leq_m^P A$. We say A is **NP**-*complete* if A is **NP**-hard and, in addition, $A \in$ **NP**. That is, an **NP**-complete problem A is one of the hardest problems in **NP** with respect to the reduction $\leq_m^P$. For an optimization problem A, we also say A is **NP**-hard (or, **NP**-complete) if its (polynomial-time equivalent) decision version A_{D} is **NP**-hard (or, respectively, **NP**-complete).

It follows immediately from Proposition 1.5 that if an **NP**-complete problem is in **P**, then **P** = **NP**. Thus, in view of our inability to solve the **P** vs. **NP** question, the next best way to prove a problem intractable is to show that it is **NP**-complete (and so it is most likely not in **P**, unless **P** = **NP**).

Among all problems, SAT was the first problem proved **NP**-complete. It was proved by Cook [1971], who showed that for any polynomial-time nondeterministic Turing machine, its computation on any input x can be encoded by a Boolean formula ϕ_x of polynomially bounded length such that the formula ϕ_x is satisfiable if and only if M accepts x. This proof is called a *generic reduction*, since it works directly with the computation of a nondeterministic Turing machine. In general, it does not require a generic reduction to prove a new problem A to be **NP**-complete. Instead, by Proposition 1.5(b), we can use any problem B that is already known to be **NP**-complete and only need to prove that $B \leq_m^P A$. For instance, we can prove that KNAPSACK$_{\mathrm{D}}$ is **NP**-complete by reducing the problem SAT to it.

Theorem 1.6 KNAPSACK$_{\mathrm{D}}$ *is* **NP**-*complete.*

Proof. We have already seen that KNAPSACK$_{\mathrm{D}}$ is in **NP**. We now prove that KNAPSACK$_{\mathrm{D}}$ is complete for **NP**. In order to do this, we introduce a subproblem 3-SAT of SAT. In a Boolean formula, a variable or the negation of a variable is called a *literal*. An elementary sum of literals is called a *clause*. A Boolean formula is in 3-CNF (*conjunctive normal form*) if it is a product of a finite number of clauses, each being the sum of exactly three literals. For instance, the following is a 3-CNF formula:

$$(v_1 + \bar{v}_2 + \bar{v}_3)(\bar{v}_1 + v_3 + v_4)(v_2 + \bar{v}_3 + \bar{v}_4).$$

The problem 3-SAT asks whether a given 3-CNF Boolean formula is satisfiable. This problem is a restrictive form of the problem SAT, but it is also known to be **NP**-complete. Indeed, there is a simple way of transforming a Boolean formula ϕ into a new 3-CNF formula ψ such that ϕ is satisfiable if and only if ψ is satisfiable.

We omit the proof and refer the reader to textbooks on complexity theory. In the following, we present a proof for 3-SAT $\leq_m^P$ KNAPSACK$_D$.

Let ϕ be a 3-CNF formula that is of the form $C_1C_2\cdots C_m$, where each C_j is a clause with three literals. Assume that ϕ contains Boolean variables $v_1, v_2, \ldots, v_n$. We are going to define a list of $2n+2m$ integers $c_1, c_2, \ldots, c_{2n+2m}$, plus an integer K. All integers c_i and the integer K are of value between 0 and 10^{n+m}. These integers will satisfy the following property:

$$\phi \text{ is satisfiable} \iff (\exists x_1, x_2, \ldots, x_{2n+2m} \in \{0,1\}) \sum_{i=1}^{2n+2m} c_i x_i = K. \quad (1.4)$$

Now, let $S = K$, $s_i = c_i$ for $i = 1, 2, \ldots, 2n+2m$. Then, it follows that the formula ϕ is satisfiable if and only if the instance $(S, K, s_1, \ldots, s_{2n+2m}, c_1, \ldots, c_{2n+2m})$ to the problem KNAPSACK$_D$ has the answer YES. Therefore, this construction is a reduction function for 3-SAT $\leq_m^P$ KNAPSACK$_D$.

We now describe the construction of these integers and prove that they satisfy property (1.4). First, we note that each integer is between 0 and 10^{n+m}, and so it has a unique decimal representation of *exactly* $n+m$ digits (with possible leading zeroes). We will define each integer digit by digit, with the kth digit indicating the kth most significant digit. First, we define the first n digits of K to be 1 and the last m digits to be 3. That is,

$$K = \underbrace{11\cdots11}_{n}\underbrace{33\cdots33}_{m}.$$

Next, for each $i = 1, 2, \ldots, n$, we define the integer c_i as follows: The ith digit and the $(n+j)$th digits, for all $1 \leq j \leq m$ such that C_j contains the literal $\bar{v}_i$, of c_i are 1 and all other digits are 0. For instance, if $\bar{v}_3$ occurs in C_1, C_5, and C_m, then

$$c_3 = \underbrace{00100\cdots0}_{n}\underbrace{100010\cdots01}_{m}.$$

Similarly, for $i = 1, 2, \ldots, n$, the integer c_{n+i} is defined as follows: The ith digit and the $(n+j)$th digits, for all $1 \leq j \leq m$ such that C_j contains the literal v_i, of c_{n+i} are 1 and all other digits are 0.

Finally, for $j = 1, 2, \ldots, m$, we define $c_{2n+2j-1} = c_{2n+2j}$ as follows: Their $(n+j)$th digit is 1 and all other digits are 0. This completes the definition of the integers.

Now, we need to show that these integers satisfy property (1.4). First, we observe that for any k, $1 \leq k \leq n+m$, there are at most five integers among c_t's whose kth digit is nonzero, and each nonzero digit must be 1. Thus, to get the sum K, we must choose, for each $i = 1, 2, \ldots, n$, exactly one integer among c_t's whose ith digit is 1, and, for each $j = 1, 2, \ldots, m$, exactly three integers whose $(n+j)$th digit is 1. The first part of this condition implies that we must choose, for each $i = 1, 2, \ldots, n$, exactly one of c_i or c_{n+i}.

Now, assume that ϕ has a truth assignment τ on variables $v_1, v_2, \ldots, v_n$. We define the sequence $(x_1, x_2, \ldots, x_{2n+2m})$ as follows:

(1) For each $i = 1, 2, \ldots, n$, let $x_{n+i} = 1 - x_i = \tau(v_i)$.

(2) For each $j = 1, 2, \ldots, m$, define $x_{2n+2j-1}$ and x_{2n+2j} as follows: If τ satisfies all three literals of C_j, then $x_{2n+2j-1} = x_{2n+2j} = 0$; if τ satisfies exactly two literals of C_j, then $x_{2n+2j-1} = 1$ and $x_{2n+2j} = 0$; and if τ satisfies exactly one literal of C_j, then $x_{2n+2j-1} = x_{2n+2j} = 1$.

Then it is easy to verify that $\sum_{i=1}^{2n+2m} c_i x_i = K$.

Next, assume that there exists a sequence $(x_1, x_2, \ldots, x_{2n+2m}) \in \{0,1\}^{2n+2m}$ such that $\sum_{i=1}^{2n+2m} c_i x_i = K$. Then, from our earlier observation, we see that exactly one of x_i and x_{n+i} has value 1. Define $\tau(v_i) = x_{n+i}$. We claim that τ satisfies each clause C_j, $1 \leq j \leq m$. Since the $(n+j)$th digit of the sum $\sum_{i=1}^{2n+2m} c_i x_i$ is equal to 3, and since there are at most two integers among the last $2m$ integers whose $(n+j)$th digit is 1, there must be an integer $k \leq 2n$ such that $x_k = 1$ and the $(n+j)$th digit of c_k is 1. Suppose $1 \leq k \leq n$; then it means that $\tau(v_k) = 0$, and C_j contains the literal $\bar{v}_k$. Thus, τ satisfies C_j. On the other hand, if $n + 1 \leq k \leq 2n$, then we know that $\tau(v_{k-n}) = 1$, and C_j contains the literal v_{k-n}; and so τ also satisfies C_j. This completes the proof of property (1.4).

Finally, we remark that the above construction of these integers from the formula ϕ is apparently polynomial-time computable. Thus, this reduction is a polynomial-time reduction. □

In addition to the above two problems, thousands of problems from many seemingly unrelated areas have been proven to be **NP**-complete in the past four decades. These results demonstrate the importance and universality of the concept of **NP**-completeness. In the following, we list a few problems that are frequently used to prove a new problem being **NP**-complete.

> VERTEX COVER (VC): Given an undirected graph $G = (V, E)$ and a positive integer K, determine whether there is a set $C \subseteq V$ of size $\leq K$ such that, for every edge $\{u, v\} \in E$, $C \cap \{u, v\} \neq \emptyset$. (Such a set C is called a *vertex cover* of G.)
>
> HAMILTONIAN CIRCUIT (HC): Given an undirected graph $G = (V, E)$, determine whether there is a simple cycle that passes through each vertex exactly once. (Such a cycle is called a *Hamiltonian circuit*.)
>
> PARTITION: Given n positive integers $a_1, a_2, \ldots, a_n$, determine whether there is a partition of these integers into two parts that have the equal sum. (This is a subproblem of KNAPSACK.)
>
> SET COVER (SC): Given a family $\mathcal{C}$ of subsets of $I = \{1, 2, \ldots, n\}$ and a positive integer K, determine whether there is a subfamily $\mathcal{C}'$ of $\mathcal{C}$ of at most K subsets such that $\bigcup_{A \in \mathcal{C}'} A = I$.

For instance, from the problem HC, we can easily prove that (the decision versions of) the following optimization problems are also **NP**-complete. We leave their proofs as exercises.

TRAVELING SALESMAN PROBLEM (TSP): Given a complete graph and a distance function that gives a positive integer as the distance between every pair of vertices, find a Hamiltonian circuit with the minimum total distance.

MAXIMUM HAMILTONIAN CIRCUIT (MAX-HC): Given a complete graph and a distance function, find a Hamiltonian circuit with the maximum total distance.

MAXIMUM DIRECTED HAMILTONIAN PATH (MAX-DHP): Given a complete directed graph and a distance function, find a Hamiltonian path with the maximum total distance. (A *Hamiltonian path* is a simple path that passes through each vertex exactly once.)

1.6 Performance Ratios

As we pointed out earlier, the two most important criteria in the study of approximation algorithms are efficiency and the performance ratio. By efficiency, we mean polynomial-time computability. By performance ratio, we mean the ratio of the objective function values between the approximate and optimal solutions. More precisely, for any optimization problem Π and any input instance I, let $opt(I)$ denote the objective function value of the optimal solution to instance I, and $A(I)$ the objective function value produced by an approximation algorithm A on instance I. Then, for a minimization problem, we define the *performance ratio* of an approximation algorithm A to be

$$r(A) = \sup_I \frac{A(I)}{opt(I)}$$

and, for a maximization problem, we define it to be

$$r(A) = \sup_I \frac{opt(I)}{A(I)},$$

where I ranges over all possible input instances. Thus, for any approximation algorithm A, $r(A) \geq 1$, and, in general, the smaller the performance ratio is, the better the approximation algorithm is.

For instance, consider the maximization problem KNAPSACK again. Let $opt(I)$ be the maximum value of the objective function on input instance I, and $c_G(I)$ and $c_{GG}(I)$ the objective function values obtained by Algorithms 1.B and 1.C, respectively, on instance I. Then, by Theorems 1.1 and 1.2, the performance ratios of these two algorithms (denoted by $A_{1\text{B}}$ and $A_{1\text{C}}$) are

$$r(A_{1\text{B}}) = \sup_I \frac{opt(I)}{c_G(I)} \leq 2$$

and

$$r(A_{1\text{C}}) = \sup_I \frac{opt(I)}{c_{GG}(I)} \leq 1 + \varepsilon.$$

That is, both of these algorithms achieve a constant approximation ratio, but Algorithm 1.C has a better ratio.

As another example, consider the famous TRAVELING SALESMAN PROBLEM (TSP) defined in the last section. We assume that the distance between any two vertices is positive. In addition, we assume that the given distance function d satisfies the *triangle inequality* (abbr. Δ-*inequality*); that is,

$$d(a,b) + d(b,c) \geq d(a,c),$$

for any three vertices a, b, and c. Then, there is a simple approximation algorithm for TSP that finds a tour (i.e., a Hamiltonian circuit) with the total distance within twice of the optimum. This algorithm uses two basic linear-time algorithms on graphs:

> **Minimum Spanning-Tree Algorithm**: Given a connected graph G with a distance function d on all edges, this algorithm finds a minimum spanning tree T of the graph G. (T is a *minimum spanning tree* of G if T is a connected subgraph of G with the minimum total distance.)
>
> **Euler Tour Algorithm**: Given a connected graph G in which each vertex has an even degree, this algorithm finds an *Euler tour*, i.e., a cycle that passes through each edge in G exactly once.

Algorithm 1.G (*Approximation Algorithm for* TSP *with* Δ-*Inequality*)

Input: A complete graph $G = (V, E)$, where $V = \{1, 2, \ldots, n\}$, and a distance function $d : V \times V \to \mathbb{N}$ that satisfies the triangle inequality.

(1) Find a minimum spanning tree T of G.

(2) Change each edge e in T to two (parallel) edges between the same pair of vertices. Call the resulting graph H.

(3) Find an Euler tour P of H.

(4) Output the Hamiltonian circuit Q that is obtained by visiting each vertex once in the order of their first occurrence in P. (That is, Q is the *shortcut* of P that skips a vertex if it has already been visited. See Figure 1.4.) ■

We first note that, after step (2), each vertex in graph H has an even degree and hence the Euler Tour Algorithm can find an Euler tour of H in linear time. Thus, Algorithm 1.G is well defined. Next, we verify that its performance ratio is bounded by 2. This is easy to see from the following three observations:

(a) The total distance of the minimum spanning tree T must be less than that of any Hamiltonian circuit C, since we can obtain a spanning tree by removing an edge from C.

(b) The total distance of P is exactly twice that of T, and so at most twice that of the optimal solution.

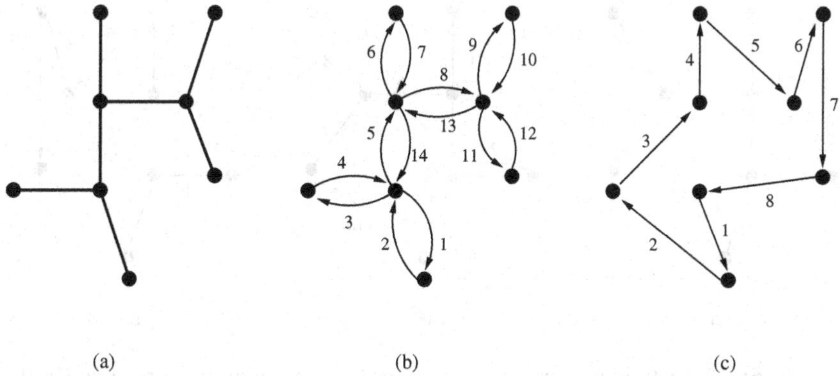

Figure 1.4: Algorithm 1.G: **(a)** the minimum spanning tree; **(b)** the Euler tour; and **(c)** the shortcut.

(c) By the triangle inequality, the total distance of the shortcut Q is no greater than that of tour P.

Christofides [1976] introduced a new idea into this approximation algorithm and improved the performance ratio to $3/2$. This new idea requires another basic graph algorithm:

> **Minimum Perfect Matching Algorithm**: Given a complete graph G of an even number of vertices and a distance function d on edges, this algorithm finds a perfect matching with the minimum total distance. (A *matching* of a graph is a subset M of the edges such that each vertex occurs in at most one edge in M. A *perfect matching* of a graph is a matching M with each vertex occurring in exactly one edge in M.)

Algorithm 1.H (*Christofides's Algorithm for* TSP *with* Δ*-Inequality*)

Input: A complete graph $G = (V, E)$, where $V = \{1, 2, \ldots, n\}$, and a distance function $d : V \times V \to \mathbb{N}$ that satisfies the triangle inequality.

(1) Find a minimum spanning tree $T = (V, E_T)$ of G.

(2) Let V' be the set of all vertices in T of odd degrees;
Let $G' = (V', E')$ be the subgraph of G induced by vertex set V';
Find a minimum perfect matching M for G';
Add the edges in M to tree T (with possible parallel edges between two vertices) to form a new graph H'.
[See Figure 1.5(b).]

(3) Find an Euler tour P' of H'.

(4) Output the shortcut Q of the tour P' as in step (4) of Algorithm 1.G. ■

It is clear that after adding the matching M to tree T, each vertex in graph H' has an even degree. Thus, step (3) of Algorithm 1.H is well defined. Now, we note

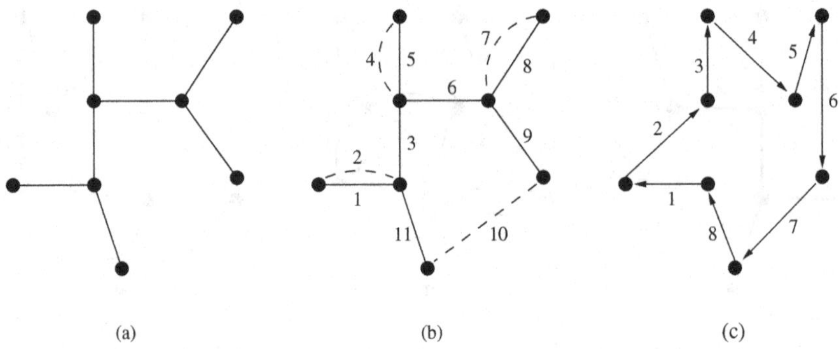

Figure 1.5: Christofides's approximation: **(a)** the minimum spanning tree; **(b)** the minimum matching (shown in broken lines) and the Euler tour; and **(c)** the shortcut.

that the total distance of the matching M is at most one half of that of a minimum Hamiltonian circuit C' in G', since we can remove alternating edges from C' to obtain a perfect matching. Also, by the triangle inequality, the total distance of the minimum Hamiltonian circuit in G' is no greater than that of the minimum Hamiltonian circuit in G. Therefore, the total distance of the tour P', as well as that of Q, is at most $3/2$ of the optimal solution. That is, the performance ratio of Algorithm 1.H is bounded by $3/2$.

Actually, the performance ratio of Christofides's approximation can be shown to be exactly $3/2$. Consider the graph G of Figure 1.6. Graph G has $2n + 1$ vertices $v_0, v_1, \ldots, v_{2n}$ on the Euclidean space $\mathbb{R}^2$, with the distance $d(v_i, v_{i+1}) = 1$ for $i = 0, 1, \ldots, 2n - 1$, and $d(v_i, v_{i+2}) = 1 + a$ for $i = 0, 1, \ldots, 2n - 2$, where $0 < a < 1/2$. It is clear that the minimum spanning tree T of G is the path from v_0 to v_{2n} containing all edges of distance 1. There are only two vertices, v_0 and v_{2n}, having odd degrees in tree T. Thus, the traveling salesman tour produced by Christofides's algorithm is the cycle $(v_0, v_1, v_2, \ldots, v_{2n}, v_0)$, whose total distance is $2n + n(1 + a) = 3n + na$. Moreover, it is easy to see that the minimum traveling salesman tour consists of all horizontal edges plus the two outside nonhorizontal edges, whose total distance is $(2n - 1)(1 + a) + 2 = 2n + 1 + (2n - 1)a$. So, if we let $A_{1\text{H}}$ denote Christofides's algorithm, we get, in this instance I,

$$\frac{A_{1\text{H}}(I)}{opt(I)} = \frac{3n + na}{2n + 1 + (2n - 1)a},$$

which approaches $3/2$ as a goes to 0 and n goes to infinity. It follows that $r(A_{1\text{H}}) = 3/2$.

Theorem 1.7 *For the subproblem of* TSP *with the triangle inequality, as well as the subproblem of* TSP *on Euclidean space, the Christofides's approximation* $A_{1\text{H}}$ *has the performance ratio* $r(A_{1\text{H}}) = 3/2$.

For simplicity, we say an approximation algorithm A is an α-approximation if $r(A) \leq \alpha$ for some constant $\alpha \geq 1$. Thus, we say Christofides's algorithm is a

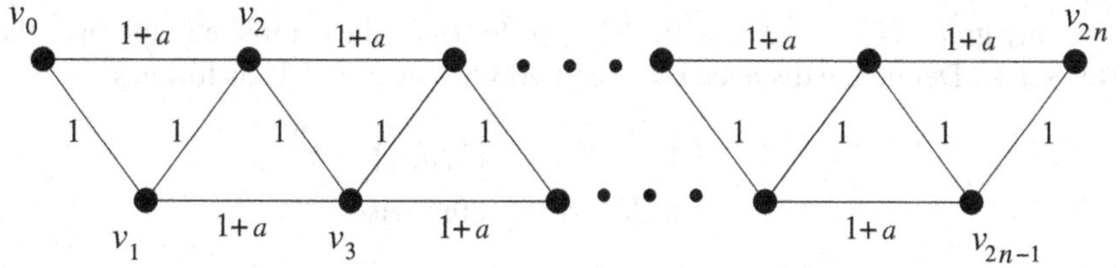

Figure 1.6: A worst case of Christofides's approximation.

(3/2)-approximation for TSP with the triangle inequality, but not an α-approximation for any $\alpha < 3/2$.

An approximation algorithm with a constant performance ratio is also called a *bounded approximation* or a *linear approximation*. An optimization problem Π is said to have a *polynomial-time approximation scheme* (*PTAS*) if, for any $k > 0$, there exists a polynomial-time approximation algorithm A_k for Π with performance ratio $r(A_k) \leq 1 + 1/k$. Furthermore, if the running time of the algorithm A_k in the approximation scheme is a polynomial function in $n+1/k$, where n is the input size, then the scheme is called a *fully polynomial-time approximation scheme* (*FPTAS*). For instance, the generalized greedy algorithm (Algorithm 1.C) is a PTAS, and the polynomial tradeoff approximation (Algorithm 1.D) is an FPTAS for KNAPSACK.

In this book, our main concern is to find efficient approximations to intractable problems with the best performance ratios. However, some optimization problems are so hard that they don't even have any polynomial-time bounded approximations. In these cases, we also need to prove that such approximations do not exist. Since most optimization problems are **NP**-complete, they hence have polynomial-time optimal algorithms if **P** = **NP**. So, when we try to prove that a bounded approximation does not exist, we must assume that **P** ≠ **NP**. Very often, we simply prove that the problem of finding a bounded approximation (or, an α-approximation for some fixed constant α) itself is **NP**-hard. The following is a simple example. We will present a more systematic study of this type of inapproximability results in Chapter 10.

Theorem 1.8 *If* **P** ≠ **NP**, *then there is no polynomial-time approximation algorithm for* TSP *(without the restriction of the triangle inequality) with a constant performance ratio.*

Proof. For any fixed integer $K > 1$, we will construct a *reduction* from the problem HC to the problem of finding a K-approximation for TSP.[5] That is, we will construct a mapping from each instance G of the problem HC to an instance (H, d) of TSP, such that the question of whether G has a Hamiltonian circuit can be determined from any traveling salesman tour for (H, d) whose total distance is within K times of the length of the shortest tour.

[5]Note that TSP is not a decision problem. So, the reduction here has a more general form than that defined in Section 1.5.

For any graph $G = (V, E)$, with $|V| = n$, let H be the complete graph over the vertex set V. Define the distance between two vertices $u, v \in V$ as follows:

$$d(u,v) = \begin{cases} 1, & \text{if } \{u,v\} \in E, \\ n(K+1), & \text{otherwise.} \end{cases}$$

Now, assume that C is a traveling salesman tour of the instance (H, d) whose total distance is at most K times of the length of the shortest tour. If the total distance of C is less than $n(K+1)$, then we know that all edges in C are of distance 1 and so they are all in E. Thus, C is a Hamiltonian circuit of G. On the other hand, if the total distance of C is greater than or equal to $n(K+1)$, this implies that the minimum traveling salesman tour has total distance at least $n(K+1)/K$ and hence greater than n. It implies that the minimum traveling salesman tour must contain an edge not in E. Thus, G has no Hamiltonian circuit.

Thus, if there is a polynomial-time K-approximation for TSP, we can then use it to solve the problem HC, which is **NP**-complete. It follows that **P** = **NP**. □

Exercises

1.1 Prove that Algorithm 1.A always finds the optimal solution for KNAPSACK. More precisely, prove by induction that if there is a subset $A \subseteq \{1, \ldots, i\}$ such that $\sum_{k \in A} c_k = j$ and $\sum_{k \in A} s_k \leq S$, then the value $a(i, j)$ obtained at the end of step (3) of Algorithm 1.A satisfies $a(i, j) \neq nil$ and $a(i, j)$ has the minimum total cost $\sum_{k \in a(i,j)} s_k$ among such sets A.

1.2 Formulate the following logic puzzles into satisfiability instances and solve them:

(a) Three men named Lewis, Miller, and Nelson fill the positions of accountant, cashier, and clerk in a department store. If Nelson is the cashier, Miller is the clerk. If Nelson is the clerk, Miller is the accountant. If Miller is not the cashier, Lewis is the clerk. If Lewis is the accountant, Nelson is the clerk. What is each man's job?

(b) Messrs. Spinnaker, Buoy, Luff, Gybe, and Windward are yacht owners. Each has a daughter, and each has named his yacht after the daughter of one of the others. Mr. Spinnaker's yacht, the *Iris*, is named after Mr. Buoy's daughter. Mr. Buoy's own yacht is the *Daffodil*; Mr. Windward's yacht is the *Jonquil*; Mr. Gybe's, the *Anthea*. Daffodil is the daughter of the owner of the yacht that is named after Mr. Luff's daughter. Mr. Windward's daughter is named Lalage. Who is Jonquil's father?

1.3 For any Boolean function f, $F_f(\boldsymbol{y})$ is defined as in Section 1.3. Prove that for $\boldsymbol{y} \in \{0,1\}^n$, $0 \leq F_f(\boldsymbol{y}) \leq 1$.

1.4 For a 3-CNF formula $\phi = C_1 C_2 \cdots C_m$ over Boolean variables $x_1, x_2, \ldots,$ x_n, let $\boldsymbol{x}$ be the vector $(x_1, x_2, \ldots, x_n)$ in $\{0,1\}^n$. For each variable x_j, $1 \leq j \leq n$, define a corresponding real variable y_j, and let $\boldsymbol{y}$ be the vector $(y_1, y_2, \ldots, y_n)$ in $\mathbb{R}^n$. Define a function $f_1 : \mathbb{R}^n \to \mathbb{R}$ as follows: First, for each pair (i, j), with $1 \le i \le m$ and $1 \le j \le n$, define a *literal function*

$$q_{ij}(y_j) = \begin{cases} (y_j - 1)^2, & \text{if } x_j \text{ is in clause } C_i, \\ (y_j + 1)^2, & \text{if } \bar{x}_j \text{ is in clause } C_i, \\ 1, & \text{neither } x_j \text{ nor } \bar{x}_j \text{ is in } C_i, \end{cases}$$

and, for each $1 \le i \le m$, define a *clause function* $c_i(\boldsymbol{y}) = \prod_{j=1}^n q_{ij}(y_j)$. Finally, define f_1 to be the sum of the clause functions: $f_1(\boldsymbol{y}) = \sum_{i=1}^m c_i(\boldsymbol{y})$.

Define a correspondence between $\boldsymbol{x}$ and $\boldsymbol{y}$ as follows:

$$x_j = \begin{cases} 1, & \text{if } y_j = 1, \\ 0, & \text{if } y_j = -1, \\ \text{undefined}, & \text{otherwise.} \end{cases}$$

Then it is clear that ϕ is satisfiable if and only if the minimum value of $f_1(\boldsymbol{y})$ is 0. Now, define $f(\boldsymbol{y}) = f_1(\boldsymbol{y}) + \sum_{j=1}^n (y_j^2 - 1)^2$, and consider the following minimization problem:

$$\text{minimize } f(\boldsymbol{y}).$$

Show that the objective function $f(\boldsymbol{y})$ satisfies the following properties:

(a) There exists $\boldsymbol{y}$ such that $f(\boldsymbol{y}) = 0$ if and only if there exists $\boldsymbol{y}$ such that $f(\boldsymbol{y}) < 1$.

(b) At every minimum point $\boldsymbol{y}^*$, $f(\boldsymbol{y}^*)$ is strictly convex.

1.5 Consider the greedy algorithm for KNAPSACK that selects the most valuable item first. That is, in Algorithm 1.B, replace the ordering $c_1/s_1 \ge c_2/s_2 \ge \cdots \ge c_n/s_n$ by $c_1 \ge c_2 \ge \cdots \ge c_n$. Show that this greedy algorithm is not a linear approximation.

1.6 Give an example to show that the performance ratio of Algorithm 1.G for TSP with the triangle inequality cannot be any constant smaller than 2.

1.7 When the distance function in TSP is allowed to be asymmetric, i.e., possibly $d(u, v) \neq d(v, u)$, the problem is called DIRECTED TSP. Give an example to show that Christofides's approximation (Algorithm 1.H) does not work for DIRECTED TSP with triangle inequality.

1.8 (a) Suppose there exists an algorithm that can compute the maximum value *opt* of the objective function for KNAPSACK. Can you use this algorithm as a subroutine to design an algorithm computing an optimal solution for KNAPSACK (i.e., the 0-1 vector $(x_1^*, x_2^*, \ldots, x_n^*)$ such that $\sum_{i=1}^n c_i x_i^* =$ *opt*) in polynomial time, provided that the time spent by the subroutine is not counted?

(b) Suppose there exists an algorithm that can compute the distance of the shortest tour for TSP. Can you use this algorithm as a subroutine to design an algorithm computing an optimal solution for TSP (i.e., the shortest tour) in polynomial time, provided that the time spent by the subroutine is not counted?

(c) Suppose there exists an algorithm that can compute a value within a factor α from the distance of the shortest tour for TSP, where α is a constant. Can you use this algorithm as a subroutine to design an algorithm computing an optimal solution for TSP in polynomial time, provided that the time spent by the subroutine is not counted?

1.9 Show that for any $\varepsilon > 0$, there exists a polynomial-time $(2 + \varepsilon)$-approximation for MAX-HC and there exists a polynomial-time 2-approximation for MAX-DHP. [Hint: Use the polynomial-time *Maximum Matching Algorithm*.]

1.10 Consider the following problem:

MINIMUM VERTEX COVER (MIN-VC): Given an undirected graph G, find a vertex cover of the minimum size.

(a) Design a polynomial-time 2-approximation for the problem [Hint: Use the polynomial-time *Maximum Matching Algorithm*.]

(b) Show that MIN-VC in bipartite graphs can be solved in polynomial time.

1.11 A subset S of vertices in a graph $G = (V, E)$ is *independent* if no edges exist between any two vertices in S.

(a) Show that I is a maximum independent set of graph $G = (V, E)$ if and only if $V - I$ is a minimum vertex cover of G.

(b) Give an example to show that if C is a vertex cover within a factor of 2 from the minimum, then $V - C$ is still an independent set but may not be within a factor of 2 from the maximum.

1.12 Find a polynomial-time 2-approximation for the following problem:

STEINER MINIMUM TREE (SMT): Given a graph $G = (V, E)$ with a distance function on E, and a subset $S \subseteq V$, compute a shortest tree interconnecting the vertices in S.

1.13 There are n jobs $J_1, J_2, \ldots, J_n$ and m identical machines. Each job J_i, $1 \leq i \leq n$, needs to be processed in a machine without interruption for a time period p_i. Consider the problem of finding a scheduling to finish all jobs with the m machines in the minimum time. Graham [1966] proposed a simple algorithm for this problem: Put n jobs in an arbitrary order; whenever a machine becomes available, assign it the next job. Show that Graham's algorithm is a polynomial-time 2-approximation.

1.14 There are n students in a late-night study group. The time has come to order pizzas. Each student has his or her own list of preferred toppings (e.g., mushroom, pepperoni, onions, garlic, sausage, etc.), and each pizza may have only one topping. Answer the following questions:

(a) If each student wants to eat at least one half of a pizza with the topping on his or her preferred list, what is the complexity of computing the minimum number of pizzas to order to make everyone happy?

(b) If everyone wants to eat at least one third of a pizza with the topping on his or her preferred list, what is the complexity of computing the minimum number of pizzas to order to make everyone happy?

1.15 Assume that $\mathcal{C}$ is a collection of subsets of a set X. We say a set $Y \subseteq X$ *hits* a set $C \in \mathcal{C}$ if $Y \cap C \neq \emptyset$. A set $Y \subseteq X$ is a *hitting set* for $\mathcal{C}$ if Y hits every set $C \in \mathcal{C}$. Show that the following problems are **NP**-hard:

(a) MINIMUM HITTING SET (MIN-HS): Given a collection $\mathcal{C}$ of subsets of a set X, find a minimum hitting set Y for $\mathcal{C}$.

(b) Given a collection $\mathcal{C}$ of subsets of a set X, find a subset Y of X of the minimum size such that all sets $Y \cap C$ for $C \in \mathcal{C}$ are distinct.

(c) Given two collections $\mathcal{C}$ and $\mathcal{D}$ of subsets of X and a positive integer d, find a subset $A \subseteq X$ of size $|A| \leq d$ that minimizes the total number of subsets in $\mathcal{C}$ not hit by A and subsets in $\mathcal{D}$ hit by A.

1.16 Show that the following problems are **NP**-hard:

(a) Given a graph $G = (V, E)$ and a positive integer m, find the minimum subset $A \subseteq V$ such that A covers at least m edges and the complement of A has no isolated vertices.

(b) Given a 2-connected graph $G = (V, E)$ and a set $A \subseteq V$, find the minimum subset $B \subseteq V$ such that $A \cup B$ induces a 2-connected subgraph.

1.17 Show that the following problem is **NP**-complete:

Given two disjoint sets X and Y, and a collection $\mathcal{C}$ of subsets of $X \cup Y$, determine whether $\mathcal{C}$ can be partitioned into two disjoint subcollections covering X and Y, respectively.

1.18 Let $k > 0$. A collection $\mathcal{C}$ of subsets of a set X is a *k-set cover* if $\mathcal{C}$ can be partitioned into k disjoint subcollections each being a set cover for X.

(a) Consider the following problem:

k-SET COVER (k-SC): Given a collection $\mathcal{C}$ of subsets of a set X, determine whether it is a k-set cover.

Show that the problem 2-SC is **NP**-complete.

(b) Show that the following problem is not polynomial-time 2-approximable unless **P** = **NP**:

> Given a collection $\mathcal{C}$ of subsets of a set X, compute the minimum k such that $\mathcal{C}$ is a k-set cover.

1.19 For each 3-CNF formula F, we define a graph $G(F)$ as follows: The vertex set of $G(F)$ consists of all clauses and all literals in F. An edge exists in $G(F)$ between a clause C and a literal x if and only if x belongs to C, and an edge exists between two literals x and y if and only if $x = \bar{y}$. A 3-CNF formula is called a *planar* formula if $G(F)$ is a planar graph. Show that the following problems are **NP**-complete:

(a) NOT-ALL-EQUAL 3-SAT: Given a 3-CNF F, determine whether F has an assignment which assigns, for each clause C, value 1 to a literal in C and value 0 to another literal in C.

(b) ONE-IN-THREE 3-SAT: Given a 3-CNF F, determine whether F has an assignment which, for each clause C, assigns value 1 to exactly one literal in C.

(c) PLANAR 3-SAT: Given a planar 3-CNF formula F, determine whether F is satisfiable.

1.20 A subset D of vertices in a graph $G = (V, E)$ is called a *dominating set* if every vertex $v \in V$ either is in D or is adjacent to a vertex in D.

(a) Show that the problem of computing the minimum dominating set for a given graph is **NP**-hard.

(b) Show that the problem of determining whether there exist two disjoint dominating sets for a given graph is polynomial-time solvable.

(c) Show that the problem of determining whether there exist three disjoint dominating sets for a given graph is **NP**-complete. [Hint: Use NOT-ALL-EQUAL 3-SAT.]

(d) Show that the problem of computing the maximum number of disjoint dominating sets for a given graph is not $(3/2)$-approximable in polynomial time unless **P** = **NP**.

1.21 A graph is said to be *k-colorable* if its vertices can be partitioned into k disjoint independent sets.

(a) Show that the problem of deciding whether a given graph is 2-colorable or not is polynomial-time solvable.

(b) Show that the problem of deciding whether a given graph is 3-colorable or not is **NP**-complete.

(c) Show that the problem of computing, for a given graph G, the minimum k such that G is k-colorable is not $(3/2)$-approximable unless **P** = **NP**.

1.22 A subset C of vertices of a graph $G = (V, E)$ is a *clique* if the subgraph of G induced by C is a complete graph. Study the computational complexity of the following problems:

(a) For a given graph, compute the maximum number of disjoint vertex covers.

(b) For a given graph, compute the minimum number of disjoint cliques such that their union contains all vertices.

Historical Notes

Graham [1966] initiated the study of approximations using the performance ratio to evaluate the approximation algorithms. However, the importance of this work was not fully understood until Cook [1971] and Karp [1972] established the notion of **NP**-completeness and its ubiquitous existence in combinatorial optimization. With the theory of **NP**-completeness as its foundation, the study of approximation algorithms took off quickly in the 1970s. Garey and Johnson [1979] gave an account of the development in this early period.

The PTAS for KNAPSACK belongs to Sahni [1975]. The first FPTAS for KNAPSACK was discovered by Ibarra and Kim [1975]. Since then, many different FPTASs, including Algorithm 1.D of Section 1.1, have been found for KNAPSACK. Christofides [1976] found a polynomial-time (3/2)-approximation for TSP with the triangle inequality. So far, nobody has found a better one in terms of the performance ratio.

2

Greedy Strategy

Someone reminded me that I once said, "Greed is good."
Now it seems that it's legal.

— Gordon Gekko (in *Wall Street: Money Never Sleeps*)

I think greed is healthy. You can be greedy
and still feel good about yourself.

— Ivan Boesky

The greedy strategy is a simple and popular idea in the design of approximation algorithms. In this chapter, we study two general theories, based on the notions of independent systems and submodular potential functions, about the analysis of greedy algorithms, and present a number of applications of these methods.

2.1 Independent Systems

The basic idea of a greedy algorithm can be summarized as follows:

(1) We define an appropriate *potential function* $f(A)$ on potential solution sets A.

(2) Starting with $A = \emptyset$, we grow the solution set A by adding to it, at each stage, an element that maximizes (or, minimizes) the value of $f(A \cup \{x\})$, until $f(A)$ reaches the maximum (or, respectively, minimum) value.

We first consider a simple setting, in which the potential function is the same as the objective function. In the following, we write $\mathbb{N}^+$ to denote the set of positive integers, and $\mathbb{R}^+$ the set of nonnegative real numbers.

Let E be a finite set and $\mathcal{I}$ a family of subsets of E. The pair $(E, \mathcal{I})$ is called an *independent system* if

(I_1) $I \in \mathcal{I}$ and $I' \subseteq I \Rightarrow I' \in \mathcal{I}$.

Each subset in $\mathcal{I}$ is called an *independent subset*. Let $c : E \to \mathbb{R}^+$ be a nonnegative function. For every subset F of E, define $c(F) = \sum_{e \in F} c(e)$. Consider the following problem:

> MAXIMUM INDEPENDENT SUBSET (MAX-ISS): Given an independent system $(E, \mathcal{I})$ and a cost function $c : E \to \mathbb{R}^+$,

$$\begin{array}{ll} \text{maximize} & c(I) \\ \text{subject to} & I \in \mathcal{I}. \end{array}$$

We remark that the family $\mathcal{I}$ has, in general, an exponential size and cannot be given explicitly (and, hence, an exhaustive search for the maximum $c(I)$ is impractical). In most applications, however, the system $(E, \mathcal{I})$ is given in such a way that the condition of whether $I \in \mathcal{I}$ can be determined in polynomial time. Under this assumption, the following greedy algorithm, which uses the objective function c as the potential function, works in polynomial time.

Algorithm 2.A (*Greedy Algorithm for* MAX-ISS)

Input: An independent system $(E, \mathcal{I})$ and a cost function $c : E \to \mathbb{R}^+$.

(1) Sort all elements in $E = \{e_1, e_2, \ldots, e_n\}$ in the decreasing order of c. Without loss of generality, assume that $c(e_1) \geq c(e_2) \geq \cdots \geq c(e_n)$.

(2) Set $I \leftarrow \emptyset$.

(3) **For** $i \leftarrow 1$ **to** n **do**
  if $I \cup \{e_i\} \in \mathcal{I}$ **then** $I \leftarrow I \cup \{e_i\}$.

(4) Output $I_G \leftarrow I$. ■

For any instance $(E, \mathcal{I}, c)$ of the problem MAX-ISS, let I^* be its optimal solution and I_G the independent set produced by Algorithm 2.A. We will see that $c(I_G)/c(I^*)$ has a simple upper bound that is independent of the cost function c.

For any $F \subseteq E$, a set $I \subseteq F$ is called a *maximal independent subset* of F if no independent subset of F contains I as a proper subset. For any set $I \subseteq E$, let $|I|$ denote the number of elements in I. Define

$$\begin{aligned} u(F) &= \min\{|I| \mid I \text{ is a maximal independent subset of } F\}, \\ v(F) &= \max\{|I| \mid I \text{ is an independent subset of } F\}. \end{aligned} \tag{2.1}$$

Theorem 2.1 *The following inequality holds for any independent system* $(E, \mathcal{I})$ *and any function* $c : E \to \mathbb{R}^+$*:*

$$1 \leq \frac{c(I^*)}{c(I_G)} \leq \max_{F \subseteq E} \frac{v(F)}{u(F)}.$$

Proof. Assume that $E = \{e_1, e_2, \ldots, e_n\}$, and $c(e_1) \geq \cdots \geq c(e_n)$. Denote $E_i = \{e_1, \ldots, e_i\}$. We claim that $E_i \cap I_G$ is a maximal independent subset of E_i. To see this, we assume, by way of contradiction, that this is not the case; that is, there exists an element $e_j \in E_i \setminus I_G$ such that $(E_i \cap I_G) \cup \{e_j\}$ is independent. Now, consider the jth iteration of the loop of step (3) of Algorithm 2.A. The set I at the beginning of the jth iteration is a subset of I_G, and so $I \cup \{e_j\}$ must be a subset of $(E_i \cap I_G) \cup \{e_j\}$ and, hence, is an independent set. Therefore, the algorithm should have added e_j to I in the jth iteration. This contradicts the assumption that $e_j \notin I_G$.

From the above claim, we see that

$$|E_i \cap I_G| \geq u(E_i).$$

Moreover, since $E_i \cap I^*$ is independent, we have

$$|E_i \cap I^*| \leq v(E_i).$$

Now, we express $c(I_G)$ and $c(I^*)$ in terms of $|E_i \cap I_G|$ and $|E_i \cap I^*|$, respectively. We note that for each $i = 1, 2, \ldots, n$,

$$|E_i \cap I_G| - |E_{i-1} \cap I_G| = \begin{cases} 1, & \text{if } e_i \in I_G, \\ 0, & \text{otherwise.} \end{cases}$$

Therefore,

$$\begin{aligned} c(I_G) &= \sum_{e_i \in I_G} c(e_i) = c(e_1) \cdot |E_1 \cap I_G| + \sum_{i=2}^{n} c(e_i) \cdot (|E_i \cap I_G| - |E_{i-1} \cap I_G|) \\ &= \sum_{i=1}^{n-1} |E_i \cap I_G| \cdot (c(e_i) - c(e_{i+1})) + |E_n \cap I_G| \cdot c(e_n). \end{aligned}$$

Similarly,

$$c(I^*) = \sum_{i=1}^{n-1} |E_i \cap I^*| \cdot (c(e_i) - c(e_{i+1})) + |E_n \cap I^*| \cdot c(e_n).$$

Denote $\rho = \max_{F \subseteq E} v(F)/u(F)$. Then we have

$$\begin{aligned} c(I^*) &\leq \sum_{i=1}^{n-1} v(E_i) \cdot (c(e_i) - c(e_{i+1})) + v(E_n) \cdot c(e_n) \\ &\leq \sum_{i=1}^{n-1} \rho \cdot u(E_i) \cdot (c(e_i) - c(e_{i+1})) + \rho \cdot u(E_n) \cdot c(e_n) \leq \rho \cdot c(I_G). \quad \square \end{aligned}$$

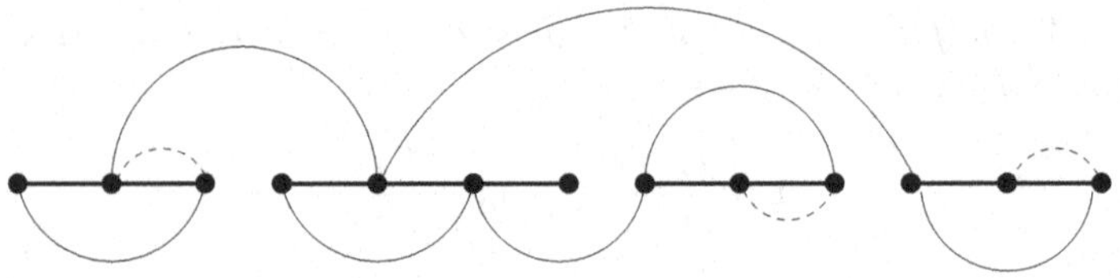

Figure 2.1: Two maximal independent subsets I and J for the problem MAX-HC (the thick lines indicate edges of I, the thin curves and dotted curves indicate the edges of J, and the dotted curves indicate edges shared by I and J).

We note that the ratio $\rho = \max_{F \subseteq E} v(F)/u(F)$ depends only on the structure of the family $\mathcal{I}$ and is independent of the cost function c. Thus, this upper bound is often easy to calculate. We demonstrate the application of this property in two examples.

First, consider the problem MAX-HC defined in Section 1.5. Each instance of this problem consists of n vertices and a distance table on these n vertices. The problem is to find a Hamiltonian circuit of the maximum total distance. Let E be the edge set of the complete graph on the n vertices. Let $\mathcal{I}$ be the family of subsets of E such that $I \in \mathcal{I}$ if and only if I is either a Hamiltonian circuit or a union of disjoint paths (i.e., paths that do not share any common vertex). Clearly, $(E, \mathcal{I})$ is an independent system and whether or not I is in $\mathcal{I}$ can be determined in polynomial time. That is, the problem MAX-HC is a special case of the problem MAX-ISS, and Algorithm 2.A runs on MAX-HC in polynomial time.

Lemma 2.2 *Let $(E, \mathcal{I})$ be the independent system defined above, and F a subset of E. Suppose that I and J are two maximal independent subsets of F. Then $|J| \leq 2|I|$.*

Proof. For $i = 1, 2$, let V_i denote the set of vertices of degree i in I. That is, V_1 is the set of end vertices in I and V_2 is the set of intermediate vertices in I. Clearly, $|I| = |V_2| + |V_1|/2$. Since I is a maximal independent subset of F, every edge in F either is incident on a vertex in V_2 or connects two endpoints of a path in I. Let J_2 be the set of edges in J incident on a vertex in V_2, and $J_1 = J \setminus J_2$. Since J is an independent set, at most two edges in J_2 could be incident on each vertex in V_2. That is, $|J_2| \leq 2|V_2|$. Moreover, every edge in J_1 must connect two endpoints in V_1 in a path of I, and at most one edge in J_1 could be incident on each vertex in V_1. Therefore, $|J_1| \leq |V_1|/2$. (Figure 2.1 shows an example of maximal independent subsets I and J.) Together, we have

$$|J| = |J_1| + |J_2| \leq \frac{|V_1|}{2} + 2|V_2| \leq 2|I|. \qquad \square$$

Theorem 2.3 *When it is applied to the problem* MAX-HC, Algorithm 2.A *is a polynomial-time 2-approximation.*

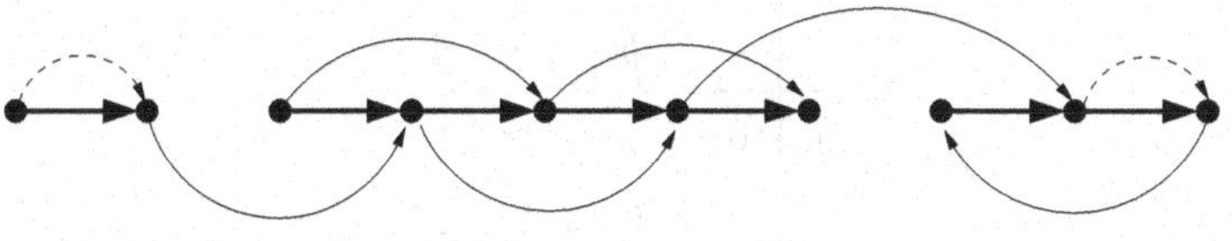

Figure 2.2: Two maximal independent subsets I and J for the problem MAX-DHP.

A similar application gives us a rather weaker performance ratio for the problem MAX-DHP, also defined in Section 1.5. An instance of this problem consists of n vertices and a directed distance table on these n vertices. The problem is to find a directed Hamiltonian path of the maximum total distance. Let E be the set of edges of the complete directed graph on the n vertices. Let $\mathcal{I}$ be the family of subsets of E such that $I \in \mathcal{I}$ if and only if I is a union of disjoint paths. Clearly, $(E, \mathcal{I})$ is an independent system, and whether or not I is in $\mathcal{I}$ can be determined in polynomial time.

Lemma 2.4 *Let $(E, \mathcal{I})$ be the independent system defined as above, and F a subset of E. Suppose that I and J are two maximal independent subsets of F. Then $|J| \leq 3|I|$.*

Proof. Since I is a maximal independent subset of F, every edge in F must have one of the following properties:

(1) It shares a head with an edge in I;

(2) It shares a tail with an edge in I; or

(3) It connects from the head to the tail of a maximal path in I.

(Figure 2.2 shows an example of two maximal independent subsets I and J.)

Let J_1, J_2, and J_3 be the subsets of edges in J that have properties (1), (2) and (3), respectively. Since J is an independent subset, each edge in I can share its head (or its tail) with at most one edge in J, and each maximal path in I can be connected from the head to the tail by at most one edge in J. That is, $|J_i| \leq |I|$, for $i = 1, 2, 3$. Thus,

$$|J| = |J_1| + |J_2| + |J_3| \leq 3|I|. \qquad \square$$

Theorem 2.5 *When it is applied to the problem* MAX-DHP, Algorithm 2.A *is a polynomial-time* 3*-approximation.*

The following simple example shows that the performance ratio given by the above theorem cannot be improved.

Example 2.6 Consider the following distance table on four vertices, in which the parameter ε is a positive real number less than 1:

	a	b	c	d
a	0	1	ε	ε
b	ε	0	1	ε
c	ε	$1+\varepsilon$	0	1
d	ε	ε	ε	0

It is clear that the longest Hamiltonian path has distance 3 and yet the greedy algorithm selects the edge (c, b) first and gets a path of total distance $1 + 3\varepsilon$. The performance ratio is, thus, equal to $3/(1+3\varepsilon)$, which approaches 3 when ε approaches zero. □

2.2 Matroids

Let E be a finite set and $\mathcal{I}$ a family of subsets of E. The pair $(E, \mathcal{I})$ is called a *matroid* if

(I_1) $I \in \mathcal{I}$ and $I' \subseteq I \Rightarrow I' \in \mathcal{I}$; and

(I_2) For any subset F of E, $u(F) = v(F)$,

where $u(F)$ and $v(F)$ are the two functions defined in (2.1). Thus, an independent system $(E, \mathcal{I})$ is a matroid if and only if, for any subset F of E, all maximal independent subsets of F have the same cardinality. From Theorem 2.1, we know that Algorithm 2.A produces an optimal solution for the problem MAX-ISS if the input instance $(E, \mathcal{I})$ is a matroid. The next theorem shows that this property actually characterizes the notion of matroids.

Theorem 2.7 *An independent system $(E, \mathcal{I})$ is a matroid if and only if for every nonnegative function $c : E \to \mathbb{R}^+$, the greedy* Algorithm 2.A *produces an optimal solution for the instance $(E, \mathcal{I}, c)$ of* MAX-ISS.

Proof. The "only if" part is just Theorem 2.1. Now, we prove the "if" part. Suppose that $(E, \mathcal{I})$ is not a matroid. Then we can find a subset F of E such that F has two maximal independent subsets I and I' with $|I| > |I'|$. Define, for any $e \in E$,

$$c(e) = \begin{cases} 1+\epsilon, & \text{if } e \in I', \\ 1, & \text{if } e \in I \setminus I', \\ 0, & \text{if } e \in E \setminus (I \cup I'), \end{cases}$$

where ϵ is a positive number less than $1/|I'|$ (so that $c(I) > c(I')$). Clearly, for this cost function c, Algorithm 2.A produces the solution set I', which is not optimal. □

The following are some examples of matroids.

Example 2.8 Let E be a finite set of vectors and $\mathcal{I}$ the family of linearly independent subsets of E. Then the size of the maximal independent subset of a subset $F \subseteq E$ is the rank of F and is unique. Thus, $(E, \mathcal{I})$ is a matroid. □

Example 2.9 Given a graph $G = (V, E)$, let $\mathcal{I}$ be the family of edge sets of acyclic subgraphs of G. Then it is clear that $(E, \mathcal{I})$ is an independent system. We verify that it is actually a matroid, which is usually called a *graph matroid*.

Consider a subset F of E. Suppose that the subgraph (V, F) of G has m connected components. We note that in each connected component C of (V, F), a maximal acyclic subgraph is just a spanning tree of C, in which the number of edges is exactly one less than the number of vertices in C. Thus, every maximal acyclic subgraph of (V, F) has exactly $|V| - m$ edges. So, condition (I_2) holds for the independent system $(E, \mathcal{I})$, and hence $(E, \mathcal{I})$ is a matroid. □

Example 2.10 Consider a directed graph $G = (V, E)$ and a nonnegative integer function f on V. Let $\mathcal{I}$ be the family of edge sets of subgraphs whose out-degree at any vertex u is no more than $f(u)$. It is clear that $(E, \mathcal{I})$ is an independent system. We verify that $(E, \mathcal{I})$ is actually a matroid.

For any subset $F \subseteq E$, let $d_F^+(u)$ be the number of out-edges at u which belong to F. Then, all maximal independent sets in F have the same size,

$$\sum_{u \in V} \min\{f(u), d_F^+(u)\}.$$

Therefore, $(E, \mathcal{I})$ is a matroid. □

In a matroid, all maximal independent subsets have the same cardinality. They are called *bases*. For instance, in a graph matroid defined by a connected graph $G = (V, E)$, every base is a spanning tree of G and they all have the same size $|V| - 1$.

There is an interesting relationship between the intersection of matroids and independent systems.

Theorem 2.11 *For any independent system $(E, \mathcal{I})$, there exist a finite number of matroids $(E, \mathcal{I}_i)$, $1 \le i \le k$, such that $\mathcal{I} = \bigcap_{i=1}^k \mathcal{I}_i$.*

Proof. Let $C_1, \dots, C_k$ be all minimal dependent sets of $(E, \mathcal{I})$ (i.e, they are the minimal sets among $\{F \mid F \subseteq E, F \not\in \mathcal{I}\}$). For each $i \in \{1, 2, \dots, k\}$, define

$$\mathcal{I}_i = \{F \subseteq E \mid C_i \not\subseteq F\}.$$

Then it is not hard to verify that $\mathcal{I} = \bigcap_{i=1}^k \mathcal{I}_i$. We next show that each $(E, \mathcal{I}_i)$ is a matroid.

It is easy to see that $(E, \mathcal{I}_i)$ is an independent system. Thus, it suffices to show that condition (I_2) holds for $(E, \mathcal{I}_i)$. Consider $F \subseteq E$. If $C_i \not\subseteq F$, then F contains a unique maximal independent set, which is itself. If $C_i \subseteq F$, then every maximal independent subset of F is equal to $F \setminus \{u\}$ for some $u \in C_i$ and hence has size $|F| - 1$. □

Theorem 2.12 *Suppose the independent system $(E, \mathcal{I})$ is the intersection of k matroids $(E, \mathcal{I}_i)$, $1 \le i \le k$; that is, $\mathcal{I} = \bigcap_{i=1}^k \mathcal{I}_i$. Then*

$$\max_{F \subseteq E} \frac{v(F)}{u(F)} \le k,$$

where $u(F)$ *and* $v(F)$ *are the two functions defined in* (2.1).

Proof. Let $F \subseteq E$. Consider two maximal independent subsets I and J of F with respect to $(E, \mathcal{I})$. For each $1 \le i \le k$, let I_i be a maximal independent subset of $I \cup J$ with respect to $(E, \mathcal{I}_i)$ that contains I. [Note that I is an independent subset of $I \cup J$ with respect to $(E, \mathcal{I}_i)$, and so such a set I_i exists.] For any $e \in J \setminus I$, if $e \in \bigcap_{i=1}^{k} (I_i \setminus I)$, then $I \cup \{e\} \in \bigcap_{i=1}^{k} \mathcal{I}_i = \mathcal{I}$, contradicting the maximality of I. Hence, e occurs in at most $k-1$ different subsets $I_i \setminus I$. It follows that

$$\sum_{i=1}^{k} |I_i| - k|I| = \sum_{i=1}^{k} |I_i \setminus I| \le (k-1)|J \setminus I| \le (k-1)|J|,$$

or

$$\sum_{i=1}^{k} |I_i| \le k|I| + (k-1)|J|.$$

Now, for each $1 \le i \le k$, let J_i be a maximal independent subset of $I \cup J$ with respect to $(E, \mathcal{I}_i)$ that contains J. Since, for each $1 \le i \le k$, $(E, \mathcal{I}_i)$ is a matroid, we must have $|I_i| = |J_i|$. In addition, for every $1 \le i \le k$, $|J| \le |J_i|$. Therefore, we get

$$k|J| \le \sum_{i=1}^{k} |J_i| = \sum_{i=1}^{k} |I_i| \le k|I| + (k-1)|J|.$$

It follows that $|J| \le k|I|$. □

Example 2.13 Consider the independent system $(E, \mathcal{I})$ for MAX-DHP defined in Section 2.1. Based on the analysis in the proof of Lemma 2.4 and Examples 2.9 and 2.10, we can see that $\mathcal{I}$ is actually the intersection of the following three matroids:

(1) The family $\mathcal{I}_1$ of all subgraphs with out-degree at most 1 at each vertex;

(2) The family $\mathcal{I}_2$ of all subgraphs with in-degree at most 1 at each vertex; and

(3) The family $\mathcal{I}_3$ of all subgraphs that do not contain a cycle when the edge direction is ignored.

Thus, Theorem 2.5 can also be derived from Theorem 2.12.

On the other hand, for the independent system $(E, \mathcal{I})$ for MAX-HC defined in Section 2.1, the analysis in the proof of Lemma 2.2 uses a more complicated counting argument and does not yield the simple property that $(E, \mathcal{I})$ is the intersection of two matroids. In fact, it can be proved that $(E, \mathcal{I})$ is *not* the intersection of two matroids. We remark that, in general, the problem MAX-ISS for an independent system that is the intersection of two matroids can often be solved in polynomial time. □

Example 2.14 Let X, Y, Z be three sets. We say two elements (x_1, y_1, z_1) and (x_2, y_2, z_2) in $X \times Y \times Z$ are *disjoint* if $x_1 \neq x_2$, $y_1 \neq y_2$, and $z_1 \neq z_2$. Consider the following problem:

> MAXIMUM 3-DIMENSIONAL MATCHING (MAX-3DM): Given three disjoint sets X, Y, Z and a nonnegative weight function c on all triples in $X \times Y \times Z$, find a collection $\mathcal{F}$ of disjoint triples with the maximum total weight.

For given sets X, Y, and Z, let $E = X \times Y \times Z$. Also, let $\mathcal{I}_X$ $(\mathcal{I}_Y, \mathcal{I}_Z)$ be the family of subsets A of E such that no two triples in any subset share an element in X $(Y, Z$, respectively). Then $(E, \mathcal{I}_X)$, $(E, \mathcal{I}_Y)$, and $(E, \mathcal{I}_Z)$ are three matroids and MAX-3DM is just the problem of finding the maximum-weight intersection of these three matroids. By Theorem 2.12, we see that Algorithm 2.A is a polynomial-time 3-approximation for MAX-3DM. □

2.3 Quadrilateral Condition on Cost Functions

Theorem 2.7 gives us a tight relationship between matroids and the optimality of greedy algorithms. It is interesting to point out that this tight relationship holds with respect to *arbitrary* nonnegative objective functions c. That is, if $(E, \mathcal{I})$ is a matroid, then the greedy algorithm will find optimal solutions for all objective functions c. On the other hand, if $(E, \mathcal{I})$ is not a matroid, then the greedy algorithm may still produce an optimal solution, but the optimality must depend on some specific properties of the objective functions. In this section, we present such a property.

Consider a directed graph $G = (V, E)$ and a cost function $c : E \to \mathbb{R}$. We say (G, c) satisfies the *quadrilateral condition* if, for any four vertices u, v, u', v' in V,

$$\begin{aligned} c(u, v) &\geq \max\{c(u, v'), c(u', v)\} \\ &\Longrightarrow c(u, v) + c(u', v') \geq c(u, v') + c(u', v). \end{aligned}$$

The quadrilateral condition is quite useful in the analysis of greedy algorithms. The following are some examples.

Let $G = (V_1, V_2, E)$ be a complete bipartite graph with $|V_1| = |V_2|$. Let $\mathcal{I}$ be the family of all matchings (recall that a *matching* of a graph is a set of edges that do not share any common vertex). Clearly, $(E, \mathcal{I})$ is an independent system. It is, however, not a matroid. In fact, for some subgraphs of G, maximal matchings may have different cardinalities (although all maximal matchings for G always have the same cardinality). A maximal matching in the bipartite graph is called an *assignment*.

> MAXIMUM ASSIGNMENT (MAX-ASSIGN): Given a complete bipartite graph $G = (V_1, V_2, E)$ with $|V_1| = |V_2|$, and an edge weight function $c : E \to \mathbb{R}^+$, find a maximum-weight assignment.

Theorem 2.15 *If the weight function c satisfies the* quadrilateral condition *for all* $u, u' \in V_1$ *and* $v, v' \in V_2$, *then* Algorithm 2.A *produces an optimal solution for the instance* (G, c) *of* MAX-ASSIGN.

Proof. Assume that $V_1 = \{u_1, u_2, \ldots, u_n\}$ and $V_2 = \{v_1, v_2, \ldots, v_n\}$. Also, assume, without loss of generality, that $M = \{(u_i, v_i) \mid i = 1, 2, \ldots, n\}$ is the assignment found by Algorithm 2.A, in the order of $(u_1, v_1), (u_2, v_2), \ldots, (u_n, v_n)$. We claim that there must be an optimal assignment that contains the edge (u_1, v_1): Let $M^* \subseteq E$ be an arbitrary optimal solution. If the edge (u_1, v_1) is not in M^*, then M^* must have two edges (u_1, v') and (u', v_1), where $v' \neq v_1$ and $u' \neq u_1$. From the greedy strategy of Algorithm 2.A, we know that $c(u_1, v_1) \geq \max\{c(u_1, v'), c(u', v_1)\}$. Therefore, by the quadrilateral condition,

$$c(u_1, v_1) + c(u', v') \geq c(u_1, v') + c(u', v_1).$$

This means that replacing edges (u_1, v') and (u', v_1) in M^* by (u_1, v_1) and (u', v') does not decrease the total weight of the assignment. This completes the proof of the claim.

Using the same argument, we can prove that for each $i = 1, 2, \ldots, n$, there exists an optimal assignment that contains all edges $(u_1, v_1), \ldots, (u_i, v_i)$. Thus, M is actually an optimal solution. □

Next, let us come back to the problem MAX-DHP.

Theorem 2.16 *For the problem* MAX-DHP *restricted to the graphs with distance functions satisfying the quadrilateral condition, the greedy* Algorithm 2.A *is a polynomial-time* 2*-approximation.*

Proof. Assume that $G = (V, E)$ is a directed graph, and $c : E \to \mathbb{R}^+$ is the distance function. Let $n = |V|$. Let $e_1, e_2, \ldots, e_{n-1}$ be the edges selected by Algorithm 2.A into the solution set H, in the order of their selection into H. They are, hence, in nonincreasing order of their length. For each $i = 1, 2, \ldots, n-1$, let P_i be a longest simple path in G that contains edges $e_1, e_2, \ldots, e_i$, and let $Q_i = P_i - \{e_1, e_2, \ldots, e_i\}$. In particular, $Q_0 = P_0$ is an optimal solution, and $Q_{n-1} = \emptyset$. For any set T of edges in G, we write $c(T)$ to denote the total length of edges in T. We claim that for $i = 1, 2, \ldots, n-1$,

$$c(Q_{i-1}) \leq c(Q_i) + 2c(e_i).$$

To prove the claim, let us consider the relationship between P_{i-1} and P_i. If $P_{i-1} = P_i$, then $Q_{i-1} = Q_i \cup \{e_i\}$, and so

$$c(Q_{i-1}) = c(Q_i) + c(e_i) \leq c(Q_i) + 2c(e_i).$$

If $P_{i-1} \neq P_i$, then we must have $e_i \notin P_{i-1}$. Assume that $e_i = (u, v)$. To add e_i to P_{i-1} to form a simple path P_i, we must remove up to three edges from P_{i-1} (and add e_i and some new edges):

(1) The edge in P_{i-1} that begins with u;

(2) The edge in P_{i-1} that ends with v; and

(3) An edge in the path from v to u if P_{i-1} contains such a subpath.

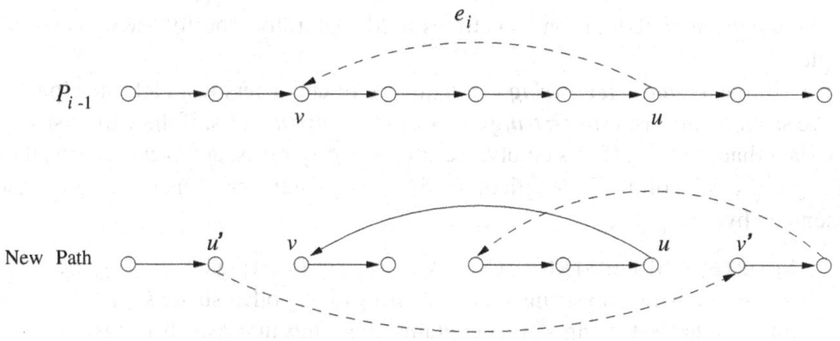

Figure 2.3: From path P_{i-1} to a new path.

In addition, these edges are all in $Q_{i-1} \setminus \{e_i\}$. Figure 2.3 shows an example of this process.

From the greedy strategy of the algorithm, we know that $c(e_i) \geq c(e)$ for any edge $e \in Q_{i-1}$. So, the total length of the edges removed is at most $3c(e_i)$. We consider two cases:

Case 1. We may form a new path passing through $e_1, \ldots, e_i$ from P_{i-1} by removing at most two edges, say, e'_j and e'_k. Then, $c((P_{i-1} \setminus \{e'_j, e'_k\}) \cup \{e_i\}) \leq c(P_i)$. Hence,

$$c(Q_{i-1}) \leq c(Q_i) + c(\{e'_j, e'_k\}) \leq c(Q_i) + 2c(e_i).$$

Case 2. We must remove three edges from P_{i-1} to form a new path passing through $e_1, e_2, \ldots, e_i$. As discussed above, these three edges must be (u, v'), (u', v), for some $u', v' \in V$, and an edge e in the subpath from v to u in P_{i-1}, and u, v, u', and v' are all distinct. This means that P_{i-1} has a subpath from u' to v', which contains these three edges. Thus, after deleting (u, v'), (u', v), and e, we can add edge (u', v') to form a new path (cf. Figure 2.3). Therefore, we have

$$\begin{aligned} c(Q_i) &\geq c(Q_{i-1}) - c(\{(u', v), e, (u, v')\}) + c(u', v') \\ &\geq c(Q_{i-1}) - c(e) - c(u, v) \\ &\geq c(Q_{i-1}) - 2c(e_i), \end{aligned}$$

where the second inequality follows from the quadrilateral condition on u, v, u', and v' and the fact that $c(u, v) \geq c(e')$ for all $e' \in Q_{i-1}$. This completes the proof of the claim.

Now, we note that $Q_{n-1} = \emptyset$, and so $c(Q_{n-1}) = 0$. Thus, we have

$$\begin{aligned} c(P_0) = c(Q_0) &\leq c(Q_1) + 2c(e_1) \\ &\leq c(Q_2) + 2c(e_1) + 2c(e_2) \\ &\leq \cdots \leq c(Q_{n-1}) + 2\sum_{i=1}^{n-1} c(e_i) = 2c(H). \end{aligned}$$ □

The quadrilateral condition sometimes holds naturally. The following is an example.

Recall that a *(character) string* is a sequence of characters from a finite alphabet Σ. We say a string s is a *superstring* of t, or t is a *substring* of s, if there exist strings u, v such that $s = utv$. If u is empty, we say t is a *prefix* of s, and if v is empty, then we say t is a *suffix* of s. The length of a string s is the number of characters in s, and is denoted by $|s|$.

> SHORTEST SUPERSTRING (SS): Given a set of strings $S = \{s_1, s_2, \ldots, s_n\}$ in which no string s_i is a substring of any other string $s_j, j \neq i$, find the shortest string s^* that contains all strings in S as substrings.

The problem SS has important applications in computational biology and data compression.

A string v is called an *overlap* of string s with respect to string t if v is both a suffix of s and a prefix of t, that is, if $s = uv$ and $t = vw$ for some strings u and w. We note that the overlap string may be an empty string. Also, the notion of overlap strings is not symmetric. That is, an overlap of s with respect to t may not be an overlap of t with respect to s. For any two strings s and t, we write $ov(s, t)$ to denote the longest overlap of s with respect to t.

To find an approximation algorithm for SS, we can transform the problem SS into the problem MAX-DHP: First, for any set $S = \{s_1, s_2, \ldots, s_n\}$ of strings, we define the *overlap graph* $G(S) = (S, E)$ to be the complete directed graph on the vertex set S, with all self-loops removed. For each edge (s_i, s_j) in E, we let its length be $c(s_i, s_j) = |ov(s_i, s_j)|$.

Suppose that s^* is a shortest superstring for S and that $s_1, s_2, \ldots, s_n$ are the strings in S in the order of occurrence from left to right in s^*. Then, for each $i = 1, \ldots, n-1$, s_i and s_{i+1} must have the maximal overlap in s^* for, otherwise, s^* could be shortened and would not be the shortest superstring. It is not hard to verify that the sequence $(s_1, s_2, \ldots, s_n)$ forms a directed Hamiltonian path H in the overlap graph $G(S)$, whose total edge length, denoted by $c(H)$, is equal to the sum of the total length of all overlap strings in s^*:

$$c(H) = \sum_{i=1}^{n-1} |ov(s_i, s_{i+1})|.$$

Next, consider an arbitrary directed Hamiltonian path $H = (s_{h(1)}, s_{h(2)}, \ldots, s_{h(n)})$ in $G(S)$. We can construct a superstring for S from H as follows: For each $i = 1, 2, \ldots, n-1$, let z_i be the prefix of $s_{h(i)}$ such that $s_{h(i)} = z_i \cdot ov(s_{h(i)}, s_{h(i+1)})$. Then, define $p(H) = z_1 z_2 \cdots z_{n-1} s_{h(n)}$. It is easy to check that $p(H)$ is a superstring of all $s_{h(i)}$, for $i = 1, 2, \ldots, n$ (cf. Figure 2.4). Clearly,

$$\begin{aligned} p(H)| &= \sum_{i=1}^{n-1} |z_i| + |s_{h(n)}| \\ &= \sum_{i=1}^{n-1} (|s_{h(i)}| - |ov(s_{h(i)}, s_{h(i+1)})|) + |s_{h(n)}| \end{aligned}$$

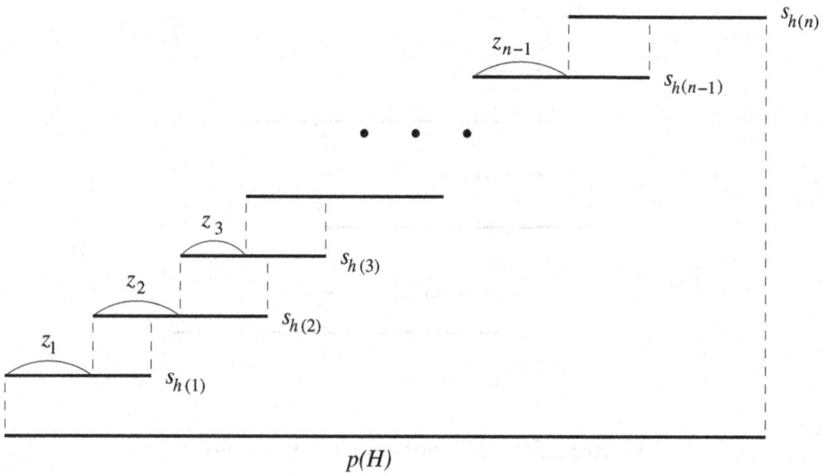

Figure 2.4: A superstring obtained from a Hamiltonian path.

$$= \sum_{i=1}^{n} |s_{h(i)}| - \sum_{i=1}^{n-1} |ov(s_{h(i)}, s_{h(i+1)})| = \sum_{i=1}^{n} |s_i| - c(H).$$

That is, the length of $p(H)$ equals the total length of the strings in S minus the total edge length of the path H. It follows that the string $p(H)$ generated from a longest directed Hamiltonian path H is a shortest superstring of S, and vice versa.

Theorem 2.17 *If H is a longest directed Hamiltonian path in the overlap graph $G(S)$, then the string $p(H)$ is a shortest superstring for S. Conversely, if s^* is a shortest superstring for S, then $s^* = p(H)$ for some longest directed Hamiltonian path H in $G(S)$.*

From this relationship, we can convert Algorithm 2.A into an approximation algorithm for the problem SS.

Algorithm 2.B (*Greedy Algorithm for* SS)

Input: A set $S = \{s_1, s_2, \ldots, s_n\}$ of strings.

(1) Set $G \leftarrow \{s_1, s_2, \ldots, s_n\}$.

(2) **While** $|G| > 1$ **do**

select s_i, s_j in G with the maximum $|ov(s_i, s_j)|$;
let $s_i \leftarrow s_i u$, where $s_j = ov(s_i, s_j)u$;
$G \leftarrow G \setminus \{s_j\}$.

(3) Output the only string s_G left in G. ∎

Tarhio and Ukkonen [1988] and Turner [1989] noticed independently that the overlap graph $G(S)$ satisfies the quadrilateral condition.

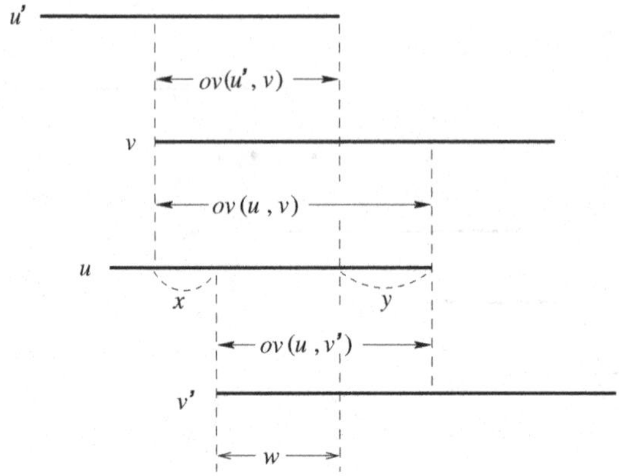

Figure 2.5: Overlaps among four strings.

Lemma 2.18 *Let $G(S)$ be the overlap graph of a set S of strings. Let u, v, u', and v' be four distinct strings in S. If $|ov(u,v)| \geq \max\{|ov(u,v')|, |ov(u',v)|\}$, then*

$$|ov(u,v)| + |ov(u',v')| \geq |ov(u,v')| + |ov(u',v)|.$$

Proof. The proof is trivial when $|ov(u,v)| \geq |ov(u,v')| + |ov(u',v)|$. Thus, we may assume that $|ov(u,v)| < |ov(u,v')| + |ov(u',v)|$.

Since both $ov(u,v)$ and $ov(u',v)$ are prefixes of v, $|ov(u',v)| \leq |ov(u,v)|$ implies that $ov(u',v)$ is a prefix of $ov(u,v)$. Similarly, we get that $ov(u,v')$ is a postfix of $ov(u,v)$ (see Figure 2.5). Because $|ov(u,v)| < |ov(u,v')| + |ov(u',v)|$, we know that the overlap of $ov(u',v)$ with respect to $ov(u,v')$ is not empty. Let $w = ov(ov(u',v), ov(u,v'))$. Then, we have $ov(u,v) = xwy$, $ov(u',v) = xw$ and $ov(u,v') = wy$ for some strings x and y (cf. Figure 2.5). That is, w is an overlap of u' with respect to v'. It follows that

$$|ov(u',v')| \geq |w| = |ov(u,v')| + |ov(u',v)| - |ov(u,v)|. \quad \square$$

Theorem 2.19 *Let s^* be a shortest superstring for S. Let $||S||$ be the total length of strings in S. Then*

$$||S|| - |s^*| \leq 2(||S|| - s_G),$$

where s_G is the superstring generated by Algorithm 2.B.

Proof. The theorem follows immediately from Lemma 2.18 and Theorem 2.16. $\square$

The following example shows that the bound on $(||S|| - s^*)/(||S|| - s_G)$ given in Theorem 2.19 is the best possible.

Example 2.20 Let $S = \{ab^k, b^{k+1}, b^k a\}$, where $k \geq 1$. The shortest superstring for S is $ab^{k+1}a$. However, Algorithm 2.B may generate a superstring $ab^k ab^{k+1}$ (by first merging the string ab^k with $b^k a$). Thus, for this example, we have $||S|| - |s_G| = k$ and $||S|| - |s^*| = 2k$. $\square$

In the above example, we also have $|s_G|/|s^*| = (2k+3)/(k+3)$. This means that the performance ratio of Algorithm 2.B cannot be better than 2. It has been conjectured that the performance ratio of Algorithm 2.B is indeed equal to 2; that is, $|s_G| \le 2|s^*|$, while the best known result is $|s_G| \le 4|s^*|$ [Blum et al., 1991].

In the above, we have seen a nice relationship between the problem SS and the problem MAX-DHP. This relationship can be extended to an interesting transformation from the problem SS to the traveling salesman problem TSP on directed graphs (called DIRECTED TSP).

Let $S = \{s_1, s_2, \ldots, s_n\}$ be an instance of the problem SS. Let s_{n+1} be the empty string. Consider a complete directed graph with vertex set $V = S \cup \{s_{n+1}\}$, and the distance function

$$d(s_i, s_j) = |s_i| - |ov(s_i, s_j)|,$$

for $s_i, s_j \in V$. [Note that $ov(s_{n+1}, s_i) = ov(s_i, s_{n+1}) = s_{n+1}$ for all $1 \le i \le n$.] It is easy to see that the shortest superstring for set S corresponds to a minimum Hamiltonian circuit with respect to the above distance function, and vice versa. Thus, a good approximation for this special case of DIRECTED TSP would also be a good approximation for the problem SS. It has also been proved that the above distance function satisfies the triangle inequality; that is, for any s_i, s_j, and s_k, with $1 \le i, j, k \le n+1$, $d(s_i, s_k) \le d(s_i, s_j) + d(s_j, s_k)$ [Turner, 1989]. Based on this relationship between the two problems DIRECTED TSP and SS, we will present, in Chapter 6, a polynomial-time 3-approximation for SS, even though no constant-ratio polynomial-time approximation for DIRECTED TSP is known.

2.4 Submodular Potential Functions

In the last three sections, we have applied the notion of independent systems to study greedy algorithms. The readers may have noticed that most applications we studied were about maximization problems. While minimization and maximization look similar, the behaviors of approximation algorithms for them are quite different. In this section, we introduce a different theory for the analysis of greedy algorithms for minimization problems.

Consider a finite set E (called the *ground* set) and a function $f : 2^E \to \mathbb{Z}$, where 2^E denotes the power set of E (i.e., the family of all subsets of E). The function f is said to be *submodular* if for any two sets A and B in 2^E,

$$f(A) + f(B) \ge f(A \cap B) + f(A \cup B). \tag{2.2}$$

Example 2.21 (a) The function $f(A) = |A|$ is submodular since

$$|A| + |B| = |A \cap B| + |A \cup B|.$$

Actually, in this case, the equality always holds, and we call f a *modular* function.

(b) Let $(E, \mathcal{I})$ be a matroid. For any $A \in 2^E$, define the *rank* of A as

$$rank(A) = \max_{I \in \mathcal{I}, I \subseteq A} |I|.$$

Then, the function *rank* is a submodular function.

To see this, consider two subsets A and B of E. Let $I_{A\cap B}$ be a maximal independent subset of $A \cap B$. Let I' be a maximal independent subset in A that contains $I_{A\cap B}$ as a subset. Since all maximal independent subsets in A have the same cardinality, we know that $|I'| = rank(A)$. Next, let I'' be a maximal independent subset in $A \cup B$ that contains I' as a subset. Similarly, we have $|I''| = rank(A \cup B)$. Let $J = I'' \setminus I'$. We note that J must be a subset of B since I' is a maximal independent subset in A. Thus, $I_{A\cap B} \cup J \subseteq I'' \cap B$ is an independent subset in B. So, $|I_{A\cap B} \cup J| = |I_{A\cap B}| + |J| \le rank(B)$. Or,

$$\begin{aligned} rank(A \cup B) &+ rank(A \cap B) - rank(A) \\ &= |I''| + |I_{A\cap B}| - |I'| = |J| + |I_{A\cap B}| \le rank(B). \end{aligned}$$ □

Assume that f is a submodular function on subsets of E. Define

$$\Delta_D f(C) = f(C \cup D) - f(C)$$

for any subsets C and D of E; that is, $\Delta_D f(C)$ is the extra amount of f value we gain by adding D to C. Then, the submodularity property (2.2) may be expressed as

$$\Delta_D f(A \cap B) \ge \Delta_D f(B), \tag{2.3}$$

where $D = A \setminus B$. When $D = \{x\}$ is a singleton, we simply write $\Delta_x f(C)$ instead of $\Delta_{\{x\}} f(C)$.

To see the role of submodular functions in the analysis of greedy algorithms, let us study a specific problem:

> MINIMUM SET COVER (MIN-SC): Given a set S and a collection $\mathcal{C}$ of subsets of S such that $\bigcup_{C\in\mathcal{C}} C = S$, find a subcollection $\mathcal{A} \subseteq \mathcal{C}$ with the minimum cardinality such that $\bigcup_{C\in\mathcal{A}} C = S$.

For any subcollection $\mathcal{A} \subseteq \mathcal{C}$, let $\cup\mathcal{A}$ denote the union of sets in $\mathcal{A}$; i.e., $\cup\mathcal{A} = \bigcup_{C\in\mathcal{A}} C$, and define $f(\mathcal{A}) = |\cup\mathcal{A}|$. Then f is a submodular function. To see this, we verify that, for any two subcollections $\mathcal{A}$ and $\mathcal{B}$ of $\mathcal{C}$, $f(\mathcal{A}) + f(\mathcal{B}) - f(\mathcal{A} \cup \mathcal{B})$ is equal to the number of elements in both $\cup\mathcal{A}$ and $\cup\mathcal{B}$. Moreover, every element in $\cup(\mathcal{A} \cap \mathcal{B})$ must appear in both $\cup\mathcal{A}$ and $\cup\mathcal{B}$. Therefore,

$$f(\mathcal{A}) + f(\mathcal{B}) - f(\mathcal{A} \cup \mathcal{B}) \ge f(\mathcal{A} \cap \mathcal{B}).$$

A function g on 2^E is said to be *monotone increasing* if, for all $A, B \subseteq E$,

$$A \subseteq B \Longrightarrow g(A) \le g(B).$$

It is easy to check that the above function f is monotone increasing. We can use this function f as the potential function to design a greedy approximation for MIN-SC as follows:

Algorithm 2.C (*Greedy Algorithm for* MIN-SC)

Input: A set S and a collection $\mathcal{C}$ of subsets of S.

(1) $\mathcal{A} \leftarrow \emptyset$.

(2) **While** $f(\mathcal{A}) < |S|$ **do**

Select a set $C \in \mathcal{C}$ to maximize $f(\mathcal{A} \cup \{C\})$;

Set $\mathcal{A} \leftarrow \mathcal{A} \cup \{C\}$.

(3) Output $\mathcal{A}$. ■

This approximation algorithm can be analyzed as follows:

Theorem 2.22 *Greedy* Algorithm 2.C *is a polynomial-time* $(1 + \ln \gamma)$*-approximation for* MIN-SC, *where* γ *is the maximum cardinality of a subset in the input collection* $\mathcal{C}$.

Proof. Let $A_1, \ldots, A_g$ be the solution found by Algorithm 2.C, in the order of their selection into the collection $\mathcal{A}$. Denote $\mathcal{A}_i = \{A_1, \ldots, A_i\}$, for $i = 0, 1, \ldots, g$. Let $C_1, C_2, \ldots, C_m$ be a minimum set cover (i.e., $m = opt$ is the number of subsets in a minimum set cover). By the greedy strategy, we know that A_{i+1} covers the maximum number of elements that are not yet covered by $\mathcal{A}_i$. Let U_i denote the set of elements in S that are not covered by $\mathcal{A}_i$. Then the total number of elements in U_i is $|U_i| = |S| - f(\mathcal{A}_i)$. The set U_i can be covered by the m subsets in the minimum set cover $\{C_1, \ldots, C_m\}$. By the pigeonhole principle, there must be a subset C_j that covers at least $(|S| - f(\mathcal{A}_i))/m$ elements in U_i. Therefore,

$$f(\mathcal{A}_{i+1}) - f(\mathcal{A}_i) \geq \frac{|S| - f(\mathcal{A}_i)}{m}. \tag{2.4}$$

Or, equivalently,

$$|S| - f(\mathcal{A}_{i+1}) \leq (|S| - f(\mathcal{A}_i)) \cdot \left(1 - \frac{1}{m}\right).$$

By a simple induction, we get

$$|U_i| = |S| - f(\mathcal{A}_i) \leq |S| \cdot \left(1 - \frac{1}{m}\right)^i \leq |S| \cdot e^{-i/m}.$$

We note that the size of U_i decreases from $|S|$ to 0, and so there must be an integer $i \in \{1, 2, \ldots, g\}$ such that $|U_{i+1}| < m \leq |U_i|$. That is, after $i + 1$ iterations of the while-loop of step (2) of Algorithm 2.C, there are at most $m - 1$ elements left uncovered, and so the greedy Algorithm 2.C will halt after at most $m - 1$ more iterations. That is, $g \leq i + m$. In addition, we have $m \leq |U_i| \leq |S|e^{-i/m}$, and so

$$i \leq m \cdot \ln\left(\frac{|S|}{m}\right) \leq m \cdot \ln \gamma$$

and

$$g \leq i + m \leq m(1 + \ln \gamma). \qquad \square$$

In the above, we used the pigeonhole principle to prove inequality (2.4). It may appear that the submodularity of the potential function f is not required in the proof. It is important to point out that the above proof actually used the submodularity property of f implicitly. To clarify this point, we present, in the following, an alternative proof that uses the submodularity property of f explicitly, and avoids the use of the specific meaning of f about set coverings.

Alternative Proof for (2.4). Recall that $\{C_1, \ldots, C_m\}$ is a minimum set cover. For each $j = 1, 2, \ldots, m$, let $\mathcal{C}_j = \{C_1, \ldots, C_j\}$. By the greedy strategy, we have, for each $1 \leq j \leq m$,

$$f(\mathcal{A}_{i+1}) - f(\mathcal{A}_i) = \Delta_{A_{i+1}} f(\mathcal{A}_i) \geq \Delta_{C_j} f(\mathcal{A}_i),$$

and so

$$f(\mathcal{A}_{i+1}) - f(\mathcal{A}_i) \geq \frac{1}{m} \cdot \sum_{j=1}^{m} \Delta_{C_j} f(\mathcal{A}_i).$$

On the other hand, we note that

$$|S| - f(\mathcal{A}_i) = f(\mathcal{A}_i \cup \mathcal{C}_m) - f(\mathcal{A}_i) = \sum_{j=1}^{m} \Delta_{C_j} f(\mathcal{A}_i \cup \mathcal{C}_{j-1}).$$

Therefore, to get (2.4), it suffices to have

$$\Delta_{C_j} f(\mathcal{A}_i) \geq \Delta_{C_j} f(\mathcal{A}_i \cup \mathcal{C}_{j-1}),$$

which follows from the submodularity and monotone increasing properties of the function f. □

The second proof above illustrates that the submodularity and monotone increasing properties of the potential function are sufficient conditions for inequality (2.4). In particular, for $m = 2$, inequality (2.4) is equivalent to

$$\Delta_{C_2} f(\mathcal{A}_i) \geq \Delta_{C_2} f(\mathcal{A}_i \cup C_1).$$

We will show, in the following, that this is equivalent to the condition that f is submodular and monotone increasing.

Lemma 2.23 *Let f be a submodular function on 2^E. Then, for all sets $A, C \subseteq E$,*

$$\Delta_C f(A) \leq \sum_{x \in C} \Delta_x f(A).$$

Proof. Note that if $x \in A$, then $\Delta_x f(A) = 0$. Thus, without loss of generality, we may assume that $A \cap C = \emptyset$. For any $x \in C$, set $X = A \cup \{x\}$ and $Y = A \cup (C - \{x\})$. Then, by the definition of submodular functions, we have

$$f(C \cup A) + f(A) = f(X \cup Y) + f(X \cap Y)$$
$$\leq f(X) + f(Y) = f(A \cup \{x\}) + f(A \cup (C - \{x\})).$$

It follows that

$$\Delta_C f(A) \leq \Delta_x f(A) + \Delta_{C-\{x\}} f(A).$$

The lemma can now be derived easily from this inequality. □

Lemma 2.24 *Let f be a function on all subsets of a set E. Then f is submodular if and only if, for any two subsets $A \subseteq B$ of E and any element $x \notin B$,*

$$\Delta_x f(A) \geq \Delta_x f(B). \tag{2.5}$$

Proof. From $A \subseteq B$ and $x \notin B$, we know that $(A \cup \{x\}) \cup B = B \cup \{x\}$ and $(A \cup \{x\}) \cap B = A$. Therefore, if f is submodular, then

$$f(A \cup \{x\}) + f(B) \geq f(A) + f(B \cup \{x\}).$$

That is,

$$\Delta_x f(A) \geq \Delta_x f(B).$$

Conversely, suppose (2.5) holds for all subsets $A \subseteq B$ and all $x \notin B$. Consider two arbitrary subsets A, B of E. Let $D = A \backslash B$, and assume that $D = \{x_1, \ldots, x_k\}$. Then

$$\Delta_D f(A \cap B) = \sum_{i=1}^{k} \Delta_{x_i} f((A \cap B) \cup \{x_1, \ldots, x_{i-1}\})$$
$$\geq \sum_{i=1}^{k} \Delta_{x_i} f(B \cup \{x_1, \ldots, x_{i-1}\}) = \Delta_D f(B).$$

(Note that $D = A \setminus B$, and so $x_i \notin B$ for all $i = 1, 2, \ldots, n$.) That is, inequality (2.3) holds and hence f is submodular. □

Lemma 2.25 *Let f be a function on all subsets of a set E. Then f is submodular and monotone increasing if and only if, for any two subsets $A \subseteq B$ and any element $x \in E$,*

$$\Delta_x f(A) \geq \Delta_x f(B).$$

Proof. We note that f is monotone increasing if and only if, for any subset $A \subseteq E$ and any $x \in E$, $\Delta_x f(A) \geq 0$. Now, assume that f is also submodular. Then, for any subsets $A \subseteq B \subseteq E$ and any $x \in E \setminus B$, we have, by Lemma 2.24, $\Delta_x f(A) \geq \Delta_x f(B)$; and for $x \in B$, we also have, by monotonicity of f, $\Delta_x f(A) \geq 0 = \Delta_x f(B)$.

Conversely, assume that $\Delta_x f(A) \geq \Delta_x f(B)$ for any subsets $A \subseteq B \subseteq E$ and any $x \in E$. Then, by Lemma 2.24, we know that f is submodular. In addition, set

$B = E$; we get $\Delta_x f(A) \geq \Delta_x f(E) = 0$ for all $x \in E$, which implies that f is monotone increasing. □

A submodular function is *normalized* if $f(\emptyset) = 0$. Every submodular function f can be normalized by setting $g(A) = f(A) - f(\emptyset)$. We note that if f is a normalized, monotone increasing submodular function, then $f(A) \geq 0$ for every set $A \subseteq E$. A normalized, monotone increasing, submodular function f is also called a *polymatroid function*. If f is defined on 2^E, then (E, f) is called a *polymatroid*. There are close relationships among polymatroids, matroids, and independent systems; see Exercises 2.18–2.24.

Consider a submodular function f on 2^E. Let $\Omega_f = \{C \subseteq E \mid (\forall x \in E)\ \Delta_x f(C) = 0\}$. Intuitively, Ω_f contains the *maximal sets* C under function f; that is, $f(C \cup B) = f(C)$ for all sets B.

Lemma 2.26 *Let f be a monotone increasing, submodular potential function on 2^E. Then, $\Omega_f = \{C \mid f(C) = f(E)\}$.*

Proof. If $C \in \Omega_f$, then

$$0 \leq f(E) - f(C) = \Delta_{E-C} f(C) \leq \sum_{x \in E-C} \Delta_x f(C) = 0.$$

Therefore, $f(C) = f(E)$.

Conversely, if $f(C) = f(E)$, then, for any $x \in E$, $f(C) \leq f(C \cup \{x\}) \leq f(E)$, and so $f(C) = f(C \cup \{x\})$. That is, for any $x \in E$, $\Delta_x f(C) = 0$. □

We are now ready to present a general result about greedy approximations which use a monotone increasing, submodular function as the potential function. Consider the following minimization problem.

> MINIMUM SUBMODULAR COVER (MIN-SMC): Given a finite set E, a normalized, monotone increasing, submodular function f on 2^E, and a nonnegative cost function c on E,

$$\begin{aligned} &\text{minimize} && c(A) = \sum_{x \in A} c(x), \\ &\text{subject to} && A \in \Omega_f. \end{aligned}$$

This minimization problem is a general form for many problems. In most applications, the submodular function f is not given explicitly in the form of the input/output pairs, but its value at any set $A \subseteq E$ is computable in polynomial time.

Example 2.27 Consider the weighted version of the problem MIN-SC.

> MINIMUM-WEIGHT SET COVER (MIN-WSC): Given a set S, a collection $\mathcal{C}$ of subsets of S with $\cup\mathcal{C} = S$, and a weight function w on all sets $C \in \mathcal{C}$, find a set cover with the minimum total weight.

Following the discussion on MIN-SC, let the input collection $\mathcal{C}$ be the ground set, and define, for any subcollection $\mathcal{A}$ of $\mathcal{C}$, $f(\mathcal{A}) = |\cup\mathcal{A}|$. Then, f is a submodular function. Moreover, f is apparently monotone increasing. With this function f, $\Delta_C f(\mathcal{A}) = 0$ if and only if $C \subseteq \cup\mathcal{A}$. This means that a subcollection $\mathcal{A}$ belongs to Ω_f if and only if $\mathcal{A}$ is a set cover of $S = \cup\mathcal{C}$. Thus, the problem MIN-WSC is just the problem MIN-SMC with respect to this potential function f. □

Example 2.28 A *hypergraph* $H = (V, \mathcal{C})$ is a pair of sets V and $\mathcal{C}$, where $\mathcal{C}$ is a family of subsets of V. Each element in V is called a *vertex* and each subset in $\mathcal{C}$ is called an *edge* (and sometimes, to emphasize that it is an edge of a hypergraph, called a *hyperedge*). The *degree* of a vertex is the number of edges that contain the vertex.

A subset A of vertices is called a *hitting set* of the hypergraph $H = (V, \mathcal{C})$ if every edge in $\mathcal{C}$ contains at least one vertex from A. The following problem is the weighted version of MIN-HS defined in Exercise 1.15:

> MINIMUM-WEIGHT HITTING SET (MIN-WHS): Given a hypergraph $H = (V, \mathcal{C})$ and a nonnegative weight function c on vertices in V, find a hitting set $A \subseteq V$ of the minimum total weight.

Let V be the ground set, and define, for each $A \subseteq V$, $E(A)$ to be the collection of sets $C \in \mathcal{C}$ such that $C \cap A \neq \emptyset$, and let $f(A) = |E(A)|$. Then it is easy to see that $E(A \cup B) = E(A) \cup E(B)$ and $E(A \cap B) \subseteq E(A) \cap E(B)$. Thus, we have

$$\begin{aligned} |E(A)| + |E(B)| &= |E(A) \cup E(B)| + |E(A) \cap E(B)| \\ &\geq |E(A \cup B)| + |E(A \cap B)|. \end{aligned}$$

That is, function f is a submodular function. Furthermore, it is easy to check that $E(\emptyset) = \emptyset$, and if $A \subseteq B$, then $E(A) \subseteq E(B)$. Thus, f is a normalized, monotone increasing, submodular function.

Now, what is Ω_f? It is not hard to verify that $A \in \Omega_f$ if and only if A is a hitting set. Thus, the problem MIN-WHS is just the problem MIN-SMC with respect to this submodular potential function f. □

The problem MIN-SMC has a natural greedy algorithm: In each iteration, we add an element x to the solution set A to maximize the value $\Delta_x f(A)$, relative to the cost $c(x)$.

Algorithm 2.D (*Greedy Algorithm* for MIN-SMC)

Input: A finite set E, a submodular function f on 2^E, and a function $c : E \to \mathbb{R}^+$.

(1) Set $A \leftarrow \emptyset$.

(2) **While** there exists an $x \in E$ such that $\Delta_x f(A) > 0$ **do**

  select a vertex x that maximizes $\Delta_x f(A)/c(x)$;

  $A \leftarrow A \cup \{x\}$.

(3) **Return** $A_G \leftarrow A$. ■

The following theorem gives an estimation of the performance of this algorithm. We write $H(n)$ to denote the *harmonic function* $H(n) = \sum_{i=1}^{n} 1/i$. Note that $H(n) \leq 1 + \ln n$ (see Exercise 2.6).

Theorem 2.29 *Let f be a normalized, monotone increasing, submodular function. Then* Algorithm 2.D *produces an approximate solution within a factor of $H(\gamma)$ from the optimal solution to the input (E, f, c), where $\gamma = \max_{x \in E} f(\{x\})$.*

Proof. Let A be the approximate solution obtained by Algorithm 2.D. Assume that $x_1, x_2, \ldots, x_k$ are the elements of A, in the order of their selection into the set. Denote $A_i = \{x_1, x_2, \ldots, x_i\}$; in particular, $A_0 = \emptyset$. Let A^* be an optimal solution to the same instance.

For any set $B \subseteq E$, we write $c(B)$ to denote the total cost of B: $c(B) = \sum_{x \in B} c(x)$. We are going to prove that

$$c(A) \leq c(A^*) \cdot H(\gamma)$$

by a weight-decomposition counting argument. That is, we decompose the total cost $c(A)$ of the approximate solution and distribute it to the elements of the optimal solution A^* through a weight function $w(y)$ on $y \in A^*$. Then we calculate the weight decomposition according to the optimal solution A^* and show that each element $y \in A^*$ can pick up at most weight $c(y) \cdot H(\gamma)$. It follows, therefore, that $c(A^*)$ is at least $c(A)/H(\gamma)$.

In other words, we need to assign weight $w(y)$ to each element y of A^* so that it satisfies the following properties:

(a) $c(A) \leq \sum_{y \in A^*} w(y)$; and

(b) $w(y) \leq c(y) \cdot H(\gamma)$.

Property (b) implies that $\sum_{y \in A^*} w(y) \leq c(A^*)H(\gamma)$. Thus, properties (a) and (b) together establish the desired result.

First, to simplify the notation, we let $r_i = \Delta_{x_i} f(A_{i-1})$ and $z_{y,i} = \Delta_y f(A_{i-1})$. Now, we define, for each $y \in A^*$,

$$w(y) = \sum_{i=1}^{k} (z_{y,i} - z_{y,i+1}) \frac{c(x_i)}{r_i}.$$

Before we prove properties (a) and (b), we observe that

$$\sum_{i=1}^{k} (z_{y,i} - z_{y,i+1}) = z_{y,1} - z_{y,k+1} = \Delta_y f(A_0) - \Delta_y f(A_k) = f(\{y\}).$$

[In the above, $\Delta_y f(A_0) = f(\{y\})$ because f is normalized, and $\Delta_y f(A_k) = 0$ because $A_k = A \in \Omega_f$.] Therefore,

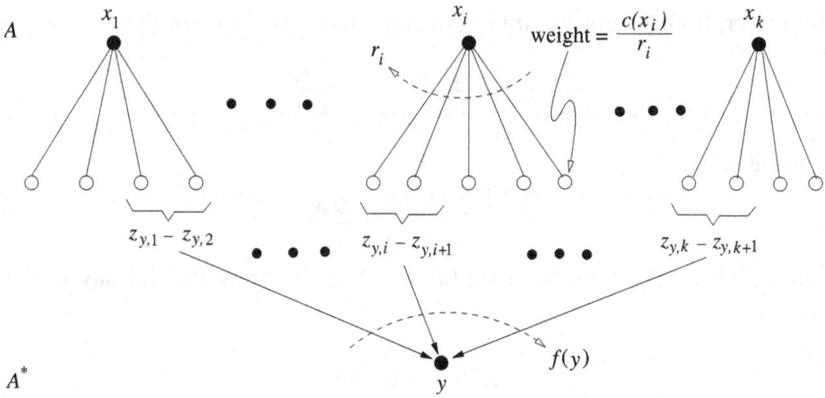

Figure 2.6: The weight decomposition.

$$\sum_{y\in A^*}\sum_{i=1}^{k}(z_{y,i} - z_{y,i+1}) = \sum_{y\in A^*} f(\{y\})$$
$$\geq f(A^*) = f(A) = \sum_{i=1}^{k}\Delta_{x_i} f(A_{i-1}) = \sum_{i=1}^{k} r_i,$$

since both A^* and A are in Ω_f. This relationship provides some intuition about how the weight-decomposition function is defined: As illustrated in Figure 2.6, we divide each element x_i into r_i parts, each of weight $c(x_i)/r_i$, so that the total weight of all parts, over all $x_i \in A$, is $c(A)$. Then each $y \in A^*$ picks up $z_{y,i} - z_{y,i+1}$ parts from the element x_i. The total number of parts picked up by y, disregarding the different weight, is $f(\{y\})$. Our goal here is to distribute part of each $x_i \in A$ to some $y \in A^*$, while each $y \in A^*$ does not take too much weight.

We now proceed to prove properties (a) and (b). For property (a), we can write weight $w(y)$ in the following form:

$$w(y) = \sum_{i=1}^{k}(z_{y,i} - z_{y,i+1})\frac{c(x_i)}{r_i}$$
$$= \frac{c(x_1)}{r_1} z_{y,1} + \sum_{i=2}^{k}\left(\frac{c(x_i)}{r_i} - \frac{c(x_{i-1})}{r_{i-1}}\right) z_{y,i}.$$

[Note that $z_{y,k+1} = \Delta_y f(A_k) = 0$.] In addition, $c(A)$ can also be expressed in a similar form:

$$c(A) = \sum_{i=1}^{k}\frac{r_i}{r_i}c(x_i) = \sum_{i=1}^{k}\left(\sum_{j=i}^{k} r_j - \sum_{j=i+1}^{k} r_j\right)\frac{c(x_i)}{r_i}$$
$$= \frac{c(x_1)}{r_1}\sum_{j=1}^{k} r_j + \sum_{i=2}^{k}\left(\frac{c(x_i)}{r_i} - \frac{c(x_{i-1})}{r_{i-1}}\right)\sum_{j=i}^{k} r_j.$$

Moreover, from the greedy strategy of Algorithm 2.D, we know that

$$\frac{r_1}{c(x_1)} \geq \frac{r_2}{c(x_2)} \geq \cdots \geq \frac{r_k}{c(x_k)};$$

or, equivalently,

$$\frac{c(x_i)}{r_i} - \frac{c(x_{i-1})}{r_{i-1}} \geq 0,$$

for all $i = 1, \ldots, k$. Thus, to prove (a), it suffices to prove that for any $i = 1, 2, \ldots, k$,

$$\sum_{j=i}^{k} r_j \leq \sum_{y \in A^*} z_{y,i}.$$

This inequality holds since, by Lemmas 2.23 and 2.26,

$$\begin{aligned}
\sum_{j=i}^{k} r_j &= \sum_{j=i}^{k} \Delta_{x_j} f(A_{j-1}) = \sum_{j=i}^{k} (f(A_j) - f(A_{j-1})) \\
&= f(A) - f(A_{i-1}) = f(A^*) - f(A_{i-1}) \\
&= f(A^* \cup A_{i-1}) - f(A_{i-1}) = \Delta_{A^*} f(A_{i-1}) \\
&\leq \sum_{y \in A^*} \Delta_y f(A_{i-1}) = \sum_{y \in A^*} z_{y,i}.
\end{aligned}$$

Next, we prove property (b). Let y be a fixed element in A^*. From the greedy strategy of Algorithm 2.D, we know that if $z_{y,i} > 0$, then

$$\frac{c(x_i)}{r_i} \leq \frac{c(y)}{z_{y,i}},$$

for all $i = 1, 2, \ldots, k$. In addition, we know from Lemma 2.25 that $z_{y,i} \geq z_{y,i+1}$. Let $\ell = \max\{i \mid 1 \leq i \leq k, z_{y,i} > 0\}$. We have

$$\begin{aligned}
w(y) &= \sum_{i=1}^{\ell} (z_{y,i} - z_{y,i+1}) \frac{c(x_i)}{r_i} \\
&\leq \sum_{i=1}^{\ell} (z_{y,i} - z_{y,i+1}) \frac{c(y)}{z_{y,i}} = c(y) \sum_{i=1}^{\ell} \frac{z_{y,i} - z_{y,i+1}}{z_{y,i}}.
\end{aligned}$$

Note that for any integers $p > q > 0$, we have

$$\frac{p-q}{p} = \sum_{j=q+1}^{p} \frac{1}{p} \leq \sum_{j=q+1}^{p} \frac{1}{j} = H(p) - H(q).$$

So, we have

$$w(y) \le c(y) \sum_{i=1}^{\ell-1} (H(z_{y,i}) - H(z_{y,i+1})) + c(y)\, H(z_{y,\ell}) = c(y)\, H(z_{y,1}).$$

Note that $z_{y,1} = f(\{y\}) \le \gamma$ for all $y \in A^*$. Therefore, we have proved property (b) and, hence, the theorem. □

2.5 Applications

Now we present some applications of the greedy Algorithm 2.D.

First, from Example 2.27, we get the upper bound for the performance ratio of the greedy algorithm for MIN-WSC immediately. More specifically, the submodular potential function f for the problem MIN-WSC is defined to be $f(\mathcal{A}) = |\cup \mathcal{A}|$. Therefore, when applied to MIN-WSC, the greedy strategy for Algorithm 2.D is to select, at each stage, the set $C \in \mathcal{C}$ with the highest value of

$$\frac{|\cup(\mathcal{A} \cup \{C\})| - |\cup \mathcal{A}|}{c(C)},$$

where $c(C)$ is the weight of set C, and add C to the solution collection $\mathcal{A}$. Also, the parameter γ in the performance ratio $H(\gamma)$ of Theorem 2.29 is equal to the maximum value of $f(\{C\}) = |C|$ over all $C \in \mathcal{C}$. Therefore, we have the following result:

Corollary 2.30 *When it is applied to the problem* MIN-WSC, Algorithm 2.D *is a polynomial-time $H(m)$-approximation, where m is the maximum cardinality of subsets in the input collection $\mathcal{C}$.*

From Example 2.28, we know that the function $f(A) = |E(A)|$ is monotone increasing and submodular for the problem MIN-WHS. With respect to this potential function f, Algorithm 2.D selects, at each stage, the element $x \in S$ with the highest value of

$$\frac{|E(A \cup \{x\})| - |E(A)|}{c(x)},$$

and adds x to the solution set A. We note that in the setting of the problem MIN-WHS, the parameter γ in the performance ratio $H(\gamma)$ of Theorem 2.29 is just the maximum degree over all vertices. So, we get the following result:

Corollary 2.31 *When it is applied to the problem* MIN-WHS, Algorithm 2.D *is a polynomial-time $H(\delta)$-approximation, where δ is the maximum degree of a vertex in the input hypergraph.*

Note that if all edges in the input hypergraph $H = (V, \mathcal{C})$ have exactly two elements, then this subproblem of MIN-WHS is actually the weighted version of the vertex cover problem MIN-VC (see Exercise 1.10).

> MINIMUM-WEIGHT VERTEX COVER (MIN-WVC): Given a graph $G = (V, E)$, with a nonnegative weight function $c : V \to \mathbb{R}^+$, find a vertex cover of the minimum total weight.

We prove that the bound $H(\delta)$ of Corollary 2.31 is actually tight, even for the nonweighted version of MIN-VC on bipartite graphs.

Theorem 2.32 *For any $n \geq 1$, there exists a bipartite graph G with degree at most n and a minimum vertex cover of size $n!$ such that* Algorithm 2.D *produces a vertex cover of size $H(n) \cdot (n!)$ on graph G.*

Proof. Let $V_1, V_{2,1}, V_{2,2}, \ldots, V_{2,n}$ be $n+1$ pairwisely disjoint sets of size $|V_1| = n!$ and $|V_{2,i}| = n!/i$, for each $i = 1, 2, \ldots, n$. The bipartite graph G has the vertex sets V_1 and $V_2 = \bigcup_{i=1}^{n} V_{2,i}$. To define the edges in G, we perform the following process for each $1 \leq i \leq n$: We partition V_1 into $n!/i$ disjoint subsets, each of size i, and build a one-to-one correspondence between these $n!/i$ subsets and $n!/i$ vertices in $V_{2,i}$. Then, for each subset A of V_1, we connect every vertex in A to the vertex in $V_{2,i}$ that corresponds to subset A.

Thus, in the bipartite graph G, each vertex in V_1 has degree n and each vertex in $V_{2,i}$ has degree $i \leq n$. Clearly, V_1 is a minimum hitting set, which has size $n!$. However, the greedy Algorithm 2.D on graph G may produce V_2 as the hitting set, which has size $\sum_{i=1}^{n} (n!)/i = H(n) \cdot (n!)$. □

The above result indicates that Algorithm 2.D is not a good approximation for the nonweighted MIN-VC, as MIN-VC actually has a polynomial-time 2-approximation, and MIN-VC in bipartite graphs can be solved in polynomial time (see Exercise 1.10). On the other hand, Algorithm 2.D is probably the best approximation for the nonweighted hitting set problem, unless certain complexity hierarchies collapse (see Historical Notes).

Our next example is the problem of subset interconnection design. Recall that for any graph $G = (V, E)$ and any set $S \subseteq V$, $G|_S$ denotes the subgraph of G induced by set S; i.e., $G|_S$ is the graph with vertex set S and edge set $E|_S = \{\{x, y\} \in E \mid x, y \in S\}$. For any subsets $S_1, S_2, \ldots, S_m$ of V, we say a subgraph $H = (V, F)$ of G is a *feasible graph* for $S_1, S_2, \ldots, S_m$ if, for each $i = 1, 2, \ldots, m$, the subgraph $H|_{S_i}$ induced by S_i is connected.

> WEIGHTED SUBSET INTERCONNECTION DESIGN (WSID): Given a complete graph $G = (V, E)$ with a nonnegative edge weight function $c : E \to \mathbb{R}^+$, and m vertex subsets $S_1, S_2, \ldots, S_m \subseteq V$, find a feasible subgraph $H = (V, F)$ for $S_1, S_2, \ldots, S_m$, with the minimum total edge weight.

Example 2.33 Let $V = \{v_1, v_2, \ldots, v_5\}$, and consider the five subsets $S_1 = \{v_1, v_2\}$, $S_2 = \{v_1, v_2, v_3\}$, $S_3 = \{v_3, v_4, v_5\}$, $S_4 = \{v_1, v_2, v_4\}$, and $S_5 = \{v_2, v_4, v_5\}$. These subsets form a hypergraph on V, as shown in Figure 2.7, together with a cost function c. Figure 2.8 shows two feasible graphs for these subsets. With respect to the cost function c given in Figure 2.7, the graph in Figure 2.8(b) is a minimum-cost feasible graph. □

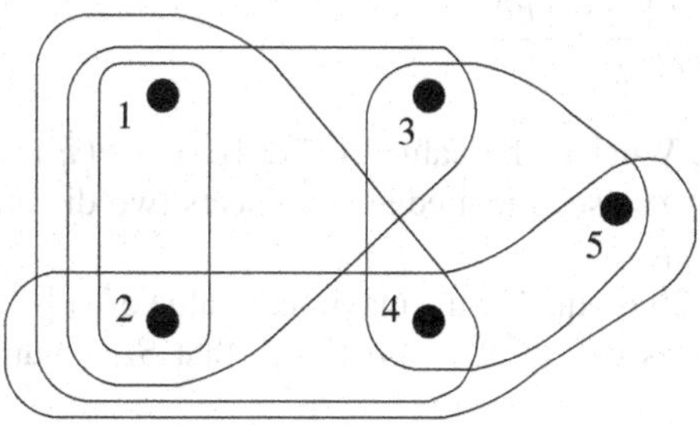

$c(i,j)$	1	2	3	4	5
1		5	6	7	8
2			5	6	7
3				5	6
4					5
5					

Figure 2.7: A hypergraph and its cost function.

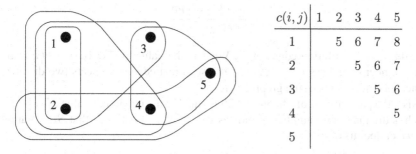

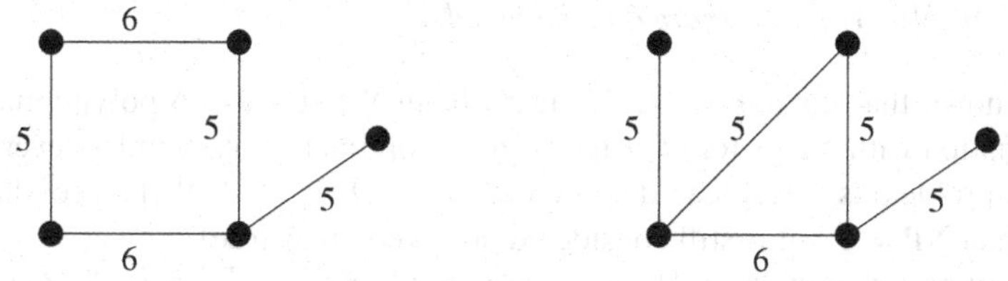

(a) (b)

Figure 2.8: Feasible graphs for the input of Figure 2.7.

In the following, we define a submodular function r on subsets of the edge set E. Consider the graph matroid of the induced subgraph $G|_{S_i} = (V, E_i)$ (see Example 2.9), where $E_i = E|_{S_i}$. In this graph matroid, a set $I \subseteq E_i$ is an independent subset if (S_i, I) is an acyclic subgraph of $G|_{S_i}$. Let r_i be the rank function of the graph matroid of graph $G|_{S_i}$ (see Example 2.21(b)). That is, for any $A \subseteq E$, $r_i(A) =$ the size of the largest edge set $I \subseteq A \cap E_i$ such that (S_i, I) is an acyclic subgraph of $G|_{S_i}$. Equivalently,

$$r_i(A) = |S_i| - \text{the number of connected components of the graph } (S_i, A \cap E_i).$$

By Example 2.21(b), r_i is a submodular function.

Now, define $r(A) = \sum_{i=1}^{m} r_i(A)$. Note that the sum of submodular functions is submodular. Therefore, r is a submodular function. Furthermore, it is not hard to check that r is monotone increasing and normalized.

For this submodular function r, the set Ω_r is the collection of sets $A \subseteq E$ such that $r(A \cup \{e\}) = r(A)$ for all edges e in E. It is not hard to see that Ω_r is just the set of all feasible graphs. Thus, the problem WSID is actually the minimization problem MIN-SMC with respect to the submodular potential function r. So, Algorithm 2.D and Theorem 2.29 can be applied to it.

To be more precise, the greedy criterion of Algorithm 2.D for the problem WSID is to select, at each stage, an edge $\{e\}$ with the maximum ratio

$$\frac{r(F \cup \{e\}) - r(F)}{c(e)}$$

and add it to the solution edge set F. What is the value $r(F \cup \{e\}) - r(F)$? It is the number of indices $i \in \{1, 2, \ldots, m\}$ such that edge e connects two distinct connected components of the graph $G|_{F \cap S_i}$.

Also, the parameter γ of Theorem 2.29 is equal to the maximum value of $r(\{e\})$, which is the maximum number of indices $i \in \{1, 2, \ldots, m\}$ such that S_i contains the two endpoints of e.

Corollary 2.34 *When it is applied to the problem* WSID, Algorithm 2.D *is a polynomial-time $H(K)$-approximation, where K is the maximum number of induced subgraphs $G|_{S_i}$ that share a common edge.*

It is known that for $0 < \rho < 1$, the problem WSID has no polynomial-time approximation within a factor of $\rho \ln n$ from the optimal solution unless every **NP**-complete problem is solvable in deterministic time $O(n^{polylog n})$[1] (this condition is weaker than **NP** = **P** but is still considered not likely to be true).

For a connected graph $G = (V, E)$, we say a subset $C \subseteq V$ is a *connected vertex cover* if C is a vertex cover for G and the induced subgraph $G|_C$ is connected. Consider the following problem:

> MINIMUM-WEIGHT CONNECTED VERTEX COVER (MIN-WCVC): Given a connected graph $G = (V, E)$ and a nonnegative vertex weight function $c : V \to \mathbb{R}^+$, find a connected vertex cover with the minimum total weight.

For a graph $G = (V, E)$ and a subset $C \subseteq V$, let $g(C)$ be the number of edges in E that are not covered by C, and $h(C)$ the number of connected components of $G|_C$. Define $p(C) = |E| - g(C) - h(C)$. Clearly, $p(\emptyset) = |E| - g(\emptyset) - h(\emptyset) = 0$.

We are going to prove that p is a monotone increasing, submodular function, using a new characterization of submodular functions. In the following, we write $\Delta_x \Delta_y f(A)$ to denote $\Delta_y f(A \cup \{x\}) - \Delta_y f(A)$. For the proofs of the following two lemmas, see Exercise 2.14.

Lemma 2.35 *Let f be a function on 2^E. Then f is submodular if and only if for any $A \subseteq E$ and any two distinct elements $x, y \notin A$,*

$$\Delta_x \Delta_y f(A) \leq 0.$$

Lemma 2.36 *Let f be a function on 2^E. Then f is monotone increasing and submodular if and only if for any $A \subseteq E$ and $x, y \in E$,*

$$\Delta_x \Delta_y f(A) \leq 0.$$

---

[1]The notation *polylog* n denotes the class of functions $(\log n)^k$, for all $k \geq 1$.

Now, we apply this characterization to show that p is a monotone increasing, submodular function.

Lemma 2.37 *p is monotone increasing and submodular.*

Proof. Consider a vertex subset C and a vertex $u \notin C$. Then $\Delta_u p(C) = -\Delta_u g(C) - \Delta_u h(C)$. We observe that $-\Delta_u g(C)$ is just the number of edges incident on u in graph G that are not covered by C. It follows that $-\Delta_u g(C) = |N(u) \setminus C|$, where $N(u)$ is the set of vertices in G that are adjacent to u. Moreover, $-\Delta_u h(C)$ is equal to the number of connected components in $G|_C$ that are adjacent to u *minus* 1. Therefore, we always have $-\Delta_u g(C) \geq 0$ and $-\Delta_u h(C) \geq -1$.

By Lemma 2.36, it is sufficient to prove that for any vertex subset C and two vertices u and v,

$$\Delta_v \Delta_u p(C) \leq 0.$$

Note that if $u \in C$, then both $\Delta_u p(C \cup \{v\})$ and $\Delta_u p(C)$ are equal to 0, and hence $\Delta_v \Delta_u p(C) = 0$. Also, if $v \in C$, then we have $\Delta_u p(C \cup \{v\}) = \Delta_u p(C)$, and hence $\Delta_v \Delta_u p(C) = 0$. Thus, we may assume that neither u nor v belongs to C.

We consider three cases.

Case 1: $u = v$. Since $\Delta_u p(C \cup \{v\}) = 0$, it suffices to show $\Delta_u p(C) \geq 0$. If $C \cap N(u) = \emptyset$, then $-\Delta_u g(C) = \deg(u)$ and $\Delta_u h(C) = -1$, which implies that $\Delta_u p(C) = \deg(u) - 1 \geq 0$, because G is connected and so $\deg(u)$ is at least 1. If $C \cap N(u) \neq \emptyset$, then u is adjacent to at least one connected component of $G|_C$ and hence $-\Delta_u h(C) \geq 0$, which also implies that $\Delta_u p(C) \geq 0$.

Case 2: $u \neq v$ and u is not adjacent to v. Then $N(u) \setminus (C \cup \{v\}) = N(u) \setminus C$, and hence $-\Delta_u g(C \cup \{v\}) = -\Delta_u g(C)$. Consider an arbitrary connected component of $G|_{C \cup \{v\}}$ that is adjacent to u. If it does not contain v, then it is also a connected component of $G|_C$ adjacent to u. If it contains v, then it must contain at least one connected component of $G|_C$ adjacent to u. Thus, the number of connected components of $G|_{C \cup \{v\}}$ adjacent to u is no more than the number of connected components of $G|_C$ adjacent to u; that is, $-\Delta_u h(C \cup \{v\}) \leq -\Delta_u h(C)$. So $\Delta_u p(C \cup \{v\}) \leq \Delta_u p(C)$.

Case 3: $u \neq v$ but u is adjacent to v. Then $N(u) \setminus (C \cup \{v\}) = (N(u) \setminus C) \setminus \{v\}$, and hence $-\Delta_u g(C \cup \{v\}) = -\Delta_u g(C) - 1$. Also, among all connected components of $G|_{C \cup \{v\}}$ that are adjacent to u, exactly one contains v and all others are connected components of $G|_C$ adjacent to u. Hence, $-\Delta_u h(C \cup \{v\}) \leq -\Delta_u h(C) + 1$. Therefore, $\Delta_u p(C \cup \{v\}) \leq \Delta_u p(C)$. □

It can be verified that with respect to this submodular function p, the set Ω_p is exactly the collection of connected vertex covers of G.

Lemma 2.38 *Let $G = (V, E)$ be a connected graph with at least three vertices. For any subset $C \subseteq V$, C is a connected vertex cover if and only if, for any vertex $x \in V$, $\Delta_x p(C) = 0$.*

Proof. If C is a connected vertex cover, then it is clear that $p(C) = |E| - g(C) - h(C) = |E| - 0 - 1 = |E| - 1$, reaching the maximum value of p.

Conversely, suppose that for any vertex $x \in V$, $\Delta_x p(C) = 0$. It is clear that $C \neq \emptyset$, for otherwise we can find a vertex $x \in V$ of degree ≥ 2 and get $\Delta_x p(C) = -\Delta_x g(\emptyset) - \Delta_x h(\emptyset) \geq 2 - 1 = 1$. Now, assume, for the sake of contradiction, that C is not a connected vertex cover. Let $B = \{x \in V \mid x \text{ is adjacent to some } v \in C\}$, and $A = V \setminus (B \cup C)$. Consider two cases.

Case 1: There exists an edge in E that is not covered by C. Then there must be an edge e in E not covered by C such that one of its endpoints x is in B (otherwise, A forms a nonempty connected component of G, contradicting the assumption that G is connected). Now, we note that $C \cup \{x\}$ covers at least one extra edge e than C, and so $-g(C \cup \{x\}) > -g(C)$. In addition, since x is in B and is adjacent to at least one vertex in C, adding x to C does not increase the number of connected components. Therefore, $-h(C \cup \{x\}) \geq -h(C)$. Together, we get $\Delta_x p(C) > 0$, which is a contradiction.

Case 2: C covers every edge, but $G|_C$ is not connected. Since G is connected, there must be a path in G connecting two connected components of $G|_C$. Furthermore, such a shortest path must contain exactly two edges $\{u, x\}$ and $\{x, v\}$ with $u, v \in C$ and $x \in B$, for otherwise it would contain an edge whose two endpoints are not in C. But then we have $-h(C \cup \{x\}) > -h(C)$ but $-g(C \cup \{x\}) = -g(C) = 0$, and hence $\Delta_x p(C) > 0$, a contradiction again. □

Corollary 2.39 *When it is applied to the problem* MIN-WCVC *on connected graphs of at least three vertices, with respect to the potential function* p, Algorithm 2.D *is a polynomial-time* $H(\delta - 1)$*-approximation, where* δ *is the maximum vertex degree of the input graph* G.

Proof. It follows from Theorem 2.29 and the facts that the maximum value of $|E| - g(\{x\})$ is equal to δ and that $-h(\{x\}) = -1$ for all $x \in V$. □

The next example is a 0–1 integer programming problem.

GENERAL COVER (GC): Given nonnegative integers a_{ij}, b_i, and c_j, for $i = 1, 2, \ldots, m$ and $j = 1, 2, \ldots, n$,

$$\begin{aligned} \text{minimize} \quad & \sum_{j=1}^{n} c_j x_j \\ \text{subject to} \quad & \sum_{j=1}^{n} a_{ij} x_j \geq b_i, && i = 1, 2, \ldots, m, \\ & x_j \in \{0, 1\}, && j = 1, 2, \ldots, n. \end{aligned}$$

We define a function $f : 2^{\{1,\ldots,n\}} \to \mathbb{N}$ as follows: For any $J \subseteq \{1, \ldots, n\}$,

$$f(J) = \sum_{i=1}^{m} \min\left\{b_i, \sum_{\ell \in J} a_{i\ell}\right\}.$$

Let $I(J) = \{i \mid \sum_{\ell \in J} a_{i\ell} < b_i\}$. Then it is clear that for any $j, k \in \{1, 2, \ldots, n\}$,

$$\Delta_j f(J) = \sum_{i \in I(J)} \min\left\{a_{ij}, b_i - \sum_{\ell \in J} a_{i\ell}\right\}, \quad \text{and}$$

$$\Delta_j f(J \cup \{k\}) = \sum_{i \in I(J \cup \{k\})} \min\left\{a_{ij}, b_i - \sum_{\ell \in J} a_{i\ell} - a_{ik}\right\}.$$

Moreover, it is not hard to verify that for any $1 \le k \le n$, $I(J \cup \{k\}) \subseteq I(J)$. Thus, $\Delta_j f(J \cup \{k\}) \le \Delta_j f(J)$ for all sets $J \subseteq \{1, 2, \ldots, n\}$ and all $j, k \in \{1, 2, \ldots, n\}$. Thus, by Lemma 2.36, f is a monotone increasing, submodular function.

The collection Ω_f consists of all sets $J \subseteq \{1, 2, \ldots, n\}$ with the maximum value $f(J) = \sum_{i=1}^{n} b_i$. So, Algorithm 2.D and Theorem 2.29 are applicable to problem GC. In particular, the greedy criterion of Algorithm 2.D adds, at each stage, the index j with the maximum value of

$$\frac{1}{c_j} \sum_{i \in I(J)} \min\left\{a_{ij}, b_i - \sum_{\ell \in J} a_{i\ell}\right\}$$

to the solution set J. Also, the parameter γ of the performance ratio $H(\gamma)$ is no more than the maximum value of $\sum_{i=1}^{m} a_{ij}$, $j = 1, 2 \ldots, n$.

Corollary 2.40 *When it is applied to the problem* GC, Algorithm 2.D *produces an* $H(\gamma)$*-approximation in polynomial time, where* $\gamma = \max_{1 \le j \le n} \sum_{i=1}^{m} a_{ij}$.

Finally, we consider a problem about matroids. Recall that a base of a matroid $(E, \mathcal{I})$ is just a maximal independent set. Consider the following problem:

MINIMUM-COST BASE (MIN-CB):

Given a matroid $(E, \mathcal{I})$ and a nonnegative function $c : E \to \mathbb{R}^+$,

$$\begin{array}{ll} \text{minimize} & c(I) \\ \text{subject to} & I \in \mathcal{B}, \end{array}$$

where $\mathcal{B}$ is the family of all bases of the matroid $(E, \mathcal{I})$.

Recall the function *rank* on a matroid $(E, \mathcal{I})$ defined in Example 2.21(b). Then *rank* is a normalized, monotone increasing, submodular function, and it has $\Omega_{rank} = \mathcal{B}$. Therefore, MIN-CB is a special case of MIN-SMC with the potential function $rank$. Note that the corresponding parameter γ in Theorem 2.29 is $\gamma = \max_{x \in E} rank(\{x\}) = 1$, and hence $H(\gamma) = 1$. In other words, the greedy Algorithm 2.D for MIN-CB actually gives the optimal solutions.

Corollary 2.41 *When it is applied to the problem* MIN-CB, *the greedy* Algorithm 2.D *produces a minimum solution in polynomial time.*

2.6 Nonsubmodular Potential Functions

When the associated potential function is not submodular, Theorem 2.29 for the greedy algorithm no longer holds. In such circumstances, how do we analyze the performance of the greedy algorithm? We study this problem in this section.

A *dominating set* of a graph $G = (V, E)$ is a subset $D \subseteq V$ such that every vertex is either in D or adjacent to a vertex in D. A *connected dominating set* C is a dominating set with an additional property that it induces a connected subgraph. The following problem has many applications in wireless communication.

> MINIMUM CONNECTED DOMINATING SET (MIN-CDS): Given a connected graph $G = (V, E)$, find a connected dominating set of G with the minimum cardinality.

Consider a graph G and a subset C of vertices in G. Divide vertices in G into three classes with respect to C, and assign different colors to them: Vertices that belong to C are colored in *black*; vertices that are not in C but are adjacent to C are colored in *gray*; and vertices that are neither in C nor adjacent to C are colored in *white*.

Clearly, C is a connected dominating set if and only if there does not exist a white vertex and the subgraph induced by black vertices is connected. This observation suggests that we use the function $g(C) = p(C) + h(C)$ as the potential function in the greedy algorithm, where $p(C)$ is the number of connected components of the subgraph $G|_C$ induced by C, and $h(C)$ is the number of white vertices. It is clear that C is a connected dominating set if and only if $g(C) = 1$. However, the function g is not really a good candidate for the potential function, because a set C may not be a connected dominating set even if $\Delta_x g(C) = 0$ for all vertices x. Figure 2.9 shows such an example, in which $g(C) = p(C) + h(C) = 2 + 0 = 2 > 1$, but $g(C \cup \{x\}) = g(C)$ for all vertices x. This means that if we apply Algorithm 2.D to MIN-CDS with this potential function g, its output is not necessarily a connected dominating set.

In general, we observe that the graph shown in Figure 2.9 is a typical case resulting from Algorithm 2.D with respect to the potential function g.

Lemma 2.42 *Let $G = (V, E)$ be a connected graph, and $C \subseteq V$. If the subgraph $G|_C$ induced by black vertices is not connected but $\Delta_x g(C) = 0$ for all $x \in V$, then*

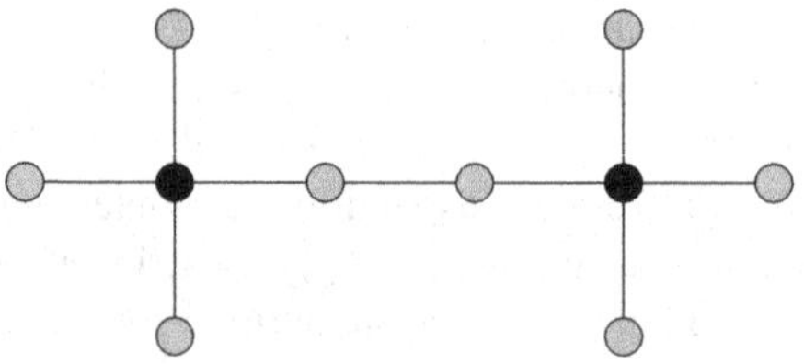

Figure 2.9: $\Delta_x g(C) = 0$ for all vertices x, but C is not a connected dominating set.

all black connected components of $G|_C$ can be connected together through chains of gray vertices, with each chain having exactly two vertices.

Proof. We first note that if $\Delta_x g(C) = 0$ for all $x \in V$, then G has no gray vertex that is adjacent to two black components, since coloring such a gray vertex in black would reduce the value of $g(C)$. In addition, G also has no white verex, for otherwise, by the connectivity of G, there must be a gray vertex adjacent to some white vertex, and coloring this gray vertex in black would reduce the value of $g(C)$, too. Now, suppose, for the sake of contradiction, that some black component cannot be connected to another black component through chains of two adjacent gray vertices. Then, we can divide all black vertices into two parts such that the distance between the two parts is more than 3. Consider a shortest path $\pi = (u, x_1, x_2, \ldots, x_k, v)$ between the two parts, with u and v belonging to the two different parts and $x_1, x_2, \ldots, x_k$ are gray vertices with $k \geq 3$. Since x_2 is gray, it must be adjacent to a black vertex w. If w and u are in the same part, then the path from w to v is a path between the two parts of black vertices shorter than π, which is a contradiction. On the other hand, if w and v are in the same part, then the path from u to w is a path between the two parts shorter than π, also a contradiction. So, the lemma is proven. □

From this lemma, a simple idea of an approximation algorithm works as follows: First, apply the greedy algorithm with the potential function g until $\Delta_x g(C) = 0$ for all $x \in V$. Then, add extra vertices to connect components of $G|_C$. A careful analysis using the pigeonhole principle shows that this modified greedy algorithm achieves the performance ratio $H(\delta) + 3$, where δ is the maximum degree of G (see Section 6.2).

In the following, we take a different approach by choosing a different potential function. Namely, we replace $h(C)$ by $q(C)$, the number of connected components of the subgraph with vertex set V and edge set $D(C)$, where $D(C)$ is the set of all edges incident on some vertices in C. Define $f(C) = p(C) + q(C)$.

Lemma 2.43 *Suppose G is a connected graph with at least three vertices. Then C is a connected dominating set if and only if $f(C \cup \{x\}) = f(C)$ for every $x \in V$.*

Proof. If C is a connected dominating set, then $f(C) = 2$, which reaches the minimum value. Therefore, $f(C \cup \{x\}) = f(C)$ for every $x \in V$.

Conversely, suppose $f(C \cup \{x\}) = f(C)$ for every $x \in V$. First, C cannot be the empty set. In fact, if $C = \emptyset$, then we can pick a vertex x of degree ≥ 2 and get $f(C \cup \{x\}) \leq |V| - 1 < |V| = f(C)$.

So, we may assume $C \neq \emptyset$. Consider a connected component of the subgraph induced by C. Let B denote its vertex set, which is a subset of C, and A be the set of vertices in $V - B$ that are adjacent to a vertex in B. We claim that $V = B \cup A$ (and hence $C = B$ is a connected dominating set for G).

To prove this claim, suppose, by way of contradiction, that $V \neq B \cup A$. Then, since G is connected, there must be a vertex x not in $B \cup A$ that is adjacent to a vertex $y \in B \cup A$. Since all vertices adjacent to B are in A, we know that y must be in A. Now, if x is white or gray, then we must have $p(C \cup \{y\}) \leq p(C)$

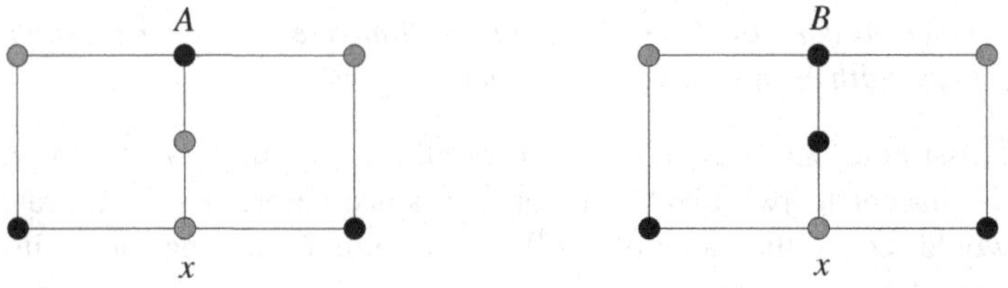

Figure 2.10: A counterexample showing f not supmodular.

and $q(C \cup \{y\}) < q(C)$. If x is black, then we have $p(C \cup \{y\}) < p(C)$ and $q(C \cup \{y\}) \leq q(C)$. In either case, we get $f(C \cup \{y\}) < f(C)$, a contradiction to our assumption. So, the claim, and hence the lemma, is proven. □

This lemma shows that the greedy Algorithm 2.D for MIN-CDS with respect to the potential function f will produce a connected dominating set.

A function $f : 2^E \to \mathbb{R}$ is *supmodular* if $-f$ is submodular. Clearly, all results about monotone increasing, submodular functions can be converted into the results about the corresponding monotone decreasing, supmodular functions. It is easy to see that f is monotone decreasing. Therefore, if f is a supmodular function, then we could directly employ Theorem 2.29 to get the performance ratio of the greedy Algorithm 2.D with respect to f. Unfortunately, as shown in the counterexample of Figure 2.10, f is not supmodular. More specifically, in this example, $A \subseteq B$ but $\Delta_x f(A) = -1 > -2 = \Delta_x f(B)$, and so $-f$ does not satisfy the condition of Lemma 2.36 and is not submodular.

Actually, f is the sum of two functions p and q, where q is supmodular but p is not.

Lemma 2.44 *If $A \subseteq B$, then $\Delta_y q(A) \leq \Delta_y q(B)$.*

Proof. Note that $-\Delta_y q(B) =$ the number of the connected components of the graph $(V, D(B))$ that are adjacent to y but do not contain y. Since each connected component of graph $(V, D(B))$ is constituted by one or more connected components of graph $(V, D(A))$, the number of connected components of $(V, D(B))$ adjacent to y is no more than the number of connected components of $(V, D(A))$ adjacent to y. Thus, we get $-\Delta_y q(B) \leq -\Delta_y q(A)$. □

How do we analyze the performance of the greedy Algorithm 2.D with respect to a nonsubmodular potential function? Let us look at the proof of Theorem 2.22 about the greedy algorithm for MIN-SC again, and see where the submodularity property of the potential function is used. It turns out that it was used only once, when we proved the inequality

$$\Delta_{C_j} f(\mathcal{A}_i) \geq \Delta_{C_j} f(\mathcal{A}_i \cup \mathcal{C}_{j-1}) \tag{2.6}$$

to get (2.4). An important observation about this inequality is that the incremental variables C_j, $1 \leq j \leq m$, are sets of the optimal solution, arranged in an arbitrary order. Therefore, although for nonsubmodular functions f this inequality may not hold for an arbitrary ordering of sets in the optimal solution, a carefully arranged ordering on these sets might still satisfy, or almost satisfy, this inequality. In the following, we will implement this idea for the problem MIN-CDS.

Let the vertices $x_1, \ldots, x_g$ be the elements of the solution found by Algorithm 2.D with respect to the potential function f, in the order of their selection into the solution set. Denote $C_i = \{x_1, x_2, \ldots, x_i\}$ and consider $f(C_i)$. Initially, $f(C_0) = n$, where n is the number of vertices in G. Let C^* be a minimum connected dominating set for G. Assume that $|C^*| = m$.

Lemma 2.45 *For* $i = 1, 2, \ldots, g$,

$$f(C_i) \le f(C_{i-1}) - \frac{f(C_{i-1}) - 2}{m} + 1. \tag{2.7}$$

Proof. First, consider the case of $i \ge 2$. We note that

$$f(C_i) = f(C_{i-1}) + \Delta_{x_i} f(C_{i-1}).$$

Since C^* is a connected dominating set, we can always arrange the elements of C^* in an ordering $y_1, y_2, \ldots, y_m$ such that y_1 is adjacent to a vertex in C_{i-1} and, for each $j \ge 2$, y_j is adjacent to a vertex in $\{y_1, \ldots, y_{j-1}\}$. Denote $C_j^* = \{y_1, y_2, \ldots, y_j\}$. Then

$$\Delta_{C^*} f(C_{i-1}) = \sum_{j=1}^{m} \Delta_{y_j} f(C_{i-1} \cup C_{j-1}^*).$$

For each $1 \le j \le m$, we note that y_j can dominate at most one additional connected component in the subgraph $G|_{C_{i-1} \cup C_{j-1}^*}$ than in $G|_{C_{i-1}}$, which is the one that contains C_{j-1}^*, since all vertices $y_1, \ldots, y_{j-1}$ in C_{j-1}^* are connected. Since $-\Delta_y p(C)$ is equal to the number of connected components of $G|_C$ that are adjacent to y *minus* 1, it follows that

$$-\Delta_{y_j} p(C_{i-1} \cup C_{j-1}^*) \le -\Delta_{y_j} p(C_{i-1}) + 1.$$

Moreover, by Lemma 2.44,

$$-\Delta_{y_j} q(C_{i-1} \cup C_{j-1}^*) \le -\Delta_{y_j} q(C_{i-1}).$$

So we have

$$-\Delta_{y_j} f(C_{i-1} \cup C_{j-1}^*) \le -\Delta_{y_j} f(C_{i-1}) + 1.$$

[Note that this inequality is close to our desired inequality (2.6).] From this inequality, we get

$$\begin{aligned} f(C_{i-1}) - 2 &= -\Delta_{C^*} f(C_{i-1}) \\ &= \sum_{j=1}^{m} (-\Delta_{y_j} f(C_{i-1} \cup C_{j-1}^*)) \le \sum_{j=1}^{m} (-\Delta_{y_j} f(C_{i-1}) + 1). \end{aligned}$$

By the pigeonhole principle, there exists an element $y_j \in C^*$ such that

$$-\Delta_{y_j} f(C_{i-1}) + 1 \geq \frac{f(C_{i-1}) - 2}{m}.$$

By the greedy strategy of Algorithm 2.D,

$$-\Delta_{x_i} f(C_{i-1}) \geq -\Delta_{y_j} f(C_{i-1}) \geq \frac{f(C_{i-1}) - 2}{m} - 1.$$

Or, equivalently,

$$f(C_i) \leq f(C_{i-1}) - \frac{f(C_{i-1}) - 2}{m} + 1.$$

For the case of $i = 1$, the proof is essentially identical, with the difference that y_1 could be an arbitrary vertex in C^*. □

Theorem 2.46 *When it is applied to the problem* MIN-CDS *with respect to the potential function $-f$, the greedy* Algorithm 2.D *is a polynomial-time* $(2 + \ln \delta)$-*approximation, where δ is the maximum degree of the input graph.*

Proof. If $g \leq 2m$, then the proof is already done. So we assume that $g > 2m$.

Rewrite the inequality (2.7) as

$$f(C_i) - 2 \leq (f(C_{i-1}) - 2)\Big(1 - \frac{1}{m}\Big) + 1.$$

Solving this recurrence relation, we have

$$\begin{aligned} f(C_i) - 2 &\leq (f(C_0) - 2)\Big(1 - \frac{1}{m}\Big)^i + \sum_{k=0}^{i-1}\Big(1 - \frac{1}{m}\Big)^k \\ &= (f(C_0) - 2)\Big(1 - \frac{1}{m}\Big)^i + m\Big(1 - \Big(1 - \frac{1}{m}\Big)^i\Big) \\ &= (f(C_0) - 2 - m)\Big(1 - \frac{1}{m}\Big)^i + m. \end{aligned}$$

From the greedy strategy of Algorithm 2.D, we reduce the value $f(C_{i-1})$ in each stage $i \leq g$. Therefore, $f(C_i) \leq f(C_{i-1}) - 1$. In addition, $f(C_g) = 2$. So we have $f(C_{g-2m}) \geq 2m + 2$. Set $i = g - 2m$, and observe that

$$2m \leq f(C_i) - 2 \leq (n - 2 - m)\Big(1 - \frac{1}{m}\Big)^i + m,$$

where n is the number of vertices in G. Since $(1 - 1/m)^i \leq e^{-i/m}$, we obtain

$$i \leq m \cdot \ln \frac{n - 2 - m}{m}.$$

Note that each vertex has at most δ neighbors and so can dominate at most $\delta + 1$ vertices. Hence, $n/m \leq \delta + 1$. It follows that $g = i + 2m \leq m(2 + \ln \delta)$. □

Now, let us consider another simple idea for designing greedy algorithms with respect to a nonsubmodular potential function. In the greedy Algorithm 2.C for the problem MIN-SC, we add, in each iteration, one subset C to the solution $\mathcal{A}$. Suppose we are allowed to add two or more subsets to $\mathcal{A}$ in each iteration. Does this give us a better performance ratio? It is easy to see that the answer is no. In general, does this idea work for the greedy Algorithm 2.D with respect to a submodular potential function f? The answer is again no, since a submodular function satisfies the property of Lemma 2.23. On the other hand, if the potential function f is not submodular, then this idea may actually work. In the following, we show that the greedy algorithm based on this idea actually gives a better performance ratio for MIN-CDS than Algorithm 2.D. More precisely, the performance ratio of the following greedy algorithm for MIN-CDS approaches $1 + \ln \delta$, as k tends to ∞.

Algorithm 2.E (*Greedy Algorithm for* MIN-CDS)

Input: A connected graph $G = (V, E)$ and an integer $k \geq 2$.

(1) $C \leftarrow \emptyset$.

(2) **While** $f(C) > 2$ **do**

Select a set $X \subseteq V$ of size $|X| \leq 2k - 1$ that maximizes $\dfrac{-\Delta_X f(C)}{|X|}$;

Set $C \leftarrow C \cup X$.

(3) Output $C_g \leftarrow C$. ■

To analyze greedy Algorithm 2.E, we note the following property of the potential function $-f$.

Lemma 2.47 *Let A, B, and X be three vertex subsets. If both $G|_B$ and $G|_X$ are connected, then*

$$-\Delta_X f(A \cup B) + \Delta_X f(A) \leq 1.$$

Proof. Since q is supmodular, we have $\Delta_X q(A) \leq \Delta_X q(A \cup B)$.

For function p, we note that, since $G|_X$ is connected, $-\Delta_X p(A)$ is equal to the number of black components dominated by X in graph $G|_A$ minus 1. Since the subgraph $G|_B$ is connected, the number of black components dominated by X in $G|_{A \cup B}$ is at most one more than the number of black components dominated by X in $G|_A$. Therefore, we have $-\Delta_X p(A \cup B) \leq -\Delta_X p(A) + 1$. It follows that $-\Delta_X f(A \cup B) \leq -\Delta_X f(A) + 1$. □

Let C^* be a minimum solution to MIN-CDS. We show two properties of C^* in the following two lemmas.

Lemma 2.48 *For any integer $k \geq 2$, C^* can be decomposed into $Y_1, Y_2, \ldots, Y_h$, for some $h \geq 1$, such that*

(a) $C^* = Y_1 \cup Y_2 \cup \cdots \cup Y_h$;

(b) For each $1 \leq i \leq h$, both $G|_{Y_1 \cup Y_2 \cup \cdots \cup Y_i}$ and $G|_{Y_i}$ are connected;

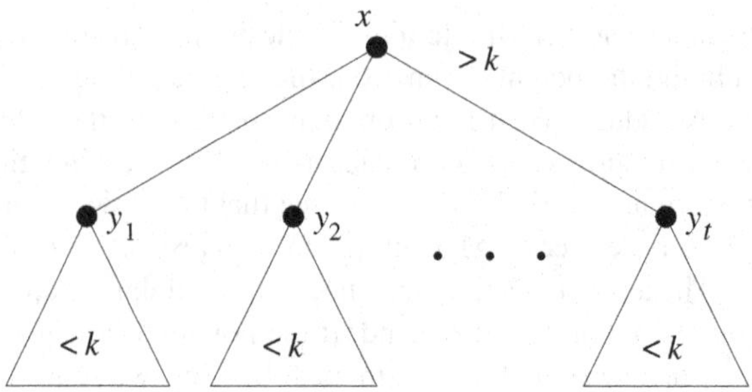

Figure 2.11: Case 2 in proof of Lemma 2.48.

(c) *For each* $1 \leq i \leq h$, $1 \leq |Y_i| \leq 2k - 1$; *and for all but one* $1 \leq i \leq h$, $k + 1 \leq |Y_h|$; *and*

(d) $|Y_1| + |Y_2| + \cdots + |Y_h| \leq |C^*| + h - 1$.

Proof. We can construct sets $Y_1, \ldots, Y_h$ recursively.

Let T be a subtree of $G|_{C^*}$ that contains all vertices in C^*. Choose an arbitrary vertex $r \in C^*$ as the root of T. For any vertex $x \in C^*$, let $T(x)$ denote the subtree of T rooted at x, and $|T(x)|$ the number of vertices in $T(x)$.

If $|T| \leq 2k - 1$, then let $Y_1 = C^*$ and the lemma holds with $h = 1$. If T contains more than $2k - 1$ vertices, then there must exist a vertex $x \in C^*$ such that $|T(x)| \geq k + 1$ and for every child y of x, $|T(y)| \leq k$. Now, consider two cases.

Case 1. There is a child y of x such that $|T(y)| = k$. Let Y_1 consist of all vertices of $T(y)$ together with x and delete all vertices of $T(y)$ from T.

Case 2. For every child y of x, $|T(y)| \leq k-1$. Suppose $y_1, \ldots, y_t$ are all children of x (cf. Figure 2.11). There must exist an integer $1 \leq j \leq t - 1$ such that

$$|T(y_1)| + \cdots + |T(y_j)| \leq k - 1$$

and

$$|T(y_1)| + \cdots + |T(y_j)| + |T(y_{j+1})| \geq k.$$

Since $|T(y_{j+1})| \leq k - 1$, we have

$$|T(y_1)| + \cdots + |T(y_j)| + |T(y_{j+1})| \leq 2k - 2.$$

Let Y_1 consist of all vertices in $T(y_1) \cup \cdots \cup T(y_{j+1})$ together with x and delete $Y_1 - \{x\}$ from T.

Repeating the above process on the remaining T, and rearranging the order of the sets $Y_1, \ldots, Y_h$, we will obtain a required decomposition. □

Lemma 2.49 *Let* δ *be the maximum degree of* $G = (V, E)$. *Then we have* $|V| \leq (\delta - 1)|C^*| + 2$.

Proof. We prove by induction on $|C|$ that a subset C of V with connected $G|_C$ can dominate at most $(\delta-1)|C|+2$ vertices. For $|C|=1$, it is trivially true. For $|C|\geq 2$, choose a vertex $x\in C$ such that $G|_{C-\{x\}}$ is still connected. Since x has at most δ neighbors, and at least one of them is in $C-\{x\}$, we see that C dominates at most $\delta-1$ more vertices than $C-\{x\}$ does. By the induction hypothesis, $C-\{x\}$ can dominate at most $(\delta-1)(|C|-1)+2$ vertices. Therefore, C can dominate at most $(\delta-1)|C|+2$ vertices. □

Theorem 2.50 *For any $\varepsilon>0$, there exists a polynomial-time approximation with performance ratio $(1+\varepsilon)\ln(\delta-1)$ for* MIN-CDS, *where δ is the maximum degree of the input graph.*

Proof. Let $G=(V,E)$ be a connected graph with the maximum degree δ. We can find easily a minimum connected dominating set of G if $\delta\leq 2$: If $\delta=1$, then G contains only one edge, and either vertex of the edge is a minimum connected dominating set. If $\delta=2$, G is either a path or a cycle, and a minimum connected dominating set of G can be obtained by deleting, respectively, either the two leaves or any two adjacent vertices.

For graphs with $\delta\geq 3$, we consider Algorithm 2.E on G. Let $X_1,\ldots,X_g$ be the sets chosen by greedy Algorithm 2.E on graph G, in the order of their selection into set C. Denote $C_i=X_1\cup\cdots\cup X_i$, for $0\leq i\leq g$ (in particular, C_g is the output of Algorithm 2.E). Let C^* be a minimum connected dominating set for G, and $m=|C^*|$. Decompose C^* into $Y_1,Y_2,\ldots,Y_h$, satisfying conditions given in Lemma 2.48. Denote $C_j^*=Y_1\cup\cdots\cup Y_j$, for $0\leq j\leq h$.

From Lemma 2.48, we know that $G|_{Y_j}$ and $G|_{C_j^*}$ are connected for each $1\leq j\leq h$. Thus, we have, by Lemma 2.47,

$$-\Delta_{Y_j}f(C_i\cup C_{j-1}^*)\leq -\Delta_{Y_j}f(C_i)+1,$$

for $0\leq i\leq g$ and $1\leq j\leq h$. By the greedy rule of Algorithm 2.E, we get

$$\frac{-\Delta_{X_{i+1}}f(C_i)}{|X_{i+1}|}\geq\frac{-\Delta_{Y_j}f(C_i)}{|Y_j|},$$

for $0\leq i\leq g$ and $1\leq j\leq h$. Note that $f(C^*)=2$ and, hence, for $0\leq i\leq g-1$,

$$\begin{aligned}\frac{-\Delta_{X_{i+1}}f(C_i)}{|X_{i+1}|}&\geq\frac{-\sum_{j=1}^{h}\Delta_{Y_j}f(C_i)}{\sum_{j=1}^{h}|Y_j|}\\&\geq\frac{-(h-1)-\sum_{j=1}^{h}\Delta_{Y_j}f(C_i\cup C_{j-1}^*)}{\sum_{j=1}^{h}|Y_j|}\\&\geq\frac{-(h-1)-(f(C_i\cup C^*)-f(C_i))}{m+h-1}\\&=\frac{f(C_i)-(h+1)}{m+h-1}.\end{aligned}$$

Denote $a_i = f(C_i) - (h+1)$. Then the above inequality can be rewritten as

$$\frac{a_i - a_{i+1}}{|X_{i+1}|} \geq \frac{a_i}{m+h-1}, \qquad \text{for } 0 \leq i \leq g-1.$$

That is, for each $0 \leq i \leq g-1$,

$$\begin{aligned} a_{i+1} &\leq a_i\Big(1 - \frac{|X_{i+1}|}{m+h-1}\Big) \leq a_i \cdot \exp\Big(\frac{-|X_{i+1}|}{m+h-1}\Big) \\ &\leq a_0 \cdot \exp\Big(\frac{-(|X_{i+1}| + |X_i| + \cdots + |X_1|)}{m+h-1}\Big). \end{aligned} \tag{2.8}$$

Fix the index i, $0 \leq i \leq g-1$, such that

$$a_i \geq m > a_{i+1},$$

and let $b = a_i - m$ and $b' = m - a_{i+1}$. Write $|X_{i+1}| = d + d'$ such that

$$\frac{b}{d} = \frac{b'}{d'} = \frac{a_i - a_{i+1}}{|X_{i+1}|} \geq \frac{a_i}{m+h-1}.$$

(In case of $b = 0$, just let $d' = |X_{i+1}|$.) We now divide the greedy solution $|C_g|$ into two parts, $|X_1| + \cdots + |X_i| + d$, and $d' + |X_{i+2}| + \cdots + |X_g|$, and bound them separately.

For the first part, we note that

$$\frac{a_i - m}{d} = \frac{b}{d} \geq \frac{a_i}{m+h-1},$$

and so

$$m \leq a_i\Big(1 - \frac{d}{m+h-1}\Big) \leq a_i \cdot e^{-d/(m+h-1)}.$$

Combining this with (2.8), we get

$$m \leq a_0 \cdot e^{-(d+|X_i|+\cdots+|X_1|)/(m+h-1)}.$$

Note that $a_0 = f(\emptyset) - (h+1) = |V| - (h+1)$. Thus,

$$|X_1| + \cdots + |X_i| + d \leq (m+h-1)\ln\frac{|V| - (h+1)}{m}.$$

For the second part, we note that $-\Delta_{X_{j+1}} f(C_j)/|X_{j+1}| \geq 1$ for all $0 \leq j \leq g-1$, since we can, by Lemma 2.43, always find a vertex v to make $-\Delta_{\{v\}} f(C_j) \geq 1$. That is,

$$|X_{j+1}| \leq f(C_j) - f(C_{j+1}),$$

for $0 \leq j \leq g-1$. Thus,

$$\begin{aligned} d' + |X_{i+2}| + \cdots + |X_g| &\leq b' + f(C_{i+1}) - f(C_g) \\ &= m - a_{i+1} + f(C_{i+1}) - f(C^*) = m + h - 1. \end{aligned}$$

Together, we have

$$|X_1| + \cdots + |X_g| \le (m + h - 1)\Big(1 + \ln \frac{|V| - (h+1)}{m}\Big).$$

From conditions (c) and (d) of Lemma 2.48, we know that

$$(h-1)(k+1) + 1 \le |Y_1| + |Y_2| + \cdots + |Y_h| \le m + h - 1,$$

and hence

$$h - 1 \le \frac{m}{k}.$$

Moreover, by Lemma 2.49, $|V| \le (\delta - 1)m + 2$. Since $h \ge 1$, we have

$$\frac{|V| - (h+1)}{m} \le \delta - 1.$$

Therefore,

$$|X_1| + \cdots + |X_g| \le m\Big(1 + \frac{1}{k}\Big)\Big(1 + \ln(\delta - 1)\Big).$$

Now, the theorem follows by choosing k such that $1/k < \varepsilon$. □

Exercises

2.1 Let $(E, \mathcal{I})$ be an independent system. Suppose that all maximal independent subsets of E have cardinality k. Define

$$p = \max_{F \subseteq E} \frac{v(F)}{u(F)},$$

where $u(F)$ and $v(F)$ are the functions defined in (2.1). Let $c : E \to \mathbb{R}^+$ be a nonnegative cost function on E. Also, let I^* be a maximal independent subset of E with the minimum cost, and I_G an independent subset obtained by greedy Algorithm 2.A on the problem MAX-ISS. Prove that

$$c(I^*) \le c(I_G) \le \frac{1}{p} \cdot c(I^*) + \frac{p-1}{p} \cdot k \cdot M,$$

where $M = \max_{e \in E} c(e)$.

2.2 For a complete directed graph $G = (V, E)$, let $\mathcal{I}_G$ be the family of the edge sets of all acyclic subgraphs of G. Show that for any integer $k > 0$, there exists a complete directed graph $G = (V, E)$ such that for the independent system $(E, \mathcal{I}_G)$,

$$\max_{F \subseteq E} \frac{v(F)}{u(F)} \ge k.$$

2.3 Show that for every integer $k \geq 1$, there exists an independent system $(E, \mathcal{I})$ that is an intersection of k matroids but not an intersection of less than k matroids, such that

$$\max_{F \subseteq E} \frac{v(F)}{u(F)} = k.$$

2.4 Prove that an independent system $(E, \mathcal{I})$ is a matroid if and only if, for any cost function $c : E \to \mathbb{N}^+$, the greedy Algorithm 2.D produces a minimum solution for MIN-CB.

2.5 Prove that the distance function defined in the transformation from the problem SS to the problem TSP, as described at the end of Section 2.3, satisfies the triangle inequality.

2.6 Prove that for every positive integer m, $\sum_{i=1}^{m} 1/i \leq 1 + \ln m$.

2.7 In terms of the notion of hypergraphs, the problem MIN-SC asks for a minimum-size hyperedge set that is incident on each vertex of the input hypergraph. A k-matching in a hypergraph H is a sub-hypergraph of degree at most k. Let m_k be the maximum number of edges in a k-matching. Prove that

(a) $m_k \leq k \cdot |C^*|$, where C^* is a minimum set cover of H, and

(b) $|C_G| \leq \sum_{i=1}^{d} m_i/(i(i+1)) + m_d/d$, where C_G is the output of the greedy Algorithm 2.C, and d is the maximum degree of H.

2.8 Use Exercise 2.7 to give another proof to Theorem 2.22.

2.9 Let $G = (V, E)$ be a graph and $c : E \to 2^{\mathbb{N}}$ a color-set function (i.e., $c(e)$ is a *color set* for edge e). A *color-covering* of the graph G is a color set $C \subseteq \mathbb{N}$ such that the set of edges e with $c(e) \cap C \neq \emptyset$ contains a spanning tree of G. Prove that the following problem has a polynomial-time $(1 + \ln |V|)$-approximation:

> For a given graph G and a given color-set function $c : E \to 2^N$, find a color-covering of the minimum cardinality.

2.10 Show that the following problem has a polynomial-time $(2 + \ln |V|)$-approximation:

> Given a graph $G = (V, E)$ and a color-set function $c : E \to 2^{\mathbb{N}}$, find the subset $C \subseteq V$ of the minimum cardinality such that all colors of the edges incident upon the vertices in C form a color-covering of G.

2.11 A function $g : \mathbb{N} \to \mathbb{R}^+$ is a *concave function* if, for any $m, r, n \in \mathbb{N}$, with $m < r < n$, $g(r) \geq tg(m) + (1 - t)g(n)$, where $t = (n - r)/(n - m)$. Let E be a finite set, and let f be a real function defined on 2^E such that $f(A) = g(|A|)$ for all $A \subseteq E$. Show that f is submodular if and only if g is concave.

2.12 Consider a graph $G = (V, E)$. Let $\bar{\delta}(X)$ for $X \subseteq V$ denote the set of edges between X and $V - X$. Show that $|\bar{\delta}(X)|$ is a submodular function.

2.13 Show that a function f on 2^E is modular (both submodular and supmodular) if and only if f is linear.

2.14 Prove Lemmas 2.35 and 2.36.

2.15 Suppose f and c are two polymatroid functions on 2^E, and f is an integer function. Consider the problem MIN-SMC with a possibly nonlinear cost function c; i.e., the problem of minimizing $c(A)$ over $\{A \subseteq E \mid f(A) = f(E)\}$. Show that the greedy Algorithm 2.D for MIN-SMC is a $(\rho \cdot H(\gamma))$-approximation, where $\gamma = \max\{f(\{x\}) \mid x \in E\}$ and ρ is the *curvature* of c, defined by

$$\rho = \min\left\{\frac{\sum_{e \in S} c(e)}{c(S)} \;\middle|\; f(S) = f(E)\right\}.$$

2.16 Consider a digraph $G = (V, E)$. For $X \subseteq V$, let $\bar{\delta}_+(X)$ ($\bar{\delta}_-(X)$) denote the set of edges going out from (coming into, respectively) X. Show that $|\bar{\delta}_+(X)|$ and $|\bar{\delta}_-(X)|$ are submodular functions.

2.17 Let r be a function mapping 2^E to $\mathbb{N}$. Show that the following statements are equivalent:

(a) $\mathcal{I} = \{I \subseteq E \mid r(I) = |I|\}$ defines a matroid $(E, \mathcal{I})$ and r is its rank function.

(b) For all $A, B \subseteq E$, r satisfies the following conditions:

(i) $r(A) \le |A|$;

(ii) if $A \subseteq B$, then $r(A) \le r(B)$; and

(iii) r is submodular.

2.18 Show that a polymatroid (E, r) is a matroid if and only if $r(\{x\}) = 1$ for every $x \in E$.

2.19 Suppose $(E, r_1), (E, r_2), \ldots, (E, r_k)$ are matroids. Show that $(E, \sum_{i=1}^k r_i)$ is a polymatroid.

2.20 Let $(E, \mathcal{I})$ be a matroid, and *rank* its rank function. Consider a collection $\mathcal{C}$ of subsets of E. For $\mathcal{A} \subseteq \mathcal{C}$, define

$$f(\mathcal{A}) = rank\big(\textstyle\bigcup_{A \in \mathcal{A}} A\big).$$

Show that (E, f) is a polymatroid.

2.21 Show that for any polymatroid $(\mathcal{E}, f)$, there exist a matroid (E, r) and a one-to-one mapping $\phi : \mathcal{E} \to 2^E$ such that

$$f(\mathcal{A}) = r\big(\textstyle\bigcup_{A \in \phi(\mathcal{A})} A\big).$$

2.22 For any polymatroid (E, f), define f^d on 2^E with

$$f^d(S) = \sum_{j \in S} f(\{j\}) - f(E) - f(E - S).$$

Show that (E, f^d) is still a polymatroid. [It is called the *dual polymatroid* of (E, f).]

2.23 For any polymatroid (E, f), let $\mathcal{I} = \{A \mid f(A) = |A|, A \subseteq E\}$. Show that $(E, \mathcal{I})$ is an independent system.

2.24 Let $(E, \mathcal{I})$ be an independent system. Define $r(A) = \max\{|I| \mid I \in \mathcal{I}, I \subseteq A\}$. Give an example of $(E, \mathcal{I})$ for which r is not a polymatroid function.

2.25 Let (E, f) be a polymatroid and c a nonnegative cost function on E. Show that the problem of computing $\min\{c(A) \mid f(A) \geq k, A \subseteq E\}$ has a greedy approximation with performance ratio $H(\min\{k, \gamma\})$, where $\gamma = \max_{x \in E} f(\{x\})$.

2.26 Consider the application of Algorithm 2.D to MIN-CDS with the potential function $f(C) = p(C) + q(C)$. Find a graph G on which the algorithm produces an approximate solution of size $g \leq 2|C^*|$.

2.27 Given a hypergraph $H = (V, S)$ and a function $f : S \to \mathbb{N}^+$, find a minimum vertex cover C such that for every hyperedge $s \in S$, $|C \cap s| \geq f(s)$. Prove that this problem has a polynomial-time $(1 + \ln d)$-approximation, where d is the maximum vertex degree in H.

2.28 Let $f : 2^E \to \mathbb{R}$ be a normalized submodular function. We associate a weight $w_i \geq 0$ with each $i \in E$. Consider the following linear program:

$$\begin{aligned} \text{maximize} \quad & \sum_{i \in E} w_i x_i \\ \text{subject to} \quad & \sum_{i \in A} x_i \leq f(A), \quad A \subseteq E. \end{aligned}$$

Show that this problem can be solved by the following greedy algorithm:

(1) Sort elements of E and rename them so that $w_1 \geq w_2 \geq \cdots \geq w_n$.

(2) $A_0 \leftarrow \emptyset$; **for** $k \leftarrow 1$ **to** n **do** $A_k \leftarrow \{1, 2, \ldots, k\}$.

(3) **For** $k \leftarrow 1$ **to** n **do** $x_i \leftarrow f(A_i) - f(A_{i-1})$.

2.29 Let E be a finite set and $p : E \to \mathbb{R}^+$ a positive function on E. For every subset A of E, define

$$g(A) = \left(\sum_{i \in A} p(i)\right)^2 + \sum_{i \in A} p(i)^2.$$

Show that g is a supmodular function.

2.30 Show that the following greedy algorithm for the problem MIN-CDS has performance ratio $2(1 + H(\delta))$, where δ is the maximum vertex degree:

> Grow a tree T starting from a vertex of the maximum degree. At each iteration, add one or two adjacent vertices to maximize the increase in the number of dominated vertices.

2.31 In the proof of Lemma 2.45, a simple argument has been suggested as follows:

> Since $m = |C^*|$ vertices are able to reduce the total number of connected components in the two subgraphs from $f(C_{i-1})$ to 2, there must exist a vertex that is able to reduce at least $\lceil (f(C_{i-1}) - 2)/m \rceil - 1$ components (here, the term -1 comes from considering the increase in the number of black components). Therefore, $-\Delta_{x_i} f(C_{i-1}) \geq (f(C_{i-1}) - 2)/m - 1$, and hence the lemma holds.

Find the error of this argument and explain why with a counterexample to the above statement.

2.32 Give a counterexample to show that Lemma 2.47 does not hold if $G|_X$ is not connected.

2.33 A dominating set A in a graph is said to be *weakly connected* if all edges incident upon vertices in A induce a connected subgraph. Show that there exists a greedy $H(\delta)$-approximation for the problem of finding the minimum-size weakly connected dominating set of a given graph, where δ is the maximum vertex degree of the input graph.

2.34 Consider a hypergraph $(V, \mathcal{E})$, where $\mathcal{E}$ is a collection of subsets of V. A subcollection $\mathcal{C}$ of $\mathcal{E}$ is called a *connected set cover* if $\mathcal{C}$ is a set cover of V and $(V, \mathcal{C})$ is a connected sub-hypergraph. Show that the problem of finding a connected set cover with the minimum cardinality has a greedy $H(\delta)$-approximation, where δ is the maximum vertex degree of the input hypergraph.

2.35 Consider a hypergraph $(V, \mathcal{E})$, where $\mathcal{E}$ is a collection of subsets of V. A subset A of V is called a *dominating set*, if every vertex is either in A or adjacent to A. Furthermore, A is said to be *connected* if A induces a connected sub-hypergraph. Design a greedy approximation for computing the minimum connected dominating set in hypergraphs. Could you reach approximation ratio $(1 - \varepsilon)(1 + \ln \delta)$ for any $\varepsilon > 0$, where δ is the maximum vertex degree of the input hypergraph?

2.36 A set S of sensors is associated with a graph $G = (S, E)$, and each sensor $s \in S$ can monitor a set T_s of targets. Let T be the collection of all targets; i.e., $T = \bigcup_{s \in S} T_s$. Consider the following problem:

> CONNECTED TARGET COVERAGE (CTC): Given a sensor graph $G = (S, E)$ and, for each sensor $s \in S$, a target set T_s, find a minimum-cardinality subset A of S such that A can monitor all targets in T and such that A also induces a connected subgraph of G.

Design a greedy approximation for CTC and analyze the performance ratio of your algorithm.

Historical Notes

The analysis of the greedy algorithm for independent systems was first reported by Jenkyns [1976] and Korte and Hausmann [1978]. Hausmann, Korte, and Jenkyns [1980] further studied algorithms of this type. Submodular set functions play an important role in combinatorial optimization. Some of the results presented in Section 2.4 can be found in Wolsey [1982a].

Lund and Yannakakis [1994] proved that for any $0 < \rho < 1/4$, there is no polynomial-time approximation algorithm with performance ratio $\rho \ln n$ for MIN-SC unless **NP** $\subseteq$ **DTIME**$(n^{poly \log n})$. Feige [1998] improved this result by relaxing ρ to $0 < \rho < 1$. This means that it is unlikely for MIN-SC to have a constant-bounded polynomial-time approximation. Johnson [1974] and Lovász [1975] independently discovered a polynomial-time greedy $H(\delta)$-approximation for MIN-SC. Chvátal [1979] extended the greedy approximation to the weighted case. The greedy algorithm for MIN-SC can be analyzed in many ways. Slavik [1997] presented a tight one. The problem WSID was proposed by Du and Miller [1988]. Prisner [1992] presented a greedy approximation for it and claimed that it has performance ratio $1 + \ln K$. Unfortunately, his proof contained an error. Du, Wu, and Kelley [1998] fixed this error. They also showed, based on a reduction from the problem MIN-SC, a lower bound on the performance ratio for WSID. It is known that the problem MIN-CDS is **NP**-hard [Garey and Johnson, 1978]. Guha and Khuller [1998a] presented a greedy algorithm for it with performance ratio $3 + \ln \delta$. Ruan et al. [2003] gave a new one with performance $2 + \ln \delta$. The $(1 + \varepsilon)(1 + \ln \delta)$-approximation can be found in Du et al. [2008].

3 Restriction

Success is restricted only from those who
restrict themselves from success.
— Gillis Triplett

When we design an approximate algorithm by the restriction method, we add some constraints on an optimization problem to shrink the feasible domain so that the optimization problem on the resulting domain becomes easier to solve or approximate. We may then use the optimal or approximate solutions for this restricted problem to approximate the original problem.

When we analyze the performance of the algorithms designed with the restriction method, we often reverse the process. Namely, for a minimization problem $\min_{\boldsymbol{x}\in\Omega} f(\boldsymbol{x})$, assume that we restrict the solutions to $\boldsymbol{x} \in \Gamma \subseteq \Omega$ and find the optimal solution $\boldsymbol{y}^* \in \Gamma$. For the analysis, we consider an optimal solution $\boldsymbol{x}^*$ to the original problem, and modify it to a solution $\boldsymbol{y}$ that satisfies the restriction. The difference $f(\boldsymbol{y}) - f(\boldsymbol{x}^*)$ between the costs of these two solutions then can be used to determine the performance ratio of this approximation. More precisely, as explained in Section 1.2 (see Figure 1.3), the performance ratio of the algorithm can be estimated by

$$\frac{f(\boldsymbol{y}^*)}{f(\boldsymbol{x}^*)} \leq \frac{f(\boldsymbol{y})}{f(\boldsymbol{x}^*)} = 1 + \frac{f(\boldsymbol{y}) - f(\boldsymbol{x}^*)}{f(\boldsymbol{x}^*)}.$$

For a maximization problem $\max_{\boldsymbol{x}\in\Omega} f(\boldsymbol{x})$, the approach is similar. Here, the performance ratio is $f(\boldsymbol{x}^*)/f(\boldsymbol{y}^*)$, and it can be bounded as follows:

$$\frac{f(\boldsymbol{x}^*)}{f(\boldsymbol{y}^*)} \leq \frac{f(\boldsymbol{x}^*)}{f(\boldsymbol{y})} = \left(1 - \frac{f(\boldsymbol{x}^*) - f(\boldsymbol{y})}{f(\boldsymbol{x}^*)}\right)^{-1}.$$

In this and the next two chapters, we will apply the restriction method and this analysis technique to a number of optimization problems.

3.1 Steiner Trees and Spanning Trees

The Steiner tree problem is a classical intractable problem with many applications in the design of computer circuits, long-distance telephone lines, and mail routing, etc. Given a set of points, called *terminals*, in a metric space, any minimal tree interconnecting all terminals is called a *Steiner tree* (by "minimal," we mean that no edge can be deleted). The Steiner tree problem asks, for a given set of terminals, find a shortest Steiner tree, called a *Steiner minimum tree (SMT)*, for them.

In a Steiner tree, the nonterminal vertices are called *Steiner points* or *Steiner vertices*, and the terminals are also called *regular points*. If there is a terminal with degree more than 1, then the tree can be decomposed at this terminal. In this way, a Steiner tree can be decomposed into smaller subtrees such that every terminal in a subtree is a leaf. These smaller subtrees are called *full components*. The size of a full component is the number of terminals in it. Figure 3.1 shows an example of a full component of size 5. A Steiner tree with only one full component is called a *full* tree.

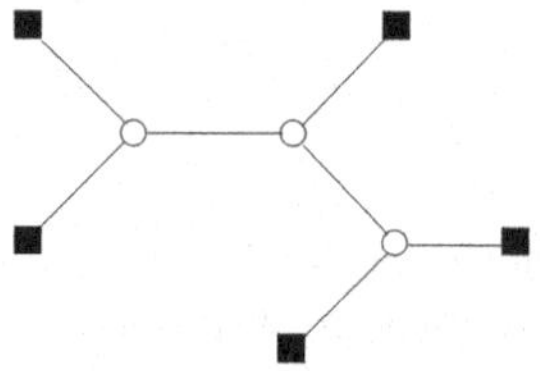

Figure 3.1: A full component (■ indicates a terminal, and ○ indicates a Steiner point).

Depending on the specific metric spaces on which the trees are defined, the Steiner tree problem may assume different forms. The following are three classical Steiner tree problems.

> EUCLIDEAN STEINER MINIMUM TREE (ESMT): Given a finite set P of terminals in the Euclidean plane, find a shortest network interconnecting all terminals in P.

> RECTILINEAR STEINER MINIMUM TREE (RSMT): Given a finite set P of terminals in the rectilinear plane, find a shortest network interconnecting all terminals in P.[1]

[1]The rectilinear plane is the plane with the distance function $d(\langle x_1, x_2\rangle, \langle y_1, y_2\rangle) = |x_1 - y_1| + |x_2 - y_2|$.

NETWORK STEINER MINIMUM TREE (NSMT): Given an edge-weighted graph (called a *network*) $G = (V, E)$ and a subset $P \subseteq V$ of terminals, find a subgraph of G with the minimum total weight interconnecting all vertices in P.

All of the above three versions of the Steiner tree problem are **NP**-hard, and we need to look for approximations for them. A simple, natural idea is to *restrict* the solutions to spanning trees, and use a *minimum spanning tree (MST)* to approximate the Steiner minimum tree.

A spanning tree is a Steiner tree with the restriction that no Steiner points exist, or, equivalently, a Steiner tree in which all full components are of size 2. In general, an MST can be computed in time $O(n^2)$. In addition, in the Euclidean or the rectilinear plane, an MST can be computed in time $O(n \log n)$.

For any set P of terminals, we let $mst(P)$ denote the length of the MST for set P, and $smt(P)$ the length of the SMT for set P. When we use the MST as the approximate solution to the Steiner tree problem, the performace ratio of this algorithm is then the maximum of $mst(P)/smt(P)$ over all input instances P. In the following, we show some results on the MST approximation to the Steiner tree problem.

We first consider the problem NSMT. In the problem NSMT, we usually assume that the input graph is a complete graph, and that the edge weight satisfies the triangle inequality. In fact, if an input graph is not complete, we can construct a complete graph on the same set of vertices and let the weight of each edge $\{u, v\}$ be the cost of the shortest path connecting u and v. Thus, the network SMT in the new graph is equivalent to the original one.

Theorem 3.1 *For the problem* NSMT, *the performance ratio of the MST approximation is equal to* 2.

Proof. Consider an SMT T interconnecting the terminals in P. Note that there exists an Euler tour T_1 of T, which uses each edge in T twice. Since we are working in a metric space that satisfies the triangle inequality, the length of an Euler tour must be greater than that of an MST. This means that $mst(P) \leq length(T_1) \leq 2 \cdot smt(P)$, and so the performance ratio of the MST approximation is at most 2.

Next, to show that the performance ratio of the MST approximation cannot be better than 2, consider a star graph G of $n+1$ vertices, each edge of length 1. More precisely, G is the complete graph with $n+1$ vertices $\{0, 1, \ldots, n\}$ and has distance $d(0, i) = 1$ for all $i = 1, 2, \ldots, n$, and $d(i, j) = 2$ for all $i \neq j \in \{1, 2, \ldots, n\}$. For the subset $P = \{1, 2, \ldots, n\}$, it is clear that $smt(P) = n$ and $mst(P) = 2(n-1)$. Therefore,

$$\frac{mst(P)}{smt(P)} = \frac{2(n-1)}{n} = 2 - \frac{2}{n}.$$

As n approaches infinity, this ratio approaches 2. This means that the performance ratio of the MST approximation cannot be less than 2. □

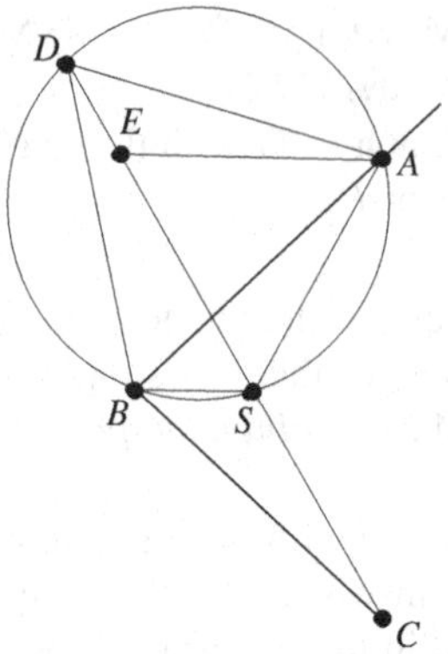

Figure 3.2: The angle between two edges of an SMT cannot be less than 120°.

For the problem ESMT, the performance ratio of the MST approximation is smaller than the general case of NSMT, due to some special properties of the SMTs in the Euclidean plane.

Lemma 3.2 *An SMT in the Euclidean plane has the following properties:*

(a) Every angle formed by two adjacent edges is at least $120°$.

(b) Every vertex has degree at most 3.

(c) Every Steiner point has degree exactly 3 *with an angle of* $120°$ *between the three edges.*

Proof. Note that (a) implies (b) and (c). To show (a), we assume, for the sake of contradiction, that there exist two edges forming an angle less than $120°$ at point B. Furthermore, assume that A and C are two points on the two edges of the angle, respectively, such that $|\overline{AB}| = |\overline{BC}|$. Draw an equilateral triangle $\triangle ABD$ with D on the opposite side of $\overline{AB}$ from C, and then draw a circle passing through the three points A, B, and D. Since $\angle ABC < 120°$, the line segment $\overline{CD}$ must intersect the circle at a point S between A and B (see Figure 3.2). We claim that $|\overline{SA}| + |\overline{SB}| = |\overline{SD}|$. To see this, let E be a point on $\overline{SD}$ such that $|\overline{DE}| = |\overline{SB}|$. Note that $\angle ADE = \angle ABS$ and $|\overline{AD}| = |\overline{AB}|$. Therefore, $\triangle ADE \cong \triangle ABS$. This implies that $\angle EAD = \angle SAB$, and so $\angle SAE = \angle BAD = 60°$. Since $\angle DSA = \angle DBA = 60°$, we see that $\triangle ASE$ is an equilateral triangle. It follows that $|\overline{SE}| = |\overline{SA}|$, and the claim is proven.

Now, if we replace the two edges $\overline{AB}$ and $\overline{BC}$ by the three edges $\overline{SA}$, $\overline{SB}$, and $\overline{SC}$, we can shorten the tree, because

$$\begin{aligned}|\overline{SA}| + |\overline{SB}| + |\overline{SC}| &= |\overline{SD}| + |\overline{SC}| \\ &= |\overline{CD}| < |\overline{CB}| + |\overline{BD}| = |\overline{AB}| + |\overline{BC}|.\end{aligned}$$

This leads to a contradiction. □

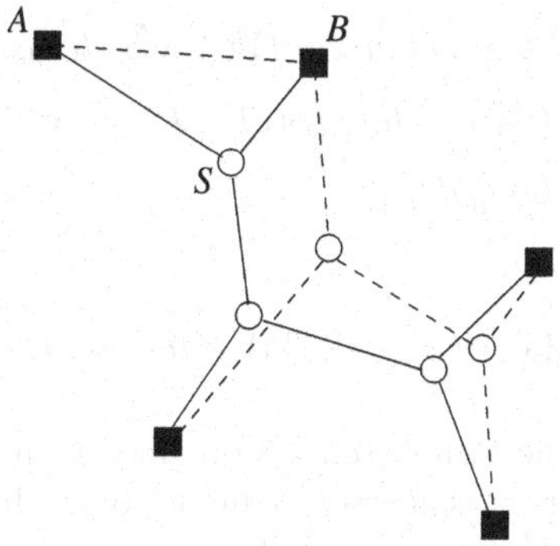

Figure 3.3: Proof of Theorem 3.3. In the above, the solid lines denote SMT(F), and the dotted lines denote SMT($V(F) - \{A\}) \cup \overline{AB}$.

Theorem 3.3 *For the problem* ESMT, *the performance ratio of the MST approximation is at most* $\sqrt{3}$.

Proof. Following the approach outlined at the beginning of this chapter, we consider a Euclidean SMT T on a set P of n terminals, and modify it into a spanning tree T' as follows:

> **While** T contains a Steiner point **do**
> find a full component F of T with two terminals A and B
> connected to a Steiner point S;
> **if** $|\overline{AS}| \geq |\overline{BS}|$
> **then** $T \leftarrow (T \setminus F) \cup \{\overline{AB}\} \cup \text{SMT}(V(F) - \{A\})$
> **else** $T \leftarrow (T \setminus F) \cup \{\overline{AB}\} \cup \text{SMT}(V(F) - \{B\})$.

[In the above, we write SMT(Q) to denote the Euclidean SMT of terminals in set Q, $V(F)$ to denote the set of terminal points in a tree F, and $\overline{AB}$ to denote the edge connecting points A and B.]

We show by induction on the number of Steiner points in T that the spanning tree T' has length at most $\sqrt{3} \cdot \mathit{length}(T)$. If T contains no Steiner point, then this holds trivially. Assume that T contains a Steiner point. Then there must exist a full component F of T with a Steiner point S adjacent to two terminals A and B. Without loss of generality, assume that $|\overline{AS}| \geq |\overline{BS}|$. From Lemma 3.2(c), we know that $\angle ASB = 120°$. It follows that

$$\begin{aligned}|\overline{AB}| &= \left(|\overline{AS}|^2 + |\overline{BS}|^2 - 2\cos 120° \cdot |\overline{AS}| \cdot |\overline{BS}|\right)^{1/2} \\ &= \left(|\overline{AS}|^2 + |\overline{BS}|^2 + |\overline{AS}| \cdot |\overline{BS}|\right)^{1/2} \leq \left(3 \cdot |\overline{AS}|^2\right)^{1/2} = \sqrt{3} \cdot |\overline{AS}|.\end{aligned}$$

Note that $(T \setminus F) \cup \text{SMT}(V(F) - \{A\})$ contains two connected components, say, T_1 and T_2. By the induction hypothesis, for $i = 1, 2$, the spanning tree T_i' obtained from T_i has length at most $\sqrt{3} \cdot \mathit{length}(T_i)$. Therefore, the spanning tree T', which is the union of T_1', T_2', and $\overline{AB}$, has length

$$\begin{aligned} length(T') &\leq |\overline{AB}| + \sqrt{3} \cdot length(T_1) + \sqrt{3} \cdot length(T_2) \\ &\leq \sqrt{3} \cdot \big(|\overline{AS}| + length((T \setminus F) \cup \mathrm{SMT}(V(F) - \{A\}))\big) \\ &\leq \sqrt{3} \cdot length(T), \end{aligned}$$

since

$$length(\mathrm{SMT}(V(F) - \{A\})) \leq length(F - |\overline{AS}|). \qquad \square$$

In each metric space, the *Steiner ratio* is the maximum ratio of the lengths between SMT and MST for the same set of input points. In other words, it is the inverse of the performance ratio of the MST approximation for SMT. For instance, Theorem 3.3 means that the Steiner ratio in the Euclidean plane is at least $1/\sqrt{3}$.

Determining the Steiner ratio in various metric spaces is a classical mathematical problem. The famous Gilbert and Pollak conjecture states that the Steiner ratio in the Euclidean plane is equal to $\sqrt{3}/2$ [Gilbert and Pollak, 1968]. This conjecture was resolved positively by Du and Hwang [1990]. That is, the performance ratio of the MST approximation for ESMT is exactly $2/\sqrt{3}$. For the problem RSMT, Hwang [1972] proved that the Steiner ratio in the rectilinear plane is equal to $2/3$.

3.2 k-Restricted Steiner Trees

We say a Steiner tree is a *k-restricted* Steiner tree if all of its full components have size at most k. In particular, a spanning tree is a 2-restricted Steiner tree. A naive idea of improving the MST approximation to the Steiner minimum-tree problems is to consider k-restricted Steiner trees, for $k \geq 3$, as approximations. Intuitively, as k gets larger, the minimum tree among all k-restricted Steiner trees, called the *k-restricted Steiner minimum tree*, gets closer to the Steiner minimum tree. In other words, the larger the parameter k is, the better the performance ratio is. In the following, we present an estimation of the performance of the k-restricted Steiner minimum tree as an approximation to the Steiner minimum tree.

Following the general approach of the analysis of an approximation designed with the restriction method, we consider an SMT and modify it into a k-restricted Steiner tree. To do so, we work on a full component T of size more than k, and perform the modification in two steps: We first express the full component T as a regular binary tree.[2] Then, we divide this tree into the union of smaller trees, each of size k.

To express T as a regular binary tree, we first modify it into a tree with the property that every Steiner point has degree exactly 3. This can be done by adding zero-length edges and new Steiner points to T. Next, we choose a root r in the middle of an edge, and convert the tree into a regular binary weighted tree, which is still called T (see Figure 3.4). In this regular binary tree, the weight of each edge is its length in the metric space.

[2] A *regular* binary tree is a binary tree in which each internal vertex has exactly two child vertices.

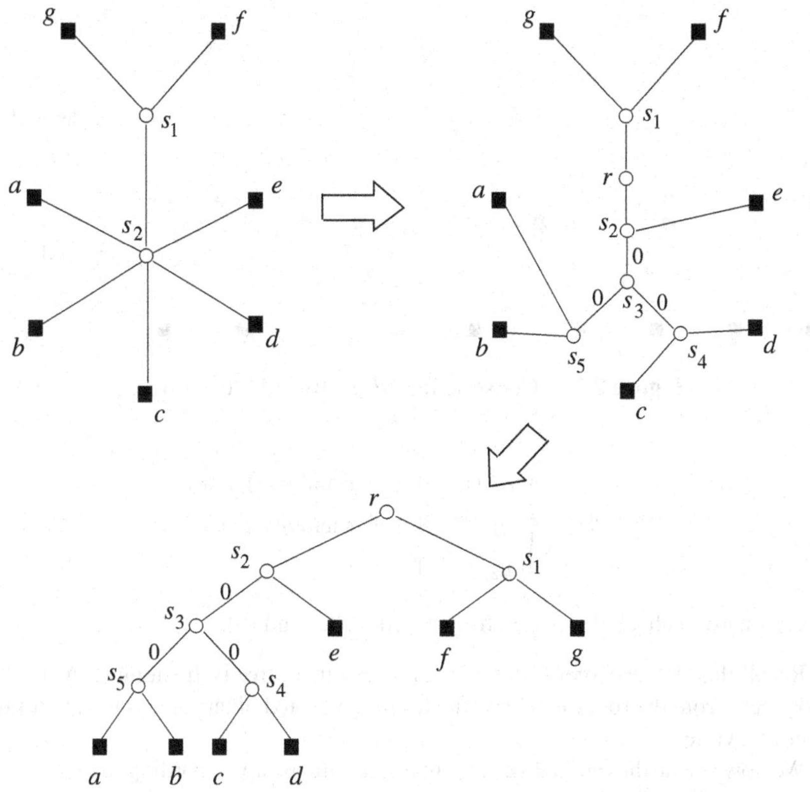

Figure 3.4: Constructing a regular binary tree from a Steiner tree.

Next, we modify this regular binary weighted tree T into a k-restricted Steiner tree. To do so, we need a lemma about regular binary trees.

Lemma 3.4 *For any regular binary tree T, there exists a one-to-one mapping f from internal vertices to leaves, such that*

(a) For any internal vertex u, $f(u)$ is a descendant of u; and

(b) All tree paths $p(u)$ from u to $f(u)$ are edge-disjoint.

Proof. We will construct, by induction on the number of internal vertices in T, a mapping f satisfying conditions (a) and (b).

If T has only one internal vertex, the lemma is obviously true. So, we assume that T has more than one internal vertex. Consider an internal vertex x both of whose two children are leaves. Let its children vertices be y_1 and y_2. Let T' be the tree T with y_1 and y_2 deleted (and so x becomes a leaf of T'). By the inductive hypothesis, there is a one-to-one mapping g from the internal vertices of T' to the leaves of T', satisfying conditions (a) and (b). Now, define f on the internal vertices of T as follows:

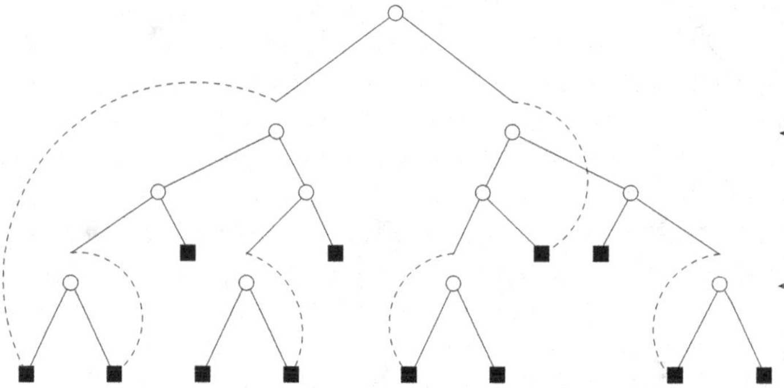

Figure 3.5: Constructing a k-restricted Steiner tree.

$$f(u) = \begin{cases} g(u), & \text{if } u \neq x \text{ and } g(u) \neq x, \\ y_1, & \text{if } u \neq x \text{ and } g(u) = x, \\ y_2, & \text{if } u = x. \end{cases}$$

It is not hard to check that f satisfies conditions (a) and (b). □

Recall that the level of a vertex u in a rooted binary tree is the number of vertices in the path from the root to u. Also, the level of a rooted binary tree is the maximum level of a vertex.

We now divide the internal vertices in tree T into groups according to their levels. Denote $q = \lfloor \log_2 k \rfloor$. For each $i \geq 1$, let I_i be the set of internal vertices at the ith level of the tree T, and $U_d = \bigcup_{i \equiv d \pmod{q}} I_i$. It is clear that sets $U_1, U_2, \ldots, U_q$ are pairwisely disjoint. Let f be the mapping found in Lemma 3.4 for the tree T. Denote by $\ell(p(u))$ the length of path $p(u)$ from an internal vertex u to the leaf $f(u)$. Let $\ell_d = \sum_{u \in U_d} \ell(p(u))$. From Lemma 3.4(b), we know that $\ell_1 + \ell_2 + \cdots + \ell_q \leq smt(P)$, where P is the set of terminals of T. By the pigeonhole principle, we can choose an integer d, $1 \leq d \leq q$, such that $\ell_d \leq smt(P)/q$.

For each nonroot vertex $u \in T$, let its parent vertex be $\pi(u)$. We construct a new tree T_d as follows: For each nonroot vertex $u \in U_d$, we replace edge $(\pi(u), u)$ by a new edge $(\pi(u), f(u))$ (see Figure 3.5). We will show that the new tree T_d is a k-restricted Steiner tree with length at most $smt(P)(q+1)/q$.

First, we prove that T_d is connected. To see this, we note that each replacement of an edge $(\pi(v), v)$ by the new edge $(\pi(v), f(v))$ keeps v and $\pi(v)$ connected, since $f(v)$ is a descendant of v. Therefore, during each step of the construction of T_d from T, all vertices are connected together.

Next, we show that T_d is k-restricted; that is, each full component of T_d has size at most k. To see this, we note that each full component of T_d must contain either the root r or a Steiner point $u \in U_d$, because each of the other Steiner points in T_d must belong to the same full component as its parent vertex (in the binary tree T). In addition, any two vertices in $U_d \cup \{r\}$ must belong to two different full components of T_d, because the edge-replacement operations divide each vertex $u \in U_d$ and its parent $\pi(u)$ into two different full components. Now, consider a full component C that contains a Steiner point $u \in U_d \cup \{r\}$. Each terminal in C can be reached from u through a path whose edges, other than the last one, are all in T. Therefore, such a path contains at most q edges. It means that if we consider C as a binary tree rooted at u, then it has at most $q+1$ levels, and so the number of terminals in C is at most $2^q \leq k$.

Finally, we check that the total length of T_d is bounded by $smt(P)(q+1)/q$. We note that, during the construction of T_d from T, each edge replacement increases the edge length by at most $\ell(p(u))$ for some $u \in U_d$. Thus, the total increase is at most ℓ_d, which is bounded by $mst(P)/q$ from our choice of d. Thus, the total length of T_d is at most $smt(P)(1+1/q)$.

Theorem 3.5 *For $k \geq 2$, the k-restricted SMT is a $(1 + 1/\lfloor \log k \rfloor)$-approximation to the Steiner minimum-tree problem.*

Let ρ_k be the maixmum lower bound of the ratio of the lengths between the SMT and the k-restricted SMT over the same set of terminals. That is,

$$\rho_k = \min_P \frac{smt(P)}{smt_k(P)},$$

where $smt_k(P)$ is the length of the k-restricted Steiner minimum tree over terminal points in the set P. This number ρ_k is called the *k-Steiner ratio*, which is the inverse of the performance ratio of the k-restricted SMT as an approximation to the SMT problem. For convenience and for historical reasons, we will use ρ_k, instead of its inverse, in later sections. To summarize our results in terms of ρ_k, we have

Corollary 3.6 *(a) For $k \geq 2$, $\rho_k \geq \lfloor \log k \rfloor / (\lfloor \log k \rfloor + 1)$.*
(b) $\lim_{k \to \infty} \rho_k = 1$.

In the above, we only proved an upper bound for ρ_k. The precise value of ρ_k is also known (see Borchers and Du [1995]): Write $k = 2^r + s$, with $0 \leq s < 2^r$; then we have

$$\rho_k = \frac{r \cdot 2^r + s}{(r+1)2^r + s}.$$

Theorem 3.5 indicates that, for large k, the k-restricted SMT could be a good approximation solution to SMT if it can be computed in polynomial time. Unfortunately, for $k \geq 4$, it is known that computing the k-restricted SMT is **NP**-hard, and for $k = 3$, it is still open whether the 3-restricted SMT can be computed in polynomial time or whether it is **NP**-hard. In the next section, we will study how to find good approximations to the k-restricted SMT itself, and use them to approximate the SMT.

3.3 Greedy k-Restricted Steiner Trees

Since the minimum spanning trees (i.e., the 2-restricted SMTs) can be found by greedy algorithms in polynomial time, it is natural to try to find approximate k-restricted SMTs by the greedy strategy. Before we present greedy approximations for the k-restricted SMTs, we first develop a general result for greedy Algorithm 2.D with respect to noninteger potential functions.

Recall the setting of the greedy Algorithm 2.D: Assume that f is a polymatroid on 2^E, and $\Omega_f = \{C \subseteq E \mid (\forall x \in E)\ \Delta_x f(C) = 0\}$. The problem MIN-SMC is to compute $\min_{A\in\Omega_f} c(A)$, where $c(A) = \sum_{x\in A} c(x)$. Algorithm 2.D finds an approximate solution to MIN-SMC as follows:

(1) Set $A \leftarrow \emptyset$.

(2) **While** there exists an $x \in E$ such that $\Delta_x f(A) > 0$ **do**

  select a vertex x that maximizes $\Delta_x f(A)/c(x)$;
  $A \leftarrow A \cup \{x\}$.

(3) **Return** $A_G \leftarrow A$.

Assume that A^* is an optimal solution to the problem MIN-SMC, and A_G is the approximate solution obtained by Algorithm 2.D with respect to the potential function f and cost function c. Let $x_1, x_2, \ldots, x_k$ be the elements in A_G in the order of their selection into the set A_G, and denote $A_0 = \emptyset$ and $A_i = \{x_1, x_2, \ldots, x_i\}$, for $i = 1, \ldots, k$.

Theorem 3.7 *Assume that the approximate solution A_G produced by* Algorithm 2.D *satisfies the condition $\Delta_{x_i} f(A_{i-1})/c(x_i) \geq 1$ for all $i = 1, 2, \ldots, k$. Then*

$$c(A_G) \leq \left(1 + \ln \frac{f(A^*)}{c(A^*)}\right) \cdot c(A^*).$$

Proof. Let $a_i = f(A^*) - f(A_i)$ for $i = 0, 1, \ldots, k$. Then $\Delta_{x_i} f(A_{i-1}) = a_{i-1} - a_i$, and $a_0 = f(A^*)$.

Suppose $A^* = \{y_1, y_2, \ldots, y_h\}$. Then, for each $j = 1, 2, \ldots, k$, we have, from the greedy choice of x_j and Lemma 2.23, that

$$\begin{aligned}\frac{a_{j-1} - a_j}{c(x_j)} &\geq \max_{1\leq i\leq h} \frac{\Delta_{y_i} f(A_{j-1})}{c(y_i)} \geq \frac{\sum_{i=1}^{h} \Delta_{y_i} f(A_{j-1})}{c(A^*)} \\ &\geq \frac{\Delta_{A^*} f(A_{j-1})}{c(A^*)} = \frac{f(A^*) - f(A_{j-1})}{c(A^*)} = \frac{a_{j-1}}{c(A^*)}. \end{aligned} \tag{3.1}$$

Hence, for each $j = 1, 2, \ldots, k$,

$$a_j \leq a_{j-1} \cdot \left(1 - \frac{c(x_j)}{c(A^*)}\right). \tag{3.2}$$

Note that

$$a_0 = f(A^*) = f(A_G) = \sum_{i=1}^{k} \Delta_{x_i} f(A_{i-1}) \geq \sum_{i=1}^{k} c(x_i) = c(A_G) \geq c(A^*),$$

and $a_k = f(A^*) - f(A_G) = 0$. Moreover, for each $i = 1, 2, \ldots, k$, $a_i \le a_{i-1}$ since f is monotone increasing. Thus, there exists an integer r, $0 \le r \le k$, such that $a_{r+1} < c(A^*) \le a_r$. From (3.1), we know that

$$\frac{a_r - a_{r+1}}{c(x_{r+1})} \ge \frac{a_r}{c(A^*)}.$$

We divide the numerator of the left-hand side of the above inequality into two parts: $a' = c(A^*) - a_{r+1}$, $a'' = a_r - c(A^*)$ (so that $a' + a'' = a_r - a_{r+1}$), and also divide the denominator into two parts proportionally: $c(x_{r+1}) = c' + c''$, with c' and c'' satisfying

$$\frac{a'}{c'} = \frac{a''}{c''} = \frac{a_r - a_{r+1}}{c(x_{r+1})}.$$

Then

$$\frac{a''}{c''} = \frac{a_r - a_{r+1} - a'}{c''} \ge \frac{a_r}{c(A^*)}.$$

Hence, by repeatedly applying (3.2), we get

$$\begin{aligned} c(A^*) &= a_{r+1} + a' \le a_r\Big(1 - \frac{c''}{c(A^*)}\Big) \\ &\le a_0\Big(1 - \frac{c(x_1)}{c(A^*)}\Big) \cdots \Big(1 - \frac{c(x_r)}{c(A^*)}\Big)\Big(1 - \frac{c''}{c(A^*)}\Big) \\ &\le a_0 \cdot \exp\Big(-\frac{c'' + \sum_{i=1}^{r} c(x_i)}{c(A^*)}\Big), \end{aligned}$$

since $1 + x \le e^x$. It follows that

$$c'' + \sum_{i=1}^{r} c(x_i) \le c(A^*) \cdot \ln \frac{a_0}{c(A^*)}.$$

Note that

$$\sum_{i=r+2}^{k} c(x_i) \le \sum_{i=r+2}^{k} \Delta_{x_i} f(A_{i-1}) = f(A) - f(A_{r+1}) = a_{r+1}.$$

Also, $a'/c' \ge a_r/c(A^*) \ge 1$. Therefore,

$$\begin{aligned} c(A) &\le c(A^*) \cdot \ln \frac{a_0}{c(A^*)} + c' + a_{r+1} \\ &\le c(A^*) \cdot \ln \frac{a_0}{c(A^*)} + a' + a_{r+1} = c(A^*)\Big(1 + \ln \frac{f(A^*)}{c(A^*)}\Big). \end{aligned}$$ □

In many cases, the potential function f is closely related to the cost function c and satisfies the condition $\Delta_{x_i} f(A_{i-1})/c(x_i) \ge 1$ of Theorem 3.7, as the cost $c(x_i)$ is usually no more than the *savings* from $\Delta_{x_i} f(A_{i-1})$.

Indeed, we can verify that this condition is satisfied by the potential function f of the following natural greedy algorithm for the k-restricted SMT problem. For a given set P of terminals, let Q_k be the set of all full components of size at most k (over all possible Steiner trees) on P. For any $A \subseteq Q_k$, let $\mathrm{MST}(P:A)$ be the minimum spanning tree on P after every edge in every component of A is contracted into a single point, and let $mst(P:A)$ denote its length. Then the greedy algorithm for the k-restricted SMT problem can be described as follows:

(1) $A \leftarrow \emptyset$; $T \leftarrow \mathrm{MST}(P)$.

(2) **While** A does not connect all terminals in P **do**
  find $K \in Q_k$ that miximizes $(mst(P:A) - mst(P:A \cup K))/c(K)$;
  $A \leftarrow A \cup K$;
  $T \leftarrow \mathrm{MST}(P:A)$.

(3) Output A.

In other words, this is the greedy Algorithm 2.D with respect to the potential function

$$f(A) = mst(P) - mst(P:A).$$

Lemma 3.8 $f(A) = mst(P)$ *if and only if A forms a connected graph interconnecting all terminals.*

Proof. Trivial. □

To prove that f is submodular, we will reduce the general case of $k \geq 2$ to the special case of $k = 2$. Since this reduction technique may be applied to other potential functions, we state it as a separate lemma.

Lemma 3.9 *Suppose that $g: 2^E \to \mathbb{R}$ is a monotone increasing, submodular function, and that $\mathcal{C}$ is a collection of subsets of E. Then the function $h: 2^{\mathcal{C}} \to \mathbb{R}$ induced from g by $h(\mathcal{A}) = g(\bigcup_{S \in \mathcal{A}} S)$ is also monotone increasing and submodular.*

Proof. It is clear that h is monotone increasing. To see that h is submodular, let $\mathcal{A} \subseteq \mathcal{B} \subseteq \mathcal{C}$ and $X \in \mathcal{C}$. We need to show that $\Delta_X h(\mathcal{A}) \geq \Delta_X h(\mathcal{B})$. Since g is monotone increasing and submodular, we have

$$\Delta_y g\big(\textstyle\bigcup_{S \in \mathcal{A}} S\big) \geq \Delta_y g\big(\textstyle\bigcup_{S \in \mathcal{B}} S\big),$$

for any $y \in E$, because $\mathcal{A} \subseteq \mathcal{B}$ implies $\bigcup_{S \in \mathcal{A}} S \subseteq \bigcup_{S \in \mathcal{B}} S$. This inequality can be extended so that, for any $X \subseteq E$,

$$\Delta_X g\big(\textstyle\bigcup_{S \in \mathcal{A}} S\big) \geq \Delta_X g\big(\textstyle\bigcup_{S \in \mathcal{B}} S\big).$$

It follows that

$$\Delta_X h(\mathcal{A}) = \Delta_X g\big(\textstyle\bigcup_{S \in \mathcal{A}} S\big) \geq \Delta_X g\big(\textstyle\bigcup_{S \in \mathcal{B}} S\big) = \Delta_X h(\mathcal{B}).$$ □

Lemma 3.10 *The function f is a polymatroid function on 2^{Q_k}.*

Proof. Clearly, f is normalized and monotone increasing. To see that it is submodular, we reduce the general case to the case $k = 2$. For a given set P of terminals, let E be the set of all edges connecting terminals in P, and $g : 2^E \to \mathbb{R}$ be the function defined by $g(S) = mst(P) - mst(P : S)$ (that is, g is the function f in the case $k = 2$). Now, for any $T \in Q_k$, let $e(T)$ be the set of edges in a spanning tree on the terminals in T. Then it is easy to see that $f(A) = g(\bigcup_{T \in A} e(T))$. Thus, by Lemma 3.9, we only need to prove that g is submodular.

Note that

$$
\begin{aligned}
&g \text{ is submodular and monotone increasing}\\
&\quad \iff (\forall A \subseteq B \subseteq E)\,(\forall y \in E)\,\Delta_y g(B) \le \Delta_y g(A)\\
&\quad \iff (\forall A \subseteq E)\,(\forall x, y \in E)\,\Delta_y g(A \cup \{x\}) \le \Delta_y g(A)\\
&\quad \iff (\forall A \subseteq E)\,(\forall x, y \in E)\,\Delta_{\{x,y\}} g(A) \le \Delta_x g(A) + \Delta_y g(A).
\end{aligned}
$$

From the definition of g, we have

$$\Delta_x g(A) = g(A \cup \{x\}) - g(A) = mst(P : A) - mst(P : A \cup \{x\}).$$

So, it suffices to prove, for any $A \subseteq E$ and any $x, y \in E$,

$$
\begin{aligned}
&mst(P : A) - mst(P : A \cup \{x, y\})\\
&\quad \le (mst(P : A) - mst(P : A \cup \{x\})) + (mst(P : A) - mst(P : A \cup \{y\})).
\end{aligned}
$$

Let $T = \mathrm{MST}(P : A)$. This tree T contains a path π_x connecting two endpoints of x and a path π_y connecting two endpoints of y. Let e_x (and e_y) be a longest edge in π_x (in π_y, respectively). Then we have

$$
\begin{aligned}
mst(P : A) - mst(P : A \cup \{x\}) &= length(e_x),\\
mst(P : A) - mst(P : A \cup \{y\}) &= length(e_y).
\end{aligned}
$$

In addition, the value of $mst(P : A) - mst(P : A \cup \{x, y\})$ can be computed as follows: Choose a longest edge e' from $\pi_x \cup \pi_y$. Notice that $T \cup \{x, y\} - \{e'\}$ contains a unique cycle C. Choose a longest edge e'' from $(\pi_x \cup \pi_y) \cap C$. Then we have

$$mst(P : A) - mst(P : A \cup \{x, y\}) = length(e') + length(e'').$$

Now, to show the submodularity of g, it suffices to prove

$$length(e_x) + length(e_y) \ge length(e') + length(e''). \tag{3.3}$$

Case 1. Neither e_x nor e_y is in $\pi_x \cap \pi_y$. Without loss of generality, assume $length(e_x) \ge length(e_y)$. Then we have $length(e') = length(e_x)$. So, if we choose $e' = e_x$, then $(\pi_x \cup \pi_y) \cap C = \pi_y$. Hence, we have $length(e'') = length(e_y)$. It follows that the two sides of (3.3) are equal.

Case 2. $e_x \notin \pi_x \cap \pi_y$ and $e_y \in \pi_x \cap \pi_y$. Clearly, $length(e_x) \geq length(e_y)$. Hence, we may choose $e' = e_x$ so that $(\pi_x \cup \pi_y) \cap C = \pi_y$, and $length(e'') = length(e_y)$. Again, the two sides of (3.3) are equal.

Case 3. $e_x \in \pi_x \cap \pi_y$ and $e_y \notin \pi_x \cap \pi_y$. Similar to Case 2.

Case 4. Both e_x and e_y are in $\pi_x \cap \pi_y$. In this case, $length(e_x) = length(e_y) = length(e') \geq length(e'')$. Hence, inequality (3.3) holds. □

Lemma 3.11 *Each element x_i, $1 \leq i \leq k$, selected by* Algorithm 2.D, *with respect to the potential function f, must satisfy the condition $\Delta_{x_i} f(A_{i-1})/c(x_i) \geq 1$.*

Proof. It is clear that $\Delta_e f(A_{i-1})/c(e) = 1$ for any edge e of MST$(P : A_{i-1})$. It follows that the value $\Delta_{x_i} f(A_{i-1})/c(x_i)$ of the best choice x_i, which is greater than or equal to this value, must be at least 1. □

Let $c(T)$ denote the length of tree T. The following theorem follows from Theorem 3.7.

Theorem 3.12 *Suppose A is the approximate solution produced by* Algorithm 2.D *with respect to the potential function f defined above. Then*

$$\frac{c(A)}{smt_k(P)} \leq 1 + \ln \frac{mst(P)}{smt_k(P)}.$$

Corollary 3.13 *Suppose A is the approximate solution produced by* Algorithm 2.D. *Then*

$$\frac{c(A)}{smt(P)} \leq \rho_k^{-1}\left(1 + \ln \frac{\rho_k}{\rho_2}\right).$$

Proof. By Theorem 3.12,

$$\frac{c(A)}{smt(P)} \leq \frac{smt_k(P)}{smt(P)}\left(1 + \ln \frac{smt(P)/smt_k(P)}{smt(P)/mst(P)}\right).$$

Note that

$$\frac{smt(P)}{smt_k(P)} \geq \rho_k \quad \text{and} \quad \frac{smt(P)}{mst(P)} \geq \rho_2.$$

Now, the corollary follows from the observation that the function $(1 + \ln(x/a))/x$ is monotone decreasing when $x \geq a$. □

Note that $\lim_{k\to\infty} \rho_k = 1$. Thus, when k goes to ∞, the greedy Algorithm 2.D produces approximate solutions with performance ratio close to $1 - \ln \rho_2$.

In the above analysis, the condition in Theorem 3.7 that the selected element x always satisfies $\Delta_x f(A_{i-1})/c(x) \geq 1$ is critical. Suppose this condition does not hold; can we still get a good estimate of the performance ratio of the greedy Algorithm 2.D? The answer is *yes*, but we may need to modify the potential function

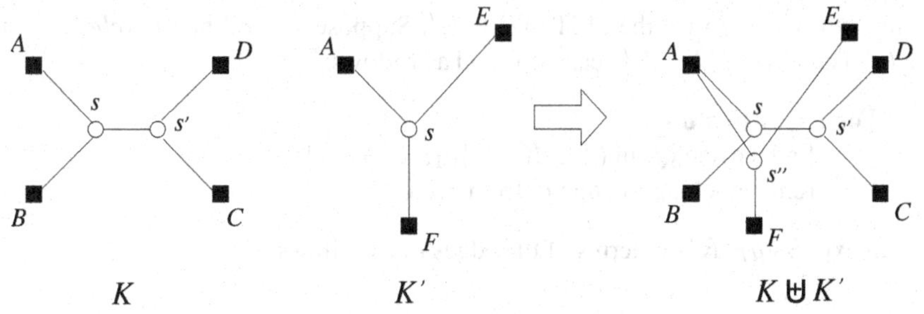

Figure 3.6: Operation $K \uplus K'$.

f and/or the cost function c so that a property similar to the condition of Theorem 3.7 still holds. In the following, we present such an example, which gives a better approximation for NSMT.

The idea of this greedy algorithm is as follows: It again begins with $T = \mathrm{MST}(P)$. At each iteration, it selects a full component K in Q_k, replaces T by the union of T and K, and then eliminates edges from the union until it does not have a cycle. The greedy strategy suggested by Algorithm 2.D would select K to maximize the saving of this process relative to the cost $c(K)$. However, since the saving here is not necessarily greater than or equal to $c(K)$, Theorem 3.7 cannot be applied directly, and so we need to modify this strategy.

Before we describe how to modify this algorithm, we first define the notion of the *union* of two Steiner trees. For $A, B \in Q_k$, we let the union $A \uplus B$ be the graph obtained from A and B by identifying the same terminals in A and B, but keeping separate copies of the same Steiner vertices (see Figure 3.6). More precisely, suppose A has terminals T_A, Steiner vertices S_A, and edges E_A; and B has terminals T_B, Steiner vertices S_B, and edges E_B. Then $A \uplus B$ has terminals $T_A \cup T_B$, Steiner vertices $S_{A \uplus B} = \{s^A \mid s \in S_A\} \cup \{s^B \mid s \in B\}$, and edges $E_{A \uplus B} = E_A \cup E_B$.[3] This definition of operation $\uplus$ can also be extended to two subgraphs A and B.

Now we can define the potential function g for this greedy algorithm. For convenience, we define $\Delta_K g(T)$ directly and denote it by $g_T(K)$: For $A \subseteq Q_k$ and a Steiner tree T on P, let

$$g_T(A) = c(T) - mst\big(T \uplus \big(\biguplus_{K \in A} K\big)\big).$$

Lemma 3.14 *Let T be a Steiner tree on terminal set P. Then, for $K, K' \in Q_k$,*

$$g_T(K \uplus K') \le g_T(K) + g_T(K').$$

Proof. It suffices to show that

$$mst(T \uplus K) - mst(T \uplus K \uplus K') \le g_T(K'). \tag{3.4}$$

[3]Note that if $|T_A| > 2$, then all edges in E_A must have a Steiner vertex as an endpoint. This implies that $E_A \cap E_B = \emptyset$ unless $T_A = T_B$ has size 2.

We first study how to get the MST of $T \uplus K'$. Suppose $T \uplus K'$ has a *cycle base* of size h.[4] Then, $\mathrm{MST}(T \uplus K')$ can be found as follows:

> **For** $i \leftarrow 1$ **to** h **do**
> find a cycle Q_i in $(T \uplus K') \setminus \{e_1, \ldots, e_{i-1}\}$;
> remove a longest edge e_i from cycle Q_i.

We can express $g_T(K')$ in terms of the edges e_i as follows:

$$g_T(K') = \sum_{i=1}^{h} c(e_i) - c(K').$$

Next, we consider the MST of graph $H = \mathrm{MST}(T \uplus K) \uplus K'$. Again, H has a cycle base of size h, and we can find $\mathrm{MST}(H)$ by finding h cycles Q'_i, $1 \leq i \leq h$, in H and removing a longest edge e'_i from each cycle Q'_i. In order to prove (3.4), we need to show that the total cost of the removed edges is no more than the total cost of $c(e_i)$, $1 \leq i \leq h$. This property can be proved by modifying, at each stage, cycle Q_i to form a new cycle Q'_i so that each edge in Q'_i is no longer than $c(e_i)$. More precisely, we can find $\mathrm{MST}(H)$ as follows:

> **For** $i \leftarrow 1$ **to** h **do**
> find, from Q_i, a cycle Q'_i in $H \setminus \{e'_1, \ldots, e'_{i-1}\}$ with the property
> that all edges in Q'_i are no longer than e_i;
> delete a longest edge e'_i from Q'_i.

To see how to find Q'_i from Q_i with the desired property, let $H_1 = \mathrm{MST}(T \uplus K)$. If Q_i is a cycle in H, then let $Q'_i = Q_i$. On the other hand, if Q_i is not a cycle in H, that is, if there is an edge $\{u, v\}$ in $Q_i \setminus H$, then this edge must be in T and hence in $(T \uplus K) \setminus H_1$. Thus, H_1 must contain a path $\pi_{u,v}$ from u to v, which, together with $\{u, v\}$, forms a cycle in $T \uplus K$. In addition, since H_1 is a minimum spanning tree of $T \uplus K$, $\{u, v\}$ must be a longest edge in this cycle. (Note that this cycle cannot be identical to Q_i, since Q_i must contain at least one edge in K'.) Thus, for each edge $\{u, v\}$ in Q_i that is not in $H \setminus \{e'_1, \ldots, e'_{i-1}\}$, we can replace it by a path $\pi_{u,v}$ in H in which each edge is no longer than $\{u, v\}$. (This is also true for edges in $Q_i \cap \{e'_1, \ldots, e'_{i-1}\}$, since each e'_j, with $j < i$, was deleted from a cycle Q'_j in H.) Repeating this on all edges in $Q_i \setminus H$, we obtain a cycle Q'_i in H with the required property.

This implies that

$$\begin{aligned} g_{\mathrm{MST}(T \uplus K)}(K') &= mst(T \uplus K) - mst(T \uplus K \uplus K') \\ &= \textstyle\sum_{i=1}^{h} c(e'_i) - c(K') \leq \sum_{i=1}^{h} c(e_i) - c(K') = g_T(K'), \end{aligned}$$

and the lemma is proven. □

---

[4] A cycle base in a graph is a minimal set of cycles from which all cycles in the graph can be generated.

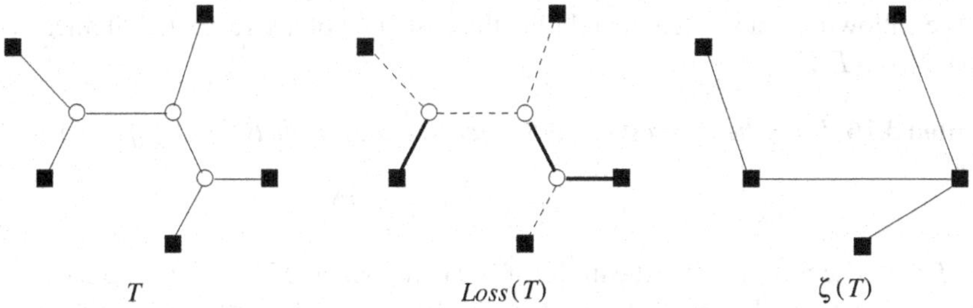

Figure 3.7: An example of Loss(T) and $\zeta(T)$.

Note that we write $g_T(K)$ to denote $\Delta_K g(T)$. So the potential function of the following greedy algorithm is submodular.

Algorithm 3.A (*Greedy Algorithm for* NSMT)
Input: A complete graph $G = (V, E)$ with edge cost c, and $P \subseteq V$.

(1) Set $T \leftarrow \text{MST}(P)$.

(2) **While** there exists a $K \in Q_k$ such that $g_T(K) > 0$ **do**
  select $K \in Q_k$ that maximizes $g_T(K)/c(K)$;
  $T \leftarrow \text{MST}(T \uplus K)$.

(3) Output $T_G \leftarrow T$. ■

As we pointed out earlier, the function g_T, unfortunately, does not necessarily satisfy the condition of Theorem 3.7, and so the performance of the above algorithm is hard to estimate. To resolve this problem, Robin and Zelikovsky [2000] introduced a new technique based on the notion of loss of a Steiner tree. The *loss* of a Steiner tree T, denoted by Loss(T), is the shortest forest connecting all Steiner points to terminals. We write $loss(T)$ to denote its length. In addition, we define $\zeta(T)$ to be the tree obtained from T by contracting every edge in Loss(T) into a point. We show Loss(T) and $\zeta(T)$ in Figure 3.7. Note that although $\zeta(T)$, as shown in the figure, looks like a spanning tree of T, the length of its edges may be shorter than the original edge length.

Proposition 3.15 *For any Steiner tree T, $loss(T) \leq length(T)/2$.*

Proof. We can construct recursively a forest L connecting each Steiner point in T to a terminal as follows:

While there is a Steiner point **do**
  find a Steiner point S adjacent to two terminals A and B;
  add to L the shorter of the two edges $\overline{SA}$ and $\overline{SB}$;
  reset S as a terminal point.

It is clear that this forest L has length at most one half of $length(T)$. □

The following is a key lemma relating the cost $c(T)$ of a k-restricted Steiner tree T with $loss(T)$.

Lemma 3.16 *Let T be a k-restricted Steiner tree. If, for all $K \in Q_k$, $g_T(K) \le 0$, then*

$$c(T) \le smt_k(P) + loss(T).$$

Proof. Suppose $\mathrm{SMT}_k(P)$ is the union of full components $K_1, \ldots, K_p$, each of size at most k. Then, by Lemma 3.14, we have

$$g_T(K_1 \uplus \cdots \uplus K_p) \le \sum_{i=1}^{p} g_T(K_i) \le 0.$$

That is,

$$c(T) \le mst(T \uplus K_1 \uplus \cdots \uplus K_p).$$

Note that $\mathrm{MST}(T \uplus K_1 \uplus \cdots \uplus K_p)$ is a shortest tree connecting all vertices in $T \uplus K_1 \uplus \cdots \uplus K_p$, including terminals and all Steiner vertices in $T, K_1, \ldots, K_p$, using the edges in $T, K_1, \ldots, K_p$. But $\mathrm{SMT}_k(P) \cup \mathrm{Loss}(T)$ is just such a tree. It follows that $c(T) \le smt_k(P) + loss(T)$. □

This lemma suggests that we can use $loss(K)$ instead of $c(K)$ as the cost of K in Algorithm 3.A. In addition, since we changed the cost to $loss(K)$, the saving $g_T(K)$ needs to be adjusted accordingly. That is, at each iteration, we only add $\zeta(K)$, instead of K, to the current Steiner tree T, to calculate $g_T(K)$ (in the following algorithm, we call this new tree H).

Algorithm 3.B (*Robin–Zelikovsky Algorithm for* NSMT)
Input: A complete graph $G = (V, E)$ with edge cost c, and $P \subseteq V$.

(1) Set $E^* \leftarrow \{K \in Q_k \mid loss(K) > 0\}$; $T \leftarrow \mathrm{MST}(P)$; $H \leftarrow \mathrm{MST}(P)$.

(2) **While** there exists a $K \in E^*$ such that $g_H(K) > 0$ **do**

select a smallest $K \in E^*$ that maximizes $g_H(K)/loss(K)$;

$T \leftarrow \mathrm{MST}(T \uplus K)$;

$H \leftarrow \mathrm{MST}(H \uplus \zeta(K))$.

(3) Output $T_G \leftarrow T$. ■

To analyze the performance of Algorithm 3.B, we observe the following properties of the tree H. In the following, for $i \ge 1$, we let K_i denote the full component K selected at the ith iteration, and H_i the Steiner tree H at the end of the ith iteration.

Lemma 3.17 *For each $i \ge 1$, $\mathrm{MST}(H_{i-1} \uplus K_i)$ must contain all edges of K_i.*

Proof. For the sake of contradiction, suppose e is an edge in K_i that is not in $\mathrm{MST}(H_{i-1} \uplus K_i)$. Then we claim that there must be a $\mathrm{Loss}(K_i)$ that does not contain e.

To see this, let us consider how to find Loss(K_i). In general, for $A \in Q_k$, we can find Loss(A) as follows: Let $Z(A)$ be the complete graph on the terminals in A, with edge cost equal to zero for all edges. Let $B = \mathrm{MST}(Z(A) \uplus A)$. Then we observe that the edges in $A \cap B$ must be a Loss(A), since all terminals are connected in B by edges in $Z(A)$. Now, consider the specific case of Loss(K_i) here. We can add $Z(K_i)$ to $H_{i-1} \uplus K_i$, and consider $B = \mathrm{MST}(H_{i-1} \uplus K_i \uplus Z(K_i))$. From the above observation, we see that the edges in $K_i \cap B$ form a Loss(K_i). Now, since e is not in $\mathrm{MST}(H_{i-1} \uplus K_i)$, there is a minimum spanning tree $B = \mathrm{MST}(H_{i-1} \uplus K_i \uplus Z(K_i))$ that does not contain e. (We can find such a tree B by adding, one by one, an edge $e' \in Z(K_i)$ to $\mathrm{MST}(H_{i-1} \uplus K_i)$ and then removing a longest edge from the cycle that resulted from the addition of e'.) It follows that the corresponding forest Loss(K_i) does not contain e. This completes the proof of the claim.

Now, we note that e divides K_i into two parts C and D. Since $e \notin \mathrm{MST}(H_{i-1} \uplus K_i)$, we have $g_{H_{i-1}}(K_i) = g_{H_{i-1}}(C \uplus D)$. By Lemma 3.14,

$$g_{H_{i-1}}(K_i) \le g_{H_{i-1}}(C) + g_{H_{i-1}}(D).$$

If e connects a terminal to a Steiner vertex, then either C or D is a single terminal point, and the other is $K_i' = K_i \setminus \{e\} \in Q_k$; and we have $g_{H_{i-1}}(K_i) = g_{H_{i-1}}(K_i')$. Moreover, $loss(K_i) = loss(K_i')$. Hence,

$$\frac{g_{H_{i-1}}(K_i)}{loss(K_i)} = \frac{g_{H_{i-1}}(K_i')}{loss(K_i')}.$$

However, K_i' is smaller than K_i, and this contradicts the greedy choice of K_i in Algorithm 3.B.

On the other hand, if both endpoints of e are Steiner vertices, then we have $loss(K_i) = loss(C) + loss(D)$, and so

$$\frac{g_{H_{i-1}}(K_i)}{loss(K_i)} \le \frac{g_{H_{i-1}}(C) + g_{H_{i-1}}(D)}{loss(C) + loss(D)} \le \max\left\{\frac{g_{H_{i-1}}(C)}{loss(C)}, \frac{g_{H_{i-1}}(D)}{loss(D)}\right\}.$$

Again, this is a contradiction to the greedy choice of K_i. So, the lemma is proven. □

Lemma 3.18 *For each $i \ge 1$,*

$$g_{H_{i-1}}(K_i) + loss(K_i) = c(H_{i-1}) - c(H_i).$$

Proof. From Lemma 3.17, we know that $A_i = \mathrm{MST}(H_{i-1} \uplus K_i)$ contains K_i. In addition, if we change the cost of each edge in Loss(K_i) to zero, we obtain the tree $\zeta(K_i)$, and since the edge cost of $\zeta(K_i)$ is no more than that of K_i, $H_i = \mathrm{MST}(H_{i-1} \uplus \zeta(K_i))$ must also contain $\zeta(K_i)$. Therefore, the edges in trees $A_i \setminus K_i$ and $H_i \setminus \zeta(K_i)$ are identical. Thus, the difference between the costs of the two trees A_i and H_i is just $c(K_i) - c(\zeta(K_i)) = loss(K_i)$. That is,

$$c(H_i) = mst(H_{i-1} \uplus \zeta(K_i)) = mst(H_{i-1} \uplus K_i) - loss(K_i).$$

In addition, by the definition of $g_{H_{i-1}}$, we know that

$$g_{H_{i-1}}(K_i) = c(H_{i-1}) - mst(H_{i-1} \uplus K_i).$$

It follows that

$$g_{H_{i-1}}(K_i) + loss(K_i) = c(H_{i-1}) - c(H_i). \qquad \square$$

Now, we are ready to estimate the performance ratio of the greedy Algorithm 3.B. The analysis is similar to that of Theorem 3.7.

Theorem 3.19 *The greedy* Algorithm 3.B *produces an approximate solution for* NSMT *with cost at most*

$$smt_k(P) + loss_k \cdot \ln\left(1 + \frac{mst(P) - smt_k(P)}{loss_k}\right),$$

where $loss_k = loss(\mathrm{SMT}_k(P))$.

Proof. Assume that greedy Algorithm 3.B halts after m iterations. For $1 \le i \le m$, let K_i denote the full component K selected at the ith iteration in Algorithm 3.B, and H_i the tree H at the end of the ith iteration. For convenience, we also let $l_i = loss(K_i)$ and $g_i = g_{H_{i-1}}(K_i)$. By Lemma 3.18,

$$c(H_{i-1}) - c(H_i) = g_i + l_i.$$

Let $Y_1, \ldots, Y_h$ be all full components of $\mathrm{SMT}_k(P)$. Then, by the greedy strategy and Lemma 3.14,

$$\begin{aligned} \frac{g_i}{l_i} &\ge \max_{1\le j\le h} \frac{g_{H_{i-1}}(Y_j)}{loss(Y_j)} \ge \frac{\sum_{j=1}^h g_{H_{i-1}}(Y_j)}{\sum_{j=1}^h loss(Y_j)} \\ &\ge \frac{g_{H_{i-1}}\left(\biguplus_{j=1}^h Y_j\right)}{loss_k} = \frac{c(H_{i-1}) - smt_k(P)}{loss_k}. \end{aligned}$$

Hence,

$$\frac{c(H_{i-1}) - c(H_i)}{l_i} = \frac{g_i + l_i}{l_i} \ge 1 + \frac{c(H_{i-1}) - smt_k(P)}{loss_k}.$$

Denote $a_i = c(H_i) + loss_k - smt_k(P)$. Then we can rewrite the above inequality as

$$\frac{a_{i-1} - a_i}{l_i} \ge \frac{a_{i-1}}{loss_k};$$

that is,

$$a_i \le a_{i-1}\left(1 - \frac{l_i}{loss_k}\right) \le a_{i-1} \cdot \exp\left(-\frac{l_i}{loss_k}\right). \tag{3.5}$$

We note that by Lemma 3.16, $c(H_m) \le smt_k(P)$ and, hence, $a_m = c(H_m) + loss_k - smt_k(P) \le loss_k$. Moreover, $a_0 = mst(P) + loss_k - smt_k(P) \ge loss_k$. Therefore, we can find an integer i such that $a_{i+1} < loss_k \le a_i$. (If $a_m = loss_k$, then set $i = m$.) Divide $a_i - a_{i+1}$ into a' and a'' by $a' = a_i - loss_k$ and $a'' = loss_k - a_{i+1}$. Also, divide l_{i+1} into c' and c'' proportionally so that $c' + c'' = l_{i+1}$ and

$$\frac{a'}{c'} = \frac{a''}{c''} = \frac{a_i - a_{i+1}}{l_{i+1}}.$$

Note that

$$\frac{a_i - loss_k}{c'} = \frac{a'}{c'} = \frac{a_i - a_{i+1}}{l_{i+1}} \geq \frac{a_i}{loss_k}.$$

Thus,

$$loss_k \leq a_i\Big(1 - \frac{c'}{loss_k}\Big) \leq a_i \cdot \exp\Big(-\frac{c'}{loss_k}\Big).$$

Applying (3.5) recursively to the above inequality, we get

$$loss_k \leq a_0 \cdot \exp\Big(-\frac{c' + l_i + \cdots + l_1}{loss_k}\Big),$$

or

$$l_1 + \cdots + l_i + c' \leq loss_k \cdot \ln \frac{a_0}{loss_k} = loss_k \cdot \ln\Big(1 + \frac{mst(P) - smt_k(P)}{loss_k}\Big).$$

Now let us estimate the cost of the output approximation T_G of Algorithm 3.B. Since the cost of the approximate Steiner tree T in each iteration is decreasing, $c(T_G)$ is at most $mst(H_0 \uplus K_1 \uplus \cdots \uplus K_{i+1})$. To estimate this value, we can construct a spanning tree S for $H_0 \uplus K_1 \uplus \cdots \uplus K_{i+1}$ as follows: We first put $L = \text{Loss}(K_1) \cup \cdots \cup \text{Loss}(K_{i+1})$ into S; then we contract each edge of L into a single point, find an MST of the resulting graph, and add it to S. It follows that

$$\begin{aligned} c(T_G) &\leq mst(H_0 \uplus K_1 \uplus \cdots \uplus K_{i+1}) \\ &\leq c(S) = mst(H_0 \uplus \zeta(K_1) \uplus \cdots \uplus \zeta(K_{i+1})) + l_1 + \cdots + l_{i+1} \\ &= c(H_{i+1}) + l_1 + \cdots + l_{i+1}. \end{aligned}$$

Furthermore, we know that

$$c(H_{i+1}) = c(H_i) - (a_i - a_{i+1}) = c(H_i) - a' - a'',$$

and that

$$\frac{a''}{c''} = \frac{a_i - a_{i+1}}{l_{i+1}} \geq \frac{a_i}{loss_k} \geq 1.$$

So, we have

$$\begin{aligned} c(T_G) &\leq c(H_{i+1}) + l_1 + \cdots + l_{i+1} \\ &= c(H_i) - a' - a'' + l_1 + \cdots + l_i + c' + c'' \\ &= (c(H_i) - a') + (l_1 + \cdots + l_i + c') + (c'' - a'') \\ &\leq smt_k(P) + loss_k \cdot \ln\Big(1 + \frac{mst(P) - smt_k(P)}{loss_k}\Big). \end{aligned}$$

□

Since the value of

$$loss_k \cdot \ln\left(1 + \frac{mst(P) - smt_k(P)}{loss_k}\right)$$

is increasing with respect to $loss_k$, we get, from Proposition 3.15,

$$loss_k \cdot \ln\left(1 + \frac{mst(P) - smt_k(P)}{loss_k}\right) \le \frac{smt_k(P)}{2} \ln\left(1 + \frac{mst(P) - smt_k(P)}{smt_k(P)/2}\right).$$

Therefore, the performance ratio of Algorithm 3.B is bounded by

$$\begin{aligned}
&\frac{smt_k(P)}{smt(P)}\left(1 + \frac{1}{2}\ln\left(1 + 2 \cdot \frac{mst(P)/smt(P) - smt_k(P)/smt(P)}{smt_k(P)/smt(P)}\right)\right) \\
&\quad \le \rho_k^{-1}\left(1 + \frac{1}{2}\ln\left(1 + 2 \cdot \frac{2 - \rho_k^{-1}}{\rho_k^{-1}}\right)\right) = \rho_k^{-1}\left(1 + \frac{\ln(4\rho_k - 1)}{2}\right).
\end{aligned}$$

When $k \to \infty$, we have $\rho_k \to 1$, and hence $\rho_k^{-1}(1 + \ln(4\rho_k - 1)/2)$ tends to $1 + (\ln 3)/2 < 1.55$.

Corollary 3.20 *The greedy* Algorithm 3.B *produces a* (1.55)*-approximation for* NSMT.

3.4 The Power of Minimum Spanning Trees

Minimum spanning trees play an important role in the design of approximation algorithms for network optimization problems. They are a natural candidate for approximation when the objective function is a function of the total edge length. In some cases, they might be a good approximation even if the objective function is not a function of edge length. This is due to many special properties of minimum spanning trees. The analysis of such approximation algorithms often depends on these special properties. We present three examples in this section.

First, consider the following problem:

> STEINER TREES WITH MINIMUM STEINER POINTS (ST-MSP): Given n terminals in the Euclidean plane and a number $r > 0$, find a Steiner tree interconnecting all terminals with the minimum number of Steiner points such that the length of each edge is at most r.

The problem ST-MSP arises from the design of networks in which there are limits on the edge length. For instance, in a wavelength-division multiplexing (WDM) optical network, each node has a limited transmission power, and signals can only travel a limited distance r. Then, finding the optimal networks under this restriction is just the problem ST-MSP.

A Steiner tree as a feasible solution for ST-MSP may contain a Steiner point of degree 2. We can obtain a Steiner tree T' with only Steiner points of degree 2 by adding Steiner points on the edges of a spanning tree T. We call such a tree a *Steinerized spanning tree* (induced from the spanning tree T). In the following, we will reserve the term "minimum spanning tree" for a spanning tree with the minimum length, and use the term "minimum Steinerized spanning tree" for a Steinerized spanning tree with the minimum number of Steiner points.

A simple heuristic for the problem ST-MSP is to use a minimum Steinerized spanning tree as an approximate solution. The following lemma shows that the Steinerized spanning tree induced from a minimum spanning tree is, in fact, a minimum Steinerized spanning tree.

Lemma 3.21 *Let T be a minimum spanning tree on a set P of terminals, and r a positive real number. Suppose, for each edge e in T, we break it into shorter edges of length at most r by adding the minimum number of Steiner points on e. Then the resulting tree is a minimum Steinerized spanning tree.*

Proof. Let T^* be an MST on P and T' an arbitrary spanning tree on P. Let $E(T^*)$ and $E(T')$ be their corresponding edge sets. Then there is a one-to-one, onto mapping f from $E(T^*)$ to $E(T')$ such that

$$length(e) \leq length(f(e)),$$

for all $e \in E(T^*)$ (see Exercise 3.16). The lemma follows immediately from this fact. □

Theorem 3.22 *Suppose that, for any set of terminals as an input to the problem* ST-MSP, *there always exists a minimum spanning tree with vertex degree at most d. Then the minimum Steinerized spanning tree is a $(d-1)$-approximation for* ST-MSP.

Proof. Let P be a set of terminals and $r > 0$ a given real number. Let S^* be an optimal tree on input P for ST-MSP with respect to the edge-length limit r. Suppose S^* contains k Steiner points $s_1, s_2, \ldots, s_k$, in the order of their occurrence in a breadth-first search starting from a terminal point of S^*. Let $N(Q)$ denote the number of Steiner points in a minimum Steinerized spanning tree on Q. We claim that, for $0 \leq i \leq k-1$,

$$N(P \cup \{s_1, \ldots, s_i\}) \leq N(P \cup \{s_1, \ldots, s_i, s_{i+1}\}) + d - 1. \quad (3.6)$$

In other words, we claim that we can eliminate Steiner points $s_k, s_{k-1}, \ldots, s_1$, one by one, and convert S^* into a Steinerized spanning tree, adding at most $d-1$ new Steiner points in each step.

To prove this claim, consider a minimum spanning tree T for $P \cup \{s_1, \ldots, s_i, s_{i+1}\}$, with degree at most d. Suppose s_{i+1} is adjacent to vertices $v_1, \ldots, v_j$, where $j \leq d$, in T. Write $d(x, y)$ to denote the Euclidean distance between two points x and y. Then we must have $d(v_\ell, s_{i+1}) \leq r$ for some $1 \leq \ell \leq j$, because, by the ordering of Steiner points $s_1, \ldots, s_k$, we know that one of the vertices in $P \cup$

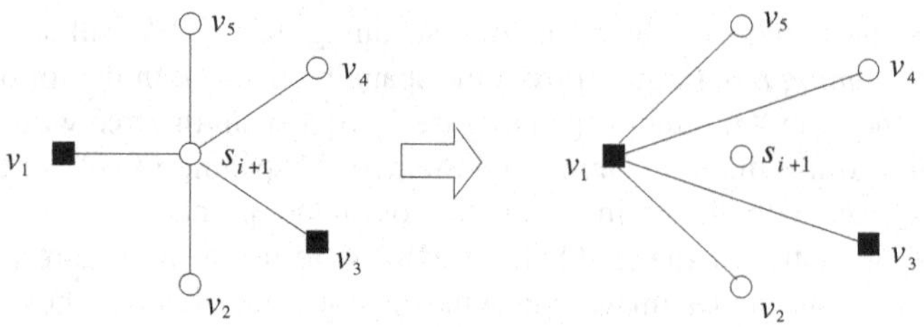

Figure 3.8: Proof of Theorem 3.22.

$\{s_1, \ldots, s_i\}$ has distance at most r from s_{i+1}. Without loss of generality, assume that $d(v_1, s_{i+1}) \le r$.

Now, we can get a spanning tree T' on $P \cup \{s_1, \ldots, s_i\}$ by deleting j edges $\{s_{i+1}, v_1\}, \ldots, \{s_{i+1}, v_j\}$, and adding $j-1$ edges $\{v_1, v_2\}, \ldots, \{v_1, v_j\}$ (see Figure 3.8). Note that, for each $2 \le \ell \le j$,

$$d(v_1, v_\ell) \le d(v_1, s_{i+1}) + d(s_{i+1}, v_\ell) \le r + d(s_{i+1}, v_\ell).$$

Thus, we only need one more degree-2 Steiner point to break the edge $\{v_1, v_\ell\}$ into shorter edges of length $\le r$ than to break the edge $\{s_{i+1}, v_\ell\}$. This means that the minimum Steinerized spanning tree induced from T' contains at most $j - 1$ more Steiner points than that induced from T. Now, (3.6) follows from Lemma 3.21.

Finally, by applying (3.6) repeatedly, we get

$$N(P) \le N(P \cup \{s_1, \ldots, s_k\}) + k(d-1) = k(d-1). \qquad \square$$

Note that for any set P of terminals in the Euclidean plane, there is a minimum spanning tree of P with degree at most 5 (see Exercise 3.19). Therefore, we have the following result:

Corollary 3.23 *The minimum Steinerized spanning tree is a* 4*-approximation for* ST-MSP *in the Euclidean plane.*

Next, we consider a problem closely related to ST-MSP.

> BOTTLENECK STEINER TREE (BNST): Given a set P of terminals in the Euclidean plane and a positive integer k, find a Steiner tree on P with at most k Steiner vertices which minimizes the length of the longest edge.

A simple approach to this problem is to use Steinerized spanning trees to approximate it. The following algorithm, called the *Optimal Cut*, applies the greedy strategy to obtain a Steinerized spanning tree from a given spanning tree T.

Algorithm 3.C (*Optimal Cut for the Steinerized spanning tree*)
Input: A spanning tree T on a set P of terminals in the Euclidean plane and an integer $k > 0$.

(1) **For** each edge $e \in T$ **do** $n(e) \leftarrow 0$.

(2) **For** $i \leftarrow 1$ **to** k **do**

select an edge $e \in T$ with the maximum $\dfrac{\mathit{length}(e)}{n(e)+1}$;

set $n(e) \leftarrow n(e) + 1$.

(3) **For** each edge $e \in T$ **do**

cut e evenly with $n(e)$ Steiner points. ∎

The following two lemmas show that Algorithm 3.C gives the best Steinerized spanning tree if we start with an MST T.

Lemma 3.24 *Among the Steinerized spanning trees induced by T with at most k Steiner points, the optimal cut tree produced by* Algorithm 3.C *has the minimum value of the longest edge length.*

Proof. Let $e_1, e_2, \ldots, e_t$ be all edges of T. Let $\mathcal{T}$ be the collection of trees that can be obtained from T by adding k Steiner points on edges $e_1, e_2, \ldots, e_t$, and let $\mathit{opt}(k; e_1, \ldots, e_t)$ be the minimum value of the longest edge length of T', among all possible trees T' in $\mathcal{T}$. We will prove the lemma by induction on k.

For $k = 0$, it is trivial. For the general case, we assume that, after adding k Steiner points to T according to Algorithm 3.C,

$$\mathit{opt}(k; e_1, \ldots, e_t) = \max_{1 \le i \le t} \frac{\mathit{length}(e_i)}{n(e_i)+1}.$$

Without loss of generality, assume that

$$\frac{\mathit{length}(e_1)}{n(e_1)+1} = \max_{1 \le i \le t} \frac{\mathit{length}(e_i)}{n(e_i)+1}.$$

From Algorithm 3.C, we need to prove

$$\mathit{opt}(k+1; e_1, \ldots, e_t) = \max\left\{\max_{2 \le i \le t} \frac{\mathit{length}(e_i)}{n(e_i)+1}, \frac{\mathit{length}(e_1)}{n(e_1)+2}\right\}. \tag{3.7}$$

We first observe that in Algorithm 3.C, on input $e_1, e_2, \ldots, e_t$, if we ignore the steps of adding points on e_1, then the remaining steps are exactly those steps in the algorithm on input $e_2, \ldots, e_t$. Therefore, by the induction hypothesis, we have

$$\mathit{opt}(k - n(e_1); e_2, \ldots, e_t) = \max_{2 \le i \le t} \frac{\mathit{length}(e_i)}{n(e_i)+1}. \tag{3.8}$$

Furthermore, as the right-hand side of Equation (3.7) is derived from a specific way of putting $k+1$ Steiner points on tree T, we see that it is greater than or equal to $\mathit{opt}(k+1; e_1, \ldots, e_t)$. Thus, it suffices to prove

$$\mathit{opt}(k+1; e_1, \ldots, e_t) \ge \max\left\{\mathit{opt}(k - n(e_1); e_2, \ldots, e_t), \frac{\mathit{length}(e_1)}{n(e_1)+2}\right\}.$$

Suppose, for the sake of contradiction,

$$opt(k+1; e_1, \ldots, e_t) < \max\left\{opt(k - n(e_1); e_2, \ldots, e_t), \frac{length(e_1)}{n(e_1)+2}\right\}. \quad (3.9)$$

Let $n^*(e_1)$ denote the number of Steiner points on e_1 in an optimal solution for $opt(k+1; e_1, \ldots, e_t)$. Thus,

$$opt(k+1; e_1, \ldots, e_t) = \max\left\{opt(k+1-n^*(e_1); e_2, \ldots, e_t), \frac{length(e_1)}{n^*(e_1)+1}\right\}.$$

Consider three cases:

Case 1. $n^*(e_1) \le n(e_1)$. Note that

$$opt(k+1; e_1, \ldots, e_t) \ge \frac{length(e_1)}{n^*(e_1)+1} \ge \frac{length(e_1)}{n(e_1)+1} = opt(k; e_1, \ldots, e_t).$$

However, from (3.8), we know that the right-hand side of (3.9) is no greater than $opt(k; e_1, \ldots, e_t)$. This is a contradiction.

Case 2. $n^*(e_1) = n(e_1) + 1$. Then,

$$opt(k+1-n^*(e_1); e_2, \ldots, e_t) = opt(k - n(e_1); e_2, \ldots, e_t),$$

and

$$\frac{length(e_1)}{n^*(e_1)+1} = \frac{length(e_1)}{n(e_1)+2}.$$

So, the two sides of (3.9) are equal. This is also a contradiction.

Case 3. $n^*(e_1) > n(e_1) + 1$. From the induction hypothesis and (3.8), we know that the right-hand side of (3.9) is no greater than $opt(k; e_1, \ldots, e_t)$. So, we have

$$opt(k+1-n^*(e_1); e_2, \ldots, e_t) \le opt(k+1; e_1, \ldots, e_t) < opt(k; e_1, \ldots, e_t).$$

Also, from $n^*(e_1) > n(e_1) + 1$, we get

$$\frac{length(e_1)}{n^*(e_1)} < \frac{length(e_1)}{n(e_1)+1} = opt(k; e_1, \ldots, e_t).$$

Hence,

$$\max\left\{opt(k+1-n^*(e_1); e_2, \ldots, e_t), \frac{length(e_1)}{n^*(e_1)}\right\} < opt(k; e_1, \ldots, e_t).$$

In other words, there is a Steinerized spanning tree T' induced by T with $n^*(e_1) - 1$ Steiner points on e_1, and $k - (n^*(e_1) - 1)$ Steiner points on other edges such that the longest edge length of T' is less than $opt(k; e_1, \ldots, e_t)$. This is again a contradiction. □

Lemma 3.25 *Among the optimal cut Steinerized spanning trees, the one induced by a minimum spanning tree has the minimum value of the longest edge length.*

Proof. Let T be a spanning tree and T^* a minimum spanning tree. By Exercise 3.16, there is a one-to-one, onto mapping f from edges in T to edges in T^* such that

$$length(e) \geq length(f(e)),$$

for all e in T. Suppose, in the optimal cut for tree T, there are $n(e)$ Steiner points on each edge e of T. Then, by putting $n(e)$ Steiner points on each edge $f(e)$ of T^*, we get a Steinerized spanning tree induced from T^* whose longest edge length is no longer than that of the optimal cut for T. By Lemma 3.24, we see that the longest edge length of the optimal cut for T^* is no longer than that of the optimal cut for T. □

Theorem 3.26 *The optimal cut Steinerized spanning tree induced by a minimum spanning tree is a 2-approximation for* BNST.

Proof. The optimal cut tree is the optimal solution to BNST with the restriction on Steinerized spanning trees. Following the general approach on the analysis of algorithms based on the restriction method, we will convert an optimal solution T to BNST to a Steinerized spanning tree with the longest edge length at most twice that of T.

Without loss of generality, it suffices to consider the case that T is a full Steiner tree with k Steiner points. Assume that the length of the longest edge length in T is R. We arbitrarily select a Steiner point s as the root. Call a path from the root to a leaf a *root-leaf path*. The length of a root-leaf path is the number of edges on the path or, equivalently, the number of Steiner points on the path. Let h be the length of a shortest root-leaf path in T, and d the length of a longest root-leaf path in T (called the *depth* of T). We will show by induction on the depth d of T that there exists a Steinerized spanning tree for all terminals in T with at most $k - h$ Steiner points such that each edge has length at most $2R$.

For $d = 0$, T contains only one terminal so it is trivial. For $d = 1$, T contains only one Steiner point. We directly connect the terminals without any Steiner points. By the triangle inequality, the distance between two terminals is at most $2R$. Thus, the induction statement holds for $d = 1$.

Next, we consider the general case of $d \geq 2$. Suppose s has m children $s_1, \ldots, s_m$. For each s_i, $1 \leq i \leq m$, there is a subtree T_i rooted at s_i with depth $\leq d - 1$. Let k_i be the number of Steiner points in T_i and h_i the length of a shortest root-leaf path in T_i, from s_i to a leaf v_i (see Figure 3.9). By the induction hypothesis, there exists, for each $1 \leq i \leq m$, a Steinerized spanning tree S_i for the terminals in T_i with at most $k_i - h_i$ Steiner points such that each edge has length at most $2R$. Without loss of generality, assume that $h_1 \geq h_2 \geq \cdots \geq h_m = h - 1$. Now, we connect all trees S_i, for $1 \leq i \leq m$, into a Steinerized spanning tree S with edges $\{v_1, v_2\}, \{v_2, v_3\}, \ldots, \{v_{m-1}, v_m\}$, and add, for each $i = 1, \ldots, m-1$, h_i Steiner points on the edge $\{v_i, v_{i+1}\}$. Note that S contains

$$\sum_{i=1}^{m}(k_i - h_i) + \sum_{i=1}^{m-1} h_i = \sum_{i=1}^{m} k_i - h_m = k - 1 - h_m = k - h$$

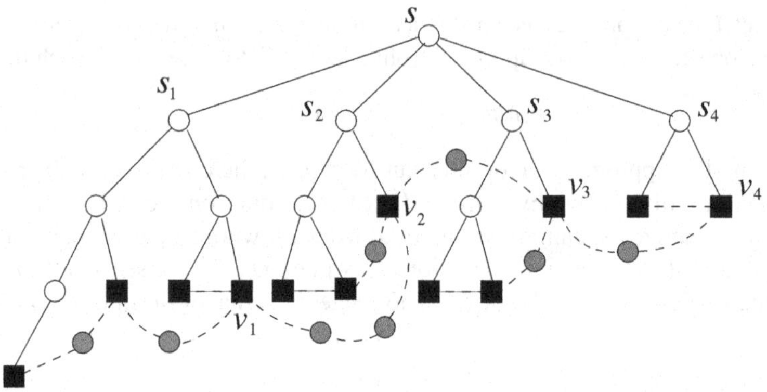

Figure 3.9: Proof of Theorem 3.26. Here, a dark square ■ denotes a terminal, a circle ○ denotes a Steiner point in the optimal solution, a dashed line denotes an edge of the approximate solution, and a shaded circle denotes a Steiner point in the approximate solution.

Steiner points. Moreover, we note that for each $1 \leq i \leq m-1$, the path between v_i and v_{i+1} in T contains $h_i + h_{i+1} + 2$ edges. By the triangle inequality, the distance between v_i and v_{i+1} is at most $(h_i + h_{i+1} + 2)R \leq 2(h_i + 1)R$. Therefore, the h_i Steiner points on the edge $\{v_i, v_{i+1}\}$ break it into $h_i + 1$ shorter edges each of length at most $2R$. Thus, all edges in S have length $\leq 2R$, and the induction proof is complete. □

Our third example is about a broadcasting problem in a wireless network. We represent a wireless network by a directed graph in the Euclidean plane. In a wireless network, a *broadcasting routing* from a source node s is an out-arborescence T rooted at s (i.e., a directed, rooted tree T with root s and with edge directions going from parents to children). Assume that a node u in T has k out-edges, (u, v_i), $i = 1, \ldots, k$. Then the energy consumption of u in the routing is

$$\max_{1 \leq i \leq k} c \cdot d(u, v_i)^{\alpha},$$

where d is the Euclidean distance function, and c and α are two positive constants with $\alpha \geq 2$. The energy consumption of a broadcasting routing T is the sum of energy consumptions over all nodes in T.

> MINIMUM-ENERGY BROADCASTING (MIN-EB): Given a set S of points in the Euclidean plane and a source node $s \in S$, find a broadcasting routing from s with the minimum total energy consumption.

A simple idea for an approximation to MIN-EB is to turn a minimum spanning tree T into a broadcasting routing. Its total energy consumption is at most

$$c \sum_{e \in T} \|e\|^{\alpha},$$

where $\|e\|$ denotes the Euclidean length of the edge e. To establish the performance ratio of this MST-approximation, we first prove the following.

Lemma 3.27 *Let C be a disk with center x and radius R, and P a set of points inside C, including the center x. Let T be a minimum spanning tree on P. Then, for $\alpha \geq 2$,*

$$\sum_{e \in T} \|e\|^\alpha \leq 8R^\alpha.$$

Proof. Since $x \in P$, the edge length of T cannot exceed R. For any $0 \leq r < R$, let T_r be the subgraph of T with vertex set P and all edges in T of length at most r. Let $n(T, r)$ denote the number of connected components in T_r.

We can rewrite $\sum_{e \in T} \|e\|^\alpha$ as

$$\sum_{e \in T} \|e\|^\alpha = \sum_{e \in T} \int_0^{\|e\|} dr^\alpha = \sum_{e \in T} \int_0^R \chi_e(r) dr^\alpha = \int_0^R \sum_{e \in T} \chi_e(r) dr^\alpha,$$

where

$$\chi_e(r) = \begin{cases} 1, & \text{if } 0 \leq r < \|e\|, \\ 0, & \text{if } \|e\| \leq r. \end{cases}$$

Note that, for fixed r, $\sum_{e \in T} \chi_e(r)$ is equal to the number of edges in T that are longer than r, or, equivalently, $n(T, r) - 1$. Therefore, we have

$$\sum_{e \in T} \|e\|^\alpha = \int_0^R \sum_{e \in T} \chi_e(r) dr^\alpha = \alpha \int_0^R (n(T, r) - 1) r^{\alpha - 1} dr.$$

For any $r \leq R$, let us associate each node $u \in P$ with a disk $D(u; r/2)$ with center u and radius $r/2$. Then these disks have the following properties: For each connected component C of T_r, the corresponding disks form a connected region. In addition, since T is a minimum spanning tree, two regions formed by disks corresponding to two different connected components of T are disjoint. Furthermore, since each of these regions contains at least one disk with radius $r/2$, its area is at least $\pi(r/2)^2$. Hence, the boundary of each region has length at least πr, because, among all connected regions of the same area, circles have the shortest boundary.

For any $r \leq R$, define $a(P, r)$ to be the total area covered by disks $D(u; r/2)$, for all $u \in P$. Then we have

$$\begin{aligned} a(P, R) &= \int_0^R d(a(P, r)) \geq \int_0^R n(T, r) \pi r \, d\Big(\frac{r}{2}\Big) \\ &= \frac{\pi}{2} \int_0^R (n(T, r) - 1) r dr + \frac{\pi R^2}{4} = \frac{\pi}{4} \sum_{e \in T} \|e\|^2 + \frac{\pi R^2}{4}. \end{aligned}$$

Note that $a(P, R)$ is contained in a disk centered at x with radius $3R/2$. Therefore,

$$\frac{\pi}{4}\sum_{e\in T}\|e\|^2 + \frac{\pi R^2}{4} \le a(P,R) \le \pi\Big(\frac{3R}{2}\Big)^2,$$

and so

$$\sum_{e\in T}\|e\|^2 \le 8R^2.$$

Finally, we note that for every $e \in T$, $\|e\| \le R$. Thus, for $\alpha \ge 2$,

$$\sum_{e\in T}\Big(\frac{\|e\|}{R}\Big)^{\alpha} \le \sum_{e\in T}\Big(\frac{\|e\|}{R}\Big)^{2} \le 8,$$

and the lemma holds for all $\alpha \ge 2$. □

Theorem 3.28 *The minimum spanning tree provides an 8-approximation for the problem* MIN-EB.

Proof. Let T^* be a minimum-energy broadcasting routing. For each node u of T^*, we draw a smallest disk to cover all out-edges from u. Let $R(D)$ be the radius of disk D, and $\mathcal{D}$ the set of all such disks. Then disks in $\mathcal{D}$ cover all points in the input set S, and the total energy consumption of T^* is

$$\sum_{D\in\mathcal{D}} c(R(D))^{\alpha}.$$

For each disk D, construct an MST T_D connecting all points in D. These MSTs form an MST T connecting all points in S. By Lemma 3.27, the energy consumption of T is at most

$$\sum_{e\in T} c\|e\|^{\alpha} \le 8\sum_{D\in\mathcal{D}} c(R(D))^{\alpha}.$$

Now, from Exercise 3.16, we see that the MST routing is an 8-approximation to MIN-EB. □

We remark that the bound $8R^{\alpha}$ of Lemma 3.27 can be improved to $6R^{\alpha}$ [Ambühl, 2005]. Thus, the minimum spanning tree is actually a 6-approximation to MIN-EB.

3.5 Phylogenetic Tree Alignment

In this section, we study a simple application of the restriction method to a problem in bioinformatics. We first give some definitions.

Let Σ be a set of finite symbols and "$-$" a special *blank* symbol not in Σ. Assume that there is a metric distance $\sigma : (\Sigma \cup \{-\})^2 \to \mathbb{N}$ between these symbols that satisfies the triangle inequality. For any two strings $s = s_1 s_2 \cdots s_n$, $s' = s'_1 s'_2 \cdots s'_n$ in $(\Sigma \cup \{-\})^*$ that are of the same length, where each s_i or s'_j denotes a symbol in $\Sigma \cup \{-\}$, the *score* between them is

$$score(s, s') = \sum_{i=1}^{n} \sigma(s_i, s'_i).$$

For k strings $s_1, \ldots, s_k \in \Sigma^*$, we can *align* them by inserting the blank symbols into them to make them of the same length. More precisely, an *alignment* of $s_1, s_2, \ldots, s_k \in \Sigma^*$ is a mapping from $(s_1, \ldots, s_k)$ to $(s'_1, \ldots, s'_k)$, where $s'_i \in (\Sigma \cup \{-\})^*$ for $1 \le i \le k$, such that

(1) $|s'_1| = |s'_2| = \cdots = |s'_k|$,

(2) Each string s'_i, $1 \le i \le k$, is generated from s_i with insertion of blanks, and

(3) At any position j, $1 \le j \le |s'_1|$, at least one string of $s'_1, \ldots, s'_k$ has a non-blank symbol.

Often, we use images $(s'_1, \ldots, s'_k)$ or a matrix with rows $s'_1, \ldots, s'_k$ to represent this alignment. For instance, the following matrix represents an alignment of strings $AGGTC$, $GTTCG$, and $TGAAC$:

$$\begin{pmatrix} A\,G\,G\,T - C - \\ - G - T\,T\,C\,G \\ T\,G - A\,A\,C - \end{pmatrix}.$$

The *score* of an alignment $(s'_1, \ldots, s'_k)$ is defined to be

$$\sum_{1 \le i < j \le k} score(s'_i, s'_j).$$

The function *score* induces a metric distance D between strings in Σ^*:

$$D(s, s') = \text{ the minimum score of an alignment of } (s, s').$$

It is not hard to see that the distance function D and the corresponding minimum score alignment can be computed by dynamic programming.

Lemma 3.29 *The minimum score alignment of two strings s and s' in Σ^* can be computed by dynamic programming in time $O(|s| \cdot |s'|)$.*

Proof. Assume that $s = s_1 s_2 \cdots s_n$ and $s' = s'_1 s'_2 \cdots s'_m$, where each s_i or s'_j denotes a symbol in Σ. Denote $V(i, j) = D(s_1 \cdots s_i, s'_1 \cdots s'_j)$. Then it is easy to see that $V(0, 0) = 0$, $V(1, 0) = \sigma(s_1, -)$, $V(0, 1) = \sigma(-, s'_1)$; and, for $i, j \ge 0$,

$$\begin{aligned} V(i+1, j+1) = \min\{&V(i, j) + \sigma(s_{i+1}, s'_{j+1}), \\ &V(i, j+1) + \sigma(s_{i+1}, -), V(i+1, j) + \sigma(-, s'_{j+1})\}. \end{aligned}$$

There are $O(nm)$ entries of $V(i, j)$'s, and each entry $V(i+1, j+1)$ can be computed in time $O(1)$ from $V(i, j)$, $V(i+1, j)$, and $V(i, j+1)$. Therefore, $V(n, m)$ can be computed in time $O(nm)$. $\square$

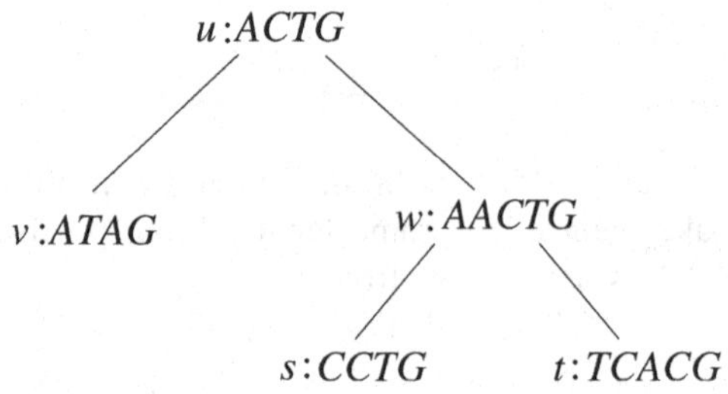

Figure 3.10: A tree with labels.

Consider a tree $T = (V, E)$ in which each vertex v is assigned a label $s_v \in \Sigma^*$. An *alignment* of T is a tree T' with the same vertex set and edge set, and possibly different labels s'_v for $v \in V$ such that the set $\{s'_v \mid v \in V\}$ is an alignment of the set $\{s_v \mid v \in V\}$. The score of the alignment tree T' is defined to be $\sum_{\{u,v\} \in E} score(s'_u, s'_v)$.

The following lemma shows that the minimum-score alignment tree can be found in polynomial time.

Lemma 3.30 *The minimum-score alignment of tree T has the score value*

$$\sum_{\{u,v\} \in E} D(s_u, s_v)$$

and can be found in time $O(nm(n+m))$, where n is the number of edges in T and m is the length of the longest label in T.

Proof. First, we note that the score of an alignment of T cannot be smaller than $\sum_{\{u,v\} \in E} D(s_u, s_v)$. Moreover, from alignments for each edge, we can induce an alignment for the whole tree, preserving score values for every edge. Thus, the minimum-score alignment of T can reach the lower bound $\sum_{\{u,v\} \in E} D(s_u, s_v)$.

More precisely, we can grow the tree T' and adjust the labels iteratively. Let $T' = (V', E')$. Initially, V' contains a single vertex v, with a label $s'_v = s_v$, and $E = \emptyset$. At each iteration, we select an edge $\{u, w\} \in E$, with $u \in V'$ and $w \notin V'$, and add w to V' and $\{u, w\}$ to E'. We follow Lemma 3.29 to find the minimum score alignment (s''_u, s'_w) of (s_u, s_w). Let t_u be the alignment of s_u such that the number of blanks between any two nonblank symbols in t_u is equal to the maximum number of blanks between them in s'_u and s''_u. String t_u may have more blanks than s'_u or s''_u. For each extra blank in t_u that is not in s''_u, we insert a blank, at the corresponding position, into s'_w. For each extra blank in t_u that is not in s'_u, we insert a blank, at the corresponding position, into each s'_v in T' (including s'_u, so that s'_u now is equal to t_u).

To make this process clear, let us look at a simple example. Consider the tree T in Figure 3.10. Assume that the minimum pairwise alignments of labels are

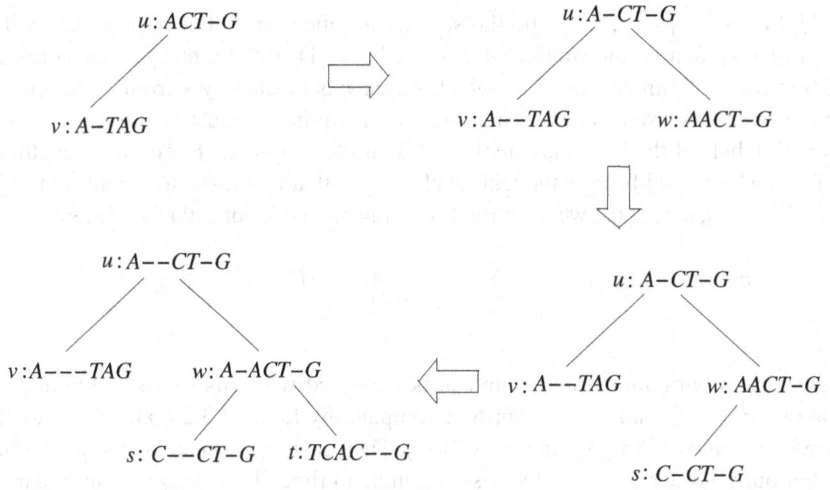

Figure 3.11: Constructing the minimum score alignment of a tree.

(u,v):	(u,w):	(w,s):	(w,t):
$A\,C\,T - G$	$A - C\,T\,G$	$A\,A\,C\,T\,G$	$A - A\,C\,T\,G$
$A - T\,A\,G$	$A\,A\,C\,T\,G$	$C - C\,T\,G$	$T\,C\,A\,C - G.$

Then the minimum-score alignment T' of T can be found as in Figure 3.11.

We note that at each iteration we added blanks to both labels of an edge at the same positions, and so did not increase its score. Thus, the total score of T' remains equal to $\sum_{\{u,v\}\in E} D(s_u, s_v)$. It is also easy to see that each iteration takes time $O(m^2 + nm)$, and so the total running time is $O(nm(n+m))$. □

Now we consider the following problem.

> Phylogenetic Tree Alignment (PTA): Given a rooted tree T with k leaves labeled with k distinct strings $s_1, \ldots, s_k \in \Sigma^*$, respectively, find string labels for internal vertices which minimize the total alignment score of the tree.

The problem PTA is known to be **NP**-hard. To find an approximation to this problem, we study a restricted version of PTA, which requires that an internal vertex must have the same label as one of its children. A tree alignment satisfying this restriction is called a *lifted alignment*. The following lemma shows that the optimal lifted alignment can be found in polynomial time; thus, it can be used as an approximation to the problem PTA.

Lemma 3.31 *The optimal lifted alignment of a tree T can be computed by dynamic programming in time $O(m^2 + k^3)$, where k is the number of leaves in T and m is the total length of leaf labels in T.*

Proof. Let $S = \{s_1, \ldots, s_k\}$ be the set of leaf labels in T. For each vertex v in tree T, let T_v denote the subtree of T rooted at v. Denote by $c(v, s)$ the score of the best lifted alignment for T_v in which vertex v is labeled by s from S. Suppose the label for v is fixed to be s. Then one of its children x must also have label s. Since all labels of the leaves are distinct, this child x is unique. For each other child y of v, the best label for y is the leaf label s' in T_y that minimizes the total score of $D(s, s') + c(y, s')$. Thus, we have the following recursive formula for $c(v, s)$:

$$c(v, s) = c(x, s) + \sum_{\substack{y \in child(v) \\ y \neq x}} \min_{s' \in leaf(T_y)} [D(s, s') + c(y, s')].$$

A dynamic programming algorithm can be designed with this formula running in time $O(m^2 + k^3)$. Indeed, we can first compute, by Lemma 3.29, all $k(k-1)/2$ pairwise distances $D(s_i, s_j)$ in time $O(m^2)$. Then each $c(v, s)$ can be computed, in the bottom-up order, from the recurisve formula in time $O(k)$. There are altogether $O(k^2)$ entries of $c(v, s)$'s. Therefore, the total running time is $O(m^2 + k^3)$. □

Next, we need to estimate the performance ratio of the optimal lifted alignment as an approximation to PTA. By Lemma 3.30, the objective function of the problem PTA is $\sum_{\{u,v\} \in E(T)} D(s_u, s_v)$, where s_u is the label of vertex u.

Following the general approach for the analysis of approximations based on the restriction method, we consider a tree T^* with the optimal assignment of labels s_v^* for internal vertices and modify it into a lifted alignment tree T_L. The modification is a bottom-up process according to the following formula:

$$s_v = \operatorname*{argmin}_{\substack{s_x \\ x \in child(v)}} D(s_v^*, s_x).$$

That is, initially, we let $s_v = s_v^*$ for all leaves $v \in T_L$. Then, in each iteration, we select a vertex v in T_L with all labels of its children already defined, and choose a child vertex x of v with the minimum $D(s_v^*, s_x)$ and set label $s_v = s_x$.

For each edge $\{v, w\}$, where w is a child of v, if $s_v \neq s_w$, then we have, by the triangle inequality,

$$D(s_v, s_w) \leq D(s_v, s_v^*) + D(s_v^*, s_w) \leq 2D(s_v^*, s_w).$$

Note that there is a *lifted path* π_w from w to a leaf z in which all vertices have the same label s_w. In particular, the leaf z of π_w in the optimal tree T^* has label $s_z^* = s_z = s_w$ (see Figure 3.12). Applying the triangle inequality to the path $\{v, w\} \cup \pi_w$ in T^*, we get

$$D(s_v^*, s_w) = D(s_v^*, s_z^*) \leq D(s_v^*, s_w^*) + \text{the score of } \pi_w \text{ in } T^*.$$

That is, we can charge the score $D(s_v, s_w)$ of T_L to the edges in the path $\{v, w\} \cup \pi_w$ in T^*, with each edge $\{x, y\}$ in this path charged with the score $2 \cdot D(s_x^*, s_y^*)$. Note that every lifted path π_w is uniquely determined by its lowest vertex w. Moreover, all lifted paths are disjoint, and all edges $\{x, y\}$ in the lifted paths have score zero

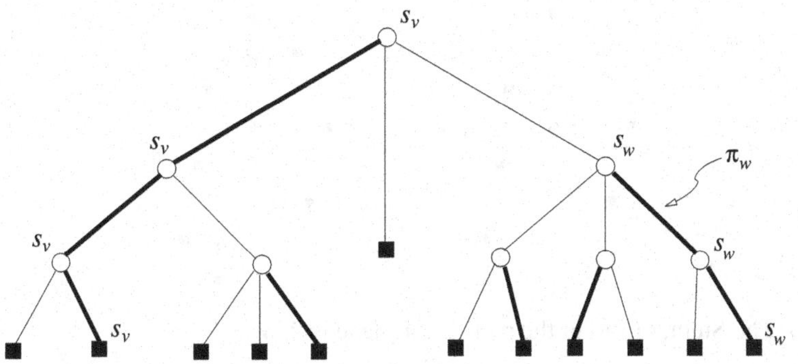

Figure 3.12: A lifted alignment tree.

in T_L. Therefore, each edge $\{v, w\}$ in T^* can be charged at most once: If $s_v \neq s_w$, then it can only be charged by $D(s_v, s_w)$, since it is not in a lifted path; otherwise, it is in a lifted path π_t, and it can only be charged by $D(s_u, s_t)$, where u is the parent of t. It follows that

$$\sum_{\{v,w\} \in E} D(s_v, s_w) \leq \sum_{\substack{\{v,w\} \in E \\ s_v \neq s_w}} 2D(s_v^*, s_w) \leq 2 \sum_{\{x,y\} \in E} D(s_x^*, s_y^*).$$

That is, the performance ratio of the optimal lifted alignment is bounded by 2.

Theorem 3.32 *The optimal lifted alignment is a polynomial-time* 2*-approximation for the problem* PTA.

Exercises

3.1 Prove the following properties of Steiner minimum trees in the d-dimensional Euclidean space, for $d \geq 3$:

(a) Every Steiner point is on the two-dimensional plane determined by the three adjacent vertices.

(b) An angle between any two adjacent edges at a vertex is at least 120°.

(c) Every Steiner point has degree 3 and the three angles at a Steiner point are all equal to 120°.

3.2 Prove the following properties about rectilinear SMTs:

(a) For any set P of terminal points, there exists a rectilinear SMT in which every maximal vertical or horizontal segment contains a terminal.

(b) For any set P of terminal points, there exists a rectilinear SMT in which every full component is in one of the following forms:

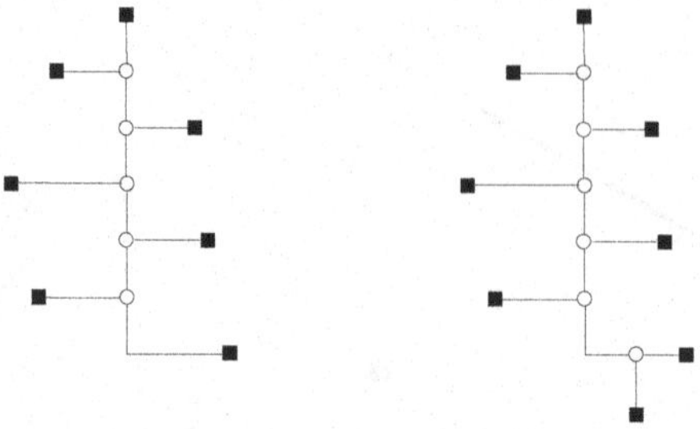

(c) The Steiner ratio in the rectilinear plane is $2/3$.

3.3 Show that for any rooted tree T, there is a mapping f from the leaves to the internal vertices such that the paths from leaves v to $f(v)$ form an edge-disjoint decomposition of tree T.

3.4 Show that for $k = 2^r + s$, where $0 \le s < 2^r$, the k-Steiner ratio for network Steiner trees is

$$\rho_k = \frac{r2^r + s}{(r+1)2^r + s}.$$

3.5 Determine whether or not the following argument is correct: Assume that f is a potential function in greedy Algorithm 2.D. Set $g(A) = f(A) + c(A)$. Then $\Delta_x g(A)/c(x) = 1 + \Delta_x f(A)/c(x)$. This means that, using $g(A)$ as a potential function, greedy Algorithm 2.D would generate the same solution as using $f(A)$. However, with the potential function $g(A)$, we always have $\Delta_x g(A)/c(x) \ge 1$. By Theorem 3.7, we conclude that greedy Algorithm 2.D generates a solution within a factor of $1 + \ln(1 + f(A^*)/c(A^*))$ from the optimal solution A^*.

3.6 Consider the following greedy algorithm for the problem NSMT: Grow a tree T starting with the empty set. At each iteration, choose a Steiner point $v \notin T$ that maximizes the number of terminals in $G \setminus T$ adjacent to v, relative to the edge-weight. In other words, let E consist of all stars in G that contain a Steiner vertex at the center and terminals as leaves. For each $T \subseteq E$, define $f(T) = r - 1$, where r is the number of leaves in T. Show that the greedy Algorithm 2.D with the potential function f is a 2-approximation for NSMT and, in addition, the performance ratio 2 is tight for this approximation.

3.7 Consider the problem NSMT. Let T be a minimum spanning tree on terminal set P. Show that if, for any full component K of size at most k, $g_T(K) \le 0$, then T is a k-restricted Steiner minimum tree.

3.8 Consider the problem NSMT. For a Steiner tree T on the terminal set P and a full component K in Q_k, define

$$gain_T(K) = mst(T) - mst(T \cup \zeta(K)) - c(K),$$

and for a subset A of full components,

$$gain_T(A) = mst(T) - mst(T \cup (\bigcup_{K \in A} \zeta(K))) - \sum_{K \in A} c(K).$$

Show the following:

(a) For any two full components K, K' of tree T, $gain_T(\{K, K'\}) \leq gain_T(K) + gain_T(K')$.

(b) If $gain_T(K) \leq 0$ for every full component K of size at most k, then T is a k-restricted Steiner minimum tree.

(c) If we replace $g_H(K)$ with $gain_H(K)$ in Algorithm 3.B, it will also give us a (1.55)-approximation for NSMT. Furthermore, when there are more than one $K \in E^*$ having the maximum value of $gain_H(K)/loss(K)$, the choice of K can be arbitrary; in other words, the condition "smallest" for K in step (2) of greedy Algorithm 3.B can be deleted.

3.9 Show that $g_T(K) = c(T) - mst(T \uplus K)$ is a submodular function, but is not a polymatroid function.

3.10 Suppose f and c are polymatroid functions on 2^E in the problem MIN-SMC. Suppose it is hard to compute the values $\max_{y \in E} \Delta_y f(A)/c(y)$. Therefore, in greedy Algorithm 2.D, instead of choosing an element $x \in E$ to maximize $\Delta_x f(A)/c(x)$, we choose an x such that

$$\alpha \cdot \frac{\Delta_x f(A)}{c(x)} \geq \max_{y \in E} \frac{\Delta_y f(A)}{c(y)},$$

for some constant $\alpha \geq 1$. Show that if the element x selected in step (2) always satisfies $\Delta_x f(A)/c(x) \geq 1$, then this modified greedy algorithm produces a solution within a factor of $1 + \alpha \cdot \ln(f(A^*)/c(A^*))$ from the optimal solution $c(A^*)$ of MIN-SMC.

3.11 Consider a rooted tree $T = (V, E)$ of n leaves, with edge cost $c : E \to \mathbb{R}^+$, and any integer $k > 0$. Let $s(v)$ be the number of leaves in the subtree rooted at v, and for $i = 0, \ldots, k$,

$$V_i = \{v \in V \mid s(v) \geq n^{(k-i)/k} \text{ and } s(v') < n^{(k-i)/k} \text{ for any child } v' \text{ of } v\}.$$

Construct a new k-level tree T^k with vertex set V, and edge set $\{(u, v) | u \in V_i, v \in V_{i+1}$, for some $i = 0, 1, \ldots, k-1$; and v is a descendant of u in $T\}$, with the cost $cost(u, v)$ equal to the total cost of the path from u to v in T. Show that

$$cost(T^k) \leq n^{1/k} \cdot cost(T).$$

3.12 Consider the following problem:

ACYCLIC DIRECTED STEINER TREE (ADST): For a given acyclic digraph $G = (V, E)$ satisfying the transitive relation, i.e., $(u, v), (v, w) \in E$ implying $(u, w) \in E$, with an edge cost function $c : E \to \mathbb{R}^+$ satisfying the triangle inequality, a given set $P \subseteq V$, and a point $r \in V$, find a minimum-cost outward-directed tree from r to all vertices in P.

(a) Let $\mathcal{A}_k$ be the set of full Steiner components of at most k levels. For a subset $A \subseteq \mathcal{A}_k$, let $f(A) = mst(P \cup \{r\}) - mst(P \cup \{r\} : A)$, where $mst(P \cup \{r\} : A)$ is the length of the minimum spanning tree for $P \cup \{r\}$ after contracting every component in A into a terminal point. Show that for $k = 1, 2$, and any $A \subseteq \mathcal{A}_k$, $\max_{T \in \mathcal{A}_k} \Delta_T f(A)/c(T)$ is polynomial-time computable.

(b) For any set $S \subseteq V$, let $U_S(s) = \{v \in V \mid s = \text{argmin}_{s \in S} c(s, v)\}$. For any $A \subseteq \mathcal{A}_k$, and any $T \in \mathcal{A}_k$, define $g_A(T) = \Delta_T f(A)/c(T)$. For $u \in V$ and $k \geq 3$, compute k-level trees $T^k(u)$ recursively as follows.

(1) Let $s_0 \leftarrow \text{argmin}_{s \in P \cup \{r\}} c(s, u)$, and $T^k(u) \leftarrow (s_0, u)$.
(2) Set $S \leftarrow P \cup \{u\}$.
(3) **While** $(\exists v \in U_S(u))\ g_{T^k(u)}(T^{k-1}(v)) \geq 0$ **do**

$v^* \leftarrow \text{argmax}_{v \in U_S(u)} g_{T^k(u)}(T^{k-1}(v))$;
$T^k(u) \leftarrow T^k(u) \cup T^{k-1}(v^*)$.

Let $T^* = \text{argmax}_{T \subseteq \mathcal{A}_k} f(T)$ and u the unique child of the root of T^*. Show that

$$f(T^k(u)) \cdot (2 + \log n)^{k-2} \geq f(T^*).$$

(c) Show that there is a polynomial-time approximation for ADST with performance ratio $n^{1/k}(1 + \log n)^{k-1}$ for any $k \geq 1$.

3.13 Let V be n stations (points) in the Euclidean plane. Each station $v \in V$ has a communication range with radius r_v, which depends on its energy consumption E_v according to the formula $E_v = c r_v^\alpha$ for some constant $\alpha \geq 2$. These communication ranges induce a digraph $G = (V, E)$ such that $(u, v) \in E$ if and only if $r_u > dist(u, v)$. They also induce an undirected graph $G' = (V, E')$, where $\{u, v\} \in E'$ if and only if both r_u and r_v are greater than $dist(u, v)$.

(a) Show that the minimum spanning tree is a 2-approximation for the problem of minimizing the total energy $\sum_v E_v$ subject to the condition that the communication ranges induce a connected undirected graph over all stations.

(b) Show that the minimum spanning tree is a 2-approximation for the problem of minimizing the total energy $\sum_v E_v$ subject to the condition that the communication ranges induce a strongly connected directed graph over all stations.

(c) Find an approximation of a constant performance ratio for the problem of minimizing the total energy $\sum_v E_v$ subject to the condition that the communication ranges induce a weakly connected directed graph over all stations.

3.14 Consider the following problem:

TERMINAL STEINER TREE (TST): Given a complete graph $G = (V, E)$ with an edge-weight function $w : E \to \mathbb{R}^+$, which satisfies

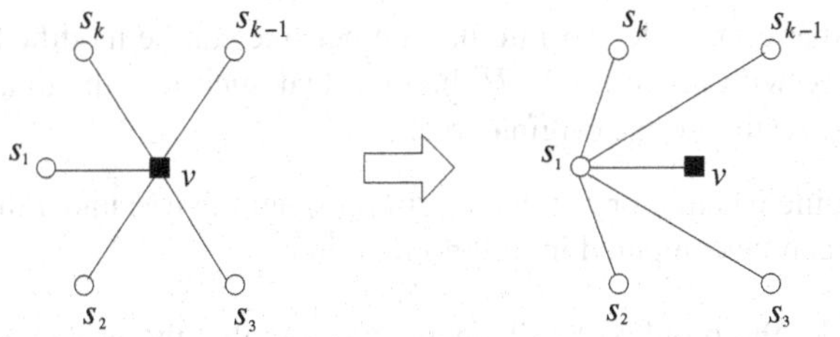

Figure 3.13: Step (3) of the algorithm in Exercise 3.15.

the triangle inequality, and a subset $P \subseteq V$ of terminals, find a shortest Steiner tree interconnecting all terminals such that all terminals are leaves.

Let *opt* denote the length of a minimum solution to this problem. Show the following results:

(a) For each terminal v, denote by $c(v)$ the closest nonterminal vertex to v. Then the total length of edges $\{v, c(v)\}$, for $v \in V$, is at most *opt*.

(b) The length of the network SMT on all $c(v)$'s is at most $2 \cdot opt$.

(c) All edges $\{v, c(v)\}$ together with a ρ-approximation of the problem NSMT on all $c(v)$'s form a $(1 + 2\rho)$-approximation for TST.

3.15 Consider the problem TST again. Assume that the problem NSMT is ρ-approximable. Show that the following algorithm is a (2ρ)-approximation for TST:

(1) $G' \leftarrow G \setminus \{\{u, v\} \mid u, v \in P\}$.

(2) In graph G', find a ρ-approximation T for NSMT on terminals P.

(3) **For** each $v \in P$ with $deg(v) > 1$ **do**

assume v's neighbors are $s_1, \ldots, s_k$, and $d(v, s_1) = \min_{1 \le i \le k} d(v, s_i)$;

for $i \leftarrow 2$ **to** k **do** $T \leftarrow T \cup \{s_1, s_i\} \setminus \{v, s_i\}$ (see Figure 3.13).

3.16 Show that for a minimum spanning tree T^* and any spanning tree T of a graph G, there exists a one-to-one, onto mapping f between their edge sets $E(T^*)$ and $E(T)$ such that $length(e) \le length(f(e))$ for each $e \in E(T^*)$.

3.17 Consider the following problem:

SELECTED-INTERNAL STEINER TREE (SIST): Given a complete graph $G = (V, E)$ with an edge-cost function $c : E \to \mathbb{R}^+$ and two vertex subsets P and P' with $P' \subsetneqq P \subseteq V$, find a shortest tree interconnecting all vertices (terminals) in P under the constraint that no vertex in P' can be a leaf.

Any tree satisfying the constraint given above is called a *selected-internal Steiner tree*.

(a) Show that every selected-internal Steiner tree can be modified into a spanning tree with no vertex in P' being a leaf such that the total length is at most twice that of the original tree.

(b) Determine whether or not the minimum spanning tree under the above constraint can be computed in polynomial time.

3.18 Consider the problem SIST again. Assume that the problem NSMT is ρ-approximable. Show that the following algorithm gives a (2ρ)-approximation for SIST.

(1) Compute a ρ-approximation T for NSMT on subset P.

(2) **For** each leaf v of T that is in P' **do**

find the closest internal vertex m_v to v such that either
$m_v \notin P'$ or $deg(m_v) \geq 3$;
choose a vertex t_v adjacent to m_v, but not in the path from v to m_v;
replace edge $\{m_v, t_v\}$ by edge $\{v, t_v\}$.

3.19 Show that for any finite set of points in the Euclidean plane, there exists a minimum spanning tree with degree at most 5.

3.20 Show that for ST-MSP in the rectilinear plane, the minimum Steinerized spanning tree is a 3-approximation to it.

3.21 Consider the following problem:

MULTIPLE SEQUENCE ALIGNMENT (MSA): Given k strings $s_1, \ldots,$ s_k, find their minimum score alignment.

(a) Show that the optimal solution to MSA can be computed by dynamic programming in time $O(k2^k m^k)$, where m is the total length of the given strings.

(b) Choose s_i to minimize $\sum_{j \neq i} D(s_i, s_j)$. Show that if $(s'_1, \ldots, s'_k)$ is an alignment of $(s_1, \ldots, s_k)$ such that $score(s'_i, s'_j) = D(s_i, s_j)$ for all $j \neq i$, then $\sum_{1 \leq j < h \leq k} score(s'_j, s'_h) \leq 2 \cdot opt$, where *opt* is the score of the optimal solution to MSA.

(c) Use fact (b) to design a 2-approximation for MSA.

3.22 Show that the optimal lifted alignment can actually be computed in time $O(m^2 + k^2)$ by dynamic programming, where k is the number of leaves and m is the total length of leaf labels.

3.23 Consider a binary tree T in which each leaf is labeled with a string. An alignment of T is *uniformly lifted* if, at each level j, either every internal vertex is assigned by the label of its left child or every internal vertex is assigned by the label of its right child.

(a) Show that the best uniformly lifted alignment can be computed faster than the best lifted alignment.

(b) Show that the best uniformly lifted alignment is a 2-approximation for PTA.

3.24 Show that, for a binary tree, at least $1/2^{d-1}$ of all lifted alignments have cost less than twice that of the optimal solution to PTA, where d is the depth of the tree T.

3.25 Show that the average cost of all lifted alignments for a binary tree is less than twice that of the optimal solution to PTA.

Historical Notes

The Steiner tree problem for three terminal points, that is, the problem of finding a point connecting three given points on the Euclidean plane with the shortest total distance, was first proposed by Fermat (see, e.g., Wesolowsky [1993]). This problem has two generalizations to the cases with more than three terminal points. The first one is to find a single point connecting all given terminals with the shortest total distance. This is commonly called the *Fermat problem*. The second one is to find a shortest network interconnecting all given terminals. This was called, for unknown reasons, the Steiner tree problem by Courant and Robbins [1941], although Gauss in 1836 had already studied this problem.

In a letter to Gauss dated on March 19, 1836, Schumacher mentioned a paradox about the Fermat problem: For four vertices of a convex quadrilateral, the solution to the Fermat problem is the intersection point of the two diagonals. When two of the neighboring vertices of the quadrilateral move toward a same point, the intersection point of the two diagonals would also move to this point. However, this point is not the solution to the Fermat problem when the quadrilaterals converge to a triangle. Two days later, Gauss wrote back to Schumacher and explained the paradox. He suggested another generalization of the Fermat problem, which aims at the network structure instead of a single point position. Gauss also discussed in the letter all possible topologies of the Steiner minimum trees (SMTs) for four terminal points. (See Schreiber [1986].)

It is well known that the Steiner tree problems in many different topologies are NP-hard [Karp, 1972; Garey and Johnson, 1977; Garey, Graham, and Johnson, 1977; Foulds and Graham, 1982]. Much effort has been devoted to find good approximate solutions. For the minimum spanning tree (MST) approximation, Hwang [1972] determined its performance ratio in the rectilinear plane. For the case in the Euclidean plane, Gilbert and Pollak [1968] conjectured that the performance ratio is exactly $2/\sqrt{3}$. This conjecture remained open for more than 20 years, and was finally proved by Du and Hwang [1990], who adopted many ideas from previous works in their proof, including Chung and Gilbert [1976], Chung and Graham [1985], Chung and Hwang [1978], Graham and Hwang [1976], and Rubinstein and Thomas [1991]. The first approximation with the performance ratio better than that of the MST approximation was found by Zelikovsky [1993] for NSMT. Later,

Du, Zhang and Feng [1991] showed that such approximations exist in all metric spaces as long as SMTs for a fixed number of points are computable in polynomial time. Recently, a (1.55)-approximation has been found for NSMT [Robin and Zelikovsky, 2000], and various PTAS algorithms have been designed for ESMT and RSMT (see Chapter 5). The performance ratios of those approximations for NSMT are determined through the estimate of the k-Steiner ratio, which was established by Borchers and Du [1995].

Steiner trees have many variations arising from various applications, such as terminal Steiner trees [Lin and Xue, 2002; Drake and Hougardy, 2004], Steiner trees with the minimum number of Steiner points [Lin and Xue, 1999, Mandoiu and Zelikovsky, 2000], acyclic directed Steiner trees [Zelikovsky, 1997], bottleneck Steiner trees [Wang and Du, 2002], and selected-internal Steiner trees [Hsieh and Yang, 2007]. In a way, the phylogenetic tree alignment can also be considered as a Steiner tree problem with a given topology in a special metric space [Ravi and Kececioglu, 1995; Wang and Gusfield, 1996].

4
Partition

But it's important that we all pull together
to reduce the strain on the grid.
— Gray Davis

The basic idea of partition is to divide the input object into smaller parts so that each part has a simple solution, and a feasible solution to the input instance can be constructed by combining the solutions of the smaller parts. The method of partition can be seen as a special form of restriction; that is, we restrict our attention to the feasible solutions that can be constructed through partitions.

The partition technique may be divided into two types: nonadaptive partition and adaptive partition. In nonadaptive partition, the input object is divided into smaller parts in one round, and the solutions to the smaller parts can be found independently from each other. In adaptive partition, the input object is divided into smaller parts by a sequence of subdivision operations recursively, and the solution to each part is also to be found recursively from the solutions of its own subproblems. We study, in this chapter, applications of nonadaptive partition to a number of geometric optimization problems. The technique of adaptive partition will be studied in the next chapter.

4.1 Partition and Shifting

We begin with a simple example to demonstrate the basic techniques of partition. In the following, by a *unit disk* we mean a disk of diameter 1.

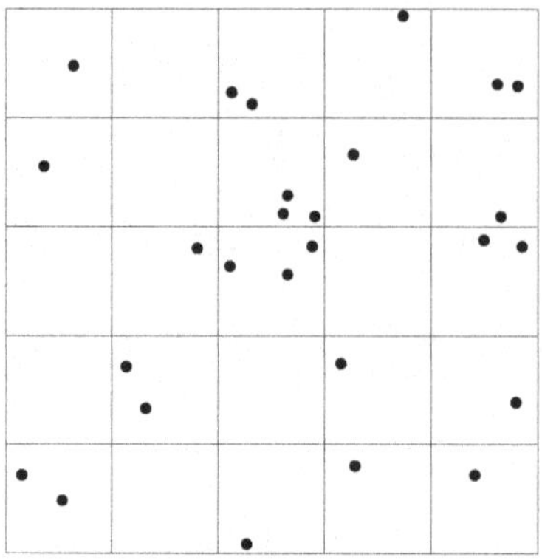

Figure 4.1: Partitioning a square into cells.

UNIT DISK COVERING (UDC): Given n points in the Euclidean plane, find a minimum number of unit disks to cover all given points.

Let P be the set of n given points in the Euclidean plane. Assume that Q is a square that covers all points in P. The idea of the partition technique for the problem UDC is as follows: First, we divide the square Q into a grid of squares, called *cells*, each of size $m \times m$ for some constant m (see Figure 4.1). Then, we solve the problem UDC for each cell. Finally, we take the union of the solutions of all cells as the solution to the original input.

Algorithm 4.A (*Partition Algorithm for* UDC)
Input: A set of points, all lying in square Q; an integer $m > 1$.

(1) Divide Q into cells, each of size $m \times m$;
Let $cell(Q) \leftarrow$ the set of all nonempty cells in Q.[1]

(2) **For** each $e \in cell(Q)$ **do**
find a minimum unit disk cover $A(e)$ for all points in e.

(3) **Output** $A \leftarrow \bigcup_{e \in cell(Q)} A(e)$. ■

To see that Algorithm 4.A runs in polynomial time, we claim that the problem UDC restricted to a single cell e can be solved in time $n^{O(m^2)}$ by an exhaustive search algorithm. Note that a unit disk can cover a $\frac{1}{\sqrt{2}} \times \frac{1}{\sqrt{2}}$ square. Since a cell of size $m \times m$ can be partitioned into at most $\lceil \sqrt{2}m \rceil^2$ such squares, at most $O(m^2)$ unit disks are needed to cover all points in a cell.

Assume that a cell e contains n_e input points. If there is a point in cell e having distance greater than 1 from any other point, then we need to use an isolated disk to cover it. If a point has a distance at most 1 from some other points, then we

[1] A cell e is *nonempty* if it contains at least one input point.

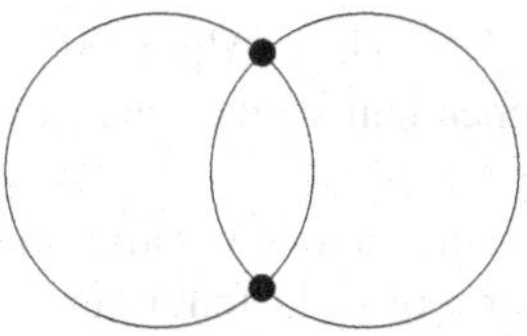

Figure 4.2: There are at most two possible positions for a unit disk with two given points on its boundary.

can use a disk D to cover it with some other points. In this case, we can move the disk to a canonical position so that at least two points covered by D lie on the boundary of D. For any two given points within distance 1, there are at most two possible canonical positions (see Figure 4.2). Therefore, for n_e given points in cell e, we need to consider at most $2\binom{n_e}{2}$ canonical positions. Together with the earlier observation that we need at most $O(m^2)$ unit disks to cover all points in a cell, we see that, in the exhaustive search algorithm, we need to inspect at most $(n_e(n_e-1))^{O(m^2)} = n_e^{O(m^2)}$ possible solutions to find a minimum disk cover for cell e.

Thus, over all nonempty cells, the total time for step (2) of Algorithm 4.A is

$$\sum_{e \in cell(Q)} n_e^{O(m^2)} \leq \left(\sum_{e \in cell(Q)} n_e \right)^{O(m^2)} = n^{O(m^2)}.$$

Next, we consider the performance of Algorithm 4.A as an approximation to UDC. Following the general approach of the analysis of approximation algorithms designed by the restriction method, we consider an optimal solution D^* to UDC and modify it to a feasible approximate solution. [Here, by a feasible approximate solution, we mean a solution that can be represented as

$$\bigcup_{e \in cell(Q)} S(e),$$

where each $S(e)$ is a unit disk cover of points in cell e.] The modification is simple: For each disk in D^* that intersects more than one cell, we make additional copies of the disk and use them to cover points in different cells. If a disk intersects k cells, $2 \leq k \leq 4$, then we make $k-1$ additional copies. Each copy is used to cover points in a different cell.

Note that if there are d disks in D^* that intersect more than one cell, then the above modification adds at most $3d$ unit disks to D^*. It follows that the solution A obtained by Algorithm 4.A satisfies

$$|A| \leq |D^*| \left(1 + \frac{3d}{|D^*|} \right),$$

since A is the optimal one among all feasible approximate solutions. In the worst case, it could happen that every disk in D^* intersects four cells (i.e., $d = |D^*|$), and we would have $|A| \leq 4|D^*|$. Thus, Algorithm 4.A is a polynomial-time 4-approximation for UDC (independent of the value of the constant m, as long as $m > 1$).

The performance ratio 4 obtained above is rather high, and we would like to improve it by reducing the number d of disks that intersect more than one cell. Before we do that, let us first look at a simple probabilistic analysis of the average-case performance of Algorithm 4.A, which might suggest some ideas for the improvement.

Suppose that the positions of the given points are evenly distributed in square Q, so that the center of each disk in D^* is evenly distributed in square Q. Note that a disk intersects more than one cell if and only if its center is within distance $1/2$ from a grid line. It follows that the probability that a disk in D^* does not intersect any grid line is equal to $(m-1)^2/m^2$, and the number of disks that intersect a grid line is binomially distributed with the success probability $p = 1-(m-1)^2/m^2 = (2m-1)/m^2$. Therefore, among $|D^*|$ disks, the expected number of disks intersecting a grid line is

$$\mu = |D^*| \cdot \frac{2m-1}{m^2} < |D^*| \cdot \frac{2}{m}.$$

Note that the median of a binomially distributed variable is equal to $\lceil \mu \rceil$ or $\lfloor \mu \rfloor$. That is, with probability $1/2$, the number of disks intersecting a grid line is at most $2|D^*|/m$. From the above analysis, we get a much better performance ratio in the average case.

Theorem 4.1 *Assume that the given input points for* UDC *are evenly distributed in a square* Q. *Then, with probability* $1/2$, *the solution* A *of* Algorithm 4.A *is a* $(1+6/m)$*-approximation for* UDC.

This probabilistic analysis suggests a randomized partition algorithm in which we choose the grid lines randomly so that the expected number d of unit disks intersecting more than one cell is close to $2|D^*|/m$. In the following, we show that this idea can actually be further improved to a deterministic partition algorithm. The basic technique here is the *shifting strategy*. That is, we shift the grid lines to find a partition with a small number of disks intersecting with grid lines.

To do this, let us examine the partition more carefully. Recall that Q is the initial square containing all n input points. Without loss of generality, assume that Q is of size $q \times q$, and that

$$Q = \{(x,y) \mid 0 \leq x \leq q, 0 \leq y \leq q\},$$

where q is a positive integer. Let $p = \lfloor q/m \rfloor + 1$. Consider the square

$$\overline{Q} = \{(x,y) \mid -m \leq x \leq mp, -m \leq y \leq mp\}.$$

Partition $\overline{Q}$ into $(p+1)^2$ cells, with each cell a square of size $m \times m$. We denote this partition of $\overline{Q}$ by $P(0,0)$. Note that the lower-left corner of the partition $P(0,0)$ is $(-m, m)$. In general, for any integers $0 \leq a < m$ and $0 \leq b < m$, we can create a new partition $P(a,b)$ for square Q by shifting the lower-left corner from $(-m,-m)$ to $(-m+a, -m+b)$ (see Figure 4.3).

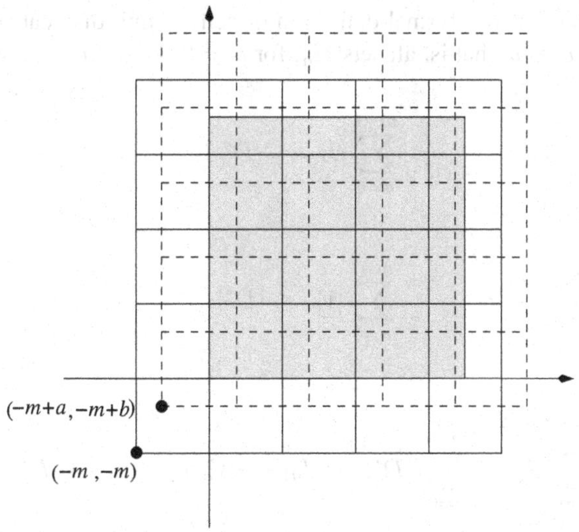

Figure 4.3: Square Q (the shaded area), partition $P(0,0)$ (the solid grid), and partition $P(a,b)$ (the dashed grid).

Consider a partition $P(a,a)$, for some integer $a \in \{0,1,\ldots,m-1\}$. Note that the cells of $P(a,a)$ cover the square Q. Let A_a denote the output of Algorithm 4.A using partition $P(a,a)$ [instead of $P(0,0)$]; that is, A_a is the union of the minimum unit disk covers for all cells in $P(a,a)$. As shown earlier, A_a can be computed by an exhaustive search algorithm in time $n^{O(m^2)}$. In the following, we show that, for at least one value of $a \in \{0,1,\ldots,m-1\}$, A_a is a $(1+3/m)$-approximation for the problem UDC.

Theorem 4.2 *For at least one value of $a \in \{0,1,\ldots,m-1\}$, $|A_a| \le (1+3/m)|D^*|$, where D^* is a minimum disk cover of* UDC *for all n input points.*

Proof. For simplicity, let us formally define that a cell is an $m \times m$ square, excluding the top and right boundaries. For each cell e in a partition $P(a,a)$, let $D^*(e)$ denote the set of unit disks in D^* that intersect cell e. Then we have

$$|A_a| \le \sum_{e \in cell(P(a,a))} |D^*(e)|,$$

where $cell(P(a,a))$ denotes the set of all cells in partition $P(a,a)$.

We call the collection of all cells in the partition $P(a,a)$ that lie along a horizontal or vertical line a *strip* of $P(a,a)$. Let H_a (or V_a) be the set of all disks in D^* that intersect two horizontal (or, respectively, vertical) strips in $P(a,a)$. Note that a unit disk can intersect at most four cells in $P(a,a)$, and if it intersects more than two cells in $P(a,a)$, it must belong to both H_a and V_a. It follows that

$$|A_a| \le \sum_{e \in cell(P(a,a))} |D^*(e)| \le |D^*| + |H_a| + 2|V_a|.$$

Now, note that, by our formal definition of cells, a unit disk cannot be in both H_a and H_b for $a \neq b$; that is, all sets H_a, for $a = 0, 1, \ldots, m-1$, are pairwisely disjoint. Thus,

$$\sum_{a=0}^{m-1} |H_a| \leq |D^*|.$$

Similarly,

$$\sum_{a=0}^{m-1} |V_a| \leq |D^*|.$$

Therefore,

$$\sum_{a=0}^{m-1} |A_a| \leq \sum_{a=0}^{m-1} (|D^*| + |H_a| + 2|V_a|) \leq (m+3)|D^*|.$$

Hence,

$$\frac{1}{m} \sum_{a=0}^{m-1} |A_a| \leq \Big(1 + \frac{3}{m}\Big)|D^*|.$$

That is, the average value of $|A_a|$ is bounded by $(1 + 3/m)|D^*|$. This implies that, for at least one value of $a \in \{0, 1, \ldots, m-1\}$, $|A_a| \leq (1 + 3/m)|D^*|$. □

Corollary 4.3 *For any* $\varepsilon > 0$*, there is a* $(1+\varepsilon)$*-approximation for* UDC *that runs in time* $n^{O(1/\varepsilon^2)}$.

Proof. Choose $m = \lceil 3/\varepsilon \rceil$. Note that computing each A_a needs at most $n^{O(m^2)} = n^{O(1/\varepsilon^2)}$ time. By Theorem 4.2, a $(1+\varepsilon)$-approximation can be obtained by computing all m solutions A_a and choosing the best one. The total running time is $mn^{O(1/\varepsilon^2)} = n^{O(1/\varepsilon^2)}$. □

In the problem UDC, we are allowed to use any unit disk in our solution. Suppose we add some restrictions on the location of the unit disks; the partition technique might still work. For instance, if we require that the center of each unit disk in the solution must be located at an input point, it is not hard to check that a similar argument for Theorem 4.2 works. In other words, the following variation of the problem UDC has, for any fixed $\varepsilon > 0$, a $(1+\varepsilon)$-approximation that runs in time $n^{O(1/\varepsilon^2)}$.

> UNIT DISK COVERING WITH RESTRICTED LOCATIONS (UDC_1): Given a set of n points in the Euclidean plane, find a minimum number of unit disks to cover all input points, with the center of each disk located at an input point.

4.2 Boundary Area

A key step in the partition technique is to combine the solutions of the smaller parts of the partition into a feasible global solution to the original input. In the approximation Algorithm 4.A for UDC, this is straightforward, as the union of the local solutions for smaller cells is naturally a feasible global solution. In general, when the relationship between local solutions and global solutions is not so simple, we may have to modify the local solutions around the boundary area of the partition to get the global solution to the original input. In this section, we study this issue through the example of the connected dominating set problem in unit disk graphs.

Recall that a *dominating* set in a graph $G = (V, E)$ is a subset D of vertices V such that every vertex is either in the set D or adjacent to some vertex in D. If, in addition, the subgraph induced by a dominating set is connected, then such a dominating set is called a *connected dominating set*. A *unit disk graph* is a graph in which each vertex is a point in the Euclidean plane and there is an edge between two points u and v if and only if the two unit disks centered at u and v have a nonempty intersection.

> CONNECTED DOMINATING SET IN A UNIT DISK GRAPH (CDS-UDG): Given a connected unit disk graph G, find a connected dominating set of G with the minimum cardinality.

First, we notice that the minimum dominating set problem in a unit disk graph can be easily converted to the problem UDC_1 and, hence, has a PTAS.

Theorem 4.4 *For any $\varepsilon > 0$, there is a $(1 + \varepsilon)$-approximation for the minimum dominating set problem in unit disk graphs that runs in time $n^{O(1/\varepsilon^2)}$.*

Proof. Let $G = (V, E)$ be a unit disk graph. Then there is an edge between two vertices u and v if and only if the distance between u and v is less than or equal to 1. Equivalently, a vertex u dominates a vertex $v \in V$ if and only if the disk $D(u, 2)$ centered at u and having diameter 2 covers the vertex v. It follows that the minimum dominating set in G has size k if and only if the set V of points can be covered by k disks, each of which is centered at an input point and has diameter 2. That is, the minimum dominating set problem in unit disk graphs is equivalent to the variation of UDC_1 in which all disks have diameter 2. The theorem now follows from Corollary 4.3. □

The following result follows easily from the simple relationship between dominating sets and connected dominating sets.

Corollary 4.5 *There is a polynomial-time* 4*-approximation for* CDS-UDG.

Proof. Let $G = (V, E)$ be a unit disk graph. Suppose that $D \subseteq V$ is a (4/3)-approximation for the minimum dominating set of G and D^* is a minimum connected dominating set. Then we must have $|D| \leq 4|D^*|/3$, since the size of a minimum dominating set cannot exceed the size of a minimum connected dominating set. Now we claim that if D is not connected, then we can reduce the number of connected components in D by one by adding one or two vertices into D. To see this, recall that the input graph G of the problem CDS-UDG is always connected. Consider a shortest path $(v_1, v_2, \ldots, v_m)$ between any two connected components of D. First, the vertex v_2 must not be in D, for otherwise the path can be shortened to $(v_2, \ldots, v_m)$. If $v_3 \in D$, then we can add v_2 to D to reduce the number of connected components in D. If $v_3 \notin D$, then v_3 must be dominated by a vertex u in D. We note that u and v_1 cannot be in the same connected component in D, for otherwise $(u, v_3, v_4, \ldots, v_m)$ would be a shorter path between two connected components of D. Therefore, we must have $m = 4$ and $u = v_m$, and adding v_2 and v_3 to D reduces the number of connected components by one. So, the claim is proven.

From the above claim, we need to add at most $2(c-1)$ vertices to D to get a connected dominating set, where c is the number of connected components in D. It is clear that $c \leq |D|$. That is, we can find a connected dominating set C of size $|C| \leq 3|D| - 2 \leq 3 \cdot (4|D^*|/3) = 4|D^*|$. □

Next, we describe how to apply the above 4-approximation algorithm to get a PTAS for CDS-UDG.

Let $G = (V, E)$ be a given connected unit disk graph. We define partitions $P(a, b)$ as in Section 4.1. That is, assume that Q is a square containing all vertices in V. Without loss of generality, assume that $Q = \{(x, y) \mid 0 \leq x \leq q, 0 \leq y \leq q\}$. Let m be an integer whose value will be determined later. Let $p = \lfloor q/m \rfloor + 1$. Consider the square $\overline{Q} = \{(x, y) \mid -m \leq x \leq mp, -m \leq y \leq mp\}$. Partition $\overline{Q}$ into $(p+1) \times (p+1)$ cells so that each cell is an $m \times m$ square, excluding the top and right boundary edges. This partition of $\overline{Q}$ is denoted by $P(0, 0)$. In general, the partition $P(a, b)$ is obtained from $P(0, 0)$ by shifting the lower-left corner of $\overline{Q}$ from $(-m, -m)$ to $(-m + a, -m + b)$.

Let h be an integer such that $2h + 2 < m$. For each cell e of size $m \times m$, we define its *central area* to be the set of points in e that have distance at least h from the boundary of e; that is, it is the $(m - h) \times (m - h)$ square that shares the same center with cell e. In addition, we define the *boundary area* of a cell e to be the set of points in e that are within distance $< h + 1$ from the boundary of e. Note that for each cell, its boundary area and central area have an overlapping area of width 1 (see Figure 4.4). Finally, we define the *boundary area* of a partition $P(a, a)$ to be the union of the boundary areas of all cells in $P(a, a)$. The idea of our algorithm is to solve the problem CDS-UDG on the central area of each cell e, and take the union of these local solutions, plus a 4-approximation global solution on the boundary area, as the solution to the input graph G.

For each cell e of a partition $P(a, a)$, let $G_c[e]$ be the subgraph of G induced by all vertices in the central area of e. This graph $G_c[e]$ may have more than one connected component. Let $C[e]$ be a minimum subset of vertices in e satisfying the following condition:

(C_1) For each connected component H of $G_c[e]$, the subgraph of G induced by $C[e]$ has a connected component dominating H.

Lemma 4.6 *For each cell e in a partition $P(a, a)$, the set $C[e]$ can be computed in time $n_e^{O(m^2)}$, where n_e is the number of vertices in e.*

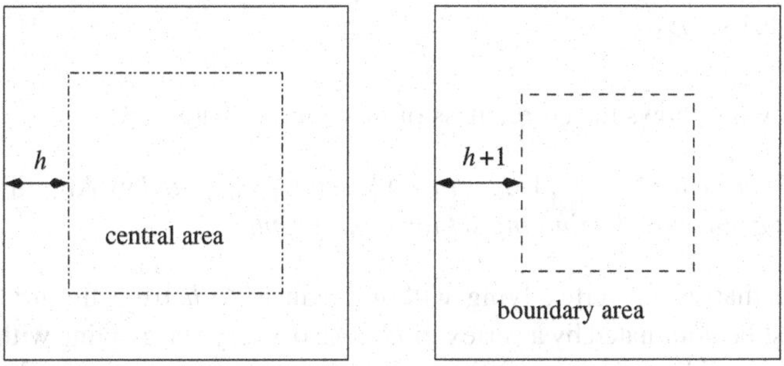

Figure 4.4: Central area and boundary area overlapping with width 1.

Proof. We note that for any square of size $\frac{1}{\sqrt{2}} \times \frac{1}{\sqrt{2}}$, the set of vertices lying inside the square induces a complete subgraph in which any vertex dominates all other vertices. It follows that the minimum dominating set S for $G_c[e]$ has size at most $\lceil \sqrt{2}m \rceil^2$. To create a connected dominating set from S, we need to add at most two vertices to reduce the number of connected components of S by 1, and so $C[e]$ has size at most $3\lceil \sqrt{2}m \rceil^2$. Thus, the number of candidates for $C[e]$ is at most

$$\sum_{k=0}^{3\lceil \sqrt{2}m \rceil^2} \binom{n_e}{k} = n_e^{O(m^2)}.$$

It is clear that for any set C' of vertices in cell e, we can check in linear time whether it holds that, for each connected component H of $G_c[e]$, the subgraph of G induced by C' has a connected component dominating H. Therefore, an exhaustive search for $C[e]$ only takes time $n_e \cdot n_e^{O(m^2)} = n_e^{O(m^2)}$. □

Now, we are ready to describe a $(1+\varepsilon)$-approximation algorithm for CDS-UDG.

Algorithm 4.B (*PTAS for* CDS-UDG)

Input: A unit disk graph $G = (V, E)$, with all vertices lying in square Q.

(1) Let $h \leftarrow 3$ and $m \leftarrow \lceil 160/\varepsilon \rceil$.

(2) Let $D \subseteq V$ be a 4-approximation to the minimum connected dominating set for G (obtained by the algorithm of Corollary 4.5).

(3) **For** $a \leftarrow 0$ **to** $m - 1$ **do**

(3.1) Let $D_a \leftarrow \{v \in D \mid v \text{ lies in the boundary area of } P(a, a)\}$;

(3.2) **For** each cell e of $P(a, a)$ **do**
compute set $C[e]$ (by exhaustive search of Lemma 4.6);

(3.3) Let $A_a \leftarrow D_a \cup \Big(\bigcup_{e \in P(a,a)} C[e] \Big)$.

(4) Let $a^* \leftarrow \operatorname{argmin}_{0 \le a < m} |A_a|$.

(5) Output $A \leftarrow A_{a^*}$. ■

The following shows the correctness of this approximation.

Lemma 4.7 *For each $a \in \{0, 1, \ldots, m-1\}$, set A_a computed by* Algorithm 4.B *in step* (3) *is a connected dominating set for input graph G.*

Proof. Note that every vertex lying within distance $\leq h$ from the grid lines of $P(a, a)$ must be dominated by a vertex in D_a. Also, every vertex lying with distance $> h$ from the grid lines of $P(a, a)$ lies in the central area of a cell e in $P(a, a)$, and hence is dominated by $C[e]$. Thus, A_a is a dominating set of G.

Next, we show that A_a is connected. Consider two connected components E_1, E_2 of D_a that are connected through a path π in D passing through the central area of a cell e of $P(a, a)$. Note that the central area and the boundary area of cell e have an overlapping area of width 1. Thus, this path π must begin with a vertex x_1 in E_1 that lies in the overlapping area and end with a vertex x_2 in E_2 that also lies in the overlapping area. Obviously, π is a subgraph of a connected component H of the graph $G_c[e]$ and hence, by the requirement on $C[e]$, is dominated by a connected component C' of the subgraph of G induced by $C[e]$. In particular, C' must dominate both x_1 and x_2. It follows that E_1 and E_2 are connected through C'. This proves that all connected components of D_a are connected through $C[e]$'s over all cells e of partition $P(a, a)$.

Moreover, a similar argument shows that every connected component C' of $C[e]$ for any cell e is connected to some vertex in D_a, and hence A_a is connected. To see this, assume, by way of contradiction, that a connected component C' of $C[e]$ is not connected to any vertex in D_a. Then every vertex of C' lies in the central area of e, for otherwise it would be dominated by a vertex of D_a. Let H be the connected component of $G_c[e]$ that contains C'. By the minimality of $C[e]$, C' dominates H. Let x be a vertex in C'. Then x is dominated by a vertex $y \in D \setminus D_a$. Since D is connected, there must be a path π in D from y to a vertex $z \in D_a$ with every vertex in π lying in the central area of e. Clearly, the path π is a subgraph of H and so is dominated by C'. In particular, $z \in D_a$ is adjacent to a vertex in C', which is a contradiction. This completes the proof of the claim and, hence, the proof of the lemma. □

Remark. In the above proof, the minimality of set $C[e]$ is not required. In fact, the proof is correct as long as, for every cell e of $P(a, a)$, $C[e]$ satisfies condition (C_1), and every connected component C' of $C[e]$ dominates some connected component H of $G_c[e]$.

To verify that Algorithm 4.B runs in polynomial time, we note that, from Lemma 4.6, each $C[e]$ can be computed in time $n_e^{O(m^2)}$, and so the total time for step (3.2) is at most

$$\sum_{e \in P(a,a)} n_e^{O(m^2)} \leq \left(\sum_{e \in P(a,a)} n_e \right)^{O(m^2)} = n^{O(m^2)}.$$

It follows that each A_a is computable in time $n^{O(m^2)}$, and so the output $A = A_{a^*}$ can be found in time $O(mn) + m \cdot n^{O(m^2)} = n^{O(m^2)}$.

Finally, we show that Algorithm 4.B is a PTAS.

Theorem 4.8 *Output* A_{a^*} *of* Algorithm 4.B *is a* $(1+\varepsilon)$*-approximation for* CDS-UDG *with computation time* $n^{O(1/\varepsilon^2)}$.

Proof. Let D^* denote a minimum connected dominating set for G. Following the general approach for analyzing the performance of an approximation constructed by the restriction method, we will modify D^* into a feasible approximate solution D'. Here, by a feasible approximate solution, we mean that, for some $a \in \{0, 1, \ldots, m-1\}$, D' contains D_a, and, for each cell e of $P(a,a)$, set $D'[e]$ of all vertices in D' lying in cell e satisfies condition (C_1) (that is, for every connected component H of $G_c[e]$, the subgraph of G induced by vertices in $D'[e]$ has a connected component dominating H). Note that A_a is such a feasible approximate solution with the minimum $C[e]$'s, and so if D' is feasible with respect to partition $P(a,a)$, then $|A_a| \le |D'|$.

The modification of D^* is divided into two steps. We first find a suitable shifting parameter $b \in \{0, 1, \ldots, m-1\}$. Then we modify D^* on each cell e of the partition $P(b,b)$ to get the required $D'[e]$.

Recall that for $a \in \{0, 1, \ldots, m-1\}$, D_a denotes the set of vertices in D that lie in the boundary area of partition $P(a,a)$; in addition, let D_a^* denote the set of vertices in D^* that lie in the boundary area of $P(a,a)$. We claim that there exists an integer $b \in \{0, 1, \ldots, m-1\}$ such that

$$6 \cdot |D_b^*| + |D_b| \le \varepsilon \cdot |D^*|.$$

To prove this, let us study how the shifting of the partition from $P(0,0)$ to $P(1,1)$, $P(2,2), \ldots, P(m-1, m-1)$ affects sets D_a^* and D_a. When the partition shifts (toward northeast), the location of the graph G relative to the grid of the partition changes. We may imagine that the grid of the partition is fixed, but actually the graph G is moving (toward southwest). For each vertex v of G, the moving of the graph leaves a trace in the grid. The trace of v consists of m points in a straight line of slope 1, with distance $\sqrt{2}$ between any two consecutive points (see Figure 4.5). Thus, the trace of v contains at most $4(h+1)$ points in the boundary area of the (fixed) partition.

In other words, for any vertex v in D^*, it belongs to at most $4(h+1)$ of the sets $D_0^*, D_1^*, \ldots, D_{m-1}^*$. Therefore, by the pigeonhole principle, we have

$$\sum_{a=0}^{m-1} |D_a^*| \le 4(h+1)|D^*|.$$

Similarly, for any vertex v in D, it belongs to at most $4(h+1)$ of the sets D_0, $D_1, \ldots, D_{m-1}$, and so

$$\sum_{a=0}^{m-1} |D_a| \le 4(h+1)|D| \le 16(h+1)|D^*|.$$

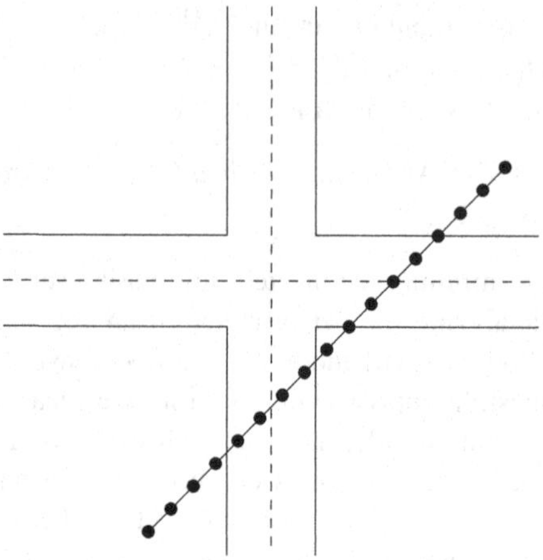

Figure 4.5: The trace of a vertex has at most $4(h+1)$ points lying in the boundary area.

It follows that

$$\sum_{a=0}^{m-1} \left(6 \cdot |D_a^*| + |D_a|\right) \le 40(h+1)|D^*|.$$

Therefore, there must exist an integer $b \in \{0, 1, \ldots, m-1\}$ such that

$$6 \cdot |D_b^*| + |D_b| \le \frac{40(h+1)}{m}|D^*| \le \varepsilon \cdot |D^*|. \tag{4.1}$$

Now, we fix this shifting parameter b. For each cell e of the partition $P(b, b)$, let $D^*[e]$ denote the set of all vertices in D^* that lie in cell e. We will modify $D^*[e]$ into a set $D'[e]$ that satisfies condition (C_1).

For a cell e of $P(b, b)$, consider a connected component H of $G_c[e]$. Clearly, $D^*[e]$ dominates H. Assume that H is dominated by k connected components $D_1, \ldots, D_k$ of $D^*[e]$. Since H is connected, these connected components can be connected together into a single component by adding at most $2(k-1)$ vertices of H. We define $D'[e]$ to be the set $D^*[e]$ plus the collection of all these connecting vertices for all connected components H of $G_c(e)$. Clearly, each $D'[e]$ satisfies condition (C_1). Therefore,

$$D' = D_b \cup \left(\bigcup_{e \in P(b,b)} D'[e] \right)$$

is a feasible solution (see the remark after Lemma 4.7), and $|A_b| \le |D'|$.

To estimate the size of D', let $D_b^*[e]$ be the set of all vertices in D_b^* that lie in the boundary area of cell e. Again, assume that a connected component H of $G_c[e]$ is dominated by k connected components $D_1, D_2, \ldots, D_k$ of $D^*[e]$. Since D^* is connected, each of its connected components D_i, $i = 1, 2, \ldots, k$, is connected to

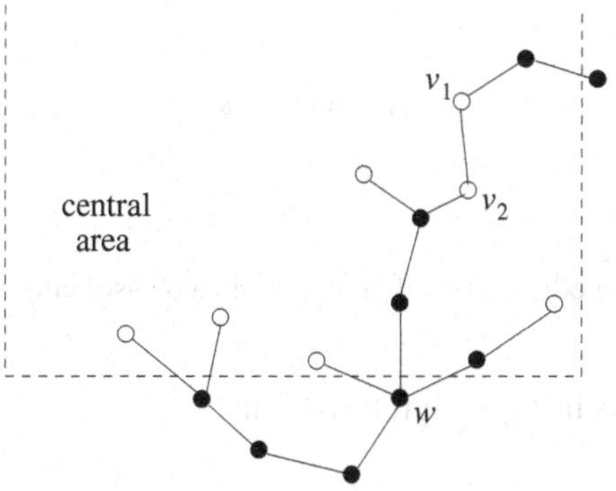

Figure 4.6: Charging v_1 and v_2 to the vertex w outside the central area. In the above, • denotes a vertex in D^*, and ∘ denotes a vertex in $G_c[e]$ that is not in D^*.

some vertices of D^* lying outside the cell e through an edge crossing a boundary edge of e (unless there is no vertex of D^* outside cell e, in which case $k = 1$). That is, each D_i must contain a vertex that lies outside the central area of cell e and, hence, belongs to $D_b^*[e]$. Now, we describe a charging method that charges the cost of vertices in $D'[e] \setminus D^*[e]$ to different vertices in $D_b^*[e]$, so that each vertex in $D_b^*[e]$ is charged at most six times.

Note that to connect the connected components $D_1, D_2, \ldots, D_k$ of $D^*[e]$ into a single component, we need to add at most $2(k-1)$ vertices in H to $D'[e]$. We charge these vertices evenly to $k-1$ of the components in such a way that

(i) When two vertices (or, one vertex) are added to connect two components D_i and D_j, they are both charged to D_i or both charged to D_j, and

(ii) Each component D_i is charged at most twice (by two vertices).

Thus, for each connected component H of $G_c[e]$, a connected component D_i of $D^*[e]$ can be charged at most twice. However, a component D_i of $D^*[e]$ may be used to dominate more than one component H of $G_c[e]$. Therefore, we need to further distribute the charges to different vertices in D_i. To be more specific, when a vertex v_1 is charged (maybe together with another vertex v_2) to a component D_i, we charge it to vertex w of D_i lying outside the central area of e that is the closest to vertex v_1 through a path of $D_i \cup \{v_1, v_2\}$ (see Figure 4.6).

Note that if we charge v_1 to w of D_i according to the above criteria, all vertices, except w, in the shortest path in $D_i \cup \{v_1, v_2\}$ between w and v_1 must lie in the central area of e. Thus, a vertex w can be charged at most 2ℓ times if it has ℓ independent neighbors inside the central area of e. (Two neighbors of w that are not independent to each other must belong to the same connected component H of the subgraph $G_c[e]$.) It is easy to see that in a unit disk graph, a vertex lying outside the central area of e can have at most three independent neighbors lying inside the central area (cf. Figure 4.6). Thus, each vertex w in D_i can be charged at most six times. Furthermore, a vertex is charged only if it lies outside the central area of e

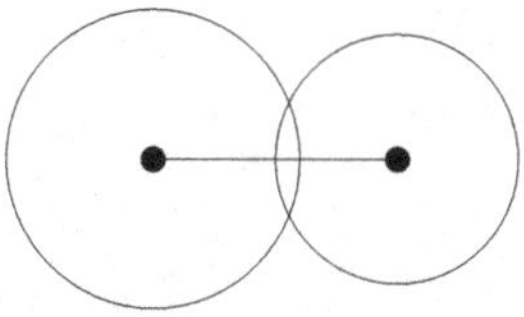

Figure 4.7: An edge exists if and only if two associated disks overlap.

and, hence, only if it is in $D_b^*[e]$. It follows that

$$|D'[e]| \le |D^*[e]| + 6 \cdot |D_b^*[e]|.$$

Now, we get, from inequality (4.1),

$$\begin{aligned} |D'| &\le |D_b| + \sum_{e \in P(b,b)} |D'[e]| \le |D_b| + \sum_{e \in P(b,b)} \left(|D^*[e]| + 6 \cdot |D_b^*[e]|\right) \\ &= |D_b| + |D^*| + 6 \cdot |D_b^*| \le (1+\varepsilon)|D^*|. \end{aligned}$$

Since the output A_{a^*} of Algorithm 4.B has the minimum size among all sets A_a, and since $|A_b| \le |D'|$, we conclude that $|A_{a^*}|$ is at most $(1+\varepsilon)|D^*|$. □

4.3 Multilayer Partition

In a unit disk graph, each vertex v is a point in the Euclidean plane and is associated with a unit disk centered at v. An edge exists between two vertices if and only if the two associated disks have a nonempty intersection. This notion can be generalized to an *intersection disk graph* in which disks may be of different sizes. More precisely, in an intersection disk graph, each vertex v is a point in the Euclidean plane and is associated with a disk centered at v, but different points may be associated with disks of different diameters; and, an edge exists between two vertices if and only if the two associated disks have a nonempty intersection (see Figure 4.7).

When we apply the partition technique to intersection disk graphs with disks of different sizes, a simple partition of a fixed size does not work well. Instead, we need to use partitions of different grid sizes to deal with disks of different sizes. We call this the *multilayer partition*. Let us look at the following example:

> MAXIMUM INDEPENDENT SET IN AN INTERSECTION DISK GRAPH (MIS-IDG): Given an intersection disk graph G, find an independent set of G with the maximum cardinality.

Clearly, a subset of vertices is independent if and only if their associated disks do not overlap. For convenience, we will identify the vertices with their associated disks and work on the disks directly. In particular, we say a set of disks is *independent* if these disks are mutually disjoint (see Figure 4.8).

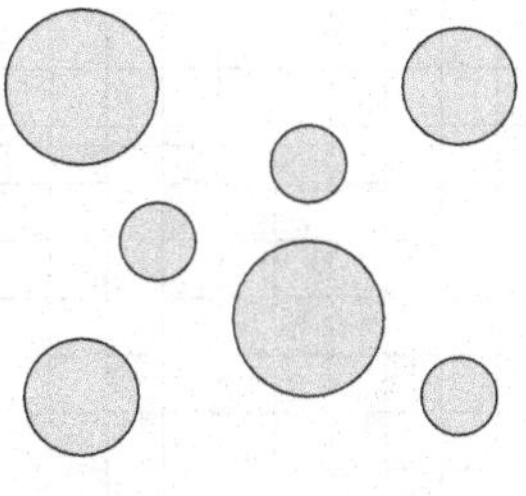

Figure 4.8: Independent disks.

In the following, we will apply the multilayer partition to the problem MIS-IDG. First, assume that all given disks are contained in the interior of a square Q. Fix an integer $k > 0$, and rescale all disks so that the largest disk has diameter $1 - 1/k$. Let $d_{\min}$ be the diameter of the smallest disk in the new scale, and let $m = \lfloor \log_{k+1}(1/d_{\min}) \rfloor$. We now divide all disks into $m+1$ layers: For $0 \le j \le m$, layer j consists of all disks with diameters d in the range $(k+1)^{-(j+1)} < d \le (k+1)^{-j}$. So, the largest disk is in layer 0, and the smallest disk is in layer m.

Next, corresponding to each layer j of disks, with $0 \le j \le m$, we define a partition of square Q. Without loss of generality, assume that

$$Q = \{(x, y) \mid 0 \le x \le q, 0 \le y \le q\}.$$

Let $p = \lfloor q/k \rfloor + 1$. We extend square Q to the square

$$\overline{Q} = \{(x, y) \mid -k \le x \le kp, -k \le y \le kp\}.$$

For each $0 \le j \le m$, partition $\overline{Q}$ into $(p+1)(k+1)^j \times (p+1)(k+1)^j$ cells so that each cell is a $k(k+1)^{-j} \times k(k+1)^{-j}$ square (excluding the top and right boundary edges). We call this the *layer-j partition* of $\overline{Q}$, and denote it by $P_j(0,0)$ (see Figure 4.9). Note that the diameter of a disk in layer j is at least $1/k$ of the grid size of a layer-$(j+1)$ partition, and is at most $1/k$ of the grid size of a layer-j partition.

Next, we describe how to apply the shifting technique to the partitions $P_j(0,0)$. The critical idea here in multilayer partition is to shift partitions in different layers with different distances. In general, for $0 \le a < k$ and $0 \le b < k$, the layer-j partition $P_j(a,b)$ can be obtained from $P_j(0,0)$ by shifting the lower-left corner of $\overline{Q}$ from $(-k,-k)$ to $(-k + a(k+1)^{-j}, -k + b(k+1)^{-j})$. Note that, for the same shifting parameters a, b, partition $P_j(a,b)$ and partition $P_{j+1}(a,b)$ have different lower-left corners. However, since we have extended the original square Q to a bigger square $\overline{Q}$, the outer square of every $P_j(a,b)$ contains the original square Q. Furthermore, inside the original square Q, the grid lines of $P_j(a,b)$ are also the grid lines of $P_{j+1}(a,b)$.

Lemma 4.9 *For any $a, b \in \{0, 1, \ldots, k-1\}$ and any $j \in \{0, 1, \ldots, m-1\}$, a grid line of the layer-j partition $P_j(a,b)$ inside the square Q is also a grid line of the layer-$(j+1)$ partition $P_{j+1}(a,b)$.*

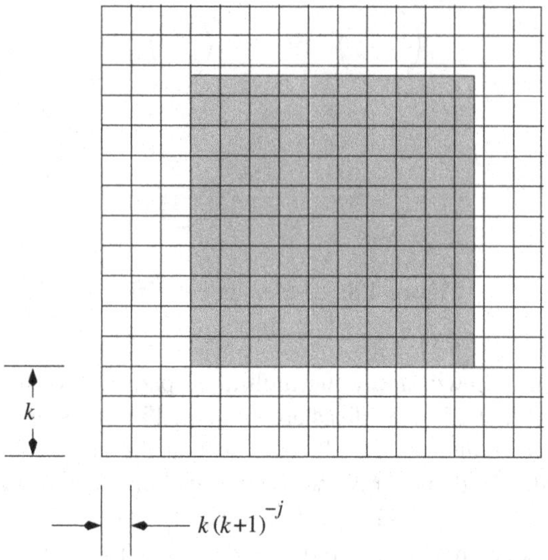

Figure 4.9: Layer-j partition $P_j(0,0)$.

Proof. From the setting of $P_j(a,b)$, the x-coordinate of a vertical grid line in $P_j(a,b)$ is of the form $-k+(a+ik)(k+1)^{-j}$ for some integer $i \geq 0$. Note that $(a+ik)(k+1)^{-j} = (a+(ik+a+i)k)(k+1)^{-(j+1)}$. Thus, every vertical grid line in $P_j(a,b)$ within Q is also a vertical grid line in $P_{j+1}(a,b)$. Similarly, every horizontal grid line in $P_j(a,b)$ within Q is also a horizontal grid line in $P_{j+1}(a,b)$. □

Let $\mathcal{D}$ be the set of the input disks, and $a,b \in \{0,1,\ldots,k-1\}$. For each j, $0 \leq j \leq m$, delete from $\mathcal{D}$ all disks in layer j that hit a grid line in the corresponding partition $P_j(a,b)$.[2] Let $\mathcal{D}(a,b)$ denote the collection of all remaining disks (in all layers).

Lemma 4.10 *The maximum independent set of disks in $\mathcal{D}(a,b)$ can be computed in time $n^{O(k^4)}$.*

Proof. For a set $\mathcal{E}$ of disks, let $opt(\mathcal{E})$ denote the maximum independent set of disks in $\mathcal{E}$. In the following, we present a dynamic programming algorithm computing $opt(\mathcal{D}(a,b))$ in time $n^{O(k^4)}$.

First, for convenience, let us call a cell in a layer-j partition a *j-cell*. A j-cell is said to be *relevant* if it contains a disk in layer j. For cells in different layers, we define a parent–child relation. For $j' > j$, we say a relevant j'-cell e' is a *child* of a relevant j-cell e if e contains e' and no other relevant j''-cell e'', with $j < j'' < j'$, satisfying $e' \subsetneq e'' \subsetneq e$. A relevant cell without a relevant parent is called a *maximal* relevant cell. Let E be the set of all maximal relevant cells. Note that for any two

[2]By hitting a grid line, we mean that the disk intersects the grid line or touches the grid line.

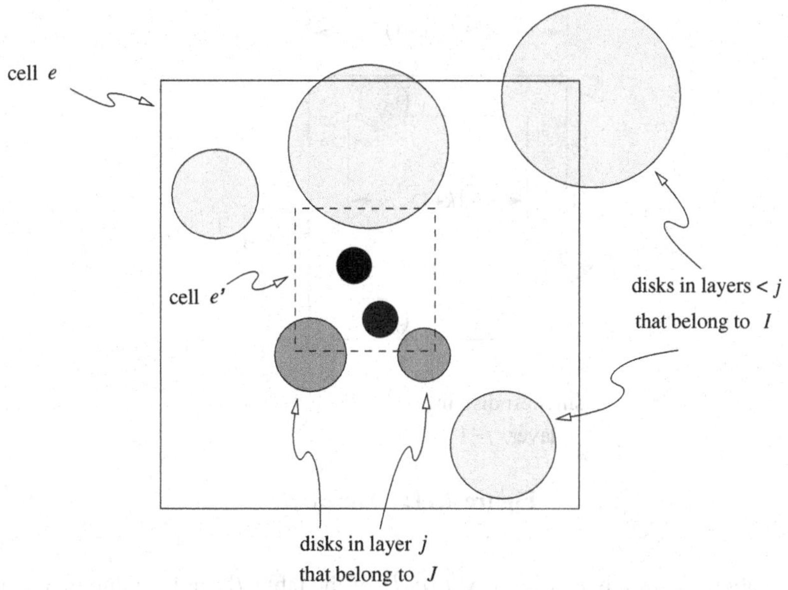

Figure 4.10: The relationships among cells e, e′ and disk sets I, J in the recursive relation.

relevant cells e and e′, we can determine, in time $O(n)$, whether e′ is a child of e and whether e is a maximal relevant cell.

In the dynamic programming algorithm for $opt(\mathcal{D}(a, b))$, we will build a table T of the following form: Let e be a relevant j-cell and I a set of independent disks in layers $< j$ that hit cell e. Then $T(e, I)$ contains the maximum independent set of disks in layers $\geq j$ that are in cell e and are disjoint from all disks in I. Clearly,

$$opt(\mathcal{D}(a, b)) = \bigcup_{e \in E} T(e, \emptyset).$$

To build table T, let $\text{IND}_j(e, I)$ denote the collection of all sets J of independent disks in layer j that are in cell e and are disjoint from all disks in I. Also, let I_e denote the set of disks in I that intersect cell e, and $child(e)$ the set of children of cell e. Then the recursive relation of the dynamic programming can be described as follows (cf. Figure 4.10):

(1) For each $J \in \text{IND}_j(e, I)$, let

$$A_J = \bigcup_{e' \in child(e)} T(e', (I \cup J)_{e'});$$

that is, A_J is the maximum independent set of disks in layers $\geq j + 1$ that are in cell e and are disjoint from $I \cup J$.

(2) Let $J^* = \text{argmax}_{J \in \text{IND}_j(e,I)} |J \cup A_J|$. Then we have $T(e, I) = J^* \cup A_{J^*}$.

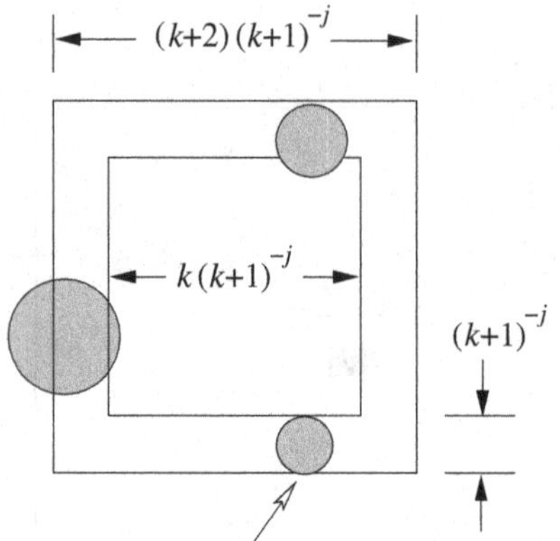

Figure 4.11: Square S.

The above shows that each entry $T(e, I)$ of the table T can be computed recursively from entries $T(e', I')$, over all children e' of e. To complete the proof, we need to verify that

(a) The computation of each entry $T(e, I)$ can be done in time $n^{O(k^4)}$, and

(b) The table size of T is bounded by $n^{O(k^2)}$.

To prove (a), we first note that disks in a set $J \in \mathrm{IND}_j(e, I)$ must be in layer j and contained in e. The cell e has size $(k(k+1)^{-j})^2$, and each disk in J has diameter $\geq (k+1)^{-(j+1)}$ and, hence, has area $\geq \pi((k+1)^{-(j+1)}/2)^2$. It means the set J contains at most

$$\frac{(k(k+1)^{-j})^2}{\pi((k+1)^{-(j+1)}/2)^2} = \frac{4}{\pi}k^2(k+1)^2 = O(k^4)$$

disks. Thus, the collection $\mathrm{IND}_j(e, I)$ has at most $n^{O(k^4)}$ sets J.

In addition, we note that there are at most n relevant cells, and, as we pointed out earlier, the parent–child relation between cells can be determined in time $O(n)$. Thus, the computation of the entry $T(e, I)$ can be done in time $n^{O(k^4)}$.

Next, we calculate the size of table T. We first count the size of I, i.e., the maximum number of independent cells in layers $< j$ that can intersect a j-cell e. To do this, we draw, as shown in Figure 4.11, a square S of size $(k+2)(k+1)^{-j} \times (k+2)(k+1)^{-j}$, that contains e in the center. We note that every disk in layer $< j$ has size at least $\pi((k+1)^{-j}/2)^2$. Thus, if it intersects cell e, then it must occupy a region of size at least $\pi((k+1)^{-j}/2)^2$ in S. Therefore, the size of set I is at most

$$\frac{((k+2)(k+1)^{-j})^2}{\pi((k+1)^{-j}/2)^2} = \frac{4(k+2)^2}{\pi} = O(k^2).$$

Therefore, for any cell e, there are at most $n^{O(k^2)}$ possible sets I to be considered, and the size of table T is bounded by $n \cdot n^{O(k^2)} = n^{O(k^2)}$. This completes the proof of (b) and, hence, the proof of the theorem. □

Now, we are ready to describe a $(1+\varepsilon)$-approximation for MIS-IDG.

Algorithm 4.C (*PTAS for* MIS-IDG)

Input: A set $\mathcal{D}$ of disks.

(1) Let $k \leftarrow 2\lceil 1 + 1/\varepsilon \rceil$.

(2) **For** $a \leftarrow 0$ **to** $k-1$ **do** compute $opt(\mathcal{D}(a,a))$.

(3) Let $a' \leftarrow \underset{0 \le a < k}{\text{argmax}}\, |opt(\mathcal{D}(a,a))|$.

(4) Output $A \leftarrow opt(\mathcal{D}(a',a'))$. ■

From Lemma 4.10, Algorithm 4.C runs in time $n^{O(1/\varepsilon^4)}$. Next, we show that it is a PTAS.

Theorem 4.11 *The output A of* Algorithm 4.C *is a $(1+\varepsilon)$-approximation to the optimal solution $opt(\mathcal{D})$.*

Proof. Let A^* be a maximum independent set of disks in $\mathcal{D}$. For each $a \in \{0, 1, \ldots, k-1\}$, let $H_j(a)$ denote the set of layer-j disks in A^* that hit a grid line in the layer-j partition $P_j(a,a)$, and let $H(a) = \bigcup_{j=0}^{m} H_j(a)$. Note that, for each $a \in \{0, 1, \ldots, k-1\}$, $A^* - H(a)$ is a feasible solution to the problem MIS-IDG with respect to the set $\mathcal{D}(a,a)$ of disks, and hence $|A| \ge |A^* - H(a)|$.

Note that a disk in layer j has a diameter $d \le (k+1)^{-j}$, and so it can appear in at most two of the sets $H(0), H(1), \ldots, H(k-1)$. Therefore,

$$\sum_{a=0}^{k-1} |H(a)| \le 2|A^*|.$$

It follows that there must exist an integer $a \in \{0, 1, \ldots, k-1\}$ such that

$$|H(a)| \le \frac{2}{k}|A^*| \le \frac{\varepsilon}{1+\varepsilon}|A^*|.$$

Now, we have

$$|A| \ge |A^* - H(a)| = |A^*| - |H(a)| \ge \frac{1}{1+\varepsilon}|A^*|;$$

or, equivalently,

$$|A^*| \le (1+\varepsilon)|A|.$$

□

4.4 Double Partition

In the previous sections, we have used the partition technique to design PTASs for some geometric problems. In these algorithms, the tradeoff between the performance ratio and the running time is straightforward. That is, in order to get a smaller performance ratio, we simply increase the cell size and spend extra time to solve the subproblems on larger cells. We note that in order for this approach to work, the running time for solving the subproblems on larger cells must remain a polynomial function in the input size—even though the degree of the polynomial function may increase along with the cell size.

For some size-sensitive problems, however, this approach may not work. That is, a problem may be easy to solve on a cell of a certain small size, but it becomes more difficult to solve (or approximate) on larger cells. For such a problem, a PTAS is difficult to get, but some kind of tradeoff between the performance ratio and running time can still be achieved. In this section, we introduce a new technique, called *double partition*, to deal with such problems. Namely, we first partition the input data into cells of a small size on which the subproblems are easy to solve; we then apply the second partition on this partitioned problem to reduce the performance ratio. To demonstrate how this technique works, we study a specific problem about unit disk graphs.

> WEIGHTED DOMINATING SET IN A UNIT DISK GRAPH (WDS-UDG): Given a unit disk graph $G = (V, E)$ with a vertex-weight function $c : V \to \mathbb{R}^+$, find a dominating set of G with the minimum total vertex weight.

We will present a polynomial-time $(6+\varepsilon)$-approximation for this problem. Since the proof of this result is quite involved, we will establish it in three steps:

(a) We find a 2-approximation for a subproblem of WDS-UDG restricted to a cell of size $\mu \times \mu$, where $\mu = 1/\sqrt{2}$.

(b) We extend result (a) to the subproblem of WDS-UDG restricted to a cell of arbitrarily large constant size, and get a 6-approximation to this subproblem.

(c) We partition the input data of the unrestricted WDS-UDG into cells of size $m\mu \times m\mu$ for constant m. We apply the 6-approximation algorithm of result (b) above to each cell (which requires a second partition), and then apply the shifting technique to get a $(6 + \varepsilon)$-approximation to the original problem.

We will present the proof of part (a) in Sections 4.4.1 and 4.4.2, that of part (b) in Section 4.4.3, and that of part (c) in Section 4.4.4.

4.4.1 A Weighted Covering Problem

To prepare for the first result (a) above, we first study a weighted unit disk covering problem (see Figure 4.12).

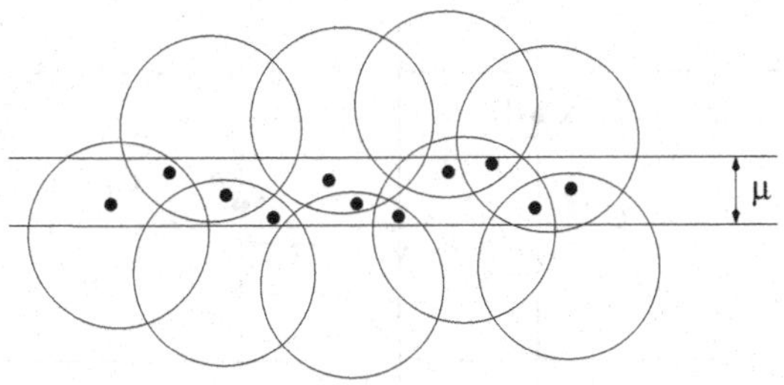

Figure 4.12: A weighted unit disk covering problem.

> WEIGHTED UNIT DISK COVERING (WUDC): Given a set P of points lying inside a horizontal strip of width μ, a set $\mathcal{D}$ of disks with radius 1 and centers lying outside the strip, and a weight function $c : \mathcal{D} \to \mathbb{R}^+$, find a minimum-weight subset $\mathcal{C} \subseteq \mathcal{D}$ of disks that cover all points in P.

The problem WUDC can be solved in polynomial time by dynamic programming.

Theorem 4.12 *The minimum-weight covering $\mathcal{C}$ for the problem* WUDC *can be computed in time $O(m^4 n)$, where $n = |P|$ and $m = |\mathcal{D}|$.*

Proof. Let $p_1, p_2, \ldots, p_n$ be all the points in P, ordered from left to right. For each $i = 1, 2, \ldots, n$, let L_i be the vertical line that passes through point p_i. We call a disk D in $\mathcal{D}$ an *upper disk* (or, a *lower disk*) if the center of D lies above (or, respectively, below) the strip. For the simplicity of the description, we add two dummy disks to $\mathcal{D}$; that is, the two boundary lines of the strip are considered as disks of weight zero, with the upper boundary an upper disk, and the lower boundary a lower disk. Note that these two dummy disks do not cover any point in P, but they always intersect line L_i for any $i = 1, 2, \ldots, n$. For any disk $D \in \mathcal{D}$ having a nonempty intersection with L_i, let $int(L_i, D)$ denote the lowest (or, highest) point in $L_i \cap D$ if D is an upper (or, respectively, lower) disk.

We will use dynamic programming to find the minimum-weight covering $\mathcal{C}$. This algorithm uses a table T with three parameters. To be more precise, for an integer $i \in \{1, 2, \ldots, n\}$, an upper disk D and a lower disk D', with $D \cup D'$ covering point p_i, we define $T_i(D, D')$ to be the set of disks with the minimum weight satisfying the following conditions:

(1) Disks in $T(D, D')$ cover points $p_1, \ldots, p_i$.

(2) D and D' are used to cover some points in $\{p_1, \ldots, p_i\}$ unless D or D' is a dummy disk.

(3) The intersection point $int(L_i, D)$ is the lowest one among all intersection points of L_i with upper disks in $T_i(D, D')$; and the intersection point $int(L_i, D')$ is the highest one among all intersection points of L_i with lower disks in $T_i(D, D')$.

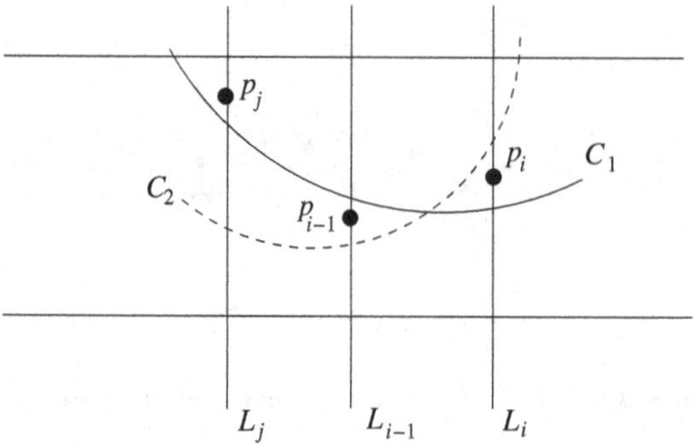

Figure 4.13: If C_2 is no lower than C_2 on line L_i and is no higher than C_1 on line L_{i-1}, then it cannot be higher than C_1 on L_j.

Let $c(T_i(D,D'))$ be the total weight of disks in $T_i(D,D')$, and $A_i(D,D') = \{(D_1,D_2) \mid D_1$ is an upper disk in $\mathcal{D}$, D_2 is a lower disk in $\mathcal{D}$, $int(L_i,D_1)$ is no lower than $int(L_i,D)$, and $int(L_i,D_2)$ is no higher than $int(L_i,D')\}$. In the following, we write, for any predicate Q, $[Q]$ to denote the truth value of Q; that is, $[Q]=1$ if Q is true, and $[Q]=0$ if Q is false. We claim that $c(T_i(D,D'))$ satisfies the following recurrence relation:

$$c(T_i(D,D')) = \min_{\substack{(D_1,D_2)\\ \in A_i(D,D')}} \{c(T_{i-1}(D_1,D_2)) + [D_1 \neq D]\, c(D) + [D_2 \neq D']\, c(D')\}. \tag{4.2}$$

Before we show the claim, we first observe a simple property between two upper (or, lower) disks (see Figure 4.13):

Property 4.13 *For any two upper disks C_1, C_2, of which C_1 is not a dummy disk, and for $1 \le j < i-1$, it is not possible that*

(i) $int(L_j,C_1)$ is lower than $int(L_j,C_2)$,

(ii) $int(L_{i-1},C_1)$ is no lower than $int(L_{i-1},C_2)$, and

(iii) $int(L_i,C_1)$ is no higher than $int(L_i,C_2)$.

A similar property holds for lower disks C_1, C_2 if C_1 is not a dummy disk.

Now we prove the claim. Let D_1 be the upper disk in $T_i(D,D')$ with the lowest intersection point $int(L_{i-1},D_1)$ among upper disks in $T_i(D,D')$, and D_2 the lower disk in $T_i(D,D')$ with the highest intersection point $int(L_{i-1},D_2)$ among lower disks in $T_i(D,D')$. Clearly, $D_1 \cup D_2$ covers p_{i-1}. Moreover, if D covers a point p_j for some $j < i-1$, then, by Property 4.13, D_1 must also cover p_j (note that D covers p_j and so is not a dummy disk). Similarly, if D' covers a point p_j for $j < i-1$, then D_2 must also cover p_j. Therefore, $(T_i(D,D') \setminus \{D,D'\}) \cup \{D_1,D_2\}$ covers points $p_1,\ldots,p_{i-1}$, and so is a candidate for $T_{i-1}(D_1,D_2)$ (i.e., they satisfy conditions (1), (2), and (3) with respect to $i-1$). It follows that

$$c(T_i(D,D')) - [D_1 \neq D]\, c(D) - [D_2 \neq D']\, c(D') \geq c(T_{i-1}(D_1,D_2)),$$

and so

$$c(T_i(D,D')) \geq \min_{\substack{(D_1,D_2)\\ \in A_i(D,D')}} \big\{ c(T_{i-1}(D_1,D_2)) + [D_1 \neq D]\, c(D) + [D_2 \neq D']\, c(D') \big\}.$$

Next, to show the "$\leq$" part of the recurrence relation (4.2), assume that the minimum value of the right-hand side of (4.2) is achieved at $(D_1^*, D_2^*) \in A_i(D,D')$; that is,

$$\begin{aligned} & c(T_{i-1}(D_1^*,D_2^*)) + [D_1^* \neq D]\, c(D) + [D_2^* \neq D']\, c(D') \\ &\quad = \min_{\substack{(D_1,D_2)\\ \in A_i(D,D')}} \big\{ c(T_{i-1}(D_1,D_2)) + [D_1 \neq D]\, c(D) + [D_2 \neq D']\, c(D') \big\}. \end{aligned}$$

Further assume that $T_{i-1}(D_1^*, D_2^*)$ contains the smallest number of disks among these minimum pairs (D_1^*, D_2^*). Then, it must be true that, for every upper disk C in $T_{i-1}(D_1^*, D_2^*)$, the intersection point $int(L_i, C)$ with L_i is no lower than $int(L_i, D)$. To see this, suppose, by way of contradiction, that there exists an upper disk $C \in T_{i-1}(D_1^*, D_2^*)$ having a lower intersection point $int(L_i, C)$ with L_i than $int(L_i, D)$. Since $(D_1^*, D_2^*) \in A_i(D, D')$, $int(L_i, D_1^*)$ is no lower than $int(L_i, D)$. So, $C \neq D_1^*$, and C must cover a point p_j for some $j < i-1$ that is not covered by D_1^* (otherwise, C can be deleted and it violates the minimality assumption about $T_{i-1}(D_1^*, D_2^*)$). However, it means that the pair (C, D_1^*) of upper disks satisfies the three conditions of Property 4.13, which is a contradiction (note that C covers p_j and so is not a dummy disk). Similarly, we can see that the intersection point $int(L_i, C')$ of every lower disk C' in $T_{i-1}(D_1^*, D_2^*)$ is no higher than $int(L_i, D')$. The above shows that the set $T_{i-1}(D_1^*, D_2^*) \cup \{D, D'\}$ satisfies conditions (1)—(3), and so is a candidate for $T_i(D, D')$.

In addition, we note that if $D \in T_{i-1}(D_1^*, D_2^*)$, then D must be identical to D_1^*, for otherwise D would cover a point p_j for some $j < i-1$ that is not covered by D_1^*, and the pair (D, D_1^*) would satisfy the three conditions of Property 4.13. Similarly, if $D' \in T_{i-1}(D_1^*, D_2^*)$, then D' must be identical to D_2^*. Together, we get

$$\begin{aligned} c(T_i(D,D')) &\leq c(T_{i-1}(D_1^*,D_2^*) \cup \{D,D'\}) \\ &= c(T_{i-1}(D_1^*,D_2^*)) + [D_1^* \neq D]\, c(D) + [D_2^* \neq D']\, c(D') \\ &= \min_{\substack{(D_1,D_2)\\ \in A_i(D,D')}} \big\{ c(T_{i-1}(D_1,D_2)) + [D_1 \neq D]\, c(D) + [D_2 \neq D']\, c(D') \big\}, \end{aligned}$$

and the proof of (4.2) is complete.

The recursive formula (4.2) induces a dynamic programming algorithm that computes all $c(T_i(D, D'))$ in time $O(nm^4)$, since the table size is $O(nm^2)$ and each entry $c(T_i(D, D'))$ can be computed from formula (4.2) in time $O(m^2)$. Finally, the minimum-weight disk cover $\mathcal{C}$ for $p_1, \ldots, p_n$ can be computed from $c(T_n(D, D'))$, over all possible $D, D' \in \mathcal{D}$, in time $O(m^2)$. □

Note that, in a unit disk graph, a vertex v dominates a vertex w if and only if the distance between v and w is at most 1. Thus, the dominating set problem in a unit disk graph can be transformed into the covering problem with disks of radius 1. In particular, Theorem 4.12 gives us the following result about a special subproblem of WDS-UDG, which will be used in the next subsection.

Corollary 4.14 *The subproblem of* WDS-UDG *that, for a given unit disk graph $G = (V, E)$ and a given strip of width $\mu = 1/\sqrt{2}$, asks for a minimum-weight set D of vertices satisfying properties (i) and (ii) below can be solved in time $O(n^5)$:*

(i) D dominates all vertices lying in the strip; and

(ii) All vertices in D lie outside the strip.

4.4.2 A 2-Approximation for WDS-UDG on a Small Cell

Now we consider the problem WDS-UDG restricted to a single cell. Let $\mu = 1/\sqrt{2}$, and consider a cell e of size $\mu \times \mu$. Let $V(e)$ denote the set of vertices in V lying in e, and $V^+(e)$ the set of vertices v in V such that v dominates some vertex in e; i.e., $V^+(e) = \{v \in V \mid v \text{ lies in } e \text{ or is adjacent to some } w \in V(e)\}$. The subproblem of WDS-UDG on a single cell e can be stated as follows:

> WDS-UDG$_1$: Given a unit disk graph $G = (V, E)$ with weight $c : V \to \mathbb{R}^+$, and a cell e of size $\mu \times \mu$, find a minimum-weight subset of $V^+(e)$ that dominates $V(e)$.

In this subsection, we show the following result:

Theorem 4.15 *There is a polynomial-time* 2*-approximation for* WDS-UDG$_1$.

To prove Theorem 4.15, let $D^*(e)$ be a minimum-weight dominating set for $V(e)$ and, for any set $U \subseteq V$, let $c(U)$ denote the total weight of set U. We consider two cases.

Case 1. $D^*(e)$ contains a vertex in $V(e)$. Since the cell e has size $\mu \times \mu$, any single vertex in $V(e)$ dominates all vertices in $V(e)$. Thus, $D^*(e)$ contains a single vertex v, which is of the minimum weight among all vertices in $V(e)$. It is easy to find this vertex in linear time.

Case 2. $D^*(e) \subseteq V^+(e) \setminus V(e)$. In this case, we will apply the algorithm of Corollary 4.14 to get a 2-approximation of $D^*(e)$ in polynomial time.

Although we do not know whether $D^*(e)$ belongs to Case 1 or Case 2 above, we can simply choose, from the two solutions obtained in the above two cases, the one with the smaller weight, and it must be a 2-approximation to $D^*(e)$.

In the following, we focus on Case 2. Let A, B, C, D be the four corners of cell e, and divide the area outside e into eight subareas, as shown in Figure 4.14. Also, let

$$\overline{N} = NW \cup CN \cup NE, \qquad \overline{S} = SW \cup CS \cup SE,$$
$$\overline{W} = NW \cup CW \cup SW, \qquad \overline{E} = NE \cup CE \cup SE.$$

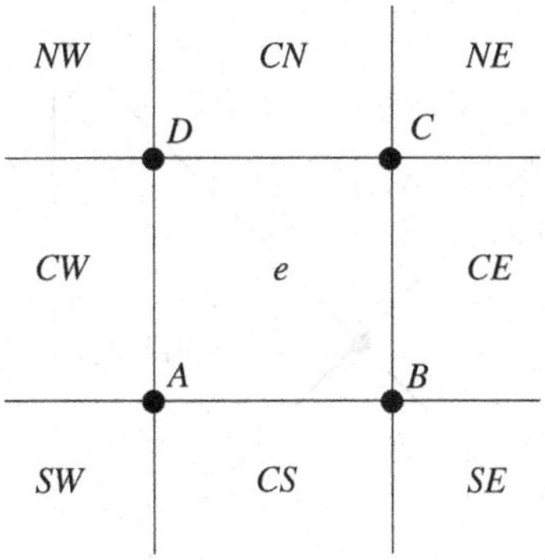

Figure 4.14: The area outside e is divided into eight subareas.

We say V_1 and V_2 form a *feasible partition* of set $V(e)$ if $V(e) = V_1 \cup V_2$, $V_1 \cap V_2 = \emptyset$, every vertex in V_1 is dominated by some vertex in $D^*(e)$ that lies in the area $\overline{N} \cup \overline{S}$, and every vertex in V_2 is dominated by some vertex in $D^*(e)$ that lies in the area $\overline{W} \cup \overline{E}$.

Suppose we are given a feasible partition (V_1, V_2) of $V(e)$; then we can apply the algorithm of Corollary 4.14 to find the minimum-weight subsets $D_1 \subseteq V^+(e) \cap (\overline{N} \cup \overline{S})$ and $D_2 \subseteq V^+(e) \cap (\overline{W} \cup \overline{E})$ that dominate vertices in V_1 and V_2, respectively. Then, $c(D_1) \leq c(D^*(e))$ and $c(D_2) \leq c(D^*(e))$. It follows that $D_1 \cup D_2$ is a 2-approximation to WDS-UDG$_1$.

Following this idea, we will develop, in the following, an algorithm that generates up to $|V(e)|^4$ different partitions of set $V(e)$ such that one of these partitions is a feasible partition. From these partitions, we can find a 2-approximation to WDS-UDG$_1$ in Case 2 by computing the optimal solutions $D_1 \cup D_2$ for each partition (V_1, V_2) of $V(e)$ and then taking the solution with the minimum weight.

For any vertex $p \in V(e)$, draw two straight lines $L_1(p)$ and $L_{-1}(p)$ passing through point p and having slopes 1 and -1, respectively. These two lines meet the boundary of the square $\square ABCD$ at an angle of 45° and divide the square $\square ABCD$ into four parts. We call them $\Delta_N(p)$, $\Delta_S(p)$, $\Delta_W(p)$, and $\Delta_E(p)$, according to their location relative to point p (see Figure 4.15).

Lemma 4.16 *If p is dominated by a vertex u in the area CS (CW, CN, or CE), then every point in the area $\Delta_S(p)$ ($\Delta_W(p)$, $\Delta_N(p)$, or $\Delta_E(p)$, respectively) is dominated by u.*

Proof. Since $\Delta_S(p)$ is a convex polygon, it suffices to show that the distance from u to every corner vertex of $\Delta_S(p)$ is at most 1.

Suppose v is a corner vertex of $\Delta_S(p)$ on line $\overline{BC}$ (cf. Figure 4.16). Draw a line L' that is perpendicular to line $\overline{pv}$ and divides $\overline{pv}$ evenly. Let $d(x, y)$ denote the distance between two points x and y. If u and v lie on the same side of line L' or if u lies on L', then we have $d(u, v) \leq d(u, p) \leq 1$. Otherwise, if u and p lie on the same side of line L', then we have $\angle uvp < \pi/2$ and, hence, $\angle uvB >$

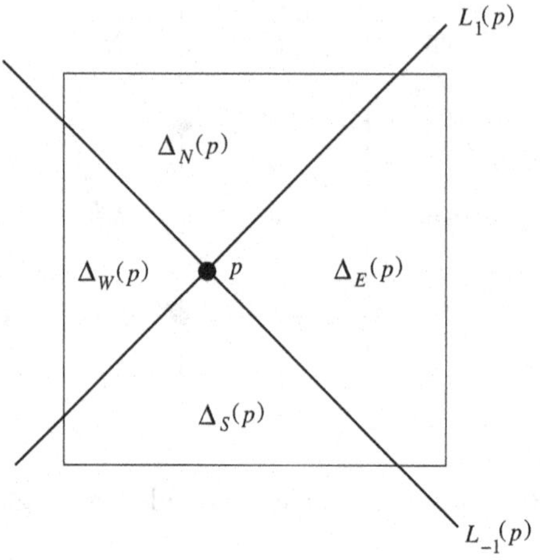

Figure 4.15: $L_1(p)$ and $L_{-1}(p)$ divide e into four parts.

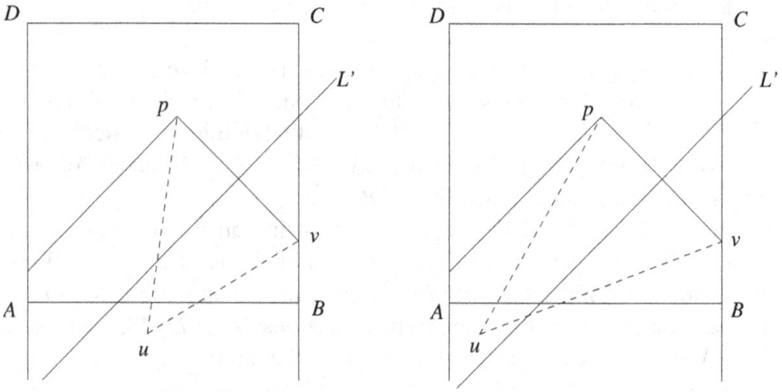

Figure 4.16: $\Delta_S(p)$ is dominated by u.

$\pi/4$ because $\angle pvC = \pi/4$. It follows that $d(u, v) \leq length(\overline{AB})/\sin(\angle uvB) < \mu/\sin(\pi/4) = 1$.

For the cases where the vertex v of $\Delta_S(p)$ lies on line $\overline{AB}$ or $\overline{AD}$, the proofs are similar. □

Next, consider two vertices $p, p' \in V(e)$. Suppose p lies to the left of p' or on the same vertical line as p'. We define $\Delta_S(p, p')$ as follows: If $\Delta_S(p') \subseteq \Delta_S(p)$, then $\Delta_S(p, p') = \Delta_S(p)$, and if $\Delta_S(p) \subseteq \Delta_S(p')$, then $\Delta_S(p, p') = \Delta_S(p')$. Otherwise, let p'' be the intersection point of lines $L_1(p)$ and $L_{-1}(p')$, and define $\Delta_S(p, p') = \Delta_S(p'')$ (see Figure 4.17). The area $\Delta_N(p, p')$ is defined in a similar way.

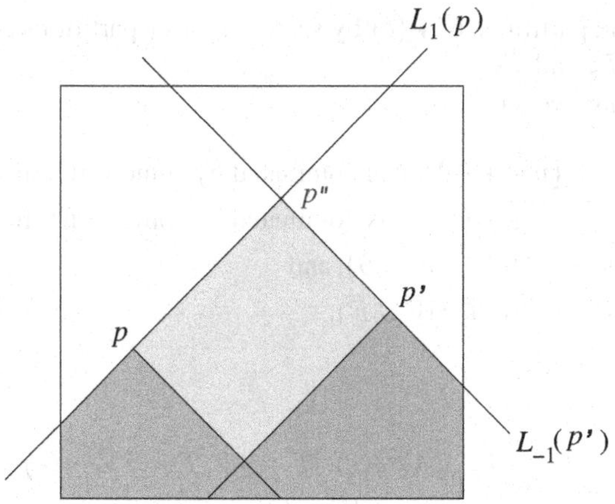

Figure 4.17: $\Delta_S(p, p')$.

Lemma 4.17 *Let K be a subset of $V^+(e) \setminus V(e)$ that dominates all vertices in $V(e)$. Assume that $p, p' \in V(e)$ and p lies to the left of p' or on the same vertical line as p'. If both p and p' are dominated by some vertices in $K \cap CS$ (or, if both are dominated by some vertices in $K \cap CN$), but neither p nor p' is dominated by any vertex in $K \cap (CW \cup CE)$, then every vertex in $\Delta_S(p, p')$ (or, respectively, $\Delta_N(p, p')$) is dominated by some vertex in $K \cap (\overline{N} \cup \overline{S})$.*

Proof. By Lemma 4.16, it suffices to consider a vertex v lying in $\Delta_S(p, p') \setminus (\Delta_S(p) \cup \Delta_S(p'))$. For the sake of contradiction, suppose v is dominated by a vertex u in $K \cap (CW \cup CE)$. If $u \in CW$, then p lies in $\Delta_W(v)$ and so, by Lemma 4.16, is dominated by u, which is a contradiction. If $u \in CE$, we can get a similar contradiction. □

In general, for a set $T \subseteq V(e)$ with $|T| \leq 2$, we define $\Delta_S(T) = \emptyset$ if $T = \emptyset$, $\Delta_S(T) = \Delta_S(p)$ if $T = \{p\}$, and $\Delta_S(T) = \Delta_S(p, p')$ if $T = \{p, p'\}$ and p lies to the left of p' or on the same vertical line as p'. We define $\Delta_N(T)$ in a similar way for a subset $T \subseteq V(e)$ with $|T| \leq 2$.

Let U_{CS} be the set of all vertices v in $V(e)$ such that v is dominated by some vertex in $D^*(e) \cap CS$, but not dominated by any vertex in $D^*(e) \cap (CW \cup CE)$. Choose p_ℓ to be the point in U_{CS} with the leftmost $L_1(p_\ell)$, and p_r to be the point in U_{CS} with the rightmost $L_{-1}(p_r)$. Similarly, let U_{CN} be the set of all vertices v in $V(e)$ such that v is dominated by some vertex in $D^*(e) \cap CN$, but not dominated by any vertex in $D^*(e) \cap (CW \cup CE)$. Choose q_ℓ to be the point in U_{CN} with the leftmost $L_{-1}(q_\ell)$, and q_r to be the point in U_{CN} with the rightmost $L_1(q_r)$.

Define $T_S = \{p_\ell, p_r\}$ if $U_{CS} \neq \emptyset$, and $T_S = \emptyset$ otherwise; and $T_N = \{q_\ell, q_r\}$ if $U_{CN} \neq \emptyset$, and $T_N = \emptyset$ otherwise. By Lemma 4.17, every vertex in $V_1(e) = \Delta_S(T_S) \cup \Delta_N(T_N)$ is dominated by $D^*(e) \cap (\overline{N} \cup \overline{S})$, and every vertex in $V_2(e) = V(e) \setminus V_1(e)$ is dominated by $D^*(e) \cap (\overline{W} \cup \overline{E})$. In other words, the partition $(V_1(e), V_2(e))$ is a feasible partition of $V(e)$. This observation suggests that we search for feasible partitions of $V(e)$ by searching over partitions corresponding to all possible sets T_S and T_N.

In the following, we let

$$\begin{aligned}
V_{CS}(e) &= \{v \in V(e) \mid v \text{ is dominated by some vertex in } CS\},\\
V_{CN}(e) &= \{v \in V(e) \mid v \text{ is dominated by some vertex in } CN\},\\
V_1^+(e) &= V^+(e) \cap (\overline{N} \cup \overline{S}), \text{ and}\\
V_2^+(e) &= V^+(e) \cap (\overline{W} \cup \overline{E}).
\end{aligned}$$

In addition, let

$$\mathcal{T}(e) = \{(T_S, T_N) \mid T_S \subseteq V_{CS}(e), |T_S| \le 2, T_N \subseteq V_{CN}(e), |T_N| \le 2\}.$$

Algorithm 4.D (*2-Approximation for* WDS-UDG$_1$)

Input: A cell e, sets $V(e)$, $V^+(e)$, a weight function $c : V^+(e) \to \mathbb{R}^+$.

(1) $u \leftarrow \text{argmin}_{v \in V(e)} c(v)$; $D \leftarrow \{u\}$.

(2) **For** each pair $(T_S, T_N) \in \mathcal{T}(e)$ **do**

(2.1) $V_1(e) \leftarrow \Delta_S(T_S) \cup \Delta_N(T_N)$;

(2,2) $V_2(e) \leftarrow V(e) \setminus V_1(e)$;

(2.3) $D_1 \leftarrow$ the minimum-weight subset of $V_1^+(e)$ that dominates $V_1(e)$ (by Corollary 4.14);

(2.4) $D_2 \leftarrow$ the minimum weight subset of $V_2^+(e)$ that dominates $V_2(e)$ (by Corollary 4.14);

(2.5) **if** $c(D) > c(D_1 \cup D_2)$ **then** $D \leftarrow D_1 \cup D_2$.

(3) Output D. ■

It is clear that step (2) is executed $O(|V(e)|^4)$ times and so, by Corollary 4.14, Algorithm 4.D runs in time $O(n^9)$.

Next, to estimate the performance of Algorithm 4.D, we note that if $D^*(e) \cap V(e) \neq \emptyset$, then $c(D) = c(D^*(e))$. On the other hand, if $D^*(e) \cap V(e) = \emptyset$, then for the sets T_S and T_N defined by points p_ℓ, p_r and q_ℓ, q_r that are chosen based on $D^*(e)$, we have

$$c(D_1) \le c(D^*(e) \cap (\overline{N} \cup \overline{S})) \quad \text{and} \quad c(D_2) \le c(D^*(e) \cap (\overline{W} \cup \overline{E})).$$

Therefore,

$$c(D) \le c(D_1 \cup D_2) \le 2c(D^*(e)).$$

This completes the proof of Theorem 4.15.

4.4.3 A 6-Approximation for WDS-UDG on a Large Cell

We first note that the 2-approximation to WDS-UDG$_1$ gives us immediately a 28-approximation to WDS-UDG (see Exercise 4.13). The performance ratio 28 is, however, too large, and we now proceed to improve this approximation algorithm. The main idea is to combine sets $V_1(e)$ of the cells along a horizontal strip and combine sets $V_2(e)$ of the cells along a vertical strip, and work on them together. This approach unfortunately only works for the graphs lying in a square of a fixed size. More precisely, we will develop the following result in this subsection.

Theorem 4.18 *For any constant $m > 0$, the subproblem of* WDS-UDG *restricted to input graphs that lie in a square of size $m\mu \times m\mu$ has a 6-approximation that runs in time $n^{O(m^2)}$.*

In the following, we assume that the input unit disk graph $G = (V, E)$ lies in the interior of a square Q of size $m\mu \times m\mu$, for some constant $m > 0$. Divide the square Q into m^2 cells with each cell of size $\mu \times \mu$. Let C be the set of the cells in Q. We collect cells in C whose lower edges lie on the same horizontal line as a *horizontal strip*, and collect the cells in C whose left edge lie on the same vertical line as a *vertical strip*. We let $H_1, H_2, \ldots, H_m$ denote all horizontal strips, and $Y_1, Y_2, \ldots, Y_m$ all vertical strips.

Intuitively, our approximate solution consists of three parts:

(1) For some cells e, we use a single vertex in e to dominate all vertices in e (like in case 1 of the Section 4.4.2).

(2) For other cells e, we get a feasible partition $(V_1(e), V_2(e))$ of $V(e)$. Then we combine sets $V_1(e)$ over all cells in a horizontal strip H_i and apply the algorithm of Corollary 4.14 to get a minimum-weight dominating set for vertices in $\bigcup_{e \in H_i} V_1(e)$.

(3) We combine sets $V_2(e)$ over all cells in a vertical strip Y_i and get a minimum-weight dominating set for vertices in $\bigcup_{e \in Y_i} V_2(e)$.

To see how this works, let us first analyze how an optimal solution can be converted to such a feasible approximate solution. Let Δ^* be a minimum-weight dominating set for G, and *opt* be its total weight; that is, $opt = c(\Delta^*)$. For each cell e, let $\Delta^+(e)$ denote the set of vertices $u \in \Delta^*$ that dominates some vertices in $V(e)$.

Recall some notations used in the last subsection: For each cell e, we let $\overline{N}$ be the area above the upper edge of e, $\overline{S}$ the area below the lower edge of e, $\overline{W}$ the area to the left of the left edge of e, and $\overline{E}$ the area to the right of the right edge of e. Let $V_1^+(e) = V^+(e) \cap (\overline{N} \cup \overline{S})$ and $V_2^+(e) = V^+(e) \cap (\overline{W} \cup \overline{E})$. Also, let $\Delta_1^+(e) = \Delta^+(e) \cap (\overline{N} \cup \overline{S})$ and $\Delta_2^+(e) = \Delta^+(e) \cap (\overline{W} \cup \overline{E})$.

Now we convert Δ^* into a feasible approximate solution. For part (1), let $C_1 = \{e \in C \mid e \cap \Delta^* \neq \emptyset\}$, and, for any $e \in C_1$, let v_e be the vertex in $\Delta^* \cap e$ of the lowest weight. Then v_e dominates all vertices in $V(e)$.

For part (2), we first let $U = \{v_e \mid e \in C_1\}$ and $Z_U = \{v \in V \mid v$ is dominated by some $v_e \in U\}$. Then for each $e \in C - C_1$, we find a feasible partition

$(V_1(e), V_2(e))$ of $V(e) \setminus Z_U$ by finding points p_ℓ, p_r, q_ℓ, q_r and sets T_S and T_N according to $\Delta^+(e)$, as described in the last subsection. Now, we define, for each $i = 1, 2, \ldots, m$, a subset $V_{H_i} = \bigcup_{e \in (C-C_1)\cap H_i} V_1(e)$. Note that the set

$$\Delta_1^+(H_i) = \Big(\bigcup_{e\in(C-C_1)\cap H_i} \Delta_1^+(e) \Big) \setminus U$$

dominates V_{H_i}.

Similarly, for part (3), we define, for each $i = 1, 2, \ldots, m$, a set $V_{Y_i} = \bigcup_{e\in(C-C_1)\cap Y_i} V_2(e)$, and observe that set

$$\Delta_2^+(Y_i) = \Big(\bigcup_{e\in(C-C_1)\cap Y_i} \Delta_2^+(e) \Big) \setminus U$$

dominates V_{Y_i}.

In summary, we divide, in the above, set V into the following (mutually disjoint) parts:

(i) $V(e)$, for $e \in C_1$, and Z_U;

(ii) V_{H_i}, for $i = 1, 2, \ldots, m$; and

(iii) V_{Y_i}, for $i = 1, 2, \ldots, m$.

For each part, a subset of Δ^* has been identified that dominates that part; namely, U dominates $\bigcup_{e\in C_1} V(e)$ and Z_U, $\Delta_1^+(H_i)$ dominates V_{H_i}, and $\Delta_2^+(Y_i)$ dominates V_{Y_i}.

The above analysis suggests that we can divide the original problem into the following three types of subproblems and use the union of the solutions to these subproblems as the approximate solution to the original problem:

A_1: For each $e \in C_1$, fix a vertex $v \in V(e)$ to dominate $V(e)$.

A_2: For each $i = 1, 2, \ldots, m$, find a minimum-weight subset of $V_1^+(H_i) = \bigcup_{e\in(C-C_1)\cap H_i} V_1^+(e)$ that dominates V_{H_i}.

A_3: For each $i = 1, 2, \ldots, m$, find a minimum-weight subset of $V_2^+(Y_i) = \bigcup_{e\in(C-C_1)\cap Y_i} V_2^+(e)$ that dominates V_{Y_i}.

We note that all these subproblems can be solved in polynomial time. In particular, each subproblem of A_2 and A_3 can be solved by the algorithm of Corollary 4.14. The only problem we have now is that all the above subproblems are defined assuming we know what sets C_1 and U are. Since C_1 and U are defined from the optimal solution Δ^*, this assumption is too strong. Instead, we will work on all possible subsets C_1 of C and all possible sets U; that is, we will solve the subproblems for all C_1 and U and use the solution of the minimum weight as the approximate solution.

To prepare for the presentation of the complete algorithm, we need some more notations. First, let C' be the set of all nonempty cells; that is, $C' = \{e \in C \mid V(e) \neq \emptyset\}$. For any $C_1 \subseteq C'$, let $\mathcal{U}_{C_1}$ be the collection of all sets U that contain exactly one vertex in each cell $e \in C_1$.

Next, let $\mathcal{T}(e)$ be the set of all possible choices of pairs (T_S, T_N), where $T_S \subseteq V_{CS}(e) \setminus Z_U$, $|T_S| \leq 2$, and $T_N \subseteq V_{CN}(e) \setminus Z_U$, $|T_N| \leq 2$. Now, for each $C_1 \subseteq C$, let $\mathcal{T}_{C_1}$ be the Cartesian product of $\mathcal{T}(e)$ over all $e \in C - C_1$; that is, if cells in $C - C_1$ are $e_1, e_2, \ldots, e_k$, then $\mathcal{T}_{C_1} = \{(T_{S,1}, T_{N,1}, T_{S,2}, T_{N,2}, \ldots, T_{S,k}, T_{N,k}) \mid (T_{S,j}, T_{N,j}) \in \mathcal{T}(e_j), j = 1, 2, \ldots, k\}$.

Algorithm 4.E (6-*Approximation for* WDS-UDG *on a Large Cell*)
Input: A unit disk graph $G = (V, E)$ on a square Q of size $m\mu \times m\mu$.

(1) **For** each $C_1 \subseteq C'$ **do**

(1.1) **for** each $U \in \mathcal{U}_{C_1}$ **do** find $A(C_1, U)$;

(1.2) let $U^*(C_1) \leftarrow \operatorname{argmin}_{U \in \mathcal{U}_{C_1}} c(A(C_1, U))$.

(2) Let $C^* \leftarrow \operatorname{argmin}_{C_1 \subseteq C'} c(A(C_1, U^*(C_1)))$.

(3) Output $A \leftarrow A(C^*, U^*(C_1^*))$. ■

In the above, each set $A(C_1, U)$ is computed by the following procedure:

Algorithm *for* Function $A(C_1, U)$:

(1) Let $Z_U \leftarrow \{u \in V \mid u \text{ is dominated by some } v \in U\}$.

(2) **For** each $T \in \mathcal{T}_{C_1}$ **do**

(2.1) **For** each cell $e \in C - C_1$ **do**

let (T_S, T_N) be the pair in T corresponding to cell e;
$V_1(e) \leftarrow \Delta_S(T_S) \cup \Delta_N(T_N)$;
$V_2(e) \leftarrow (V(e) \setminus Z_U) \setminus V_1(e)$;

(2.2) **For** $i \leftarrow 1$ **to** m **do**

$D^*(H_i) \leftarrow$ the minimum-weight subset of $V_1^+(H_i)$ that dominates V_{H_i} (by Corollary 4.14);
$D^*(Y_i) \leftarrow$ the minimum-weight subset of $V_2^+(Y_i)$ that dominates V_{Y_i} (by Corollary 4.14);

(2.3) $D(T) \leftarrow \bigcup_{i=1}^{m} (D^*(H_i) \cup D^*(Y_i))$.

(3) Let $T^* \leftarrow \operatorname{argmin}_{T \in \mathcal{T}_{C_1}} c(D(T))$.

(4) $A(C_1, U) \leftarrow D(T^*) \cup U$. ■

To prove that Algorithm 4.E is a 6-approximation, we claim that the set $A(C_1, U)$ found by Algorithm 4.E, with $C_1 = \{e \in C \mid \Delta^* \cap V(e) \neq \emptyset\}$ and $U = \{v_e \mid e \in C_1\}$, where v_e is the vertex in $\Delta^* \cap V(e)$ with the lowest weight, must have

$$c(A(C_1, U)) \leq c(U) + \sum_{i=1}^{m} c(\Delta_1^+(H_i)) + \sum_{i=1}^{m} c(\Delta_2^+(Y_i)). \tag{4.3}$$

To see this, recall that when we convert the optimal solution Δ^* to a feasible approximate solution, we constructed, for each cell $e \in C - C_1$, a pair (T_S, T_N) of subsets of $V(e)$, and used them to partition $V(e) \setminus Z_U$ into $V_1(e)$ and $V_2(e)$. We observe that in step (2) of the algorithm for function $A(C_1, U)$, we run through all possible $T \in \mathcal{T}_{C_1}$, including the one that contains, corresponding to each e, the pair (T_S, T_N) we obtained in the above conversion. Thus, for this set T, the partitions $(V_1(e), V_2(e))$ of $V(e) \setminus Z_U$ for each $e \in C - C_1$, and hence the sets V_{H_i} and V_{Y_i} for each $i = 1, 2, \ldots, m$, that we found in the algorithm for $A(C_1, U)$ are identical to those we obtained in the conversion. Thus, $c(D^*(H_i)) \leq c(\Delta_1^+(H_i))$ and $c(D^*(Y_i)) \leq c(\Delta_2^+(Y_i))$, for $i = 1, 2, \ldots, m$, and

$$\begin{aligned} c(A(C_1, U)) &\leq c(U) + c(D(T)) \leq c(U) + \sum_{i=1}^{m} c(D^*(H_i)) + \sum_{i=1}^{m} c(D^*(Y_i)) \\ &\leq c(U) + \sum_{i=1}^{m} c(\Delta_1^+(H_i)) + \sum_{i=1}^{m} c(\Delta_2^+(Y_i)). \end{aligned}$$

Next, we notice that $\Delta_1^+(H_i) \cap U = \Delta_2^+(Y_i) \cap U = \emptyset$, and so each vertex $v_e \in U$ is counted at most once on the right-hand side of (4.3). For any other vertex $u \in \Delta^*$, we note that the dominating range of u is a disk of radius 1, and so it can overlap with at most four horizontal strips. Since a vertex u lying in strip H_i cannot appear in $\Delta_1^+(H_i)$, it can appear in at most three different $\Delta_1^+(H_j)$'s. Similarly, a vertex u can appear in at most three different $\Delta_2^+(Y_j)$'s. That is, a vertex $u \in \Delta^*$ can be counted at most six times on the right-hand side of (4.3). It follows that the output A of Algorithm 4.E satisfies

$$c(A) \leq c(A(C_1, U)) \leq c(U) + \sum_{i=1}^{m} c(\Delta_1^+(H_i)) + \sum_{i=1}^{m} c(\Delta_2^+(Y_i)) \leq 6 \cdot opt.$$

Finally, let us estimate the time complexity of Algorithm 4.E. First, for any $C_1 \subseteq C'$ and any $U \in \mathcal{U}_{C_1}$, there are at most $O(n^{4m^2})$ sets $T \in \mathcal{T}_{C_1}$, and each $D^*(H_i)$ and each $D^*(Y_i)$ in step (2.2) of the algorithm for $A(C_1, U)$ can be found in time $O(n^5)$. Therefore, set $A(C_1, U)$ can be found in time $n^{O(m^2)}$. Now, we observe that there are $O(2^{m^2})$ subsets C_1 of C' and, for each $C_1 \subseteq C'$, $\mathcal{U}_{C_1}$ contains at most $n^{O(m^2)}$ sets U. Therefore, the total running time of Algorithm 4.E is $n^{O(m^2)}$. This completes the proof of Theorem 4.18

Corollary 4.19 *The subproblem of* WDS-UDG *that asks, for a given graph $G = (V, E)$, a constant $m > 0$, and an $m\mu \times m\mu$ square S, for a minimum-weight subset*

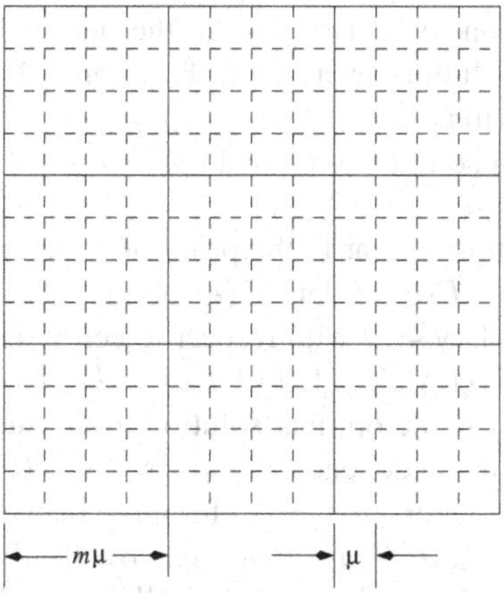

Figure 4.18: Double partition.

of vertices in G that dominate all vertices in S has a 6-approximation with running time $n^{O(m^2)}$, where $n = |V|$.

Proof. The algorithm for this problem is almost identical to Algorithm 4.E, except that we include vertices outside square S in sets $V_1^+(H_i)$ and $V_2^+(Y_i)$ in step (2.2) of the algorithm for $A(C_1, U)$. □

4.4.4 A $(6 + \varepsilon)$-Approximation for WDS-UDG

Now, we apply the double partition and shifting techniques to obtain a $(6 + \varepsilon)$-approximation to WDS-UDG.

Theorem 4.20 *For any $\varepsilon > 0$, there exists a $(6+\varepsilon)$-approximation for* WDS-UDG *with computation time $n^{O(1/\varepsilon^2)}$.*

Proof. Assume that $\varepsilon < 1$, and let $m = \lceil 72/\varepsilon \rceil$. Without loss of generality, assume that G lies in a square

$$Q = \{(x, y) \mid m\mu \le x < km\mu,\ m\mu \le y \le km\mu\}$$

for some integer $k > 1$. Let

$$\overline{Q} = \{(x, y) \mid 0 \le x < km\mu,\ 0 \le y \le km\mu\}.$$

We partition $\overline{Q}$ into k^2 cells, each of size $m\mu \times m\mu$. We call this partition $P(0, 0)$ (see Figure 4.18). For each $a \in \{0, 1, \ldots, m-1\}$, partition $P(a, a)$ is the partition $P(0, 0)$ with its lower-left corner shifted to $(a\mu, a\mu)$.

For each partition $P(a, a)$, we solve the problem WDS-UDG for each cell e in $P(a, a)$ by Corollary 4.19. Let this solution be $A_a(e)$. Then $A_a = \bigcup_{e \in P(a,a)} A_a(e)$

is an approximate solution to G. Let A_{a^*} be the one with the minimum weight $c(A_{a^*})$ among all these solutions over $a = 0, 1, \ldots, m-1$. We output $A = A_{a^*}$ as the approximate solution to G.

Let *Opt* be an optimal solution to G, and $opt = c(Opt)$. We claim that $c(A) \leq (6+\varepsilon)opt$.

The proof of the claim is similar to the proof of Theorem 4.2. For any partition $P(a,a)$ and any cell $e \in P(a,a)$, let $\Delta^*(e)$ be the optimal solution to the subproblem defined in Corollary 4.19 with respect to cell e. From Corollary 4.19, we know that $c(A_a(e)) \leq 6 \cdot c(\Delta^*(e))$. Let $Opt(e) = \{u \in Opt \mid u \text{ dominates some } v \in V(e)\}$. Since $\Delta^*(e)$ is the optimal solution to the subproblem on cell e, we have $c(\Delta^*(e)) \leq c(Opt(e))$ and, hence, $c(A_a(e)) \leq 6 \cdot c(Opt(e))$.

A vertex u in *Opt* may belong to $Opt(e)$ for more than one cell e. For any partition $P(a,a)$, let $H_a = \{u \in Opt \mid u \text{ belongs to two cells of two different horizontal strips}\}$, and $Y_a = \{u \in Opt \mid u \text{ belongs to two cells of two different vertical strips}\}$. Note that if u belongs to $Opt(e)$, then the disk D_u with center u and radius 1 has a nonempty intersection with cell e. Therefore, we have

$$c(A_a) \leq \sum_{e \in P(a,a)} c(A_a(e)) \leq 6 \sum_{e \in P(a,a)} c(Opt(e)) \leq 6(opt + c(H_a) + 2 \cdot c(Y_a)).$$

Next, we observe that a vertex u belongs to H_a only if D_u intersects a horizontal grid line of $P(a,a)$. Since the shifting distance is $1/\sqrt{2}$, a disk of radius 1 can intersect horizontal grid lines of at most four partitions. That is, a vertex can belong to at most four different sets H_a. Therefore,

$$\sum_{a=0}^{m-1} c(H_a) \leq 4 \cdot opt.$$

Similarly,

$$\sum_{a=0}^{m-1} c(Y_a) \leq 4 \cdot opt.$$

It follows that

$$\sum_{a=0}^{m-1} c(A_a) \leq 6\Big(m \cdot opt + \sum_{a=0}^{m-1} c(H_a) + 2 \sum_{a=0}^{m-1} c(Y_a)\Big) \leq 6(m+12)opt.$$

So, the minimum solution A_{a^*} has weight

$$c(A_{a^*}) \leq \frac{1}{m} \sum_{a=0}^{m-1} c(A_a) \leq \Big(6 + \frac{72}{m}\Big)opt \leq (6+\varepsilon)opt.$$

Finally, we verify that the total computation time of this algorithm is, from Corollary 4.19, $m \cdot n^{O(m^2)} = n^{O(1/\varepsilon^2)}$. □

This result can be extended to the problem of finding the minimum-weight connected dominating set in a unit disk graph.

> Weighted Connected Dominating Set in a Unit Disk Graph (WCDS-UDG): Given a unit disk graph $G = (V, E)$ with a weight function $c : V \to \mathbb{R}^+$, find a connected dominating set with the minimum total weight.

Theorem 4.21 *For any* $\varepsilon > 0$, *there exists a* $(7 + \varepsilon)$*-approximation for* WCDS-UDG *that runs in time* $n^{O(1/\varepsilon^2)}$.

Proof. We first find a dominating set D for G of total weight

$$c(D) \leq \left(6 + \frac{\varepsilon}{2}\right) c(\Delta^*),$$

where Δ^* is a minimum-weight dominating set for G. Thus, we have reduced the problem to the following subproblem:

> WCDS-UDG$_1$: Given a unit disk graph $G = (V, E)$ with a weight function $c : V \to \mathbb{R}^+$, and a dominating set $D \subseteq V$, find a minimum-weight subset $C \subseteq V - D$ that connects D.

It can be shown that WCDS-UDG$_1$ has a PTAS (see Exercise 4.17). So, we can find a $(1 + \varepsilon/2)$-approximate solution $C \subseteq V - D$ for WCDS-UDG$_1$ and use $C \cup D$ as the solution to graph G for problem WCDS-UDG.

Let C^* be a minimum-weight connected dominating set of G and C' a minimum-weight subset of V that connects D. We verify that $c(C \cup D)$ is a $(7 + \varepsilon)$-approximation to C^*:

First, it is obvious that $c(\Delta^*) \leq c(C^*)$, and so $c(D) \leq (6 + \varepsilon/2) \cdot c(C^*)$. Next, we observe that $C^* \cup D$ is a connected dominating set of G, and so $C^* \setminus D$ is a feasible solution to WCDS-UDG$_1$ for input (G, D). Therefore,

$$c(C) \leq \left(1 + \frac{\varepsilon}{2}\right) c(C') \leq \left(1 + \frac{\varepsilon}{2}\right) c(C^* \setminus D) \leq \left(1 + \frac{\varepsilon}{2}\right) c(C^*).$$

Together, we get $c(C \cup D) \leq (7 + \varepsilon) \cdot c(C^*)$. □

4.5 Tree Partition

Recall the problem Phylogenetic Tree Alignment (PTA) introduced in Section 3.5, where we showed that the optimal lifted alignment is a 2-approximation to PTA. In this section, we use a tree partition to construct a PTAS for PTA. We assume that in the input tree T to the problem PTA, every internal vertex has at least 2, and at most d, children, where d is a constant greater than 2. First, we note that Lemma 3.4 about regular binary trees can be extended to such trees.

Lemma 4.22 *For any rooted tree in which each internal vertex has at least two sons, there exists a mapping f from all internal vertices to leaves such that*

(a) For every internal vertex u, $f(u)$ is a descendant of u; and

(b) All tree paths from u to $f(u)$ are edge-disjoint.

Let T be a tree and $t > 1$ a constant. For each $i = 0, 1, \ldots, t-1$, let V_i be the set of vertices of T in any level j, with $j \equiv i \pmod{t}$. We may consider each set V_i as a partition of T into a collection of small trees of at most $t+1$ levels. To be more precise, each small tree in the partition is either rooted at the root r of T and containing all vertices in levels $j \le i$, or is rooted at a vertex $v \in V_i$ and containing all descendants of v in levels j, with $level(v) \le j \le level(v) + t$. Thus, each small tree in the partition has at most $t+1$ levels and at most $(d^{t+1} - 1)/(d-1)$ vertices. For such a small tree with $t+1$ levels, its root and leaves of level $t+1$ all belong to V_i.

Suppose T has k leaves with labels $s_1, s_2, \ldots, s_k$, respectively. The problem PTA asks for the labeling of the internal vertices of T with the minimum total alignment scores. We say a tree T' is a *phylogenetic alignment* of T if

(i) T' has the same vertex set and edge set as T,

(ii) Each vertex of T' is labeled with a string, and

(iii) The leaves of T' are labeled with the same strings as those of T.

A phylogenetic alignment tree T' is called *t-restricted* if there is an integer $i \in \{0, 1, \ldots, t-1\}$ such that the label of every vertex v in V_i is the same as the label of a descendant leaf of v.

Lemma 4.23 *Let T be a tree whose leaves have been labeled with strings s_1, s_2, $\ldots$, s_k. For any $t > 0$, there exists a t-restricted, phylogenetic alignment T' of tree T such that*

$$cost(T') \le \left(1 + \frac{3}{t}\right) opt,$$

where $cost(T')$ is the total alignment score of the tree T', and opt is the minimum cost of a phylogenetic alignment of T.

Proof. Let T^* be an optimal phylogenetic alignment of T. Assume that each internal vertex v of T^* is assigned with string s_v^*. Now, for each vertex $v \in V$, let s_v be the label of a descendant leaf of v such that $D(s_v, s_v^*)$ is minimized. Define, for each $i \in \{0, 1, \ldots, t\}$, a phylogenetic alignment T_i of T as follows: For each $v \in V_i$, label it with string s_v, and for any other internal vertex u, label it with s_u^*. Let us estimate the total alignment score $cost(T_i)$ of tree T_i.

For an internal vertex $v \in V_i$, let $\pi(v)$ denote the parent of v and $\Gamma(v)$ denote the set of the children of v. First, we observe that, for any $w \in \Gamma(v)$, s_w is the label of a descendant leaf of w, and so also the label of a descendant leaf of v. It follows that $D(s_v^*, s_v) \le D(s_v^*, s_w)$. Therefore, by the triangle inequality, we have

$$
\begin{aligned}
& D(s^*_{\pi(v)}, s_v) + \sum_{w\in\Gamma(v)} D(s_v, s^*_w) \\
&\quad \le D(s^*_{\pi(v)}, s^*_v) + \sum_{w\in\Gamma(v)} D(s^*_v, s^*_w) + (|\Gamma(v)|+1)\, D(s^*_v, s_v) \\
&\quad \le D(s^*_{\pi(v)}, s^*_v) + \sum_{w\in\Gamma(v)} D(s^*_v, s^*_w) + D(s^*_v, s_v) + \sum_{w\in\Gamma(v)} D(s^*_v, s_w) \\
&\quad \le D(s^*_{\pi(v)}, s^*_v) + 2\sum_{w\in\Gamma(v)} D(s^*_v, s^*_w) + D(s^*_v, s_v) + \sum_{w\in\Gamma(v)} D(s^*_w, s_w).
\end{aligned}
$$

Thus,

$$
\begin{aligned}
& cost(T_i) - cost(T^*) \\
&\quad \le \sum_{v\in V_i}\left[D(s^*_{\pi(v)}, s_v) + \sum_{w\in\Gamma(v)} D(s_v, s^*_w) - D(s^*_{\pi(v)}, s^*_v) - \sum_{w\in\Gamma(v)} D(s^*_v, s^*_w)\right] \\
&\quad \le \sum_{v\in V_i}\left[\sum_{w\in\Gamma(v)} D(s^*_v, s^*_w) + D(s^*_v, s_v) + \sum_{w\in\Gamma(v)} D(s^*_w, s_w)\right] \\
&\quad = \sum_{v\in V_i\cup V_{i+1}} D(s^*_v, s_v) + \sum_{v\in V_i}\sum_{w\in\Gamma(v)} D(s^*_v, s^*_w).
\end{aligned}
$$

It is clear that

$$
\sum_{i=0}^{t-1}\sum_{v\in V_i}\sum_{w\in\Gamma(v)} D(s^*_v, s^*_w) = cost(T^*).
$$

Furthermore, we note that $D(s^*_v, s_v)$ is the minimum of $D(s^*_v, s^*_z)$ over all descendant leaves z of v. Therefore, by the triangle inequality, for any descendant leaf z of v, the total cost of the path in T^* from v to z is at least as large as $D(s^*_v, s_v)$. By Lemma 4.22, there is a function f mapping each internal vertex v to a descendant leaf $f(v)$ of v such that all paths from v to $f(v)$ are edge-disjoint. Let $\Pi(v)$ denote the path in T^* from v to $f(v)$. Then

$$
\sum_{i=0}^{t-1}\sum_{v\in V_i} D(s^*_v, s_v) \le \sum_{v\in V} D(s^*_v, s^*_{f(v)}) \le \sum_{v\in V} cost\big(\Pi(v)\big) \le cost(T^*) = opt.
$$

Together, we have

$$
\sum_{i=0}^{t-1} cost(T_i) \le (t+3)opt.
$$

Therefore, there exists an integer $i \in \{0, 1, \ldots, t-1\}$ such that

$$
cost(T_i) \le \left(1 + \frac{3}{t}\right)opt. \qquad \square
$$

For any fixed integer $t > 0$, the optimal t-restricted phylogenetic alignment of a given tree T can be computed by dynamic programming in time

$$O\big(k^{d^{t-1}+2} n^{d^{t-1}+1}\big),$$

where k is the number of leaves in T and n is the total length of the leave labels (see Exercise 4.20). Therefore, we have a PTAS for the problem PTA.

Theorem 4.24 *For any* $t \geq 3$, *there exists a polynomial-time* $(1+3/t)$*-approximation for the problem* PTA.

There are a number of ways to partition trees to get approximate solutions. The reader may find more examples in the exercises.

Exercises

4.1 Find a necessary and sufficient condition for two points in the plane to have exactly one unit disk with its boundary passing through them.

4.2 Consider the problem of finding the minimum number of d-dimensional balls that cover a given set of n points in the d-dimensional Euclidean space. Show that this problem has a $(1+1/m)$-approximation with running time $O(m^d \cdot n^{2m^d+1})$.

4.3 Show that the problem of finding the minimum vertex cover in a unit disk graph has a PTAS.

4.4 A vertex cover C in a graph G is *connected* if the subgraph induced by C is connected.

(a) Show that the problem of finding the minimum connected vertex cover in a given graph has a polynomial-time 3-approximation.

(b) Show that the problem of finding the minimum connected vertex cover in a unit disk graph has a PTAS.

4.5 Show that for any connected graph G, its minimum dominating set D and minimum connected dominating set C have the following relationship: $|C| \leq 3|D| - 2$.

4.6 Can you find a polynomial-time constant approximation for the problem of finding a minimum-weight connected dominating set in a vertex-weighted unit disk graph?

4.7 An *independent set* of a graph $G = (V, E)$ is a subset $I \subseteq V$ with no edge in E connecting any two vertices in I. An independent set I is *maximal* if there is no other independent set properly contains I. Note that any maximal independent set in a graph G is a dominating set of G.

(a) Design a polynomial-time algorithm to compute a maximal independent set I for a given graph G such that $|I| \geq (|C|+1)/2$, where C is a minimum connected dominating set of G.

(b) Show that, in a unit disk graph G, any maximal independent set I and the minimum connected dominating set C have the relationship $|C| \leq 4|I|+1$.

(c) Use fact (b) above to design a polynomial-time 8-approximation for the problem of finding the minimum connected dominating set in a unit disk graph.

4.8 Consider the subproblem of ESMT with the following restriction:

(R_1) The ratio of the length of the longest edge to the length of the shortest edge in the minimum spanning tree of the terminal points is bounded above by a constant.

Show that there is a PTAS for this subproblem of ESMT.

4.9 Consider the following problem:

RECTILINEAR STEINER MINIMUM TREE WITH RECTILINEAR OBSTRUCTION (RSMTRO): Given a set T of terminal points and a set R of rectilinear rectangles in the rectilinear plane, find the Steiner minimum tree that connects terminals in T and avoids the rectilinear obstructions in R.

Show that the subproblem of RSMTRO with the restriction (R_1) defined in Exercise 4.8 has a PTAS.

4.10 Show that the problem of finding the maximum-weight independent set in a vertex-weighted intersection disk graph has a PTAS.

4.11 Show that the problem of finding the minimum-weight vertex cover in a vertex-weighted intersection disk graph has a PTAS.

4.12 In the proof of Lemma 4.10, consider a different approach in which we do not introduce the notion of relevant cells but use the following more straightforward recursive relation: For a j-cell e and a set I of independent disks in layers $< j$ that intersect e, let $J^* = \text{argmin}_{J \in \text{IND}_j(e,I)} |J \cup A_J|$ and $T(e,I) = J^* \cup A_{J^*}$, where

$$A_J = \bigcup_{e' \in C_{j+1}(e)} T(e', (I \cup J)_{e'}),$$

and $C_{j+1}(e)$ is the set of all cells e' in layer $j+1$ that are contained in e. With this recursive formula, does the corresponding dynamic programming algorithm still run in polynomial time? Justify your answer.

4.13 Show that if we divide the a square into cells of size $(1/\sqrt{2}) \times (1/\sqrt{2})$, then a unit disk can intersect at most 14 cells. Use this fact, together with the 2-approximation to the problem WDS-UDG$_1$, to get a 28-approximation to WDS-UDG.

4.14 Consider the following modification on Algorithm 4.E: We fix, for any $C_1 \subseteq C'$, $\mathcal{U}_{C_1}$ to consist of a single set $U = \{v_e \mid e \in C_1\}$, where v_e is the minimum-weight vertex in cell e. Is Algorithm 4.E still a 6-approximation to WDS-UDG on a square of size $m\mu \times m\mu$?

4.15 Suppose that in the partition for problem WDS-UDG, we use hexagonal

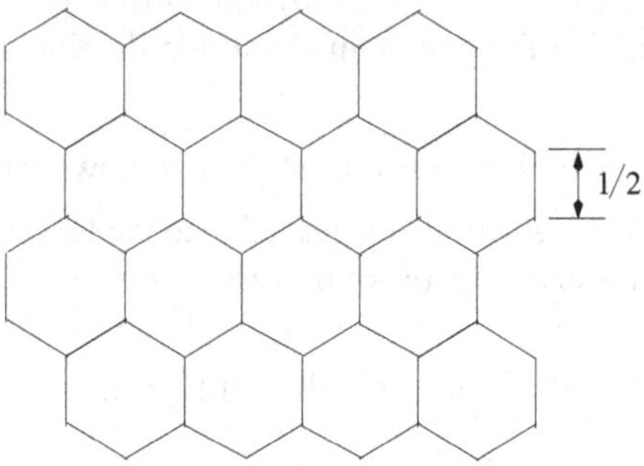

Figure 4.19: Hexagonal cells.

small cells of edge length $1/2$ (see Figure 4.19) instead of square cells of edge length μ. Can you get a polynomial-time approximation with performance ratio smaller than 2 for WDS-UDG$_1$? Can you get an approximation with performance ratio better than $(6 + \varepsilon)$?

4.16 Prove the following results about vertex-weighted unit disk graphs.

(a) Let $G = (V, E)$ be a vertex-weighted unit disk graph. For any vertex subset $U \subseteq V$, if the subgraph of G induced by U is connected, then there is a spanning tree on U with each vertex having degree at most 5.

(b) There exists a 4-approximation to the following problem: Given a weighted unit disk graph $G = (V, E)$ and a set of terminals $P \subseteq V$, find a Steiner tree on P with the minimum total vertex-weight.

4.17 Show that the following problem has a PTAS:

> Given a vertex-weighted unit disk graph $G = (V, E)$ and a dominating set $D \subseteq V$, find the minimum-weight subset $C \subseteq V$ interconnecting D.

4.18 Consider the following problem:

> MAXIMUM INDEPENDENT RECTANGLES (MAX-IR): Given a set of n rectangles in the rectilinear Euclidean plane, find the maximum subset of mutually disjoint rectangles.

(a) Show that the subproblem of MAX-IR with the following restriction has a PTAS:

(R_2) The ratio of the height to the width of every input rectangle is in the range $[a, b]$ for some constants $0 < a < b$.

*(b) Is there a constant approximation for the problem MAX-IR on rectilinear rectangles without the condition (R_2)?

4.19 Let r and s be two integers with $0 \le s < 2^r$, and let $k = 2^r + s$. For each balanced binary tree, consider the following labeling, which assigns each vertex with a set of exactly 2^r integers chosen from $L = \{1, 2, \ldots, r2^r + s\}$:

(1) For each vertex v at the ith level, $0 \le i \le r - 1$, assign v with label set $\{i2^r + 1, i2^r + 2, \ldots, (i+1)2^r\}$. In particular, the root has label set $\{1, 2, \ldots, 2^r\}$, and its two sons have label sets $\{2^r + 1, 2^r + 2, \ldots, 2^{r+1}\}$.

(2) For each $i \ge r$, assume that vertices at levels j, $0 \le j \le i - 1$, have been labeled. Let u be a vertex at level i. The label set for u is defined as follows:

(i) First, let v be the ancestor of u at level $i - r$, and let its label set be $S_v = \{\ell_1, \ell_2, \ldots, \ell_{2^r}\}$. Suppose that u is the jth level-i descendant of v; then, let $S'_u = \{\ell_{t \bmod 2^r} \mid j \le t \le j + 2^r - s - 1\}$.

(ii) Assume that the label sets of the lowest r ancestors of u are $L_1, L_2, \ldots, L_r$. Let the label set of u be

$$S_u = S'_u \cup (L - (L_1 \cup \cdots \cup L_r)).$$

Show that the above labeling induces $r2^r + s$ partitions of a balanced binary tree T such that each partition breaks tree T into smaller trees each of at most k leaves and that each vertex in T appears as a break point in at most 2^r partitions.

4.20 Let T be a tree whose leaves are labelled with strings. Let $t > 0$ be a constant integer and, for each $i \in \{0, 1, \ldots, t-1\}$, let V_i be the set of vertices in T at levels $j \equiv i \pmod{t}$. Design, by dynamic programming, a polynomial-time algorithm to find an optimal t-restricted phylogenetic alignment tree T_i of T with each vertex v in V_i labeled with the same label as one of its descendant leaves.

4.21 Consider the following problem:

VERTEX-WEIGHTED ST: Given a graph $G = (V, E)$ with vertex-weight $c : V \to \mathbb{R}^+$, and a subset P of V, find a Steiner tree interconnecting vertices in P with the minimum total vertex-weight.

For a given graph $G = (V, E)$ with vertex-weight $c : V \to \mathbb{R}^+$, and a set $P \subseteq V$, let $\pi(u, v)$ be the path between vertices u and v with the minimum total weight. Construct a complete graph K on P and assign every edge $\{u, v\}$ with weight equal to $c(\pi(u, v)) - c(u) - c(v)$. Show that the minimum spanning tree of K induces a 4-approximation for VERTEX-WEIGHTED ST in a unit disk graph.

*$\star$***4.22** Is there a constant approximation for the problem of finding the minimum dominating set in an intersection disk graph?

Historical Notes

Partition is a simple idea that has been used in the design of approximation or heuristic algorithms for a long time. Karp [1977] gave the first probabilistic analysis for partition with applications to Euclidean TSP. Komolos and Shing [1985] applied this approach to RSMT. Baker [1983, 1994] and Hochbaum and Maass [1985] introduced the shifting technique to design deterministic PTASs for a family of problems in covering and packing. This technique is used extensively to design PTASs for many problems [Min et al., 2003; Cheng et al., 2003; Zhang, Gao, Wu, and Du, 2009; Hunt et al., 1998, Vavasis, 1991; Wang and Jiang, 1996]. Cheng et al. [2003] gave the first PTAS for the minimum connected dominating set in a unit disk graph. Zhang, Gao, Wu, and Du [2009] provided a simple one that runs faster and can be extended to unit ball graphs in higher-dimensional space.

Erlebach et al. [2001] first introduced the multilayer partition technique to deal with disks with different sizes and with arbitrary squares. The maximum independent set problem in rectangle intersection graphs has interesting applications in map labeling and data mining [Agarwal et al., 1998; Berman et al., 2001; Chan, 2003; Erlebach et al., 2001]. Various partition techniques yield PTASs for this problem under the restriction (R_2) (see Exercise 4.18). For arbitrary rectangles, no constant approximation has been found. The best-known approximation has a performance ratio $O(\log n)$ [Agarwal et al., 1998; Chan, 2004; Khanna et al., 1998; Nielsen, 2000].

Ambühl et al. [2006] used the partition technique to get a polynomial-time 72-approximation for the minimum-weight dominating set and a polynomial-time 84-approximation for the minimum-weight connected dominating set in a unit disk graph. Gao et al. [2008] introduced the double-partition technique and obtained a $(6+\varepsilon)$-approximation for the minimum-weight dominating set and a $(10+\varepsilon)$-approximation for the minimum-weight connected dominating set in a unit disk graph. Dai and Yu [2009] improved the first result to a $(5+\varepsilon)$-approximation. Zou et al. [2008a] improved the second result to a $(9.85+\varepsilon)$-approximation, and Zou et al. [2008b] further lowered the performance ratio to $(6+\varepsilon)$.

Du, Zhang, and Feng [1991] proved a useful lemma for the shifting technique in tree partition when they proved a lower bound for the k-Steiner ratio. With this lemma, Jiang et al. [1994] and Wang et al. [1996] designed a PTAS for the tree alignment problem. Wang et al. [1997] introduced a new partition of balanced binary trees that results in a more efficient PTAS (see Exercise 4.19).

5
Guillotine Cut

It will be as fleeting as a cool breeze
upon the back of one's neck.

— Joseph I. Guillotine

Guillotine cut is a technique of adaptive partition that has found interesting applications in many geometric problems. Roughly speaking, a guillotine cut is a subdivision by a straight line that partitions a given area into at least two subareas. By a sequence of guillotine cut operations, we can partition the input area into smaller areas, solve the subproblems in these smaller areas, and combine these solutions to obtain a feasible solution to the original input.

In some applications of guillotine cut, there may be an exponential number of ways to form a feasible solution from the solutions of the subproblems. A few methods have been developed to reduce the number of ways of combining the solutions of the subproblems and yet still preserve good approximation. In this chapter, we study the technique of guillotine cut and the related methods for combining the solutions of subproblems.

5.1 Rectangular Partition

We start with a geometric problem MIN-RP, which has a number of applications in engineering design, such as process control, layout for integrated circuits, and architectural design.

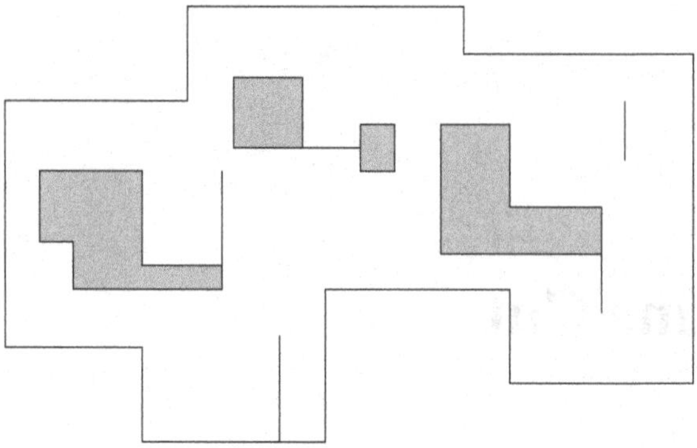

Figure 5.1: A rectilinear polygon with holes.

> MINIMUM EDGE-LENGTH RECTANGULAR PARTITION (MIN-RP): Given a rectilinear polygon, possibly with some rectilinear holes, partition it into rectangles with the minimum total edge length.

In the above definition, by a hole in the input polygon, we mean a rectilinear polygon that may be completely or partially degenerated into a line segment or a point (see Figure 5.1). The existence of holes in an input polygon makes a difference in the polynomial-time solvability of the problem: While the problem MIN-RP, in the general case, is **NP**-hard, the problem MIN-RP for hole-free inputs can be solved in time $O(n^4)$, where n is the number of vertices in the input rectilinear polygon. The polynomial-time algorithm for the hole-free MIN-RP is an application of dynamic programming, based on the following fact.

Lemma 5.1 *Suppose that the input R to* MIN-RP *is hole-free. Then there exists an optimal rectangular partition P for R in which each maximal line segment contains a vertex of the boundary.*[1]

Proof. Consider a rectangular partition P of R with the minimum total length. Suppose P has a maximal line segment $\overline{AB}$ that does not contain any vertex of the boundary. Without loss of generality, let us assume it is a vertical line segment. Then the two endpoints A and B of this line segment must lie on the interior of two horizontal line segments that are in P or in the boundary. Suppose there are r horizontal line segments touching the interior of $\overline{AB}$ from the right, and ℓ horizontal line segments touching the interior of $\overline{AB}$ from the left. We claim that r must be equal to ℓ. Indeed, if $r > \ell$ (or $r < \ell$), then we can move the line segment $\overline{AB}$ to the right (or to the left, respectively) to reduce the total length of the rectangular partition (see Figure 5.2(a)). This contradicts the optimality of P.

Since $r = \ell$, moving $\overline{AB}$ to either the right or left does not increase the total length of P. Let us keep moving $\overline{AB}$ to the left until it is not movable. Then the

[1]A maximal line segment in a partition is one that cannot be extended farther in either direction.

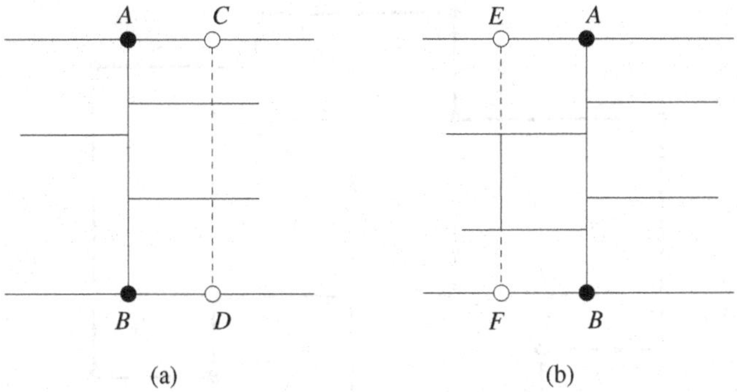

Figure 5.2: (a) Moving $\overline{AB}$ toward $\overline{CD}$ would reduce the total length of P. (b) Moving $\overline{AB}$ to the left and merging it with $\overline{EF}$ would reduce the total length of P.

line segment $\overline{AB}$, in its final position, must meet either a vertex of the boundary or another vertical line segment in P. The latter case is, however, not possible: If $\overline{AB}$ meets another line segment $\overline{EF}$ in P, then they merge into one, and the total length of the rectangular partition is reduced, contradicting the optimality of P again (see Figure 5.2(b)).

This proves that the line segment $\overline{AB}$ can be moved to meet a vertex of the boundary. We perform such movements for all line segments in P that do not contain a boundary vertex, and the resulting partition has the required property and has the same length as P. □

From the above lemma, we can see that there are only $O(n^2)$ candidates for a line segment in an optimal rectangular partition P satisfying the property of Lemma 5.1: Let us define a *grid point* to be the intersection point of any two lines that pass through a boundary vertex. Then one of the endpoints of a maximal line segment in P must be a boundary vertex, and the other one must be either a boundary vertex or a grid point (see Figure 5.3). Therefore, there are $O(n^2)$ such line segments. This observation allows us to design a dynamic programming algorithm to solve the problem MIN-RP without holes in polynomial time (see Exercise 5.1).

When the input rectilinear polygon has holes, the problem MIN-RP becomes **NP**-hard. In the following, we apply the technique of guillotine cut to approximate this problem.

A *guillotine cut* is a straight line that cuts through a connected region and breaks it into at least two subregions. A rectangular partition is called a *guillotine rectangular partition* if it can be constructed by a sequence of guillotine cuts, each cutting through a connected subregion. It is not hard to see that the minimum-length guillotine rectangular partition of a given rectilinear polygon (possibly with holes) can be found in polynomial time. First, it can be proved by the argument similar to that of Lemma 5.1 that there exists a minimum-length guillotine rectangular partition in which every maximal line segment contains a vertex of the boundary. Moreover, the restriction of using only guillotine cuts in each step allows us to apply dynamic

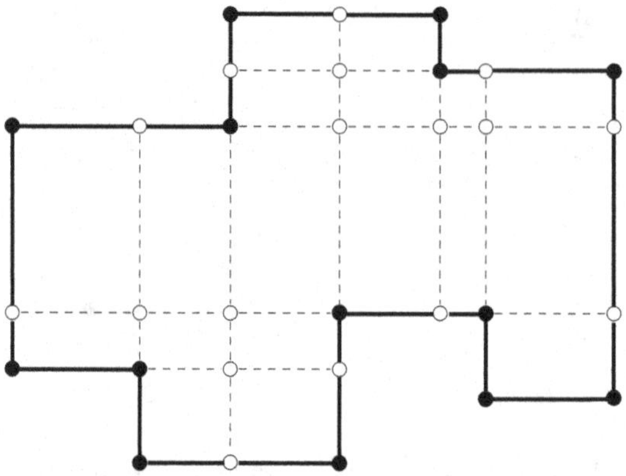

Figure 5.3: The vertices of the boundary (dark circles) and the grid points (white circles) of a rectilinear polygon.

programming to this problem. That is, we can partition a rectilinear polygon R by first using a guillotine cut to break it into two or more smaller rectilinear polygons, and then recursively partition these new rectilinear polygons. In this recursive algorithm, there are, in each iteration, $O(n)$ possible choices for the next guillotine cut. In addition, there are altogether $O(n^4)$ possible subproblems, because each subproblem's boundary is composed of pieces of the input boundary plus at most four guillotine cut edges. Therefore, the minimum-length guillotine rectangular partition can be computed by dynamic programming in time $O(n^5)$.

Since the minimum-length guillotine rectangular partition can be computed in polynomial time, it is natural to try to use it to approximate the problem MIN-RP. What is the performance ratio of this method? Unfortunately, no good bounds of the performance ratio have been found for the general case of MIN-RP. In the following, we present a special case for which this method has a nice performance ratio.

> MIN-RP$_1$: Given a rectangle R with a finite number of points inside the rectangle, find a minimum-length rectangular partition of R, treating the given points as degenerate holes.

It has been proven that the restricted version MIN-RP$_1$ is still **NP**-hard.

Theorem 5.2 *The minimum-length guillotine rectangular partition is a 2-approximation to* MIN-RP$_1$.

Proof. We follow the general approach of the analysis of approximation algorithms designed by the restriction method; that is, we will convert an optimal rectangular partition to a guillotine rectangular partition. To be more precise, let R be an input rectangle to MIN-RP$_1$, and P^* a minimum-length rectangular partition of R. We are going to construct a guillotine rectangular partition P_G from P^* such that the total edge length of P_G is at most twice the total edge length of P^*. Therefore, the total length of an optimal guillotine rectangular partition cannot exceed twice the total edge length of P^*.

In the construction, we will use guillotine cuts to divide R into smaller rectangles and recursively partition these rectangles. We will call each intermediate rectangle created by guillotine cuts a *window* (so that it will not be confused with the final rectangles created by partition P_G). For a window W, we write $int(W)$ to denote the interior area of W.

The guillotine cuts will add new edges to $P_G \setminus P^*$. In order to estimate the cost of these new edges, we will use a *charging method* to charge the cost of each new edge in $P_G \setminus P^*$ to the edges in the original partition P^*. The charging policy will be explained using the notion of *dark points* in a window W. We say a point z in W is a *vertical* (or, *horizontal*) 1-*dark point* with respect to the partition P^* and window W if each vertical (or, respectively, horizontal) half-line starting from z, but not including point z, going in either direction meets at least one horizontal (or, respectively, vertical) line segment in $P^* \cap int(W)$. (In particular, a point z on the boundary of W is not 1-dark.) In the construction, a new horizontal edge t is added to $P_G \setminus P^*$ only if all of its points are vertical 1-dark points, and so its cost can be charged to edges in P^* that lie parallel to t.

To be more precise, the guillotine rectangular partition P_G can be constructed from P^* by applying the following rules on each window W, starting with the initial window $W = R$:

(1) If $int(W)$ does not contain any edge in P^*, then do nothing.

(2) If there exists a horizontal line segment s in P^* that cuts through the whole window W, then we apply a guillotine cut to W along the line segment s.

(3) If there exists a vertical line segment s in P^* whose length in $W \geq h/2$, where h is the height of the window W, then we apply a guillotine cut to W along s. The cut extends s to a line segment at most twice as long as s.

(4) If W contains at least one edge in P^* and yet neither case (2) nor case (3) holds, then we apply to W a horizontal guillotine cut t that partitions W into two equal parts.

We note that in case (2), we did not introduce any new edge in $P_G \setminus P^*$, and so there is no extra cost. In case (3), the new edge added to $P_G \setminus P^*$ has total length $\leq length(s)$. We charge the cost of this new edge to the line segment s.

For case (4), we first claim that every point in the line segment t is a vertical 1-dark point. To see this, we assume, by way of contradiction, that there is a point z in t that is not vertical 1-dark. Then the rectangle defined by the partition P^* that contains z must have height at least $h/2$. The boundary of this rectangle must contain either a horizontal segment in P that cuts through the whole window W or a vertical segment in P of length $\geq h/2$, both cases contradicting the assumption of case (4). This proves the claim. Since each point in t is vertical 1-dark, we can charge the cost of each new edge t_1 in $t \cap (P_G \setminus P^*)$ as follows: We charge 1/2

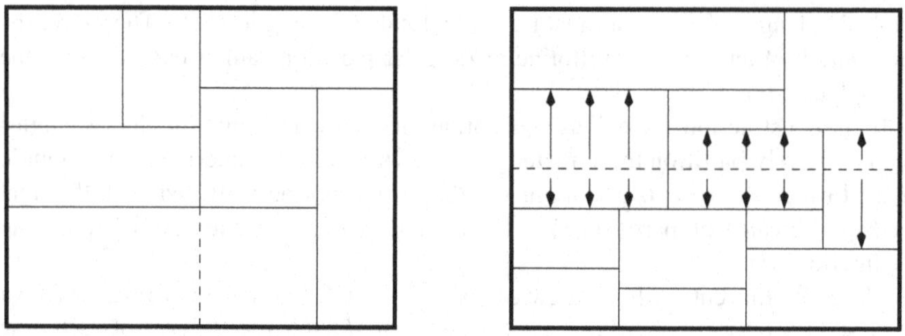

Case (3) Case (4)

Figure 5.4: Constructing a guillotine partition from a given partition. (The arrows indicate how to charge its cost.)

of the cost of t_1 to the horizontal line segments in P^* that lie immediately above the line segment t_1, and the other $1/2$ to the horizontal line segments in P^* that lie immediately below the line segment t_1 (see Figure 5.4).

In the above charging policy, each vertical line segment in P^* is charged at most once with cost less than or equal to its own length. In addition, each horizontal line segment in P^* is charged at most twice, each time with cost less than or equal to $1/2$ of its length. To see this, we note that if a horizontal line segment s has been charged once by a new line segment t_1 in P_G below it, then t_1 becomes the boundary of the new windows, and all points between t_1 and s are non-1-dark in the new window containing s. Thus, the line segment s cannot be charged again by any cut below it.

The above analysis shows that the total charge and, hence, the total length of added line segments in $P_G \setminus P^*$ cannot exceed the total length of P^*.

Finally, we observe that each time we perform a guillotine cut, each new subwindow must contain fewer line segments in P^* than the current window. Thus, after a finite number of guillotine cuts, $int(W)$ no longer contains any line segment of P^*. This means that $P^* \subseteq P_G$. In particular, P_G covers every given point in R. Thus, P_G is a guillotine rectangular partition of R. This completes the proof of the theorem. □

Since the optimal guillotine rectangular partition Q can be computed in time $O(n^5)$, and since its total length does not exceed that of P_G, we get the following conclusion:

Corollary 5.3 *The problem* MIN-RP$_1$ *has a polynomial-time* 2*-approximation.*

5.2 1-Guillotine Cut

The main idea of the proof of Theorem 5.2 is to choose, in case (4), a cut line that consists of vertical 1-dark points. This idea works for the special case of MIN-RP$_1$,

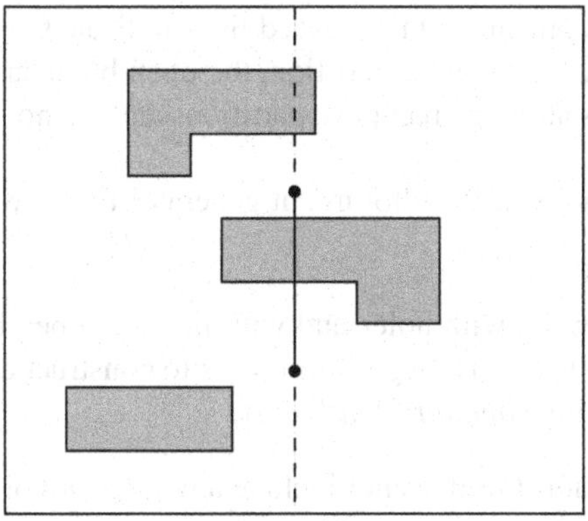

Figure 5.5: A 1-guillotine cut partitions a window into two parts with closed and open boundary segments.

but does not work directly for the general case of MIN-RP. Indeed, if a window W contains nondegenerate holes, such a line may not exist. In this section, we modify this idea to make it work for the general case. The new idea is to allow *partial cuts* that do not use the whole line segment of the cut in the partition. With the help of a technical lemma about 1-dark points, it is shown that a suitable partial cut always exists so that its length can be charged evenly to its two sides. Thus, the proof of Theorem 5.2 can be extended to the general case.

We first introduce the concept of 1-guillotine cuts, which is the simplest type of partial cuts. Consider an input rectilinear polygon R to the problem MIN-RP. We may assume that R is a rectangle, for, otherwise, we can find a rectangle that covers R, and treat the areas between R and the rectangle as holes. We let H_0 denote the holes in R, and $R_0 = R \setminus H_0$. In a guillotine rectangular partition P of R, we use a straight line L to cut a window W into two smaller windows, and add the line segments in $(L \cap W) \cap R_0$ to the partition.[2] In a 1-*guillotine rectangular partition*, we still use L to cut a window W into two smaller windows, but we do not use the whole line segment $L \cap W$ for the partition P. Instead, we select a subsegment s of $L \cap W$ and add segments in $s \cap R_0$ into the partition P. We call such a cut a 1-*guillotine cut* (more precisely, the line segment s is called a 1-guillotine cut). Figure 5.5 shows a 1-guillotine cut.

Note that, after a 1-guillotine cut s, the window W becomes two smaller windows, and the line segment $L \cap W$ becomes a common boundary edge of the new windows. This boundary edge contains a 1-guillotine cut segment s and two (possibly degenerate) line segments to the two sides of s. The 1-guillotine cut segment is called a *closed boundary segment* of the new windows and the other two line segments are called *open boundary segments* (see Figure 5.5, in which a solid line

[2] That is, the line segment $L \cap W$ may be broken into smaller segments by holes in H_0. We only add those segments in R_0 to the partition.

indicates the closed segment, and the dashed lines indicate the open segments).[3] In the construction of the rectangular partition, the open boundary segments are only temporary boundaries allowing recursive partitions and are not included in the final partition.

Thus, in each iteration, a 1-guillotine cut generates two new subproblems of the following form:

> Given a window W with holes and with possible open boundaries on each of its four sides, use 1-guillotine cuts to construct a partition P_W with the following *boundary conditions*:
>
> (1) The partition P_W does not include any interior point of the open boundary segments.
>
> (2) The partition P_W must contain the endpoints of the closed boundary segments, unless the endpoint is a corner of the window W.

With these boundary conditions, can we still use dynamic programming to find the minimum-length 1-guillotine rectangular partition? The answer is *yes*. First, it can be shown, similar to Lemma 5.1, that there exists a minimum-length 1-guillotine rectangular partition Q such that every maximal line segment in Q contains a vertex of the boundary. Therefore, if we consider only these types of 1-guillotine partitions, there are at most $O(n)$ choices for a cut line at each iteration, and at most $O(n^4)$ windows W to be considered.

In addition, we observe that each side of a window W created by a 1-guillotine cut has $O(n^2)$ choices of a line segment as the closed boundary. Therefore, there are $O(n^8)$ choices of the boundary conditions for a window W. Altogether, the total number of possible subproblems to be examined in the dynamic programming algorithm is $O(n^{12})$. For each subproblem, there are $O(n)$ choices of the cut line and, for each cut line, $O(n^2)$ choices of the closed boundary segment. So, there are $O(n^3)$ possible 1-guillotine cuts to be considered. It follows that the minimum-length 1-guillotine rectangular partition can be computed by dynamic programming in time $O(n^{15})$.

Now, we estimate the performance ratio of the minimum 1-guillotine rectangular partition as an approximation to MIN-RP. Similar to the proof of Theorem 5.2, we are going to construct, from an optimal partition P^*, a 1-guillotine partition P_1 whose total length is no more than twice the total length of P^*. To do this, we need the following interesting observation about the relationship between vertical and horizontal 1-dark points. (Here, 1-dark points only include those points in R_0.)

Lemma 5.4 (Mitchell's Lemma) *Assume that P is a rectangular partition of an instance R of* MIN-RP, *and W is a window in R. Let H (and V) be the set of all horizontal (and, respectively, vertical) 1-dark points with respect to partition P and*

[3] Note that the closed boundary segment and the open boundary segments may include points in the holes H_0.

window W. Then there exists either a horizontal cut line L_H that does not contain a line segment of P such that

$$length(L_H \cap H) \leq length(L_H \cap V),$$

or a vertical cut line L_V that does not contain a line segment of P such that

$$length(L_V \cap V) \leq length(L_V \cap H).$$

Proof. Assume that the four corners of the window W are (a, a'), (a, b'), (b, b'), and (b, a'). First, consider the case that the area of H is at least as large as the area of V. Let L_u denote the vertical line $\{(x, y) \mid x = u\}$. Then the areas of H and V can be represented by $\int_a^b length(L_u \cap H)du$ and $\int_a^b length(L_u \cap V)du$, respectively. Since

$$\int_a^b length(L_u \cap H)du \geq \int_a^b length(L_u \cap V)du,$$

and since P has only finitely many line segments, there must exist a vertical line L_u, with $u \in (a, b)$, that does not contain a line segment of P such that

$$length(L_u \cap H) \geq length(L_u \cap V).$$

The line $L_V = L_u$ is what we need.

Similarly, for the case that the area of H is smaller than the area of V, we can show that there exists a horizontal cut line L_H that does not contain a line segment of P such that

$$length(L_H \cap H) \leq length(L_H \cap V). \qquad \square$$

This lemma suggests the following strategy to construct P_1: At each iteration, we make a 1-guillotine cut through a horizontal cut line L_H (or a vertical cut line L_V) satisfying the property of Mitchell's lemma, and let the 1-guillotine cut segment s be the maximal line segment in $L_H \cap W$ (or, $L_V \cap W$) whose two endpoints are in vertical (or, respectively, horizontal) line segments in $P^* \cap int(W)$. Note that all points in $s \cap R_0$, other than the two endpoints of s, are horizontal (or, vertical) 1-dark. Actually, s is the maximal segment in line L_H (or, in L_V) with this property.

Suppose that we select a horizontal 1-guillotine cut s according to the above rule. Then this cut adds some new edges $s \cap R_0$ to $P_1 \setminus P^*$ whose total length is at most $length(L_H \cap H)$ and hence, by Mitchell's lemma, at most $length(L_H \cap V)$. This means that the total length of the horizontal edges in $P^* \cap int(W)$ that lie on each side of L_H is no less than the total length of $s \cap R_0$. Therefore, we can charge the cost of the new edges in $s \cap R_0$ to these horizontal line segments in $P^* \cap int(W)$ that lie on the two sides of L_H, with each line segment charged with at most one half of its own length. The same property holds for a vertical 1-guillotine cut.

With this analysis, the performance ratio 2 can be established for the general case of MIN-RP.

Theorem 5.5 *The minimum-length* 1*-guillotine rectangular partition is a* 2*-approximation for* MIN-RP.

Proof. Assume that P^* is a minimum-length rectangular partition of the input R (a rectangle with holes). We will construct from P^* a 1-guillotine rectangular partition P_1 by a sequence of 1-guillotine cuts such that the total length of edges in $P_1 \setminus P^*$ does not exceed the total length of P^*. At each iteration, we are given a window W with boundary conditions, and we need to find a 1-guillotine cut to divide it into two smaller windows. We select the 1-guillotine cut by the following rules.

(1) If $P^* \cap int(W) = \emptyset$, then do nothing.

(2) If $P^* \cap int(W)$ contains a line segment s that is actually a 1-guillotine cut with respect to P^* (i.e., $s \cap R_0 = P^* \cap L \cap int(W)$, where L is the cut line along s), then we perform the 1-guillotine cut s. [So the line $L \cap W$ becomes a boundary of two new windows, with the segment s being the close boundary, and the segments in $(L \cap W) \setminus s$ being the open boundaries.] We did not add any new edge to P_1.

(3) If $P^* \cap int(W) \neq \emptyset$ but it does not contain a 1-guillotine cut with respect to P^*, then the area of the set H (or set V) of horizontal (or, respectively, vertical) 1-dark points in W with respect to P^* must be greater than zero. Thus, as discussed earlier, we can select the 1-guillotine cut s by Mitchell's lemma. More precisely, we select the cut line L_H (or, L_V) with the property of Mitchell's lemma and let s be the maximal line segment in $L_H \cap int(W)$ (or, $L_V \cap int(W)$) whose two endpoints are in vertical (or, respectively, horizontal) line segments of $P^* \cap int(W)$. We perform a 1-guillotine cut s, and add all segments in $s \cap R_0$ to P_1.

We observe that in the above procedure, each new subwindow created by a 1-guillotine cut contains fewer line segments of P^* than the current window. So, after a finite number of steps, the subwindows W have no more line segments of P^* in $int(W)$. This means that $P^* \subseteq P_1$. Furthermore, we note that since the endpoints of a 1-guillotine cut must be in P^*, the endpoints of each new edge in $P_1 \setminus P^*$ must be either in P^* or on the boundary of R (including the boundary of holes). Therefore, each edge in $P_1 \setminus P^*$ must divide a rectangle created by the optimal partition P^* into two smaller rectangles. It follows that P_1 is a 1-guillotine rectangular partition of R_0.

Now, we estimate the cost of the new edges in $P_1 \setminus P^*$. In case (2), we did not add new edges to P_1. For case (3), assume that we perform a vertical 1-guillotine cut s along line L_V on window W. From the earlier analysis, the total length of edges in $s \cap R_0$ is bounded by $length(L_V \cap H)$. We charge one half of the cost of these new edges to the vertical line segments in $P^* \cap int(W)$ lying immediately to the right of L_V, and the other half to the vertical line segments in $P^* \cap int(W)$ lying immediately to the left of L_V (cf. Figure 5.6). Similar to the proof of Theorem 5.2, we see that each edge in P^* can be charged at most twice, and each time with cost at most one half its own length. So the total length of the new edges in $P_1 \setminus P^*$ is no more than the total length of P^*. This completes the proof of the theorem. □

Corollary 5.6 *The problem* MIN-RP *has a polynomial-time* 2*-approximation.*

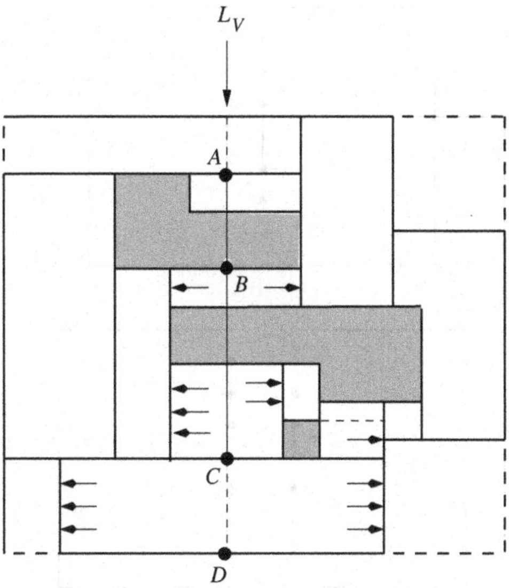

Figure 5.6: The cost of 1-guillotine cut $\overline{AC}$, excluding the points in the holes, is charged to the vertical edges lying on the two sides of $\overline{BD}$, excluding the points in the hole.

5.3 m-Guillotine Cut

The concept of 1-guillotine cut can be extended to m-guillotine cut for any $m > 1$. For any window W, an *m-guillotine cut* along a line L is a line segment s in $L \cap int(W)$, plus at most $2(m-1)$ points in $L \cap int(W)$, with at most $m-1$ points in each side of s. (Note that this includes the case when the line segment s is degenerated to a single point. Thus, any cut with at most $2m-1$ points is considered an m-guillotine cut.) After an m-guillotine cut, a window is divided into two smaller windows. The common boundary of the two new windows contains a line segment of closed boundary and up to m open boundary segments in each side of the closed boundary segment, separated by the points of the m-guillotine cut. Each new window, like that in the case of 1-guillotine cut, defines a subproblem of m-guillotine rectangular partition:

> Given a window W with holes and with up to m open boundary segments on each end of each of the four sides of W, use m-guillotine cuts to construct a partition P_W with the following boundary conditions:
>
> (1) The partition P_W does not include any interior point of the open boundary segments.
>
> (2) The partition P_W must contain the endpoints of the open and closed boundary segments, unless the endpoint is a corner of the window W.

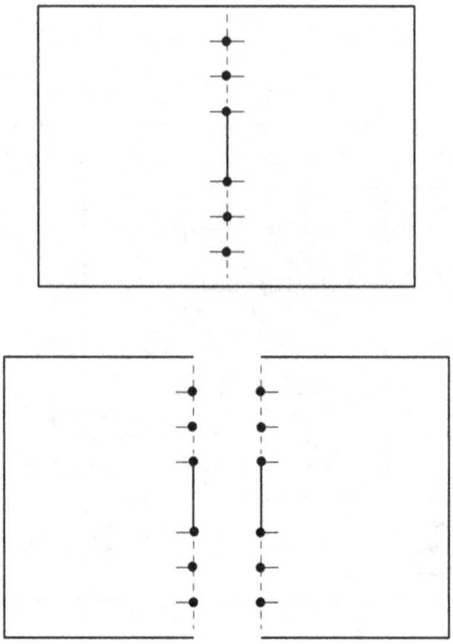

Figure 5.7: An m-guillotine cut results in $2m$ open segments on each subproblem's boundary.

Figure 5.7 shows an m-guillotine cut. (In Figure 5.7, the short horizontal lines indicate that the partition P_W must contain line segments touching the endpoints of the open and closed boundary segments.)

A rectangular partition is called an *m-guillotine rectangular partition* if it can be realized by a sequence of m-guillotine cuts. Similar to the problem of 1-guillotine rectangular partition, the problem of the minimum-length m-guillotine rectangular partition can be computed by dynamic programming in time $O(n^{10m+5})$. To see this, we observe that, at each iteration of the dynamic programming algorithm, an m-guillotine cut has at most $O(n^{2m+1})$ choices: It has $O(n)$ choices of the cut line and, at each cut line, $O(n^{2m})$ choices for the $2m$ endpoints of the open and closed boundary segments. In addition, there are $O(n^{8m+4})$ possible subproblems: There are $O(n^4)$ possible windows, each having $O(n^{8m})$ possible boundary conditions. So the total running time is $O(n^{10m+5})$.

To analyze the approximation to MIN-RP by m-guillotine cut, we need to extend the notion of 1-dark points to m-dark points, for $m > 1$. Let R be an instance of MIN-RP (i.e., a rectangle with holes), P a partition of R, and W a window of R. We say a point z in W is a *horizontal* (or, *vertical*) *m-dark point* with respect to window W and partition P if each horizontal (or, respectively, vertical) half-line starting from z, but not including z, going in either direction meets at least m vertical (or, respectively, horizontal) line segments in $P \cap int(W)$.

By an argument similar to Mitchell's lemma about 1-dark points, we can easily establish the following property about m-dark points.

Lemma 5.7 *Assume that P is a rectangular partition of an instance R of* MIN-RP, *and W is a window of R. Let $m > 1$, and H_m (and V_m) be the set of all horizontal (and, respectively, vertical) m-dark points with respect to partition P and window W. Then there exists either a horizontal cut line L_H that does not contain a line segment of P such that*

$$length(L_H \cap H_m) \le length(L_H \cap V_m),$$

or a vertical cut line L_V that does not contain a line segment of P such that

$$length(L_V \cap V_m) \le length(L_V \cap H_m).$$

Theorem 5.8 *For any $m > 1$, the minimum-length m-guillotine rectangular partition is a $(1 + 1/m)$-approximation to* MIN-RP.

Proof. Let P be a rectangular partition of an instance R of MIN-RP. Also, let H_0 be the set of the points in the holes in R and $R_0 = R \setminus H_0$. We will construct an m-guillotine rectangular partition P_m with total length bounded by $(1 + 1/m)length(P)$.

The construction is similar to that of Theorem 5.5: At each iteration with a window W, if $P \cap int(W)$ has an m-guillotine cut, then we make such a cut. Otherwise, we choose a cut line according to Lemma 5.7. Without loss of generality, assume that by Lemma 5.7, there is a vertical cut line L that does not contain a line segment in P such that $length(L \cap V_m) \le length(L \cap H_m)$. Let s be the maximal segment of L whose interior points are all vertical m-dark. Since all vertical m-dark points in L are in s, the cut line L contains exactly $m - 1$ points in $P \cap int(W)$ on each side of s. We select line segment s plus these points as the m-guillotine cut.

For such an m-guillotine cut along the cut line L, we note that the total length of $s \cap R_0$ is at most $length(L \cap V_m) \le length(L \cap H_m)$. This means that there are line segments in L, of total length at least $length(s \cap R_0)$, which have the following property: There are at least m layers of vertical edges in P lying on each side of these segments. So we can charge the length of $s \cap R_0$ to the edges in the m layers that are closest to line L, with each edge charged with at most $1/(2m)$ of its own length. Furthermore, an edge in P can be charged at most twice: After an edge t is charged by a cut s through line L, the line segment $L \cap W$ becomes a boundary edge of the two new windows. In the new window containing t, there are at most $m - 1$ layers of vertical edges of P between t and L, and so all points between L and t are not horizontal m-dark in the new window, and t can no longer be charged by any cut between t and L. It follows that the total length of m-guillotine cuts is bounded by $(1 + 1/m)length(P)$. □

Corollary 5.9 *For any constant $\varepsilon > 0$,* MIN-RP *has a polynomial-time $(1 + \varepsilon)$-approximation with running time $n^{O(1/\varepsilon)}$.*

The significance of the technique of guillotine cut stems not only from the PTAS for the problem MIN-RP, but also from wide applications to other geometric optimization problems. As another example, let us apply the technique of m-guillotine

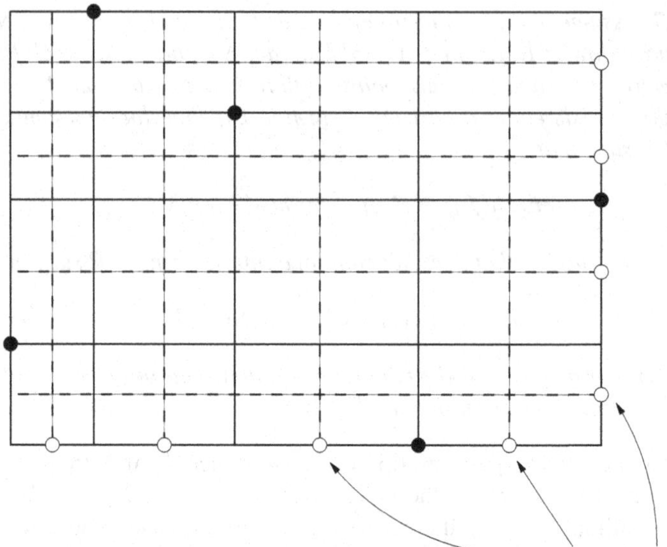

Figure 5.8: Hanan grid (solid lines) and cut lines (dashed lines) in $\mathcal{L}$.

cut to find approximation algorithms for RSMT (RECTILINEAR STEINER MINIMUM TREE) introduced in Section 3.1. Let Q be the minimal rectangle covering all n given points in the rectilinear plane. We will define the concept of m-guillotine rectilinear Steiner trees, and show that the shortest m-guillotine rectilinear Steiner tree provides a $(1 + 1/m)$-approximation to the rectilinear SMT.

First, we define the location of the cut lines. For a given set A of terminal points, let the *Hanan grid* be the set of all horizontal and vertical lines each passing through a point in A. The Hanan theorem states that for any given set A of points, there is a rectilinear SMT T^* lying on the Hanan grid. Thus, we can avoid the case of the guillotine cut lines overlapping with the edges of tree T^* by choosing the cut lines off the lines of the Hanan grid. Moreover, in order to limit the number of possible cut lines, we let $\mathcal{L}$ be the set of lines lying at the middle between two adjacent lines of the Hanan grid (see Figure 5.8), and require that each m-guillotine cut must use a line $L \in \mathcal{L}$ as a cut line.

Let W be a window of Q. Then an m-guillotine cut of W is a cut along a line $L \in \mathcal{L}$ that consists of a line segment s and up to $2m - 2$ points, with at most $m - 1$ of them on each side of s. In addition, it is required that all the cut points and the endpoints of the cut segment lie on the Hanan grid. After each cut, W is divided into two subwindows, the cut segment s is included as part of the Steiner tree, and $L \cap W$ becomes a common boundary of the two new subwindows. In the problem MIN-RP, the boundary conditions of the new subwindows are conditions about at most $2m-2$ cut points and the cut segment s. Here, our boundary conditions are those about at most $2m - 2$ cut points plus one of the endpoints of s. (These points are called *crosspoints*.) Furthermore, since the given points in A can be connected by edges passing through other windows, the boundary conditions here are more complicated than those in the problem MIN-RP: Some of the crosspoints are required to be connected to each other and some are not. Also, the cut segment s is not part of the boundary condition; instead, we choose one of its endpoints as a crosspoint and add a boundary condition on that point.

More precisely, each m-guillotine cut generates two subwindows and, for each subwindow, we need to solve a subproblem of the following form:

(1) A window W is given, together with at most four closed boundary segments and a set S of at most $8m - 4$ crosspoints on the boundary of W (with each edge of W having at most one closed segment s, $m - 1$ crosspoints on each side of s, and an endpoint of s as a crosspoint). All input points of A in window W lie in the interior of W and all crosspoints lie on the Hanan grid.

(2) A partition of the set S is given.

(3) The problem is to find a rectilinear Steiner forest F that includes the closed boundary segments of W and satisfies the following properties:

(a) All crosspoints in each part of S are connected by F;

(b) Two crosspoints in different parts of S are not connected by F;

(c) No two line segments of F cross each other (other than at a Steiner point);

(d) The Steiner forest F does not contain any point on the boundary of W other than those in set S; and

(e) If S is not empty, then each input point in A is connected by F to at least one crosspoint; otherwise, all input points are connected by F.

Note that if there is no line $L \in \mathcal{L}$ passing through the interior of a window W, then an m-guillotine cut of W is not possible. We call such a window a *minimal window*. Each minimal window W contains at most one input point in A, and each side of W contains at most one crosspoint, all lying on the Hanan grid. For a minimal window, a shortest rectilinear Steiner forest satisfying the boundary conditions is easy to find.

We say a rectilinear Steiner tree T is an *m-guillotine rectilinear Steiner tree* if it can be obtained by m-guillotine cuts so that each edge of T is either an m-guillotine cut segment or an edge in a minimal window.

Now, we show that the minimum m-guillotine rectilinear Steiner tree is a $(1 + 1/m)$-approximation to the rectilinear SMT. To do this, we will construct, from a given rectilinear Steiner tree T lying on the Hanan grid, an m-guillotine rectilinear Steiner tree T_m whose total edge length is at most $(1+1/m)$ of the total edge length of T.

We first define the notion of horizontal and vertical m-dark points with respect to T and a window W, similar to that for the problem MIN-RP. That is, a point z in W is *horizontal* (or, *vertical*) *m-dark* if each of the two horizontal (or, respectively, vertical) half-lines starting from z, but not including z, meets at least m vertical (or, respectively, horizontal) edges of T in $int(W)$. Using a similar argument as that of Mitchell's lemma, we have the following property.

Lemma 5.10 *Let A be a given set of points and T a rectilinear Steiner tree of A. Also, let Q be the minimal rectangle covering points in A, and let W be a window in Q. Then there exists either a horizontal cut line $L_H \in \mathcal{L}$ such that*

$$length(L_H \cap H_m) \leq length(L_H \cap V_m)$$

or a vertical cut line $L_V \in \mathcal{L}$ such that

$$length(L_V \cap V_m) \leq length(L_V \cap H_m),$$

where H_m (V_m) is the set of all horizontal (vertical, respectively) m-dark points with respect to T and W.

Proof. Let (a, a'), (a, b'), (b, b'), and (b, a') be the four vertices of the window W. First, assume that the area of H_m is greater than or equal to the area of V_m. Denote $L_u = \{(x, y) \mid x = u\}$. Then the areas of H_m and V_m can be represented by $\int_a^b length(L_u \cap H_m)du$ and $\int_a^b length(L_u \cap V_m)du$, respectively. Since

$$\int_a^b length(L_u \cap H_m)du \geq \int_a^b length(L_u \cap V_m)du,$$

and since there are only finitely many u's such that L_u passes through a point in A, there must exist $u \in (a, b)$ such that L_u does not pass through any point in A and

$$length(L_u \cap H_m) \geq length(L_u \cap V_m).$$

That is, line L_u must lie between two lines L_1 and L_2 of the Hanan grid. Let L_v be the line in $\mathcal{L}$ that lies between L_1 and L_2. Since T lies on the Hanan grid, the m-dark segments on L_u and L_v have the same length: A point (u, w) in L_u is vertical (or, horizontal) m-dark if and only if the point (v, w) is vertical (or, respectively, horizontal) m-dark. So $length(L_u \cap H_m) = length(L_v \cap H_m)$, and $length(L_u \cap V_m) = length(L_v \cap V_m)$. Therefore, L_v satisfies the required property.

The case when the area of H_m is smaller than the area of V_m is similar. □

We now construct tree T_m by performing a sequence of m-guillotine cuts. At each iteration of the construction, we are given a window W with boundary conditions (i.e., a set S of at most $2m-1$ crosspoints on each side of W, all on the Hanan grid, and a partition of S) and a partially constructed $T_m \cap W$ (which consists of up to four closed boundary segments) satisfying the following conditions:

(i) All crosspoints in the same part of S are connected by $(T \cup T_m) \cap W$ (note that this includes the closed boundary segments of W);

(ii) Two crosspoints in different parts of S are not connected by $(T \cup T_m) \cap W$;

(iii) All input points in W are connected by $(T \cup T_m) \cap W$ and are connected to at least one crosspoint if S is nonempty.

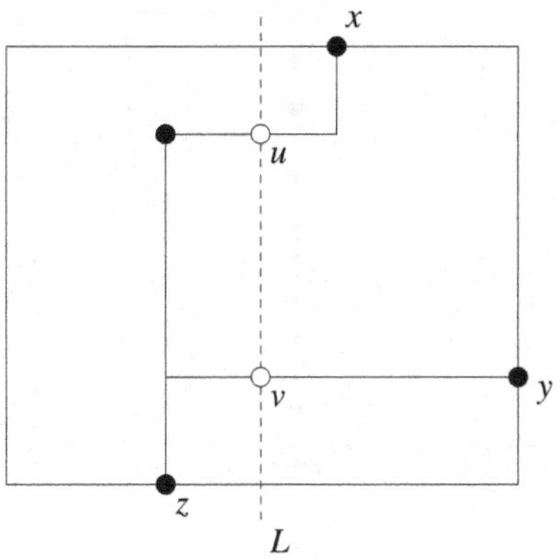

Figure 5.9: The partition of crosspoints changes.

Initially, we let $T_m = \emptyset$. Since the window Q does not have boundary conditions, T_m apparently satisfies conditions (i)–(iii) above. For a given window W, we find an m-guillotine cut by the following rules:

(1) If W is a minimal window, then do nothing.

(2) If W is not a minimal window and if there exists a cut line $L \in \mathcal{L}$ that intersects T at no more than $2m - 1$ points, then we cut W along line L, and put these intersection points as the crosspoints. (In this case, we do not add any line segment to T_m.) We partition the crosspoints of each new subwindow W_1 according to $(T \cup T_m) \cap W_1$; that is, two crosspoints are in the same part if they are connected by edges of $(T \cup T_m) \cap W_1$. Note that we need to repartition the old crosspoints on the three other boundaries of the subwindow W_1, since the connection by $T \cup T_m$ may change within W_1. This is demonstrated in Figure 5.9: In the original window W, crosspoints x and y are in the same part. After the m-guillotine cut along line L, they belong to two different parts in the right subwindow (the new partition of the right subwindow is now $\{x, u\}$, $\{y, v\}$). Also note that u and v are in different parts in the right subwindow, but they are in the same part when we consider the left subwindow.

(3) Otherwise, the area of set H_m (or, set V_m) of horizontal (or, respectively, vertical) m-dark points in W must be greater than zero. We choose a cut line L from $\mathcal{L}$ that satisfies the property of Lemma 5.10. Without loss of generality, assume that L is a vertical line. We make an m-guillotine cut of the window W along line L. This cut contains a segment s of all vertical m-dark points and $2m - 2$ points in T, with $m - 1$ points in each side of s. We add the cut segment s to T_m. For each subwindow created by this cut, we add the $2m - 2$ points of the cut as the crosspoints, and choose one endpoint of s as an additional crosspoint (see Figure 5.10). Again, we partition the crosspoints of each new subwindow W_1 according to $(T \cup T_m) \cap W_1$ so that

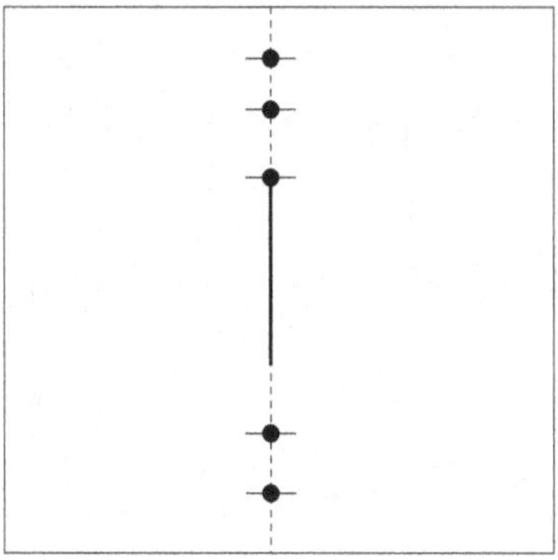

Figure 5.10: The new crosspoints.

two crosspoints are in the same part if and only if they are connected by edges of $(T \cup T_m) \cap W_1$, including the new edge s we just added to T_m.

By the above rule of partition, we see that $T \cup T_m$ satisfies boundary conditions (i)–(iii) in the new subwindows, because the boundary conditions are simply defined by $(T \cup T_m) \cap W$. Therefore, the final tree $T_m = T \cup T_m$ is an m-guillotine rectilinear Steiner tree since the initial window Q has no boundary conditions.[4]

Finally, we verify that the length of T_m is at most $(1 + 1/m)$ of the length of T. We observe that in case (3), each cut line L is chosen to satisfy the property of Lemma 5.10. Assume, without loss of generality, that L is a vertical line. Then the total length of the segments of horizontal m-dark points in $L \cap W$ is at least $length(s)$. So the cost of the cut segment s can be evenly charged to the $2m$ closest layers of vertical line segments in $T \cap W$ that lie on the two sides of L. By the same argument as in Theorem 5.8, we can see that each edge in T can be charged at most twice, each time at most $1/(2m)$ of its own length. So the total length of $T_m \setminus T$ is at most $1/m$ of the length of T. We just proved the following result:

Theorem 5.11 *For any $m > 0$, the minimum m-guillotine rectilinear Steiner tree is a $(1 + 1/m)$-approximation to* RSMT.

Next, we verify that, similar to the case of the m-guillotine rectangular partition, the minimum m-guillotine rectilinear Steiner tree can be computed by dynamic programming in polynomial time as follows:

First, at each iteration, there are $O(n)$ possible positions to choose the cut line, since the cut line must belong to $\mathcal{L}$. For each cut line, there are at most $2m - 1$ crosspoints on each cut plus a line segment s, with all crosspoints and the endpoints of s lying on the Hanan grid. Therefore, there are totally $O(n^{2m})$ possible cuts to consider in each iteration.

[4]Strictly speaking, the final set $T_m = T \cup T_m$ is not necessarily a tree since it might contain loops. This problem can be easily resolved by removing some redundant edges from the final T_m.

Next, we estimate how many subproblems may occur in the computation of the dynamic programming algorithm. Each subproblem, as defined earlier, is given by a window with four boundary lines, up to $8m - 4$ crosspoints on the boundaries, and a partition of these crosspoints. Because all boundaries of a window must be on lines in $\mathcal{L}$, and all crosspoints must lie on the Hanan grid, we see that there are $O(n^4)$ possible windows, and for each window there are $O(n^{8m-4})$ possible sets S of crosspoints. For each set S of $8m - 4$ crosspoints, there are $2^{O(m \log m)}$ different partitions of S. Note, however, that not every partition of S has a feasible solution satisfying boundary conditions (a)–(d) (for instance, in the right subwindow of Figure 5.9, the partition $\{\{x, v\}, \{u, y\}\}$ is not feasible). We will prove in Lemma 5.12 that the number of partitions of S that have a feasible solution satisfying boundary conditions (a)–(d) is $2^{O(m)}$. Therefore, the total number of subproblems that may occur in the dynamic programming algorithm is $O(n^4 \cdot n^{8m-4} \cdot 2^{O(m)}) = n^{O(m)}$. Moreover, Lemma 5.12 also shows that we can actually generate all feasible partitions of S in time $2^{O(m)}$. It follows that the dynamic programming algorithm runs in time $n^{O(m)} \cdot 2^{O(m)} = n^{O(m)}$.

Lemma 5.12 *Let W be a window and S the set of crosspoints on the boundary of W. Then the number of partitions of S that have a feasible solution satisfying the boundary conditions* (a)–(d) *is* $2^{O(m)}$.

Proof. Break the boundary of window W at a point and spread it out into a straight line. Then the problem is reduced to counting the number N_k of partitions of a set S of k ($k \leq 8m - 4$) points on a horizontal line such that there exists a forest above the line satisfying conditions (a)–(c). Let us denote the k points on the line by $p_1, p_2, \ldots, p_k$, from left to right. When point p_1 is connected to no other point, the number of required partitions is N_{k-1}. When p_1 is connected to at least one other point and p_i is the leftmost point other than p_1 that is connected to p_1, the number of required partitions is $N_{i-2}N_{k-i+1}$. (We define $N_0 = 1$.) Therefore,

$$N_k = N_{k-1} + \sum_{i=2}^{k} N_{i-2}N_{k-i+1} = \sum_{i=0}^{k-1} N_i N_{k-1-i}. \tag{5.1}$$

Let $f(x)$ be the generating function of N_k; that is, $f(x) = \sum_{k=0}^{\infty} N_k x^k$. Then we have

$$f(x)^2 = \sum_{k=0}^{\infty}\sum_{i=0}^{k} N_i N_{k-i} x^k = \sum_{k=0}^{\infty} N_{k+1} x^k.$$

Hence, $xf(x)^2 = f(x) - 1$. Thus,

$$f(x) = \frac{1 \pm \sqrt{1-4x}}{2x}.$$

Since $\lim_{x \to 0} f(x) = 1$ and $\lim_{x \to 0}(1 + \sqrt{1-4x})/(2x) = \infty$, we get

$$f(x) = \frac{1 - \sqrt{1-4x}}{2x} = -\sum_{k=1}^{\infty} \binom{1/2}{k} \frac{(-4x)^k}{2x}.$$

That is, $f(x)$ is the generating function of the well-known *Catalan numbers*, and

$$\begin{aligned} N_k &= -\binom{1/2}{k+1}\frac{(-4)^{k+1}}{2} \\ &= \frac{(1/2)(1-1/2)(2-1/2)\cdots(k-1/2)}{(k+1)!}\cdot 2^{2k+1} = 2^{O(k)}. \end{aligned}$$

In addition, we remark that we obtained the recurrence (5.1) by a simple case analysis, which may be used as a recursive algorithm to generate all feasible partitions of S. □

Corollary 5.13 *For any* $\varepsilon > 0$*, there exists a* $(1+\varepsilon)$*-approximation for the problem* RSMT *with running time* $n^{O(1/\varepsilon)}$.

5.4 Portals

In the last two sections, we have studied the technique of m-guillotine cut. We note that in the design of an approximation algorithm by guillotine cut, we often face two conflicting requirements: On the one hand, after a guillotine cut, we need to allow the two subproblems resulting from the cut to communicate through the common boundary so that we can combine the solutions of the two subproblems into a good approximate solution to the current window. On the other hand, the communication points (i.e., the crosspoints) must be limited so that the number of possible boundary conditions is polynomially bounded and, hence, an algorithm of dynamic programming can find the optimal solution (of the guillotine-cut restricted problem) in polynomial time. In the m-guillotine cut technique, this problem is resolved by allowing up to $2m-2$ crosspoints on the cut line for the communication between the two subproblems. Note, however, that the running time of the dynamic programming algorithm, though polynomially bounded, is very high even with reasonably small values of m.

In this section, we introduce a different technique to deal with these conflicting requirements. In this technique, we allow up to $O(\log n)$ crosspoints on each cut line, but the locations of the crosspoints are predetermined. That is, we define a set of $p = O(\log n)$ points on a cut line that evenly divide the cut segment (called *p-portals*) and require that the connections between the two new windows resulting from the cut can only go through these portals. Since the number of portals on a cut line is bounded by $O(\log n)$, the number of possible boundary conditions is still polynomially bounded and so the optimal guillotine-cut restricted solution can be found by dynamic programming in polynomial time.

To be more specific, let us consider the problem RSMT again. Let P be the set of n input terminals. Initially, we use a minimal square R to cover all points in P, and divide the square R, by a grid of lines, into $g \times g$ cells of equal size, where $g = \lceil (4.5)n/\varepsilon \rceil$ for a given $0 < \varepsilon < 1$. Assume the length of each side of R is L. Then each cell is a square of size $(L/g) \times (L/g)$.

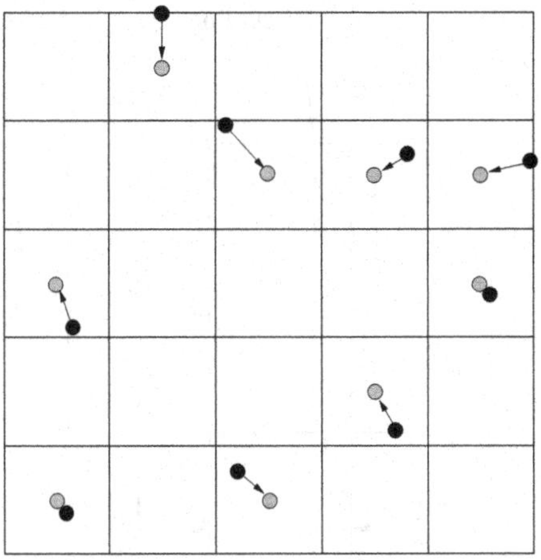

Figure 5.11: Moving each terminal to a center.

For each terminal $u \in P$, let u' be the center of the cell containing point u. Denote $P' = \{u' \mid u \in P\}$. We first show that in order to get a PTAS for P, it suffices to construct a PTAS for P' (see Figure 5.11).

Lemma 5.14 *For any $\varepsilon > 0$, if there is a polynomial-time $(1 + \varepsilon)$-approximation for* RSMT *on P', then there exists a polynomial-time $(1 + 2\varepsilon)$-approximation for* RSMT *on P.*

Proof. Let $smt(P)$ denote the length of the rectilinear SMT on P, and $mst(P)$ denote the length of the rectilinear minimum spanning tree on P. Recall that the Steiner ratio (i.e., the maximum ratio of $smt(Q)$ to $mst(Q)$ on the same input points Q) in the rectilinear plane is equal to $2/3$. Since R is the minimal square covering the input points, the length L of each side of R is no greater than $mst(P)$, and hence no greater than $(3/2)smt(P)$. In addition, we note that to move each point $u \in P$ to $u' \in P'$, we increase the length of the rectilinear Steiner tree by a value at most L/g. Thus, we have $|smt(P) - smt(P')| \leq nL/g$.

Let $T_\varepsilon(P')$ be a polynomial-time $(1 + \varepsilon)$-approximation for the rectilinear SMT on P'. That is,

$$length(T_\varepsilon(P')) \leq (1 + \varepsilon)smt(P').$$

We can construct a tree T interconnecting points in P from $T_\varepsilon(P')$ by connecting each point u' in P' to its corresponding point u in P. Then we have

$$\begin{aligned} length(T) &\leq length(T_\varepsilon(P')) + \frac{nL}{g} \\ &\leq length(T_\varepsilon(P')) + \frac{nL}{g} \\ &\leq (1 + \varepsilon)smt(P') + \frac{nL}{g} \end{aligned}$$

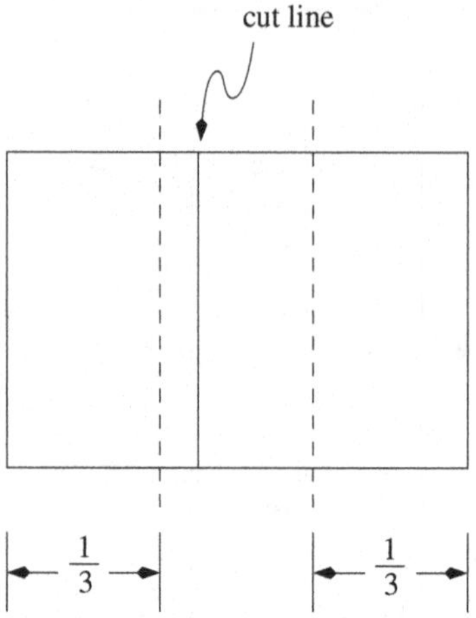

Figure 5.12: A $(1/3, 2/3)$-restricted guillotine cut.

$$\begin{aligned} &\leq (1+\varepsilon)\Big(smt(P) + \frac{nL}{g}\Big) + \frac{nL}{g} \\ &= (1+\varepsilon)smt(P) + (2+\varepsilon)\frac{nL}{g} \\ &\leq \Big(1+\varepsilon+\frac{3n}{2g}(2+\varepsilon)\Big)smt(P) \leq (1+2\varepsilon)smt(P), \end{aligned}$$

since $g \geq (4.5)n/\varepsilon$. □

Based on this lemma, we will work on set P' instead of P. That is, we will assume that all terminals lie at the centers of the cells (and we still use the name P for the set of these terminals).

Next, we apply guillotine cuts to partition the rectangle R, step by step, into smaller rectangles (called *windows*) until each rectangle contains at most one terminal (called a *minimal window*). In order to limit the depth of the cutting process, we will only choose cut lines that lie close to the middle of the window. That is, at each iteration, for a given window W, we choose a grid line parallel to the shorter edge of W which cuts through the middle $1/3$ of the longer edge of W. We call such a cut a $(1/3, 2/3)$-*restricted guillotine cut* (see Figure 5.12), and a partition made by such cuts a $(1/3, 2/3)$-*partition.*

A $(1/3, 2/3)$-partition has a natural binary tree structure. The root of the tree is the initial window R. For each window W, a guillotine cut divides W into two new windows, which are the two children of W in the tree (see Figure 5.13). This binary tree has an important property: it has depth $O(\log n)$.

Lemma 5.15 *The binary tree structure of a* $(1/3, 2/3)$*-partition of a window of* n *terminal points has* $O(\log n)$ *levels.*

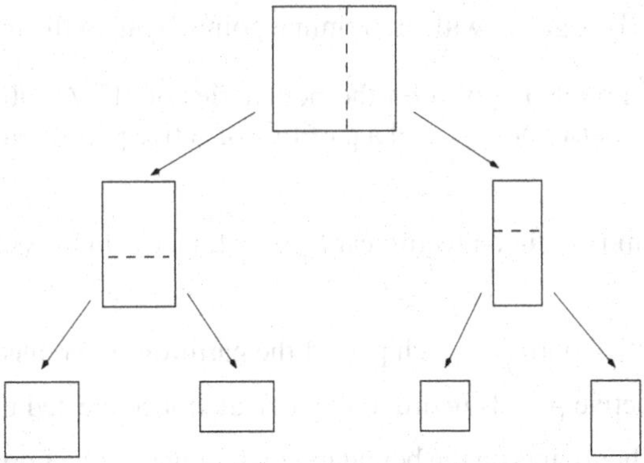

Figure 5.13: Binary tree structure of a $(1/3, 2/3)$-partition.

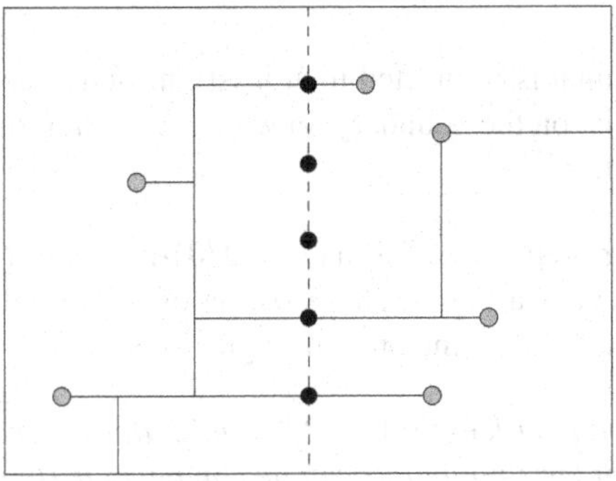

Figure 5.14: Portals.

Proof. At each node of the binary tree, a window W is divided into two smaller windows, each having area at most $2/3$ of W. Thus, a window at the ith level, for $i \geq 0$, has area at most $L^2(2/3)^i$. Since each cut runs along a grid line, the window of a leaf node has area at least $(L/g)^2$. Therefore, the level s of a leaf node satisfies $L^2(2/3)^s \geq (L/g)^2$. That is, $s = O(\log g) = O(\log n)$. □

To limit the number of crosspoints at each cut line, we fix the number and locations of portals on the line where the edges of the Steiner tree can cross the cut line. For an integer $p > 0$, we define *p-portals* on a cut line to be the p points on the line that evenly divide the cut line into $p+1$ segments. We have selected the locations of the portals independently of the input terminals. Thus, a portal only serves as a potential crosspoint, and may not actually be used in the approximate solution. Thus, in the computation of the approximation algorithm, we need to identify some portals as *active* portals, which, in the new windows resulting from a cut, must connect to the output Steiner tree; that is, active portals are the real crosspoints.

With p-portals, a subproblem in the guillotine cut algorithm for RSMT has the following form:

(1) A window W is given, with all terminal points lying in the interior of W.

(2) A set S of portals is given on the boundaries of W. A subset of portals is identified as *active* portals, and a partition of active portals on the boundary is given.

(3) The problem is to find a rectilinear Steiner forest F in W with the following properties:

   (a) All active portals in each part of the partition are connected by F;

   (b) Two active portals in different parts are not connected by F;

   (c) All other points on the boundary, including inactive portals, are not connected by F to any other portals or terminals;

   (d) No two line segments of F cross each other except at a Steiner point; and

   (e) Each terminal is connected to at least one active portal unless no active portals exist on the boundary, in which case, all terminals are connected to each other.

We say a rectilinear Steiner tree T is a $(1/3, 2/3)$-*guillotine* (p-*portal) rectilinear Steiner tree* if there exists a $(1/3, 2/3)$-partition of the initial rectangle such that each edge of T intersects a cut line only through a p-portal.

Lemma 5.16 *The minimum-length* $(1/3, 2/3)$*-guillotine* p*-portal rectilinear Steiner tree of a given set* P *of terminal points can be computed in time* $n^{11}2^{O(p)}$.

Proof. Based on the binary tree structure of the $(1/3, 2/3)$-partition, we can employ dynamic programming to find the minimum $(1/3, 2/3)$-guillotine rectilinear Steiner tree. To estimate the running time of this dynamic programming algorithm, we first note that each boundary of a window must be a grid line, and so there are $O(n^4)$ possible windows. Each window W has four sides, and one of them is the cutting line of the parent window of W and contains p portals. However, each of the three other boundary sides may contain fewer than p portals, as it may be a subsegment of a longer cutting segment from a cut on a nonparent ancestor window (see Figure 5.15). Note that there are $O(n^2)$ potential ancestor windows that may have made a cut along a side line of W. The locations of the portals on this side line resulting from a cut by different ancestor windows are different. Therefore, the number of possible sets of portals for each of these three sides is $O(n^2)$. In addition, we do not know which of the four sides is the cut segment of the parent window of W. Thus, the total number of sets of portal locations on the boundary of W is $4 \cdot O(n^6) = O(n^6)$. After the locations of the portals are fixed, we need to choose a subset of active portals and a partition of this subset. There are $2^{O(p)}$ choices for the subset of active portals and, for each subset, there are, as proved in Lemma 5.12, $2^{O(p)}$ possible choices of partitions that satisfy boundary conditions (a)–(d). Therefore, the total number of possible subproblems is $n^{10}2^{O(p)}$.

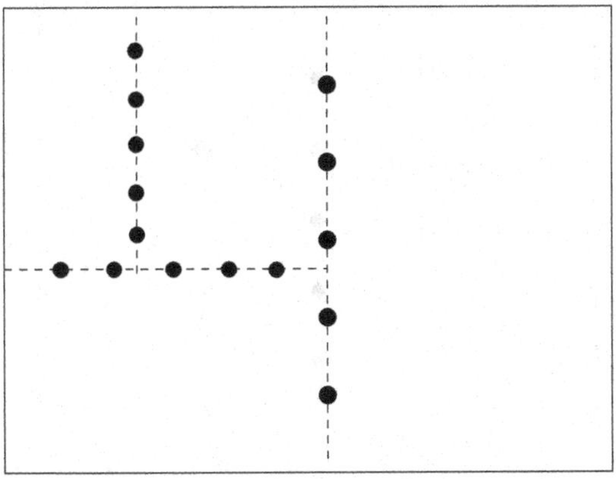

Figure 5.15: Portals defined from different cuts.

Moreover, in each iteration of the dynamic programming algorithm, the number of possible cuts is $O(n)$, since the cut must be made along a grid line. For each cut, we need to choose a set of active portals on the cutting segment and, for each subwindow created by this cut, a partition of the active portals of the subwindow. Therefore, each iteration takes time $n2^{O(p)}$ and the overall running time of the dynamic programming algorithm is $n^{11}2^{O(p)}$. □

Now, let us estimate the performance ratio of the minimum $(1/3, 2/3)$-guillotine rectilinear Steiner tree as an approximation to the rectilinear SMT. To do so, consider a rectilinear SMT T^*, and we are going to modify T^* to meet our restriction. That is, we will construct a $(1/3, 2/3)$-partition and will move all crosspoints at cut segments to portals.

More precisely, the $(1/3, 2/3)$-partition is constructed in the following way: At each step, among all possible cut lines that cut through the window W in the middle third of the longer side of W, choose the one with the minimum number of intersection points with T^*. Then, set up the p-portals on the cut segment. For each crosspoint of T^* on the cut line, move it to the nearest portal by adding a detour path (see Figure 5.16), and define these cross portals as the active portals. At last, as in the case of the construction of the m-guillotine rectilinear Steiner trees (described in the proof of Theorem 5.11), repartition the set of active portals depending on whether two active portals are connected by T^*.

The following lemma shows that moving all crosspoints to the portals does not cost much.

Lemma 5.17 *Let $i \geq 0$. The length increase that resulted from moving all crosspoints to portals in all windows at level i of a $(1/3, 2/3)$-partition is at most $(6/p) \cdot length(T^*)$.*

Proof. Let W be a window at level i. Suppose a longer edge of W has length a and a shorter edge has length b (with $0 < b \leq a$). Without loss of generality, assume that

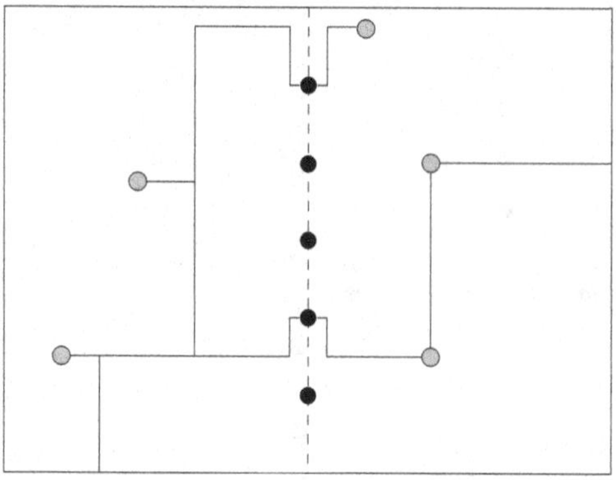

Figure 5.16: Moving crosspoints to portals.

the longer edges of W are horizontal line segments. Then the guillotine cut on W is a vertical $(1/3, 2/3)$-restricted cut of W; that is, it is a vertical line that intersects each longer edge of W in the middle third of that edge. Furthermore, this line is chosen to have, among all such vertical $(1/3, 2/3)$-restricted cuts, the minimum number of intersections (i.e., crosspoints) with tree T^*. Suppose that the chosen cut has c crosspoints with T^*. Then for every vertical line that lies in the middle third of W, it has at least c crosspoints with T^*. This means that the total length of horizontal line segments in $T_W = T^* \cap W$ is at least $ca/3$. It follows that the total length of T_W is at least $ca/3$. Moving each crosspoint to its nearest portal requires adding two edges to T^*, each of length at most $b/(p+1)$. [For the middle $p-2$ portals, each additional edge is only of length at most $b/(2(p+1))$.] So moving all c crosspoints to their respective nearest portals increases the length of the tree by at most

$$\frac{2cb}{(p+1)} \le \frac{2ca}{(p+1)} \le \frac{6}{p} \cdot \frac{ca}{3} \le \frac{6}{p} \cdot length(T_W).$$

We note that the union of T_W over all windows at level i of the $(1/3, 2/3)$-partition is just T^*, and so

$$\sum_{W \in \text{ level } i} length(T_W) = length(T^*).$$

Thus, the total length increase resulting from moving crosspoints to portals on all windows at level i is at most $(6/p) \cdot length(T^*)$. □

Theorem 5.18 *The minimum $(1/3, 2/3)$-guillotine rectilinear Steiner tree using p-portals, for some $p = O((\log n)/\varepsilon)$, is a $(1+\varepsilon)$-approximation for* RSMT. *Moreover, this tree can be computed in time $n^{O(1/\varepsilon)}$.*

Proof. Suppose that the binary tree structure of a $(1/3, 2/3)$-partition has $d \log n$ levels for some constant $d > 0$. Then the total length increase that resulted from moving crosspoints to portals on all windows of the partition is at most

$$d \log n \cdot \frac{6}{p} \cdot length(T^*) \leq \varepsilon \cdot length(T^*)$$

if we choose $p = \lceil 6d \log n/\varepsilon \rceil$. So the $(1/3, 2/3)$-guillotine rectilinear Steiner tree obtained from T^* as described above has length at most $(1+\varepsilon)length(T^*)$.

Also, note that for $p = \lceil 6d \log n/\varepsilon \rceil$, the running time of the dynamic programming algorithm is $n^{11}2^{O(p)} = n^{11+O(\lceil 6d/\varepsilon \rceil)} = n^{O(1/\varepsilon)}$. □

5.5 Quadtree Partition and Patching

We have introduced two techniques of limiting the number of crosspoints on the guillotine cut lines, namely, the m-guillotine cut technique and the portal technique. In this section, we show how to combine the two techniques to further improve the guillotine cut approximation algorithms. Let us first compare the two techniques in different applications.

First, consider geometric problems in the three- or higher-dimensional space. When we perform guillotine cuts on such a problem, a cut line needs to be replaced by a cut plane or a cut hyperplane. As a consequence, the number of portals on the cut plane or hyperplane would increase from $O(\log n/\varepsilon)$ to $O((\log n/\varepsilon)^2)$ or even higher. With so many possible crosspoints, the dynamic programming algorithm for finding the optimal guillotine cut–restricted solution may no longer run in polynomial time. On the other hand, the m-guillotine cut allows at most $2m$ crosspoints in each dimension. Since m is a constant with respect to n, the polynomial-time bound for the corresponding dynamic programming algorithm is preserved in the higher-dimensional spaces.

For some other problems, moving crosspoints to predetermined portals is difficult or even impossible. For such problems, the portal techniques cannot be applied at all. This includes the problem MIN-RP and the following problems, for which the m-guillotine cut technique works well:

> RECTILINEAR STEINER ARBORESCENCE: Given n terminals in the first quadrant of the rectilinear plane, find the minimum-length directed tree rooted at the origin, connecting to all terminals and consisting of only horizontal arcs oriented from left to right and vertical arcs oriented from bottom to top.
>
> SYMMETRIC RECTILINEAR STEINER ARBORESCENCE: Given n terminals in the first and second quadrants of the rectilinear plane, find the minimum-length directed tree rooted at the origin, connecting to all terminals and consisting of only horizontal arcs (in either orientation) and vertical arcs oriented from bottom to top.
>
> MINIMUM CONVEX PARTITION: Given a polygon with polygonal holes, partition it into convex areas with the minimum total length of cut lines.

On the other hand, the m-guillotine cut may be difficult to apply to some problems, while the portal technique works well on them. This includes the following problems:

EUCLIDEAN k-MEDIANS: Given a set P of n points in the Euclidean plane, find k *medians* in the plane such that the sum of the distances from each terminal to the nearest median is minimized.

EUCLIDEAN FACILITY LOCATION: Given n points $x_1, \ldots, x_n$ in the Euclidean plane and, for each $i = 1, \ldots, n$, a cost c_i for opening a facility at x_i, find a subset F of $\{1, 2, \ldots, n\}$ that minimizes

$$\sum_{i \in F} c_i + \sum_{i=1}^{n} \min_{j \in F} d(x_i, x_j),$$

where $d(x_i, x_j)$ is the Euclidean distance between x_i and x_j.

EUCLIDEAN GRADE STEINER TREE: Given a sequence of point sets $P_1 \subset P_2 \subset \cdots \subset P_m$ in the Euclidean plane and weights $c_1 > c_2 > \cdots > c_m$, find a network $G = (V, E)$ of the minimum total weight such that G contains a Steiner tree T_i for every P_i, where the total weight of G equals $\sum_{e \in E} length(e) \cdot \max_{e \in T_i} c_i$.

For the problems to which both techniques can be applied, such as RSMT, it is natural to ask whether the two techniques can be combined to yield a better approximation algorithm. As both techniques already produce PTASs, we mainly look for a combined method that can reduce the running time for the dynamic programming algorithm. A general idea is as follows: We may first use the portal technique to reduce the number of possible locations of crosspoints to $O((\log n)/\varepsilon)$ and then choose $2m$ portals to form a m-guillotine cut (with $m = O(1/\varepsilon)$). In this way, the dynamic programming algorithm for finding the best such partition would run in time $n^c (\log n)^{O(1/\varepsilon)}$, where c is a constant independent of ε. However, when we try to implement this idea, we might encounter troubles in the analysis of the performance ratio. More precisely, when we modify the optimal solution to meet our restriction, we first need to construct a relevant partition and, in particular, need to know how to select the cut lines for the construction. With the portal technique, we want to select the cut lines in a way that minimizes the number of crosspoints on the cut lines. On the other hand, in the m-guillotine cut technique, we need to select cut lines that satisfy the inequality in Mitchell's lemma. In general, these two selection criteria are often incompatible and would prevent us from finding a good combined partition.

How do we overcome this problem? An idea is to move our attention away from finding the local optimal guillotine cut at each step, but instead to work on the entire adaptive partition directly. To illustrate this point, let us define a family of adaptive partitions called *quadtree partitions*: Initially, we are given a square window that covers all the input points. In each subsequent step, if a square window contains

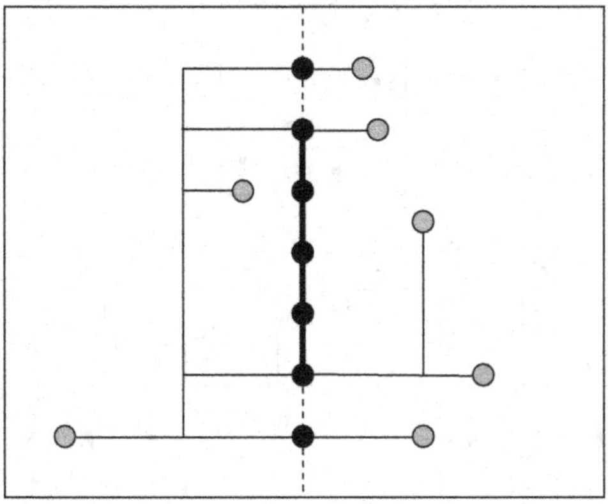

Figure 5.17: Idea of m-guillotine cut with portals.

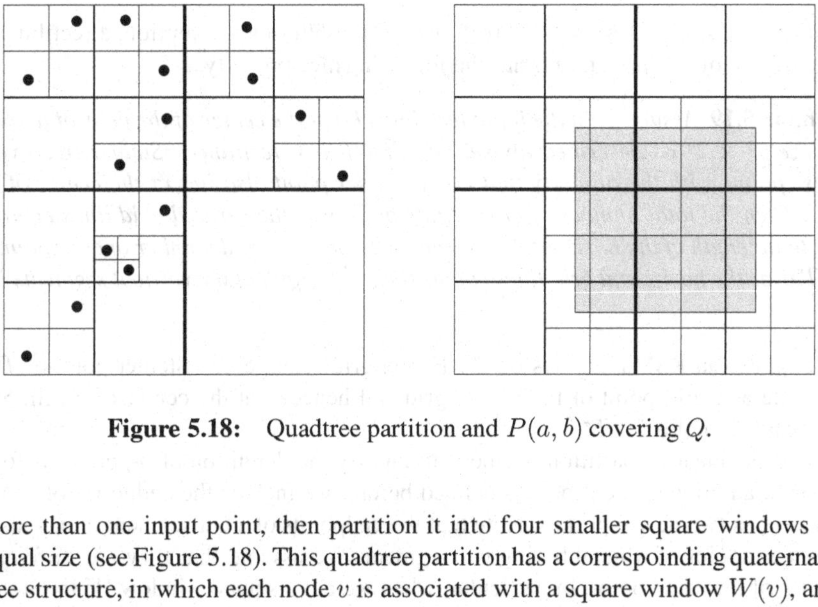

Figure 5.18: Quadtree partition and $P(a, b)$ covering Q.

more than one input point, then partition it into four smaller square windows of equal size (see Figure 5.18). This quadtree partition has a correspoinding quaternary tree structure, in which each node v is associated with a square window $W(v)$, and each internal node v has four children each associated with a smaller subsquare of $W(v)$ (see Figure 5.19).

With quadtree partitions, the cut lines are predetermined, and so there might be a large number of crosspoints on the cut segments. However, we can reduce the number of crosspoints by performing m-guillotine cuts on these cut segments. Furthermore, by the shifting technique introduced in Chapter 4, we can limit the extra cost of the m-guillotine cuts without employing Mitchell's lemma. In the following, we illustrate how this technique works on the problem RSMT.

Let Q be a square that covers all input terminals. By Lemma 5.14, we may divide Q into a $2^q \times 2^q$ grid, where $(4.5)n/\varepsilon < 2^q = O(n/\varepsilon)$, and assume that every terminal point lies at the center of a cell. We may further rescale the grid and assume

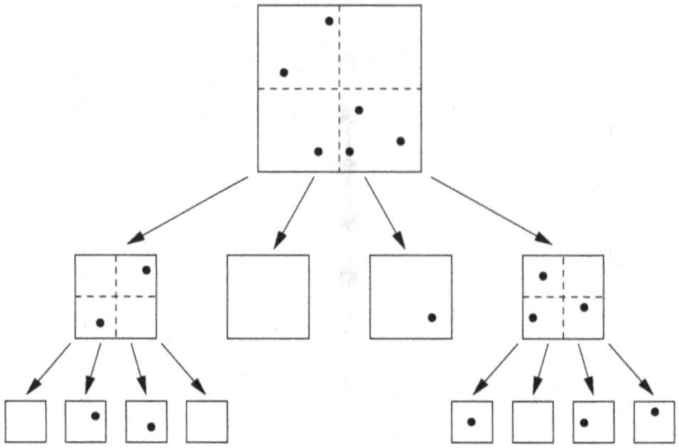

Figure 5.19: The tree structure of a quadtree partition.

that $Q = \{(x, y) \mid 0 \le x \le 2^q, 0 \le y \le 2^q\}$. With this assumption, a rectilinear Steiner tree of the input points has the following nice property:

Lemma 5.19 *Assume that the input terminals lie at the center of the cells of a grid of size $2^q \times 2^q$ as described above, and that T is a rectilinear Steiner tree over these points with the property that every Steiner point also lies at the center of a cell. Then the total number of crosspoints of T over the vertical grid lines equals the total length of the horizontal segments in T, and the total number of crosspoints of T over the horizontal grid lines equals the total length of the vertical segments in T.*

Suppose an RSMT T^* lies on the Hanan grid. Then every Steiner point of T^* must lie at a grid point of the Hanan grid and hence is at the center of a cell. So Lemma 5.19 holds for T^*.

For the quadtree partition, we need to modify the definition of p-portals as follows: In addition to the p-portals defined before, we include the endpoints of a cut segment as two new p-portals. Note that the endpoints of a cut segment s may also be an endpoint of a neighboring cut segment. We assume that they are different copies of the same point. In particular, when we cut a square window W into four square subwindows, we create four portals at the center of W, to be used as portals for the four cut segments (see Figure 5.20). We call these new portals the *endpoint portals* and the original portals the *interior portals*. Similarly, if a crosspoint on a cut segment locates at one of the endpoints of the cut segment, then we call it an *endpoint crosspoint*; otherwise, we call it an *interior crosspoint*.

Now, let $p = O((\log n)/\varepsilon)$ and $m = O(1/\varepsilon)$ be two fixed parameters. For each (a, b), with $0 \le a, b < 2^q$, we define a quadtree partition $P(a, b)$ as follows: Use (a, b) as the center to draw an initial square $\overline{Q}$ with edge length twice of that of Q (i.e., 2^{q+1}). It is obvious that $\overline{Q}$ covers Q. From this initial square $\overline{Q}$, construct a quadtree partition as described earlier, and place $p + 2$ portals on each cut segment (see Figures 5.18 and 5.20).

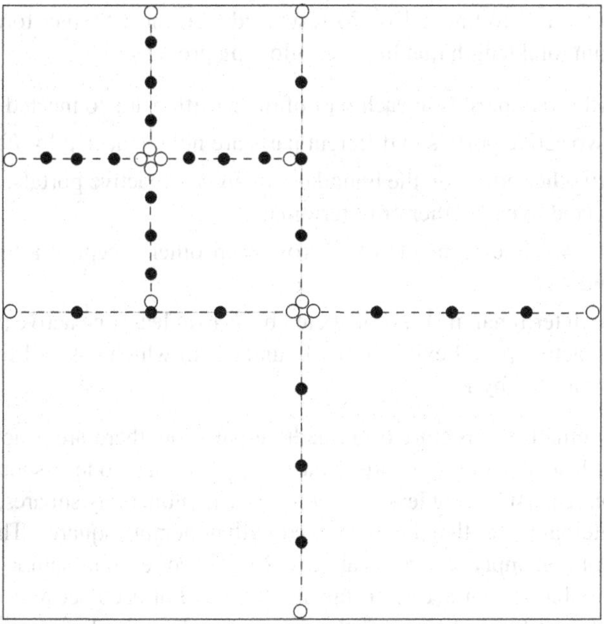

Figure 5.20: Quadtree cuts and portals on them, where a $\circ$ indicates an endpoint crosspoint.

We say a rectilinear Steiner tree T is a $P(a, b)$*-restricted rectilinear Steiner tree* (with parameters (p, m)) if, for the quadtree partition $P(a, b)$,

(1) Every edge of T crosses a cut segment at a portal, and

(2) There exist at most m interior crosspoints on every cut segment, plus possibly one or two crosspoints at the endpoints of the cut segment.

Remark. In condition (1) above, we allow an edge of T to cross a cut segment at its endpoint portal. This edge may lie on the boundary of the window, but it is treated as an edge in the interior of the window, and it can only be connected to edges in other windows through portals.

Lemma 5.20 *The minimum* $P(a, b)$*-restricted rectilinear Steiner tree with parameters* (p, m) *can be computed in time* $np^{O(m)}2^{O(m)}$.

Proof. Based on the tree structure of $P(a, b)$, we employ the method of dynamic programming to compute the minimum $P(a, b)$-restricted rectilinear Steiner tree. Each subproblem of this dynamic programming algorithm can be described as follows:

(1) A square window W is given, with all points lying in the interior of W.

(2) A set S of portals is given on the boundary of W. A subset of portals, at most m of them on the interior of each boundary, is identified as active portals, and a partition of the active portals is given.

(3) The problem is to find a $P(a, b)$-restricted rectilinear Steiner forest F of the minimum total length that has the following properties:

   (a) All active portals in each part of the partition are connected by F;
   (b) Two active portals in different parts are not connected by F;
   (c) All other points on the boundary, including inactive portals, are not connected to each other or to terminals;
   (d) No two line segments of F cross each other except at a Steiner point; and
   (e) Each terminal in W is connected by F to at least one active portal unless no active portal exists in the boundary, in which case all terminals are connected by F.

Note that in the tree structure of a quadtree partition, there are exactly n leaves associated with a nonempty square. Since each internal node associated with a nonempty square must have at least two children with nonempty squares, there are at most $n - 1$ internal nodes that are associated with nonempty squares. Therefore, the total number of nonempty squares is at most $2n - 1$. For each nonempty square, the number of possible sets of active portals is $O(p^{4m})$. For each set of active portals, the number of possible partitions of the active portals is, by Lemma 5.12, $2^{O(m)}$. Therefore, the total number of subproblems is $O(np^{4m})2^{O(m)}$.

Moreover, each iteration in the dynamic programming algorithm can be computed in time $p^{O(m)}2^{O(m)}$ since, for each cut segment, we need to consider all possible choices of the set of active portals and, for each set of active portals, all possible choices of the partition of this set. Putting them together, the dynamic programming algorithm runs in time $np^{O(m)}2^{O(m)}$. □

Choose $p = O(q/\varepsilon) = O((\log n)/\varepsilon)$ and $m = O(1/\varepsilon)$, and let T_a denote the minimum $P(a, a)$-restricted rectilinear Steiner tree with parameters (p, m). Then, by Lemma 5.20, T_a can be computed in time $n((\log n)/\varepsilon)^{O(1/\varepsilon)}$. As a result, the shortest tree among $T_0, T_1, \ldots, T_{2^q-1}$, denoted by T_{a^*}, can be computed in time $n^2(\log n)^{O(1/\varepsilon)}$.

Next, let us estimate the performance ratio of T_{a^*} as an approximation to the rectilinear SMT. To do so, consider a rectilinear SMT T^* lying on the Hanan grid so that the conclusion of Lemma 5.19 holds for T^*. For each quadtree partition $P(a, a)$, we will modify T^* to satisfy the $P(a, a)$ restriction and estimate the cost of the modification.

The modification consists of two parts. In the first part, we move each crosspoint to the nearest portal in the boundary. In the second part, we perform a *patching procedure* on cut segments to reduce the number of crosspoints such that each cut segment contains at most m interior crosspoints.

Let $\mathcal{P}$ be the family of partitions $P(a, a)$, for $a = 0, 1, \ldots, 2^q - 1$. We first estimate the total cost of the modification in the first part over all partitions in $\mathcal{P}$, instead of a single partition $P(a, a)$. That is, we calculate the total length increase resulting from moving all crosspoints to their corresponding nearest p-portals over all partitions in $\mathcal{P}$.

Lemma 5.21 *Let $c_1(P, T)$ denote the total length increase resulting from moving each crosspoint of a rectilinear Steiner tree T to the nearest p-portal in a partition P. Then, for the rectilinear SMT T^*,*

$$\frac{1}{2^q} \sum_{0 \le a < 2^q} c_1(P(a,a), T^*) \le \frac{q+1}{2(p+1)} \cdot length(T^*).$$

Proof. Consider the tree structure of the partition $P(a, a)$. As usual, we say a vertex v in this tree (or its associated square $W(v)$) is at level i, for some $i \ge 0$, if the path from the root to v has length i. In particular, the root is at level 0. A cut segment of $P(a, a)$ is also said to be at level i if it is one of the four cut segments of a level-i square W that cuts W into four squares at level $i + 1$. Note that all cut segments on a grid line must be at the same level. Thus, we may say that a grid line is at level i if all cut segments on it are at level i.

In the following, we let $\mathcal{H}$ (and $\mathcal{V}$) denote the set of all horizontal (and, respectively, vertical) grid lines in partitions in $\mathcal{P}$. Consider an arbitrary vertical grid line $\ell \in \mathcal{V}$. When we shift the partition from $P(0, 0)$ to $P(1, 1), \ldots$, and to $P(2^q - 1, 2^q - 1)$, the level of line ℓ changes along. In particular, in the family of 2^q quadtree partitions in $\mathcal{P} = \{P(a, a) \mid 0 \le a < 2^q\}$, ℓ is a level-0 cut line for exactly one partition in $\mathcal{P}$: the partition whose center vertex lies on ℓ. In addition, ℓ is also a level-1 cut line for one partition: the partition whose center vertex has distance 2^{q-1} from ℓ. In general, for each $1 \le i \le q - 1$, ℓ is a level-i cut line for a partition if the center of the partition has distance $(2j + 1)2^{q-i}$ from ℓ for some $j \ge 0$. It is easy to see that there are exactly 2^{i-1} such partitions, and hence ℓ is a level-i cut line for 2^{i-1} partitions.

Let T^*_H (and T^*_V) denote the set of all horizontal (and, respectively, vertical) line segments in tree T^*. Also, for any $\ell \in \mathcal{V}$, let $n(\ell, T^*)$ denote the number of crosspoints of T^* on line ℓ (note that this value is independent of the partitions). Note that a level-i cut segment in a partition $P(a, a)$ has edge length 2^{q-i}. Thus, for a partition $P(a, a)$ relative to which ℓ is at level i, moving a crosspoint on ℓ to a nearest p-portal on $P(a, a)$ increases the length of T^* by at most $2^{q-i}/(p + 1)$ [note that any point on ℓ has distance $\le 1/(2(p + 1))$ to the nearest portal]. Therefore, by Lemma 5.19, the total length increase for moving crosspoints at vertical cuts to portals, over all partitions in $\mathcal{P}$, is at most

$$\sum_{\ell \in \mathcal{V}} n(\ell, T^*) \cdot \left(2^q + \sum_{i=1}^{q-1} 2^{i-1} \cdot 2^{q-i}\right) \frac{1}{p+1} = length(T^*_H) \cdot \frac{2^{q-1}(q+1)}{p+1}.$$

Similarly, the total length increase resulting from moving crosspoints at all horizontal cut segments to portals, over all partitions in $\mathcal{P}$, is at most

$$length(T^*_V) \cdot \frac{2^{q-1}(q+1)}{p+1}.$$

Putting them together, we get

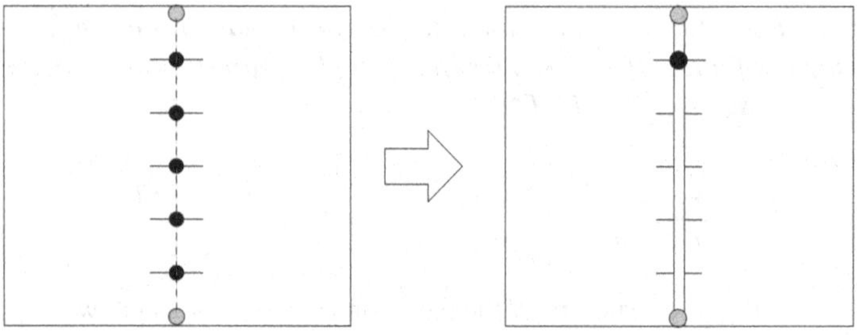

Figure 5.21: Patching.

$$\frac{1}{2^q} \sum_{0 \le a < 2^q} c_1(P(a,a), T^*) \le \frac{q+1}{2(p+1)} \cdot length(T^*). \qquad \square$$

Next, we study how to reduce the number of crosspoints so that each cut segment contains at most m interior crosspoints. An idea motivated from the m-guillotine cut is to add a guillotine cut segment to the Steiner tree and leave out at most m interior crosspoints not covered by this new segment. To simplify the operation, we will add the whole cut segment to the Steiner tree and keep only one crosspoint. More precisely, the *patching* operation on a cut segment s is as follows: If s contains more than m interior crosspoints, then we add two copies of the cut segment s to the Steiner tree (one for each subwindow resulting from the cut by s), and keep one single (interior or endpoint) crosspoint on s (see Figure 5.21).

As in the analysis of the m-guillotine cut approximation, we need to keep the total cost of patching bounded by $\varepsilon \cdot length(T^*)$. This bound is, however, difficult to get. For instance, for a cut line s at level 0, the patching operation on s would increase the length of T^* by 2^{q-1}, which by itself might be greater than $\varepsilon \cdot length(T^*)$ already. Thus, we need to modify the patching operation to avoid such expensive patchings. Intuitively, for a cut segment s having more than m interior crosspoints, we may choose a subsegment of s with a high density of crosspoints and only patch this subsegment so that the extra length added is proportional to the number of crosspoints reduced. The following procedure is an implementation of this idea.

First, let us introduce a new notation for line segments: Write $[x, y]$ to denote the line segment with endpoints x and y. For a line segment $[x, y]$, denote by $x(h)$ the point in $[x, y]$ with distance h from x. For instance, if $[x, y]$ is a line segment at level i (and hence of length 2^{q-i}), then $x = x(0)$, $y = x(2^{q-i})$, and the middle point on $[x, y]$ is $x(2^{q-i-1})$.

Iterated Patching Procedure (on a cut segment $[x, y]$ at level i):

For $k \leftarrow 0$ **to** $q - i$ **do**
  for $j \leftarrow 0$ **to** $2^{q-i-k} - 1$ **do**
    if $[x(j2^k), x((j+1)2^k)]$ has more than m interior crosspoints
      then patch the line segment $[x(j2^k), x((j+1)2^k)]$.

It is worth pointing out that although the patching operation is similar to the m-guillotine cut, it is not used in the dynamic programming algorithm to find the minimum $P(a, b)$-restricted rectilinear Steiner tree. Instead, we only use it as a tool for the analysis of the minimum $P(a, b)$-restricted rectilinear Steiner tree as an approximation to the rectilinear SMT. Since we did not use patching in the dynamic programming algorithm, we do not need to include the running time of the Iterated Patching Procedure in the construction of T_{a^*}. On the other hand, to keep the new tree T a $P(a, a)$-restricted rectilinear Steiner tree, we need to make two copies of the patching edge, one for each subwindow, so that T does not violate condition (c) given in the proof of Lemma 5.20.

Now, to reduce the number of crosspoints on a cut segment, we execute, for each partition $P(a, a)$ in $\mathcal{P}$, the Iterated Patching Procedure on every cut segment $[x, y]$ at every level in $P(a, a)$, in the order of the cut segments being generated by the quadtree partition, starting at level 0 and then moving to higher-level cut segments. The next lemma estimates the total cost of this reduction procedure over all partitions in $\mathcal{P}$.

Lemma 5.22 *Let $c_2(P, T)$ denote the total length increase resulting from executing the* Iterated Patching Procedure *on all cut segments in partition P, with respect to the crosspoints of a rectilinear Steiner tree T. Then, for the rectilinear SMT T^*,*

$$\frac{1}{2^q} \sum_{1 \le a < 2^q} c_2(P(a, a), T^*) \le \frac{2}{m} \cdot length(T^*).$$

Proof. First, we define sets $\mathcal{H}$, $\mathcal{V}$, T_H^*, and T_V^* as in the proof of Lemma 5.21.

Let ℓ be a vertical cut line in $\mathcal{V}$. Consider the procedure of Iterated Patching applied to ℓ, as if ℓ is a level-0 cut line (and hence consists of two level-0 cut segments). This procedure patches subsegments of ℓ from shorter segments to longer segments. For any k, with $0 \le k \le q$, let $g(k, \ell)$ be the number of length-2^k subsegments patched in this procedure, or, equivalently, the number of patches done by the procedure in the kth iteration. Note that this number $g(k, \ell)$ depends only on the crosspoints of T^* with line ℓ and is independent of the quadtree partitions. Indeed, for any partition $P(a, a)$ relative to which ℓ is at level $i \le q - k$, the total number of patches done in the kth iteration of the Iterated Patching Procedure, on all cut segments in ℓ, is equal to $g(k, \ell)$. (Note that when we patch a line segment, the new patch segment may intersect a cut segment s at a higher level and generate new crosspoints on s. However, all these new crosspoints locate at the endpoints of the cut segment s, and so they do not affect later patching procedures, as it only considers the interior crosspoints and ignores the endpoint crosspoints.)

Thus, if ℓ is at level i relative to a partition $P(a, a)$, then the total length increase resulting from executing the Iterated Patching Procedure on cut segments of partition $P(a, a)$ that lie in ℓ is at most

$$\sum_{k=0}^{q-i} g(k, \ell) \cdot 2^{k+1}.$$

(Note that each cut segment is doubled for patching.)

Now, consider the Iterated Patching Procedure applied to grid line ℓ over all quadtree partitions $P(a, a)$ in $\mathcal{P}$. Recall from the proof of Lemma 5.21 that ℓ is at level 0 for one partition in $\mathcal{P}$, and, for each $i \geq 1$, ℓ is at level i for 2^{i-1} partitions in $\mathcal{P}$. Therefore, the total length increase resulting from executing the Iterated Patching Procedure on all segments in ℓ, over all partitions in $\mathcal{P}$, is at most

$$\sum_{k=0}^{q} g(k,\ell)\cdot 2^{k+1} + \sum_{i=1}^{q} 2^{i-1} \sum_{k=0}^{q-i} g(k,\ell)\cdot 2^{k+1}$$
$$= \sum_{k=0}^{q} g(k,\ell)\left(2^{k+1} + \sum_{i=1}^{q-k} 2^{k+i}\right) = \sum_{k=0}^{q} g(k,\ell)\cdot 2^{q+1}.$$

Note that each patching of a cut segment reduces at least m crosspoints of T^* with ℓ. Thus,

$$\sum_{k=0}^{q} g(k,\ell) \leq \frac{n(\ell, T^*)}{m},$$

where $n(\ell, T^*)$ is the number of crosspoints of T^* on ℓ. It follows that the total length increase that resulted from Iterated Patching on ℓ, over all partitions in $\mathcal{P}$, is at most

$$2^{q+1} \cdot \frac{n(\ell, T^*)}{m}.$$

Therefore, by Lemma 5.19,

$$\frac{1}{2^q} \sum_{1\leq a<2^q} c_2(P(a,a), T^*) \leq \sum_{\ell\in\mathcal{H}\cup\mathcal{V}} \frac{2\cdot n(\ell, T^*)}{m} = \frac{2}{m}\cdot \mathit{length}(T^*). \qquad \Box$$

In summary, for each quadtree partition $P(a, a)$ in $\mathcal{P}$, we modify the rectilinear SMT T^* as follows: We first perform the Iterated Patching procedure on all cut segments of $P(a, a)$ to reduce the number of crosspoints of T^* to no more than m on each cut segment. Call the resulting tree T'_a. Then we move all crosspoints of T'_a on the grid lines of $P(a, a)$ to their corresponding nearest p-portals. Let T''_a denote the resulting Steiner tree. It is clear that T''_a is a $P(a, a)$-restricted rectilinear Steiner tree.

Lemma 5.23 *With parameters $p \geq 2(q + 1)/\varepsilon$ and $m \geq 8/\varepsilon$, at least one half of the trees T''_a, for $0 \leq a < 2^q$, have*

$$\mathit{length}(T''_a) \leq (1 + \varepsilon)\mathit{length}(T^*).$$

Proof. It is clear that

$$\mathit{length}(T''_a) \leq \mathit{length}(T^*) + c_2(P(a,a), T^*) + c_1(P(a,a), T'_a)).$$

Furthermore, we note that each crosspoint coming from an edge in $T'_a \setminus T^*$ is an endpoint portal, and need not be moved. Therefore, when we move the crosspoints of T'_a on grid lines of $P(a,a)$ to their corresponding nearest portals, only original crosspoints of T^* on a grid line of $P(a,a)$ need to be moved. Thus, $c_1(P(a,a), T'_a) \le c_1(P(a,a), T^*)$. It follows that

$$length(T''_a) \le length(T^*) + c_2(P(a,a), T^*) + c_1(P(a,a), T^*).$$

For fixed parameters p and m such that $p \ge 2(q+1)/\varepsilon$ and $m \ge 8/\varepsilon$, we have, by Lemmas 5.22 and 5.21,

$$\begin{aligned}\frac{1}{2^q} \sum_{1 \le a < 2^q} length(T''_a) &\le \left(1 + \frac{2}{m} + \frac{q+1}{2(p+1)}\right) \cdot length(T^*) \\ &\le \left(1 + \frac{\varepsilon}{2}\right) \cdot length(T^*).\end{aligned}$$

Therefore, it holds, for at least one half of $a \in \{0, 1, \ldots, 2^q - 1\}$, that $length(T''_a) \le (1+\varepsilon) \cdot length(T^*)$. □

Theorem 5.24 *There exists a* $(1+\varepsilon)$*-approximation to the problem* RSMT *that can be computed in time* $n^2(\log n)^{O(1/\varepsilon)}$. *Moreover, with probability* $1/2$, *a* $(1+\varepsilon)$*-approximation for* RSMT *can be computed in time* $n(\log n)^{O(1/\varepsilon)}$.

Proof. The first half of the corollary is a direct consequence of Lemma 5.23. For the second half, we can choose a random quadtree partition $P(a,a)$ and compute the minimum $P(a,a)$-restricted rectilinear Steiner tree T_a. □

5.6 Two-Stage Portals

In the last section, we combined the portal and m-guillotine cut techniques to get a PTAS for the problem RSMT in time $n^2(\log n)^{O(1/\varepsilon)}$. We now introduce yet another idea, called *two-stage portals*, to further improve the running time of the PTAS.

Let $[x,y]$ be a cut segment in a quadtree partition. For two integers $p_1, p_2 > 0$, we can set up two-stage (p_1,p_2)-portals on $[x,y]$ as follows: We first set up a set $\{z_0 = x, z_1, \ldots, z_{p_1}, z_{p_1+1} = y\}$ of p_1-portals on $[x,y]$. Next, we choose two points x', y' from $\{z_0, z_1, \ldots, z_{p_1+1}\}$, and set up a set $\{w_0 = x', w_1, \ldots, w_{p_2}, w_{p_2+1} = y'\}$ of p_2-portals on $[x', y']$. We call $\{w_0, w_1, \ldots, w_{p_2+1}\}$ a set of *two-stage* (p_1,p_2)*-portals* on segment $[x,y]$ (see Figure 5.22). We note that for each segment $[x,y]$, there are $O(p_1^2)$ sets of (p_1,p_2)-portals on $[x,y]$.

To apply two-stage portals to the approximation of RSMT, we first modify the notion of $P(a,b)$-restricted rectilinear Steiner trees accordingly. That is, a rectilinear Steiner tree T is a $P(a,b)$*-restricted rectilinear Steiner tree* (with parameters (p_1,p_2,m)) if it satisfies the following conditions:

(1) Each crosspoint of T on a cut segment s of $P(a,b)$ belongs to a set of (p_1,p_2)-portals on s or is an endpoint of s, and

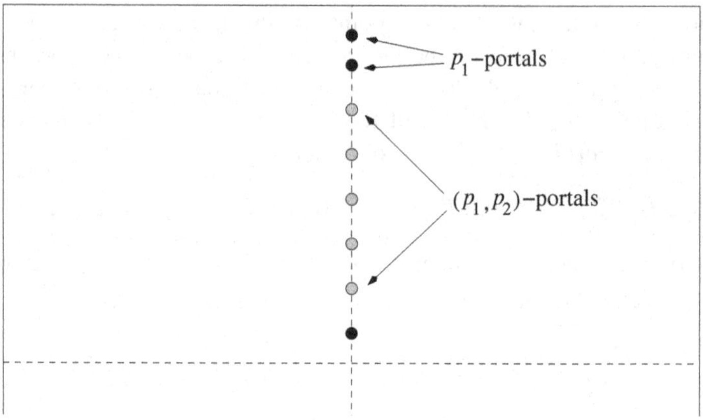

Figure 5.22: Two-stage portals.

(2) There exist at most m interior crosspoints of T on each cut segment of $P(a,b)$,

Lemma 5.25 *With parameters p_1, p_2, m, the minimum $P(a,b)$-restricted rectilinear Steiner tree can be computed in time $np_1^{10}p_2^{5m}2^{O(m)}$. In particular, if we choose $2^q = O(n/\varepsilon)$, $p_1 = O(q/\varepsilon) = O((\log n)/\varepsilon)$, $m = O(1/\varepsilon)$, and $p_2 = O(m^2)$, then the minimum $P(a,b)$-restricted rectilinear Steiner tree can be computed in time $n(\log n)^{10}(1/\varepsilon)^{O(1/\varepsilon)}$.*

Proof. Let $T(a,b)$ be the minimum $P(a,b)$-restricted rectilinear Steiner tree. We construct $T(a,b)$ by dynamic programming based on the tree structure of the quadtree partition $P(a,b)$ as in Lemma 5.20. In particular, a subproblem in the dynamic programming algorithm is the same as that in Lemma 5.20, except that the active portals are m points in a (p_1, p_2)-portal, plus possibly one or two endpoints of the cut segments. We note that for a cut segment s of the partition $P(a,b)$, there are $O(p_1^2 p_2^m)$ possible choices of active portals: First, there are $O(p_1^2)$ sets of (p_1, p_2)-portals on s; and, second, there are $O(p_2^m)$ ways to choose m active interior portals out of $p_2 + 2$ locations. Thus, following the argument in the proof of Lemma 5.20, there are $O(np_1^8 p_2^{4m}) \cdot 2^{O(m)}$ possible subproblems in the dynamic programming algorithm. Namely, there are at most $2n$ nonempty squares in the tree structure of the quadtree partition; each square has four sides, with each side having $O(p_1^2 p_2^m)$ possible choices of active portals; and, finally, there are $2^{O(m)}$ ways to partition active portals into connected parts.

Moreover, using the same argument, we can see that each iteration takes time $O(p_1^2 p_2^m) \cdot 2^{O(m)}$. So, the total running time of the dynamic programming algorithm is $np_1^{10}p_2^{5m}2^{O(m)}$. □

To estimate the average performance of $T_a = T(a,a)$ for $a \in \{0, 1, \ldots, 2^q - 1\}$, we consider a rectilinear SMT T^* lying in the Hanan grid and modify T^* to satisfy the $P(a,a)$-restriction. We first perform the Iterated Patching Procedure on each cut segment of the partition $P(a,a)$ so that each cut segment contains at most m interior crosspoints. We call the resulting tree T'_a. Then, for each cut segment s of the partition, we choose a subsegment $[x,y]$ of s with the minimum length that satisfies the following properties:

(i) Each of x, y is a p_1-portal on s;

(ii) All interior crosspoints of T'_a on s lie in $[x,y]$.

We set up the two-stage (p_1,p_2)-portals on the segment $[x,y]$, and move all interior crosspoints of T'_a to their corresponding nearest portals on this set of (p_1,p_2)-portals. We call the resulting tree T''_a. The following lemma estimates the length increase resulted from this modification.

Lemma 5.26 *With parameters $m \geq 8/\varepsilon$, $p_2 = 2m^2$, and $p_1 \geq 2(q+1)$, at least one half of the trees T''_a, for $0 \leq a < 2^q - 1$, satisfy*

$$length(T''_a) \leq (1+\varepsilon)length(T^*).$$

Proof. First, we follow the notation of the proof of Lemma 5.22 and denote the total length increase that resulted from Iterated Patching of T^* on partition $P(a,a)$ by $c_2(P(a,a),T^*)$. Also, let $c_3(P(a,a),T'_a)$ be the total length of the subsegments $[x,y]$ of cut segments s of $P(a,a)$ that we chose to form the (p_1,p_2)-portals on s. Then the length increase from moving the crosspoints of T'_a to their corresponding nearest two-stage (p_1,p_2)-portals is bounded by $(m/(p_2+1))c_3(P(a,a),T'_a)$.

Now, we claim that

$$\frac{1}{2^q}\sum_{0\leq a<2^q} c_3(P(a,a),T'_a) \leq 4\cdot length(T^*).$$

With this claim, the total length increase that resulted from modifying T^* into T''_a, over all partitions $P(a,a)$ in $\mathcal{P}$, is

$$\frac{1}{2^q}\sum_{0\leq a<2^q}\Big(c_2\left(P(a,a),T^*\right) + \frac{m}{p_2+1}\cdot c_3(P(a,a),T'_a)\Big)$$
$$\leq \Big(\frac{2}{m}+\frac{4m}{p_2+1}\Big)\cdot length(T^*) \leq \frac{\varepsilon}{2}\cdot length(T^*).$$

Therefore, for at least one half of $a \in \{0,1,\ldots,2^q-1\}$, we must have $length(T''_a) \leq (1+\varepsilon)length(T^*)$.

It remains to prove our claim.

For each cut segment $s = [x,y]$ of $P(a,a)$ of length 2^{q-i}, let

$$subseg(s) = \{[x(j2^k), x((j+1)2^k)] \mid 0 \leq k \leq q-i, 0 \leq j < 2^{q-i-k}\}.$$

For each cut segment s of $P(a,a)$ that contains at least one interior crosspoint of T'_a, let $[u(s),v(s)]$ be the shortest segment in $subseg(s)$ that contains all interior crosspoints of T'_a on s.

We note that if a cut segment s contains only one interior crosspoint, then the length of $[u(s), v(s)]$ must be 1. Thus, the total length L_1 of $[u(s), v(s)]$ over all cut segments s containing exactly one crosspoint is bounded by the number of interior crosspoints of T'_a on $P(a,a)$. Since each interior crosspoint of T'_a on $P(a,a)$ is also a crosspoint of T^* on $P(a,a)$, the length L_1 is, by Lemma 5.19, bounded by $length(T^*)$.

Next, consider a cut segment s of $P(a,a)$ that contains at least two interior crosspoints of T'_a. Suppose we apply the Iterated Patching Procedure to segment s, with parameter $m = 1$ and crosspoints of T'_a; then $[u(s), v(s)]$ would be the segment where we apply the (last) patching operation. Moreover, we note that the interior crosspoints of T'_a are obtained from the crosspoints of T^* through an Iterated Patching Procedure. Therefore, if we apply the Iterated Patching Procedure to segment s, with parameter $m = 1$ and crosspoints of T^*, the last patching operation must be on the same segment $[u(s), v(s)]$. Thus, the total length of $[u(s), v(s)]$ over all cut segments s that contain at least two crosspoints is bounded by $c_2(P(a,a), T^*)$ with respect to the parameter $m = 1$. [We write $c_2^{(1)}(P(a,a), T^*)$ to emphasize that the parameter m used in this bound is $m = 1$.]

Finally, let us move $u(s)$ and $v(s)$ to two p_1-portals $u'(s)$, $v'(s)$ of s satisfying $[u(s), v(s)] \subseteq [u'(s), v'(s)]$. This will increase the total length by at most $4c_1(P(a,a), T'_a)$. (Note that the distance between $u'(s)$ and $u(s)$ is at most $1/(p_1+1)$, but might be greater than $1/(2(p_1+1))$.) Again, since all interior crosspoints of T'_a are also crosspoints of T^*, this length increase is actually bounded by $4c_1(P(a,a), T^*)$.

Now, from Lemmas 5.21 and 5.22, we know that the total length increase $c_3(P(a,a), T'_a)$, over all partitions $P(a,a)$ in $\mathcal{P}$, can be bounded as follows:

$$\frac{1}{2^q} \sum_{0 \le a < 2^q} c_3(P(a,a), T'_a)$$

$$\le length(T^*) + \frac{1}{2^q} \sum_{0 \le a < 2^q} c_2^{(1)}(P(a,a), T^*) + \frac{1}{2^q} \sum_{0 \le a < 2^q} 4c_1(P(a,a), T^*)$$

$$\le length(T^*) + 2\Big(1 + \frac{q+1}{p_1+1}\Big) \cdot length(T^*) \le 4 \cdot length(T^*).$$

This completes the proof of the claim and, hence, the lemma. □

The following theorem follows immediately from Lemmas 5.25 and 5.26.

Theorem 5.27 *For any $\varepsilon > 0$, there exists a $(1+\varepsilon)$-approximation for the problem* RSMT *that runs in time $n^2(\log n)^{10}(1/\varepsilon)^{O(1/\varepsilon)}$. Moreover, with probability $1/2$, a $(1+\varepsilon)$-approximation for* RSMT *can be computed in time $n(\log n)^{10}(1/\varepsilon)^{O(1/\varepsilon)}$.*

So far, we have used portals together with patching to reduce the running time of the PTASs for the problem RSMT substantially. We note, however, that the cost of moving crosspoints to portals depends on the depth of the adaptive partition (cf. Theorem 5.18 and Lemma 5.21), and so it is hard to further reduce the running time of the PTAS using the portal technique. Thus, for further improvement over the running time of the PTAS, we must give up on portals and look for other techniques. One promising direction is to combine the patching technique with the graph-theoretic notions of *spanners* and *banyans*. For the problem RSMT, it has been shown, using spanners, banyans, and patching, that, for any $\varepsilon > 0$, there exist a randomized $(1 + \varepsilon)$-approximation running in time $O(n \log n)$ and a deterministic $(1 + \varepsilon)$-approximation running in time $O(n^2 \log n)$. The proof is, unfortunately, too involved to be included here. The interested reader is referred to Rao and Smith [1998] for the details.

Exercises

5.1 Show that, for any given rectilinear polygon without holes, the minimum-length rectangular partition can be found by dynamic programming in time $O(n^4)$.

5.2 A *stair* is a rectilinear polygon of the shape as shown in Figure 5.23.

(a) Show that the minimum-length rectangular partition for stairs can be computed by dynamic programming in time $O(n^2)$.

(b) Can you improve the running time of the above algorithm to $O(n \log n)$?

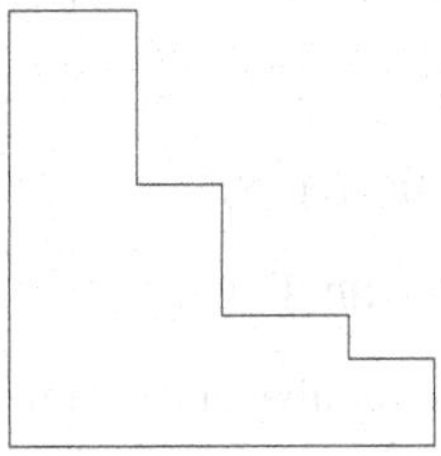

Figure 5.23: A stair.

5.3 Consider the problem MIN-RP$_1$. Prove, by constructing a counterexample, that the upper bound for the ratio of the minimum-length guillotine rectangular partition to the minimum-length rectangular partition cannot be smaller than $3/2$.

5.4 Consider a rectangular partition P of a rectilinear polygon, possibly with rectilinear holes. Let $proj_x(P)$ denote the total length of segments on a horizontal line covered by vertical projection of the partition P. Let $guil(P)$ be the set of guillotine rectangular partitions obtained from adding some segments to P. Show, by induction on the number k of segments in P, that there exists a partition $P_G \in guil(P)$ such that

$$length(P_G) \leq 2 \cdot length(P) - proj_x(P).$$

5.5 Show that for any rectilinear SMT of n points in a rectangle, and any $m \geq 1$, there exists a constant $c > 0$ such that either (i) there exists a horizontal line L, not passing through any input point, such that

$$length(L \cap H_m) + c \cdot length(L \cap H_1) \leq length(L \cap V_m) + c \cdot length(L \cap V_1),$$

or (ii) there exists a vertical line L, not passing any input point, such that

$$length(L \cap H_m) + c \cdot length(L \cap H_1) \geq length(L \cap V_m) + c \cdot length(L \cap V_1),$$

where H_m (V_m) is the set of all horizontal (vertical, respectively) m-dark points.

5.6 For each of the following problems, use both techniques of m-guillotine cut and quadtree partition with patching to design PTASs for it:

(a) RECTILINEAR STEINER ARBORESCENCE;

(b) SYMMETRIC RECTILINEAR STEINER ARBORESCENCE;

(c) MINIMUM-LENGTH CONVEX PARTITION.

In general, are the two techniques equivalent? If not, show a counterexample.

5.7 For each of the following problems, use both techniques of $(1/3, 2/3)$-partition with portals and quadtree partition with portals to design PTASs for it:

(a) EUCLIDEAN k-MEDIANS;

(b) EUCLIDEAN FACILITY LOCATION;

(c) EUCLIDEAN GRADE STEINER TREE.

In general, are the two techniques equivalent? If not, show a counterexample.

5.8 For each of the following problems, use the technique of quadtree partition with patching and portals to design a PTAS for it:

(a) ESMT;

(b) EUCLIDEAN-TSP: Given n points in the Euclidean plane, find a minimum-length tour passing through all n points;

(c) EUCLIDEAN k-SMT: Given n terminals in the Euclidean plane and an integer $1 \leq k \leq n$, find a shortest tree interconnecting at least k terminals.

5.9 Consider the following idea of combining the techniques of m-guillotine cut and portals: We first make a 1-guillotine cut, and put portals on the cut segment. Next, perform an m-guillotine cut with these portals. Apply this idea to design a $(1+\varepsilon)$-approximation for the problems RSMT, ESMT and EUCLIDEAN-TSP. Show that these approximation algorithms can be made to run in time $n^c(\log n)^{O(1/\varepsilon)}$ for some constant $c > 0$.

5.10 Design a PTAS for the following problem:

> 3-DIMENSIONAL RSMT: Given a set of n terminals in the three-dimensional rectilinear space, find a minimum-length tree interconnecting all terminals.

5.11 Show that in any rooted tree with each internal vertex having at least two children, the number of internal vertices is less than the number of leaves.

5.12 Consider the following variation of quadtree partition, called *binary-tree partition*: On an input square, at each step $2i - 1$, partition each square into two rectangles of equal size; and at each step $2i$, partition each rectangle into two squares of equal size. Show that we can use the binary-tree partition to replace the quadtree partition to get results in Sections 5.5 and 5.6.

5.13 Consider a grid on a square in the Euclidean plane with each cell a unit square. Show that for any line segment $\overline{AB}$ with the two endpoints located at the centers of two grid cells, the number of crosspoints of $\overline{AB}$ on the grid lines is bounded by $\sqrt{2} \cdot length(\overline{AB})$.

5.14 For each of the following problems, apply the technique of two-stage portals to design a PTAS for it such that a $(1 + \varepsilon)$-approximation can be found in time $n(\log n)^{10}(1/\varepsilon)^{O(1/\varepsilon)}$:

(a) ESMT;

(b) EUCLIDEAN-TSP.

5.15 Design a PTAS for the following problem:

> INTERCONNECTING HIGHWAYS: Given a set of disjoint line segments in the Euclidean plane, find a shortest tree interconnecting them.

5.16 Consider the following problem:

> RSMT WITH OBSTRUCTIONS: Given a set of terminals in the rectilinear plane with the presence of rectilinear obstructions, find a shortest tree interconnecting all terminals without passing through the obstructions.

Explain why neither the m-guillotine cut nor portal technique works for this problem. Could you find a new technique to construct a PTAS for it?

5.17 Consider the following variation of EUCLIDEAN-TSP:

> Given n disjoint regions in the Euclidean plane, find a shortest tour visiting each region at least once.

Could you find a PTAS for this problem? If you cannot, what are the difficulties when you try to apply the techniques of m-guillotine cut and portals to this problem?

5.18 Consider the following problem:

> Given a set of n sites in the Euclidean plane, a special site r, and a positive number $L > 0$, find a tour starting from r and returning to r with total length at most L that maximizes the number of visited sites.

Show that there is a 2-approximation for this problem.

5.19 Extend the m-guillotine partition technique to the approximation to polygonal partition problems, in which a partition segment is not necessarily rectilinear. In particular, show that for any polygonal partition with edge set E of total length L, there exists an m-guillotine partition of length at most

$$L + \frac{\sqrt{2}}{m}\left(L - \frac{\varsigma^{(m)}(E)}{2}\right),$$

whose edge set contains E, where $\varsigma^{(m)}(E)$ is the sum of the lengths of the four sets of one-sided m-dark points on the subsegments of E. (We say a point z is *one-sided* m*-dark* with respect to set E, in the direction $D \in \{\mathit{left}, \mathit{right}, \mathit{above}, \mathit{below}\}$, if the half-line starting from z, not including z, going in the direction D meets at least m line segments of E. In a set of one-sided m-dark points, all points in this set are one-sided m-dark in the same direction D.)

Historical Notes

Using the technique of adaptive partition to design approximation algorithms was first introduced by Du, Pan, and Shing [1986] in the study of MIN-RP. The problem MIN-RP was first proposed by Lingas et al. [1982], who showed that the general case of MIN-RP is **NP**-hard, but its hole-free subproblem can be solved in time $O(n^4)$. A naïve idea of designing approximation algorithms for the general case of MIN-RP is to use a forest connecting all holes to the boundary and then solve the resulting hole-free case. With this idea, Lingas [1983] gave the first constant-bounded approximation to MIN-RP, with the performance ratio 41. Du [1986] improved the algorithm and obtained an approximation with performance ratio 9. Meanwhile, Levcopoulos [1986] presented a faster approximation based on the greedy strategy, but with a larger performance ratio.

Motivated by the work of Du et al. [1988] on the application of dynamic programming to finding optimal routing trees, Du, Pan, and Shing [1986] initiated the idea of guillotine cut for the problem MIN-RP. They showed that the minimum-length guillotine rectangular partition can be computed in time $O(n^5)$ with dynamic programming and, as an approximation to MIN-RP, it has a performance ratio at most 2 for the special case of MIN-RP$_1$. Du, Hsu, and Xu [1987] gave a different proof for this result. They also extended the idea of guillotine cuts to the problem of MINIMUM CONVEX PARTITION. The special case of MIN-RP$_1$ was shown to be **NP**-hard by Gonzalez and Zheng [1985]. Gonzalez and Zheng [1989] improved the performance ratio 2 proved in Theorem 5.2 to 1.75 with an ad hoc case-by-case analysis.

Arora [1996] is a milestone in the study of adaptive partition. He used this technique to design PTASs for many geometric optimization problems, including the problems EUCLIDEAN-TSP, ESMT, RSMT, DEGREE-RESTRICTED SMT, k-TSP, and k-SMT. His approximation algorithms typically run in time $n^{O(1/\varepsilon)}$ for the performance ratio $1+\varepsilon$. In the meantime, an independent line of study on m-guillotine cuts had been made by Mitchell. Inspired by the work of Du, Pan, and Shing [1986], Mitchell [1996a] introduced the notion of 1-guillotine cut. Mitchell [1996b] (later published in a journal version by Mitchell [1999]) pointed out that results similar to those of Arora [1996] could be obtained by a minor modification of his work in Mitchell [1996a] (a journal version was later published as Mitchell et al. [1999]). A year later, Arora [1997] used quadtree partition and the technique of patching, which was inspired by the idea of m-guillotine cut, to reduce the running time of the PTASs from $n^{O(1/\varepsilon)}$ to $n^3(\log n)^{O(1/\varepsilon)}$. Soon later, Mitchell [1997] also improved his algorithms with the idea of two-stage portals. In Arora [1997], a family of $O((n/\varepsilon)^2)$ quadtree partitions was employed to establish the average performance of the algorithm. Du [2001] improved it to use only $O(n/\varepsilon)$ quadtree partitions, and reduced the running time of derandomization.

Interesting applications of the above techniques have been found in STEINER ARBORESCENCE by Lu and Ruan [2000], SYMMETRIC STEINER ARBORESCENCE by Cheng, DasGupta, and Lu [2001], INTERCONNECTING HIGHWAYS by Cheng, Kim, and Lu [2001], EUCLIDEAN k-MEDIANS and EUCLIDEAN FACILITY LOCATION by Arora, Raghavan, and Rao [1998], and Arkin et al. [1998]. Rao and Smith [1998] applied spanners and banyans to geometric approximation problems. Arora, Grigni et al. [1998] extended these ideas to problems in planar graphs. The application of adaptive partition to graph problems is a rich area with potential for further research.

It is an open problem whether there are $(1+\varepsilon)$-approximations that run in time $n^c(\log n)^{O(1/\varepsilon)}$ for the problems MIN-RP, RECTILINEAR STEINER ARBORESCENCE, SYMMETRIC RECTILINEAR STEINER ARBORESCENCE, EUCLIDEAN k-MEDIANS, and EUCLIDEAN FACILITY LOCATION.

6
Relaxation

Your mind will answer most questions
if you learn to relax and wait for the answer.
—William S. Burroughs

An optimization problem asks for a solution from a given feasible domain that provides the optimal value of a given objective function. The technique of relaxation is, contrary to the technique of restriction, to relax some constraints on the feasible solutions and, hence, enlarge the feasible domain so that an optimal or a good approximate solution to the relaxed version of the problem can be found in polynomial time. This optimal or approximate solution to the relaxed version is not necessarily feasible for the original problem, and we may need to modify it to get a feasible solution to the original input. This modification step often requires special tricks and is an important part of the relaxation technique.

In this chapter, we introduce various ideas about relaxation. Then, in Chapters 7, 8, and 9, we will study how to relax combinatorial optimization problems into linear programs or semidefinite programs, and how to modify their solutions to feasible solutions of the original problems.

6.1 Directed Hamiltonian Cycles and Superstrings

Depending on the nature of the problem, there are many ways of relaxing the constraints of an optimization problem. Let us first look at some simple examples about finding Hamiltonian circuits.

Example 6.1 Recall the problem TSP (TRAVELING SALESMAN PROBLEM) studied in Section 1.6. The feasible domain of an instance G of TSP consists of all Hamiltonian circuits of the input graph G. Note that a Hamiltonian circuit of G must be a spanning graph of G. As the minimum spanning tree of a graph is well known to be computable in polynomial time, we may relax the feasible domain of TSP to contain all spanning graphs, and try to use the minimum spanning tree as an approximation to the minimum Hamiltonian circuit. Note, however, that the minimum spanning tree is not a Hamiltonian circuit. Thus, we need to modify it to get a feasible solution for the original problem.

This approach was taken in Algorithms 1.G and 1.H. In Algorithm 1.G, the modification consists of three steps: We first double the edges of the minimum spanning tree so that every vertex of the tree has an even degree. Then we convert this tree into an Euler tour. Finally, we take a *shortcut* through the Euler tour to get a Hamiltonian circuit and use it as the approximate solution. When the input graph satisfies the triangle inequality, this algorithm gives us a 2-approximation to TSP.

In Algorithm 1.H (Christofides's approximation), an additional idea is used for the first step: Instead of doubling every edge of the minimum spanning tree, we only add a perfect matching on vertices of odd degrees to get a subgraph in which each vertex has an even degree. This new idea improves the performance ratio to $3/2$ when the input satisfies the triangle inequality.

For the problem of DIRECTED TSP (or, MINIMUM DIRECTED HAMILTONIAN CIRCUIT), it seems rather difficult to modify a directed spanning tree (also called an *arborescence spanning tree*) to a Hamiltonian circuit. So this approach does not work well for DIRECTED TSP. □

Example 6.2 Consider the problems MAX-HC (MAXIMUM HAMILTONIAN CIRCUIT) and MAX-DHC (MAXIMUM DIRECTED HAMILTONIAN CIRCUIT). The feasible domain of either problem is, again, the set of all Hamiltonian circuits of the input graph. Note that the objective function of these two problems is the length of a Hamiltonian circuit, which can be written as the sum of the lengths of two matchings if the number of vertices of the input graph is even, or the sum of the lengths of three matchings if the number of vertices is odd. So we may relax the feasible domain to include all pairs or triples of (independent) matchings, and in turn further relax it to simply the set of all matchings. That is, we find a maximum matching, and then modify it to a Hamiltonian circuit and use it as an approximation to MAX-HC or MAX-DHC. Note that the total length of the maximum matching of a graph G is at least one third that of the maximum Hamiltonian circuit of G. Therefore, connecting the maximum matching into a Hamiltonian circuit results in an approximation to MAX-HC or MAX-DHC with performance ratio 3. For the problem MAX-DHC, this approximation is as good as the greedy algorithm.

The above idea of relaxation can also be applied to the problems MAX-HP (MAXIMUM HAMILTONIAN PATH) and MAX-DHP (MAXIMUM DIRECTED HAMILTONIAN PATH). Since a Hamiltonian path can always be written as the sum of two matchings, the maximum matching provides an approximation to these two problems with performance ratio 2.

Theorem 6.3 *For each of the problems* MAX-HP *and* MAX-DHP, *there exists a polynomial-time* 2*-approximation.*

For the directed case (i.e., MAX-DHP), this result is better than that of the greedy algorithm (cf. Theorems 2.5 and 2.16). □

Example 6.4 For Hamiltonian circuits, there is another possible relaxation. Recall that an assignment is a maximal matching in a bipartite graph. For a complete directed graph $G = (V, E)$, we may call a collection of disjoint cycles that cover all vertices of G an *assignment*. To see this, define a bipartite graph $H = (V, V', E')$, where $V' = \{x' \mid x \in V\}$ and $E' = \{\{x, y'\} \mid \{x, y\} \in E\}$. Then a maximal matching of H is a maximal set M of disjoint edges in H. Since M is maximal and edges in M are disjoint, this matching defines a one-to-one function from V to V'. When we identify V' with V, this matching becomes a collection of disjoint cycles that cover every vertex in G.

It is clear that a Hamiltonian circuit in a directed graph is an assignment. Since maximum matching is polynomial-time computable, the above observation suggests that we relax the problem of finding directed Hamiltonian circuits to finding assignments.

For the problems MAX-DHC, this idea leads to the following approximation algorithm:

Algorithm 6.A (*Approximation Algorithm for* MAX-DHC)

Input: A complete directed graph $G = (V, E)$ without self-loops, and a weight function $w : E \to \mathbb{N}$.

(1) Find a maximum assignment $A = C_1 \cup C_2 \cup \cdots \cup C_t$ for the graph G with edge weight w, where each C_i, for $i = 1, 2, \ldots, t$, is a cycle.

(2) **For** $i \leftarrow 1$ **to** t **do**
let (u_i, v_i) be an edge in C_i with the lowest weight;
let C_i' be the path in C_i that begins at v_i and ends at u_i.

(3) Let H be the cycle formed by connecting the paths C_i', $i = 1, 2, \ldots, t$, with edges $(u_1, v_2), (u_2, v_3), \ldots, (u_{t-1}, v_t)$, and (u_t, v_1) (cf. Figure 6.1); return H. ■

Note that the maximum Hamiltonian circuit H^* of G is an assignment, and so $w(A) \geq w(H^*)$. Furthermore, since G has no self-loops, each cycle C_i in A has at least two edges. Therefore, each edge (u_i, v_i) has weight $w((u_i, v_i)) \leq w(C_i)/2$; or, equivalently, each path C_i' has weight $w(C_i') \geq w(C_i)/2$. It follows that

$$w(H) \geq \sum_{i=1}^{t} w(C_i') \geq \frac{1}{2} \sum_{i=1}^{t} w(C_i) = \frac{w(A)}{2} \geq \frac{w(H^*)}{2}.$$

Therefore, Algorithm 6.A is a 2-approximation to MAX-DHC. □

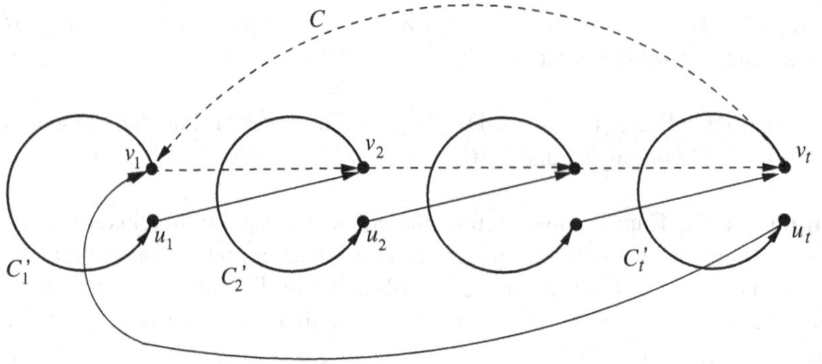

Figure 6.1: Construct a directed HC from an assignment.

We have pointed out in Example 6.1 that the relaxation of Hamiltonian circuits to spanning trees does not work well for DIRECTED TSP. Can we apply the idea of Example 6.4 to DIRECTED TSP? Unfortunately, it still looks hard. Let us see why. First, we may assume that the input graph satisfies the triangle inequality, since it is well known that finding a constant-ratio approximation to the general case of DIRECTED TSP is **NP**-hard. Next, we can modify Algorithm 6.A to the following algorithm:

Algorithm 6.B (*Approximation Algorithm for* DIRECTED TSP)

Input: A complete directed graph $G = (V, E)$ without self-loops, and a weight function $w : E \to \mathbb{N}$.

(1) Find a minimum assignment $A = C_1 \cup C_2 \cup \cdots \cup C_t$ of G, where each C_i, $1 \le i \le t$, is a cycle.

(2) **For** $i \leftarrow 1$ **to** t **do**
    Select an edge (u_i, v_i) from cycle C_i;
    Let C_i' be the path in C_i that begins at v_i and ends at u_i.

(3) Form a directed cycle C of $V' = \{v_1, v_2, \ldots, v_t\}$. Without loss of generality, assume that the cycle is $C = (v_1, v_2, \ldots, v_t, v_1)$.

(4) Let H be the cycle formed by connecting the paths C_i', $i = 1, 2, \ldots, t$, with edges $(u_1, v_2), (u_2, v_3), \ldots, (u_{t-1}, v_t), (u_t, v_1)$ (see Figure 6.1); output H. ■

It is clear that H is a directed Hamiltonian circuit, whose total length, by the triangle inequality, is no more than $w(A) + w(C)$. Since A is a minimum assignment, we have $w(A) \le w(H^*)$, where H^* is a minimum Hamiltonian circuit. Therefore, to get a constant-ratio approximation for the minimum directed Hamiltonian circuit, we only need to construct, in polynomial time, a suitable cycle C over vertices in V' with the total length bounded by $O(w(H^*))$. Unfortunately, this problem of finding the minimum cycle C is, in the general case, just DIRECTED TSP itself, and we are back to square one.

Nevertheless, for some special cases of the problem DIRECTED TSP, this relaxation approach could still produce nice approximations. In the following, we present an application of this idea to the problem SHORTEST SUPERSTRING (SS), which was first studied in Section 2.3.

First, we assume that no two strings of the input to the problem SS have the superstring–substring relationship, as we can always ignore all strings that are substrings of some other input strings. At the end of Section 2.3, we showed a natural reduction from SS to DIRECTED TSP. Recall that the overlap $ov(s,t)$ of a string s with respect to another string t is the longest string v that is a suffix of s as well as a prefix of t. Also define, for two strings s and t, $\mathit{pref}(s,t)$ to be the prefix r of s such that $r \cdot ov(s,t) = s$.[1] In the following, we reserve the name s_0 for the empty string. The overlap graph of a set $S = \{s_1, \ldots, s_n\}$ of nonempty strings is the complete directed graph $F(S) = (V, E)$ with vertex set $V = S \cup \{s_0\}$ and the following distance function:

$$d(s_i, s_j) = |s_i| - |ov(s_i, s_j)| = |\mathit{pref}(s_i, s_j)|.$$

Then strings in S and graph $F(S)$ have the following interesting relationship: The strings in S appear in a shortest superstring s^* of S in the order of $s_{i_1}, s_{i_2}, \ldots, s_{i_n}$ if and only if the cycle $H^* = (s_0, s_{i_1}, s_{i_2}, \ldots, s_{i_n}, s_0)$ is a minimum TSP tour of the directed graph $F(S)$ (note that it means we attach the empty string s_0 at the end of s^*). Furthermore, the total length of H^* is $d(H^*) = |s^*|$. From this relation, we may convert an approximation algorithm for DIRECTED TSP to an approximation algorithm for SS. In particular, we will apply the idea of relaxation of Hamiltonian circuits to assignments to the construction of an approximation algorithm for SS.

Let s be a superstring for $S = \{s_1, s_2, \ldots, s_n\}$. Assume that strings of S appear as substrings of s in the order of $s_{i_1}, s_{i_2}, \ldots, s_{i_n}$. Then we say that s is a *minimal superstring* of S with respect to the order $s_{i_1}, s_{i_2}, \ldots, s_{i_n}$ if each pair of adjacent strings s_{i_j} and $s_{i_{j+1}}$, for $j = 1, 2, \ldots, n-1$, has the maximal overlap between them in s. We write $\langle s_{i_1}, s_{i_2}, \ldots, s_{i_n} \rangle$ to denote the minimal superstring of S with respect to the order $s_{i_1}, s_{i_2}, \ldots, s_{i_n}$. We note that, for any ordering of strings in S, there is a unique minimal superstring of S with respect to this order. Also note that both the optimum superstring $Opt(S) = s^*$ and the superstring $Greedy(S)$ obtained by greedy Algorithm 2.B are minimal superstrings.

Let $A = C_1 \cup C_2 \cup \cdots \cup C_{t+1}$ be a minimum assignment in the directed graph $F(S)$, where each C_i, $1 \le i \le t+1$, is a cycle. Without loss of generality, assume that C_{t+1} contains the vertex s_0. Let $(u_1, v_1), (u_2, v_2), \ldots, (u_{t+1}, v_{t+1})$ be edges selected from cycles $C_1, C_2, \ldots, C_{t+1}$, respectively, with $v_{t+1} = s_0$. As discussed in step (3) of Algorithm 6.B, we need to find a cycle C over $V' = \{v_1, v_2, \ldots, v_{t+1}\}$. Assume that $C = (v_{i_1}, v_{i_2}, \ldots, v_{i_t}, v_{t+1}, v_{i_1})$. Let C' be the cycle C with vertex v_{t+1} removed. Then the path $C' = (v_{i_1}, v_{i_2}, \ldots, v_{i_t})$ corre-

[1] For two strings x and y, we write $x \cdot y$ or xy to denote the concatenation of x and y.

sponds to the minimal superstring $s' = \langle v_{i_1}, v_{i_2}, \ldots, v_{i_t} \rangle$. Furthermore, the length of this minimal superstring s' is equal to the length of the total distance $d(C)$ of the cycle C:

$$\begin{aligned}
&|\langle v_{i_1}, v_{i_2}, \ldots, v_{i_t} \rangle| \\
&= |pref(v_{i_1}, v_{i_2})| + |pref(v_{i_2}, v_{i_3})| + \cdots + |pref(v_{i_{t-1}}, v_{i_t})| + |v_{i_t}| \\
&= d(v_{i_1}, v_{i_2}) + d(v_{i_2}, v_{i_3}) + \cdots + d(v_{i_{t-1}}, v_{i_t}) + d(v_{i_t}, s_0) + d(s_0, v_{i_1}) \\
&= d(C).
\end{aligned}$$

We pointed out earlier that finding the minimum cycle C' covering all vertices in V' is, in general, as difficult as the problem DIRECTED TSP. However, in this case, we can prove that the greedy Algorithm 2.B for SS will actually produce a superstring of $v_1, v_2, \ldots, v_t$ with length at most $2 \cdot |Opt(S)|$.

To show this result, we need some simple properties about strings. For any nonempty string x, we write $\rho(x)$ to denote the *root* of x; that is, $\rho(x)$ is the shortest string y satisfying $x = y^k$ for some $k > 0$.

Lemma 6.5 *If y is the root of a nonempty string u [i.e., $y = \rho(u)$], then $y = \rho(y)$.*

Proof. Since $y = \rho(u)$, we know that $y^k = u$ for some $k > 0$. If $x = \rho(y) \neq y$, then $x^\ell = y$ for some $\ell > 1$. It follows that $x^{\ell k} = u$, and x is shorter than y, contradicting the assumption that y is the root of u. □

Lemma 6.6 *Suppose that u and v are two nonempty strings satisfying $uv = vu$. Then $\rho(u) = \rho(v)$.*

Proof. Without loss of generality, assume that $|u| \geq |v|$. We prove the lemma by induction on $|u|$. For $|u| = 1$, it is obvious that $\rho(v) = v = u = \rho(u)$. Now, assume $|u| > 1$. If $|u| = |v|$, then $uv = vu$ implies that $u = v$ and, hence, $\rho(u) = \rho(v)$. Suppose that $|u| > |v|$. Then $uv = vu$ implies $u = vu_1$ for some nonempty string u_1. Now, $(vu_1)v = uv = vu = v(vu_1)$ implies $u_1 v = vu_1$. By the induction hypothesis, $\rho(u_1) = \rho(v)$.

We now claim that $y = \rho(u_1) = \rho(v)$ is also the root of u. Suppose this is not true. Then the root $x = \rho(u)$ of u must be shorter than y, since $y^i = u_1$ and $y^j = v$ for some $i, j > 0$ implies that $y^{i+j} = u$. It follows that $x^k = u = y^{i+j}$ for some $k > i + j$. Now, from the relationship $x^k = y^{i+j}$, we see that x is a prefix of y, as well as a suffix of y. Let $\ell = \lfloor |y|/|x| \rfloor$, z be the suffix of y such that $y = x^\ell z$, and w be the prefix of y such that $y = wx^\ell$. Note that z is also a prefix of x, since $x^{\ell+1}$ is a prefix of y^2. Thus, both z and w are prefixes of x of the same length, and it follows that $z = w$. This means that $y = x^\ell z = zx^\ell$. If z is the empty string, then $x^\ell = y$, and this contradicts the fact that $\rho(y) = y$. On the other hand, if z is nonempty, then we have $xz = zx$. By the induction hypothesis, $\rho(x) = \rho(z)$. However, this implies $\rho(x)^p = x$ and $\rho(x)^q = z$ for some $p, q > 0$, and so $\rho(x)^{p\ell+q} = y$, again contradicting the fact that $y = \rho(y)$. This completes the proof of the claim and, hence, the lemma. □

We now consider a cycle $C = (x_1, x_2, \ldots, x_k, x_1)$ over some vertices in $F(S)$. For each $i = 1, 2, \ldots, k$, we may attach the string p_i to the edge (x_i, x_{i+1}), where $p_i = pref(x_i, x_{i+1})$ (identifying x_{k+1} with x_1). Let $s(C) = p_1 p_2 \cdots p_k$; then we have $d(C) = |s(C)|$. A string w is called a *period* of C if $w = p_i p_{i+1} \cdots p_k p_1 \cdots p_{i-1}$ for some $1 \le i \le k$, that is, if w is a cyclic shift of $s(C)$ beginning at some vertex $x_i \in C$.[2] We say that the cycle C *embeds a string* x' if x' is a substring of $s(C)^\ell$ for some sufficiently large integer ℓ. Clearly, C embeds every string x_j in the cycle C.

Lemma 6.7 *If a cycle $C = (x_1, x_2, \ldots, x_k, x_1)$ embeds, in addition to strings in C, strings $x_{k+1}, x_{k+2}, \ldots, x_m$ in S, with $m > k$, then there is another cycle C' over all vertices $x_1, x_2, \ldots, x_m$ with distance $d(C') = d(C)$.*

Proof. Let w be a period of C. Then all strings $x_1, x_2, \ldots, x_m$ occur as substrings of w^ℓ for some ℓ. We may assume, without loss of generality, that every x_i begins, as a substring of w^ℓ, in the first copy of w in w^ℓ. We can rearrange them according to the order of their occurrences in w^ℓ (since no string x_i is a substring of x_j, for $i \neq j$, this order is well defined). Now, define a cycle C' over all strings $x_1, x_2, \ldots, x_m$, according to this order. Apparently, w is still a period of C', and $d(C') = d(C)$. □

Lemma 6.8 *Assume that w is a period of a cycle $C = (x_1, x_2, \ldots, x_k, x_1)$ and $\rho(w) \neq w$. Then there exists a cycle C' over the same set of vertices in C having $\rho(w)$ as a period of C' and $d(C') = |\rho(w)|$.*

Proof. Assume that $\rho(w)^m = w$ for some $m > 1$. Then, every string x_i, for $1 \le i \le k$, occurs as a substring of $\rho(w)^\ell$ for sufficiently large ℓ. Note that the first occurrence of each x_i in $\rho(w)^\ell$ must begin within the first copy of $\rho(w)$. We can rearrange strings in C in the order of their first occurrence in $\rho(w)^\ell$, and form a cycle C' over the vertices in C in this order. Since all strings occur in $\rho(w)^\ell$ beginning in the first copy of $\rho(w)$, we have $s(C') = \rho(w)$. It follows that C' embeds all strings in C and has $d(C') = |\rho(w)|$. □

The above lemma means that the period w of a cycle C in a minimum assignment must have $\rho(w) = w$. In addition, together with Lemma 6.7, it implies that two periods w_1 and w_2 of two different cycles C_1 and C_2, respectively, in a minimum assignment must have $w_1 = \rho(w_1) \neq \rho(w_2) = w_2$.

Lemma 6.9 *Suppose C_1 and C_2 are two cycles in a minimum assignment of $F(S)$. Let x_1 and x_2 be two vertices in cycles C_1 and C_2, respectively. Then*

$$|ov(x_1, x_2)| < d(C_1) + d(C_2).$$

Proof. For contradiction, suppose $|ov(x_1, x_2)| \ge d(C_1) + d(C_2)$. Let u and v be the prefixes of $ov(x_1, x_2)$ of lengths $|u| = d(C_1)$, and $|v| = d(C_2)$, respectively.

[2] A string u is a *cyclic shift* of string v if there exist strings s, t such that $u = st$ and $v = ts$.

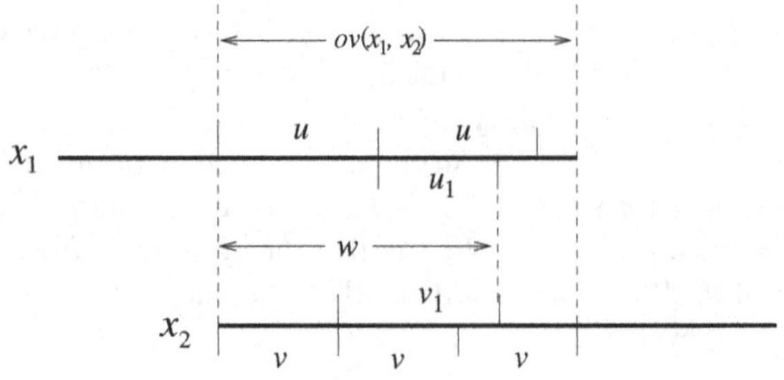

Figure 6.2: Relationships between u and v.

Then u and v are periods of C_1 and C_2, respectively. Moreover, for sufficiently large k and ℓ, u^k and v^ℓ have a prefix of length $|ov(x_1, x_2)|$ in common. We claim that $uv = vu$, and hence, by Lemma 6.6, $\rho(u) = \rho(v)$, a contradiction.

It remains to prove the claim. First, if $|u| = |v|$, then we have $u = v$, and so $uv = vu$. So, we may assume that $|u| > |v|$. Let w be the prefix of $ov(x_1, x_2)$ of length $|w| = |u| + |v|$. Since both u and v are prefixes of $ov(x_1, x_2)$, we know that v is a prefix of u. From $|ov(x_1, x_2)| \geq |u| + |v|$, we know that $w = uu_1$, where u_1 is the prefix of u of length $|v|$, and hence is equal to v (see Figure 6.2). That is, $w = uv$. On the other hand, $w = vv_1$, where v_1 is of length $|u|$. Furthermore, v_1 is the prefix of $v^{\ell-1}$ and, hence, the prefix of $ov(x_1, x_2)$. Therefore, $v_1 = u$, $w = vu$, and the claim is proven. □

Now, we come back to step (3) of Algorithm 6.B, and consider how to find a cycle over vertices in $V' = \{v_1, v_2, \ldots, v_t, v_{t+1}\}$. Let $Opt(V')$ denote the shortest supersting of strings in $V' \setminus \{v_{t+1}\}$, and $Greedy(V')$ the superstring found by the greedy Algorithm 2.B. Assume that $Opt(V') = \langle v_{i_1}, v_{i_2}, \ldots, v_{i_t} \rangle$. By Theorem 2.19, we have

$$\|V'\| - |Opt(V')| \leq 2(\|V'\| - |Greedy(V')|),$$

where $\|V'\|$ denotes the total length of strings in V'. It is clear that

$$\|V'\| = \sum_{j=1}^{t-1} \left(|pref(v_{i_j}, v_{i_{j+1}})| + |ov(v_{i_j}, v_{i_{j+1}})| \right) + |v_{i_t}|$$

and

$$|Opt(V')| = \sum_{j=1}^{t-1} |pref(v_{i_j}, v_{i_{j+1}})| + |v_{i_t}|.$$

Therefore,

$$|Greedy(V')| \leq |Opt(V')| + \frac{1}{2} \sum_{j=1}^{t-1} |ov(v_{i_j}, v_{i_{j+1}})|.$$

By Lemma 6.9,

$$\sum_{j=1}^{t-1} |ov(v_{i_j}, v_{i_{j+1}})| \le d(C_{i_1}) + 2\sum_{j=2}^{t-1} d(C_{i_j}) + d(C_{i_t}) \le 2 \cdot |Opt(S)|.$$

Moreover, $|Opt(V')| \le |Opt(S)|$. Therefore,

$$|Greedy(V')| \le 2|Opt(S)|.$$

We have just proved the following theorem:

Theorem 6.10 *Let* $A = C_1 \cup C_2 \cup \cdots \cup C_{t+1}$ *be a minimum assignment of the directed graph* $F(S)$. *Suppose that* C_{t+1} *contains the empty string* s_0, *and* $v_1, v_2, \ldots, v_t$ *are vertices chosen from cycles* $C_1, \ldots, C_t$, *respectively. Let* s' *be the superstring of vertices* $v_1, v_2, \ldots, v_t$ *found by greedy* Algorithm 2.B. *Then* $|s'| \le 2 \cdot |Opt(S)|$.

Corollary 6.11 *The problem* SS *has a polynomial-time approximation with performance ratio 3.*

Proof. Apply Algorithm 6.B to $F(S)$, using greedy Algorithm 2.B to find a cycle C in step (3). □

We remark that the performance ratio 3 of Corollary 6.11 can be further improved to a value close to 2.5 (see Historical Notes). Nevertheless, all the improvements are based on the fundamental idea we studied in this section.

6.2 Two-Stage Greedy Approximations

The algorithm used in Corollary 6.11 can be considered a two-stage approximation algorithm, in which we combine the relaxation technique with the greedy strategy to solve the problem SS. To be more precise, we relax, in the first stage, the problem SS to the minimum assignment problem, and find the minimum assignment in polynomial time. Then, in the second stage, we apply the greedy algorithm to modify the minimum assignment to an approximate superstring. In some two-stage approximations, we may also apply, in the first stage, the greedy strategy directly to the relaxed problem to find an optimal or approximate solution for the relaxed problem. Then, in the second stage, we modify the solution into the feasible region of the original problem. In the following, we study two examples in this approach.

Recall the problem MIN-CDS (MINIMUM CONNECTED DOMINATING SET) In Section 2.5, we proposed a potential function for this problem as follows: Given a graph G and a vertex subset C, we first color all vertices in three colors: a vertex in C is colored in *black*, a vertex adjacent to some black vertex is colored in *gray*, and all remaining vertices are colored in *white*. Let $p(C)$ be the number of connected components of the induced subgraph $G|_C$, and $h(C)$ the number of white vertices. Let $g(C) = p(C) + h(C)$. It is clear that C is a connected dominating set if and only if $g(C) = 1$. Therefore, we might use function g as a potential function. However, we showed, in Section 2.5, by a counterexample, that a vertex subset C may not be a connected dominating set even though $\Delta_x g(C) = 0$ for all vertices x. As a consequence, the output of a greedy algorithm using g as the potential function may not be a connected dominating set, and we did not take g as the potential function in our greedy approximation for MIN-CDS. On the other hand, if we examine this idea closely, we would find that the output of the greedy algorithm using the potential function g can be easily modified into a connected dominating set (cf. Lemma 2.42). This observation suggests the following two-stage greedy approximation.

Algorithm 6.C (*Two-Stage Greedy Algorithm for* MIN-CDS)
Input: A connected graph G.

Stage 1: Set $C \leftarrow \emptyset$;
  While there exists a vertex x such that $\Delta_x g(C) < 0$ **do**
    Choose a vertex x to maximize $-\Delta_x g(C)$;
    Set $C \leftarrow C \cup \{x\}$.

Stage 2: **While** there exists more than one black component **do**
    Find a chain of two gray vertices x and y connecting
      at least two black components;
    Set $C \leftarrow C \cup \{x, y\}$;
  Output C. ■

In this two-stage greedy approximation, Stage 1 is a greedy algorithm computing a dominating set and Stage 2 connects this dominating set into a connected set. As the value $h(C)$ is included in the potential function $g(C)$, the greedy choice based on $g(C)$ makes sure that the output of Stage 1 is a dominating set.

Lemma 6.12 *At the end of Stage* 1 *of* Algorithm 6.C, *the graph G contains no white vertex.*

Proof. Let x be a white vertex with respect to some vertex subset C. Suppose x has a white neighbor; then coloring x in black eliminates at least two white vertices, and it introduces at most one new black connected component. Therefore, we have $\Delta_x g(C) < 0$. On the other hand, if x has no white neighbor, then x must have a gray neighbor y. Then, coloring y in black does not increase the number of black connected components, but it eliminates at least one white vertex. Again, we have $\Delta_y g(C) < 0$. In either case, Stage 1 does not end at this point. □

In addition, we included value $p(C)$ in $g(C)$, and so the number of black connected components in the output of Stage 1 is kept small. As a result, we do not need to add too many vertices in Stage 2.

Theorem 6.13 *Suppose the input graph G is not a star. Then* Algorithm 6.C *is a polynomial-time* $(3 + \ln \delta)$*-approximation for* MIN-CDS, *where* δ *is the maximum vertex degree of the input graph.*

Proof. By a *piece* (with respect to a set C of black vertices), we mean a white vertex or a connected component of the subgraph induced by black vertices. A piece is said to be *touched* by a vertex x if x is either in the piece or adjacent to the piece. It is clear that, for any vertex subset C, the number of pieces with respect to C is exactly $g(C)$. Suppose $x_1, x_2, \ldots, x_t$ are the vertices selected, in this order, in Stage 1 of Algorithm 6.C. Denote $C_i = \{x_1, x_2, \ldots, x_i\}$, for $1 \leq i \leq t$, and $C_0 = \emptyset$.

Consider set C_{i-1} for some $1 \leq i \leq t$. Suppose a nonblack vertex x touches m pieces with respect to C_{i-1}. If x is white, then all pieces touched by x are white vertices. Therefore, coloring x in black would eliminate m white vertices (including x itself), and introduce one new black connected component. That is, $-\Delta_x g(C_{i-1}) = m - 1$. On the other hand, if x is gray, then it may touch k white neighbors and $m-k$ black connected components. Coloring x in black would eliminate k white pieces and connect $m - k$ black connected components into one. Again, $-\Delta_x g(C_{i-1}) = m - 1$. In other words, for any vertex x, it touches exactly $1 - \Delta_x g(C_{i-1})$ pieces with respect to C_{i-1}. Among all vertices, x_i is the vertex that touches the maximum number of pieces.

Since a piece must be touched by a vertex in the minimum connected dominating set, x_i must have touched at least $\lceil g(C_{i-1})/opt \rceil$ pieces, where *opt* is the number of vertices in a minimum connected dominating set D^*. It follows that

$$1 - \Delta_{x_i} g(C_{i-1}) \geq \frac{g(C_{i-1})}{opt};$$

or, equivalently,

$$g(C_i) \leq g(C_{i-1})\Big(1 - \frac{1}{opt}\Big) + 1.$$

Set $a_i = g(C_i) - opt$. Then we have

$$a_i \leq a_{i-1}\Big(1 - \frac{1}{opt}\Big).$$

Note that if $a_{i-1} > 0$, then $g(C_{i-1})/opt > 1$, and so $-\Delta_x g(C_{i-1}) > 0$ for some x in D^*, and hence $a_i < a_{i-1}$. It follows that $a_t \leq 0$. Choose index $j \leq t$ such that $a_j \leq 0 < a_{j-1}$. Then we must have $a_t \leq j - t$, since the value of a_i must decrease by at least one in each iteration. This implies that there are at most $opt - t + j$ pieces left when Stage 1 ends. From Lemma 6.12, we know that all these pieces are black connected components. Since we only need to add two black vertices to reduce the number of black connected components by one, at most $2(opt - t + j)$ vertices would be added in Stage 2.

Choose $i < j$ such that $a_{i+1} < opt \leq a_i$ [if no such i exists, then we have $a_0 = g(\emptyset) = n < opt$, and coloring every vertex black is a 2-approximation]. Then $j - i \leq opt$, and

$$opt \leq a_i \leq a_{i-1}\Big(1 - \frac{1}{opt}\Big) \leq a_0\Big(1 - \frac{1}{opt}\Big)^i \leq n \cdot e^{-i/opt},$$

where n is the number of vertices in the input graph. Thus,

$$i \le opt \cdot \ln\left(\frac{n}{opt}\right).$$

Note that for a nonstar graph of the maximum vertex degree δ, the size of a connected dominating set is at least n/δ. Therefore, the total number of vertices selected by Algorithm 6.C is at most

$$\begin{aligned} t + 2(opt - t + j) &\le 2 \cdot opt + j \\ &\le 3 \cdot opt + i \le opt\left(3 + \ln\left(\frac{n}{opt}\right)\right) \le opt(3 + \ln \delta). \quad \square \end{aligned}$$

Next, we study the minimum power broadcasting problem. Recall the notion of a broadcasting tree in a network introduced in Section 3.4. Let G be a network, that is, a connected, bi-directed graph with nonnegative edge weight, with the property that $w(u,v) = w(v,u)$ when both (u,v) and (v,u) are edges in G. A broadcasting tree T of G from a node s is an arborescence rooted at s over all nodes of S. The power of a nonsink node u in a broadcasting tree T is the maximum weight of out-edges in T from u, and the power of the tree T is the sum of the powers over all nonsink nodes in T.

> BROADCASTING TREE WITH MINIMUM POWER (BT-MP): Given a connected, bi-directed graph G with nonnegative edge weight and a node s, find a broadcasting tree from s with the minimum power.

A directed graph G is *weakly connected* if it is connected when direction on each edge is removed (and so G becomes a connected undirected graph). A broadcasting tree is clearly a weakly connected subgraph. This observation suggests that we may relax BT-MP to the problem of finding a weakly connected subgraph with the minimum power. Along this idea, a two-stage greedy approximation can be designed as follows: At Stage 1, use a greedy algorithm to find an approximation for the minimum-power weakly connected spanning tree; and at Stage 2, modify the weakly connected spanning tree obtained in Stage 1 to a broadcasting tree.

To design a greedy algorithm for Stage 1, we need to define a potential function. Let $G = (V, E)$ be a directed graph and w an edge-weight function on E. A *star* A (centered at a vertex v) in G is a subset of out-edges from v in G. The *weight* of a star A, denoted by $w(A)$, is the maximum weight of an edge in the star. Let F be the set of all stars A satisfying the following condition: If A contains an out-edge from v with weight w, then every out-edge from v, with weight not exceeding w, is also in A. For a directed graph G, a *weakly connected component* is a connected component of the undirected graph G' obtained from G by removing the directions of all edges in G. For every subset S of F, define $f(S)$ to be the number of weakly connected components of the subgraph $G_S = (V, \cup S)$, whose edge set consists of all edges in all stars in S. The following lemma shows that we can use $-f(S)$ as the potential function to design the greedy algorithm.

Lemma 6.14 *$f(S)$ is a monotone decreasing, supmodular function on E.*

Proof. Obvious. □

In Stage 2, we need to convert a weakly connected spanning tree into a broadcasting tree. The following lemma suggests that we can simply reverse the direction of an edge whenever it is needed.

Lemma 6.15 *Suppose B is a weakly connected spanning subgraph of G. If B does not contain a broadcasting tree from s, then there is an edge (u, v) in B such that s can reach v but cannot reach u.*

Proof. Let V_1 be the set of nodes reachable from s and V_0 the set of nodes not reachable from s. Since G is weakly connected. There must exist an edge (u, v) from V_0 to V_1. □

Algorithm 6.D (*Two-Stage Greedy Approximation for* BT-MP)

Input: A connected bi-directed graph G with a nonnegative edge-weight function w, and a node s.

Stage 1: Set $S \leftarrow \emptyset$;
While $f(S) > 1$ **do**
Choose a star $A \in F$ to maximize $-\Delta_A f(S)/w(A)$;
Set $S \leftarrow S \cup \{A\}$;
$B \leftarrow \bigcup_{A \in S} A$.

Stage 2: **While** B does not contain a broadcasting tree from s **do**
Find an edge (u, v) in B such that s can reach v but not u;
Set $B \leftarrow B \cup \{(v, u)\}$;
Output a broadcasting tree T from the graph (V, B). ■

Theorem 6.16 Algorithm 6.D *is a $2H(\delta)$-approximation for* BT-MP, *where δ is the maximum node degree of the input graph, and H is the harmonic function.*

Proof. Let *opt* be the minimum power of a broadcasting tree. Then the minimum power of a weakly connected subgraph is at most *opt*. Let B_1 be the set B at the end of stage 1. Then, by Theorem 2.29, the power of the graph $G_1 = (V, B_1)$ is at most $H(\gamma) \cdot opt$, where $\gamma = \max_{A \in F}(-f(A) + f(\emptyset)) \leq \delta$ [note that the function $g(A) = -f(A) + f(\emptyset)$ is the potential function used in Theorem 2.29]. In Stage 2, we note that for each edge (v, u) added to B, the weight of v increases to at most $w(v, u) = w(u, v)$, which does not exceed the weight of u. Therefore, the total power increase does not exceed the power of the digraph $G_1 = (V, B_1)$ at the end of Stage 1. Therefore, the power of T is at most $2H(\delta) \cdot opt$. □

6.3 Connected Dominating Sets in Unit Disk Graphs

Next, let us review the problem CDS-UDG (CONNECTED DOMINATING SET IN A UNIT DISK GRAPH). We found in Section 4.2 a PTAS for this problem. Its running time $n^{O(1/\varepsilon^2)}$ is, however, too high to be implemented for moderately large

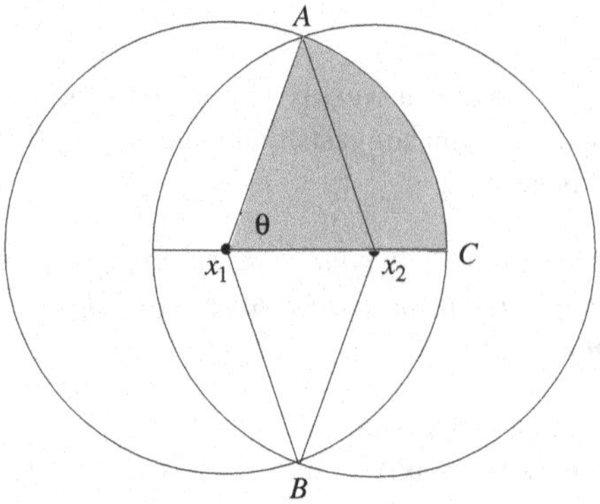

Figure 6.3: Two adjacent vertices have a common covering area.

input graphs. Therefore, one would still like to find good approximations to it with a lower running time, allowing implementation in applications in, for instance, wireless sensor networks. In this section, we follow the idea of the two-stage greedy approximations to design an approximation to this problem. Namely, we first construct, by a greedy algorithm, a dominating set of the input graph, and then connect it together.

For the construction of a dominating set, a popular way is to relax it to the maximal independent set problem.[3] We note that a maximal independent set of a graph G must be a dominating set of G. In addition, in a unit graph, the size of any maximal independent set is within a constant factor from the size of the minimum connected dominating set.

Lemma 6.17 *In a unit disk graph G, the size of a maximal independent set is upper-bounded by* $(3.74)opt + 5.26$*, where opt is the size of a minimum connected dominating set of G.*

Proof. We will bound the number of independent vertices by counting the areas of unit disks centered at these vertices. First, define the *covering area* of a vertex x in G to be the disk with center x and radius $3/2$. Two adjacent vertices in G have distance at most 1. Therefore, the covering areas of two adjacent vertices must share a common area of size at least $\frac{9}{2}\arccos\frac{1}{3} - \sqrt{2}$. To see this, we draw two circles of radius $3/2$, with centers x_1 and x_2 of distance 1 apart (see Figure 6.3). Then the angle $\theta = \angle Ax_1C$ is equal to $\arccos(1/3)$ and so the shaded area $\triangle x_1AC$ has size $s = (3/2)^2\theta/2 = (9/8)\arccos(1/3)$. The area of the intersection of the two circles is equal to $4s$ minus the area of $\Diamond x_1Ax_2B$, that is, $4s - \sqrt{2}$.

Thus, the total covering area of a minimum connected dominating set of n vertices is at most

[3]A *maximal independent* set of a graph $G = (V, E)$ is an independent set $S \subseteq V$ such that no superset $T \supseteq S$ is independent.

$$(n-1)\left[\left(\frac{3}{2}\right)^2\pi - \left(\frac{9}{2}\arccos\frac{1}{3} - \sqrt{2}\right)\right] + \left(\frac{3}{2}\right)^2\pi$$
$$\leq 2.93(n-1) + 2.25\pi.$$

Now, for every vertex y in a maximal independent set, draw a disk centered at y with radius $1/2$. Then all these disks are mutually disjoint and lie in the covering area of the minimum connected dominating set. Therefore, the size of a maximal independent set is at most

$$\frac{2.93(n-1) + 2.25\pi}{0.25\pi} \leq 3.74(n-1) + 9 = 3.74n + 5.26. \qquad \square$$

We remark that the above estimate of the upper bound for the size of a maximal independent set is not very tight and can be further improved. The best result known so far is that every maximal independent set has size at most $3.478 \cdot opt + 4.874$ [Li, Gao, and Wu, 2008]. It is conjectured in the literature that every maximal independent set has size at most $3 \cdot opt + 2$. This would be the best possible upper bound [Wan et al., 2008].

In order to have a simple connecting strategy in Stage 2, we will construct a maximal independent set D with the following property:

Π_1: For any proper subset $S \subset D$, there is a vertex x such that x is adjacent to both S and its complement $D \setminus S$.

Such a maximal independent set is easy to construct based on the white–gray–black coloring. Namely, when we add a vertex to the independent set, we always select a white vertex that has a gray neighbor.

Now, consider the connecting stage. If we consider the maximal independent set D constructed in the first stage as a set of terminals, then the problem for the connecting stage is a variation of the problem ST-MSP of Section 3.4. Since ST-MSN is **NP**-hard, we need to design an approximation for it. Here, with the maximal independent set D having the special property Π_1, a greedy approximation to this problem is easy to design. For any vertex subset C, let $p(C)$ denote the number of connected components of the subgraph of G induced by C. We can use $p(C)$ as the potential function to design a greedy algorithm to connect the connected components of D into a connected dominating set.

Algorithm 6.E (*Two-Stage Approximation Algorithm for* CDS-UDG)
Input: A connected unit disk graph G.

Stage 1: Select a vertex x; Set $D \leftarrow \{x\}$;
Color x in *black*, all its neighbors in *gray*, and all other vertices in *white*;
While there is a white vertex **do**
Choose a white vertex x with a gray neighbor;

$D \leftarrow D \cup \{x\}$;
Color x in black and its white neighbors in gray;
Return D.

Stage 2: Set $C \leftarrow D$;
While $p(C) \geq 2$ **do**
Choose a vertex x to maximize $-\Delta_x p(C)$;
$C \leftarrow C \cup \{x\}$;
Return C. ■

It is clear that the set D constructed by the end of Stage 1 has property Π_1, and hence the output C of Stage 2 is a connected dominating set. The following theorem gives us an upper bound for the performance of the second stage of Algorithm 6.E.

Theorem 6.18 *Assume that the maximal independence set D found in Stage* 1 *of* Algorithm 6.E *has size $|D| \leq \alpha \cdot opt + \beta$ for some $\alpha \geq 1$ and $\beta > 0$, where opt is the size of a minimum connected dominating set. Then the connected dominating set C found by* Algorithm 6.E *has size at most*

$$(\alpha + 2 + \ln(\alpha - 1))opt + \beta + \lfloor \beta \rfloor.$$

Proof. We follow the standard approach for the analysis of greedy algorithms. Let $x_1, \ldots, x_g$ be the vertices selected in Stage 2 of Algorithm 6.E, in the order of their selection into the set C. Also, let $\{y_1, \ldots, y_{opt}\}$ be a minimum connected dominating set for G with the property that, for each $i = 1, 2, \ldots, opt$, the set $\{y_1, \ldots, y_i\}$ induces a connected subgraph. Denote $C_0 = D$ and, for $0 \leq i \leq g-1$, $C_{i+1} = C_i \cup \{x_{i+1}\}$. In addition, for each $j = 1, 2, \ldots, opt$, we write C_j^* denote set $\{y_1, \ldots, y_j\}$.

By the greedy strategy, we know that for each $i = 0, \ldots, g-1$,

$$-\Delta_{x_{i+1}} p(C_i) \geq -\Delta_{y_j} p(C_i),$$

for all $j = 1, \ldots, opt$. In addition, since the induced subgraph $G|_{C_j^*}$ is connected, we have

$$-\Delta_{y_j} p(C_i \cup C_{j-1}^*) + \Delta_{y_j} p(C_i) \leq 1.$$

Thus, for $opt \geq 2$ and any $i = 0, 1, \ldots, g-1$,

$$\begin{aligned} -\Delta_{x_{i+1}} p(C_i) &\geq \frac{-\sum_{j=1}^{opt} \Delta_{y_j} p(C_i)}{opt} \\ &\geq \frac{-opt + 1 - \sum_{j=1}^{opt} \Delta_{y_j} p(C_i \cup C_{j-1}^*)}{opt} \\ &= \frac{-opt + 1 - p(C_i \cup C_{opt}^*) + p(C_i)}{opt} = \frac{-opt + p(C_i)}{opt}. \end{aligned}$$

That is,

$$-p(C_{i+1}) \geq -p(C_i) + \frac{-opt + p(C_i)}{opt},$$

for all $i = 0, 1, \ldots, g-1$. Denote $a_i = -opt - \beta + p(C_i)$. Then, for each $i = 0, 1, \ldots, g-1$,

$$a_{i+1} \leq a_i\Big(1 - \frac{1}{opt}\Big),$$

and so

$$a_i \leq a_0\Big(1 - \frac{1}{opt}\Big)^i \leq a_0 e^{-i/opt}.$$

First, consider the case $a_0 \geq opt$. Note that $a_g = -opt - \beta + p(C_g) = -opt - \beta + 1 < opt$. Thus, there exists an integer j, $0 \leq j < g$, such that

$$a_{j+1} < opt \leq a_j.$$

Since the values of the a_i's decrease in each iteration, we must have

$$g - (j+1) \leq a_{j+1} - a_g < opt - (-opt - \beta + 1) = 2 \cdot opt + \beta - 1;$$

or, equivalently, $g \leq j + 2 \cdot opt + \lfloor \beta \rfloor$. Now, from

$$opt \leq a_j \leq a_0 e^{-j/opt},$$

we get

$$j \leq opt \cdot \ln\Big(\frac{a_0}{opt}\Big) = opt \cdot \ln\Big(\frac{-opt - \beta + |D|}{opt}\Big) \leq opt \cdot \ln(\alpha - 1).$$

Therefore,

$$|D| + g \leq (\alpha + 2 + \ln(\alpha - 1))opt + \beta + \lfloor \beta \rfloor.$$

Next, consider the case of $a_0 < opt$. This implies that $p(C_0) < 2 \cdot opt + \beta$, and so $g \leq 2 \cdot opt - 1 + \lfloor \beta \rfloor$. Thus,

$$|D| + g \leq (\alpha + 2)opt + \beta + \lfloor \beta \rfloor - 1. \qquad \square$$

Corollary 6.19 *The connected dominating set found by* Algorithm 6.E *has size at most* $(6.7)opt + 10.26$, *where opt is the size of the minimum connected dominating set of the input graph G.*

Finally, we remark that the simple greedy strategy of Stage 2 of Algorithm 6.E works because the maximal independence set D found in Stage 1 satisfies property Π_1. However, we did not take full advantage of property Π_1 in our analysis of the algorithm. A more careful analysis using this property actually shows that the output of Algorithm 6.E has size at most $6\frac{7}{18} \cdot opt$ [Wan et al., 2008].

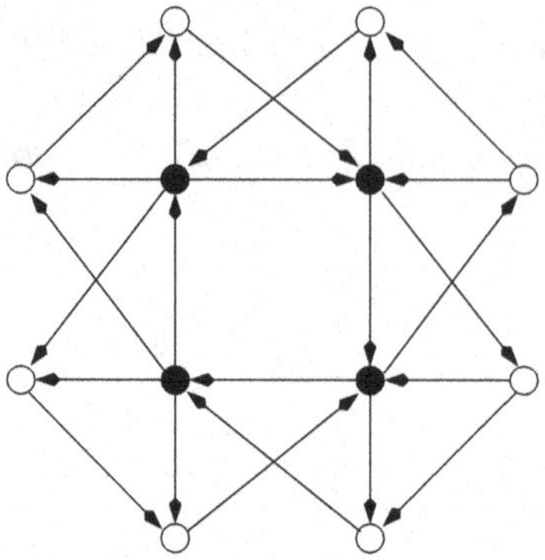

Figure 6.4: A strongly connected dominating set (dark nodes indicating the dominating set).

6.4 Strongly Connected Dominating Sets in Digraphs

Consider a digraph (i.e., a directed graph) $G = (V, E)$. A node subset $C \subseteq V$ is a *dominating set* of G if, for every node x not in C, there is an edge going from x to C and an edge coming from C to x, i.e., if there are edges $(x, y), (z, x)$ in E with $y, z \in C$. Furthermore, set C is called a *strongly connected dominating set* of G if C is a dominating set and, in addition, the subgraph $G|_C$ of G induced by C is strongly connected (see Figure 6.4). In this section, we study the following problem:

> STRONGLY CONNECTED DOMINATING SET (SCDS): Given a digraph G, find a strongly connected dominating set of the minimum cardinality.

No good direct approximations are known for this problem at this time. Here we relax it to the problem of finding broadcasting trees with the minimum number of internal nodes.

Recall that a broadcasting tree of a directed graph G is a spanning arborescence of G, i.e., a rooted directed tree in which every node is reachable from the root. For any digraph $G = (V, E)$, let G^R be the graph obtained from G by reversing the direction of each edge in E; that is, $G^R = (V, E^R)$, where $E^R = \{(y, x) \mid (x, y) \in E\}$. In a broadcasting tree, a nonleaf node is called an *internal node*. We observe that a strongly connected dominating set S of G can be viewed as the collection of internal nodes of two broadcasting trees T_1, T_2 of G and G^R, respectively, that share the same source node. Conversely, if we have two broadcasting trees T_1 and T_2 of G and G^R, respectively, sharing the same source node, then the collection of all internal nodes of T_1 and T_2 is a strongly connected dominating set. Thus, we can relax the problem SCDS to the following problem:

> BROADCASTING TREE WITH MINIMUM INTERNAL NODES (BT-MIN): Given a digraph G and a source node s, find a broadcasting tree

of G with source node s with the minimum number of internal nodes other than s.

In the following, we write $opt_{\mathrm{B}}(G,r)$ to denote the number of internal nodes in the minimum solution to BT-MIN on digraph G and source node r. Also, let $opt_{\mathrm{S}}(G)$ denote the size of the minimum strongly connected dominating set of G.

Lemma 6.20 *For any digraph G and any node r in G, we have*

$$opt_{\mathrm{B}}(G,r) \le opt_{\mathrm{S}}(G).$$

Moreover, if r belongs to an optimum solution of SCDS *on input G, then*

$$opt_{\mathrm{B}}(G,r) \le opt_{\mathrm{S}}(G) - 1.$$

Proof. Let $G = (V,E)$ be a digraph and $C^* \subseteq V$ a minimum strongly connected dominating set of G. For any node $x \in V$, there is a path from r to x that passes through only nodes in C^*. To see this, we note that C^* is a dominating set, and so there must be nodes $y, z \in C^*$ such that $(r,y),(z,x) \in E$. In addition, C^* is strongly connected and, hence, there is a path π from y to z using only nodes in C^*. So, the path $(r,y) \cup \pi \cup (z,x)$ is the desired path. This means that we can construct a broadcasting tree at source node r using only nodes in C^* as internal nodes.

When $r \in C^*$, the number of internal nodes of this broadcasting tree is at most $|C^*|$, and the value $opt_{\mathrm{B}}(G,r)$ is at most $|C^*| - 1$. □

Lemma 6.21 *Assume that there is a polynomial-time α-approximation for the problem* BT-MIN, *for some $\alpha > 1$. Then there is a polynomial-time (2α)-approximation for the problem* SCDS.

Proof. Let $G = (V,E)$ be a digraph and C^* a minimum strongly connected dominating set of G. For any node $s \in V$, apply the α-approximation algorithm for BT-MIN to find a broadcasting tree T_1 in G and a broadcasting tree T_2 in G^R, with the common source s. For $i = 1,2$, let $I(T_i)$ denote the internal nodes of T_i. We claim that $C_s = I(T_1) \cup I(T_2)$ is a strongly connected dominating set of G.

To see this, we note that for every node $x \in V$, there is a path from s to x in T_1, and so an edge from some node $y \in I(T_1)$ to x. In addition, there is a path π from s to x in T_2. Therefore, there is an edge from some node $z \in I(T_2)$ to x in G^R, which means that there is an edge from x to $z \in I(T_2)$ in G. This shows that $C_s = I(T_1) \cup I(T_2)$ is a dominating set of G. In addition, for any two nodes $x, y \in C_s$, there exists a path in G from x to s with all internal nodes in $I(T_2)$, as well as a path in G from s to y with all internal nodes in $I(T_1)$. Together, the union is a path from x to y in C_s, and so C_s is strongly connected, and the claim is proven.

Clearly,

$$\begin{aligned} |C_s| = |I(T_1) \cup I(T_2)| &\le |I(T_1) - \{s\}| + |I(T_2) - \{s\}| + |\{s\}| \\ &\le \alpha\big(opt_{\mathrm{B}}(G,s) + opt_{\mathrm{B}}(G^R,s)\big) + 1. \end{aligned}$$

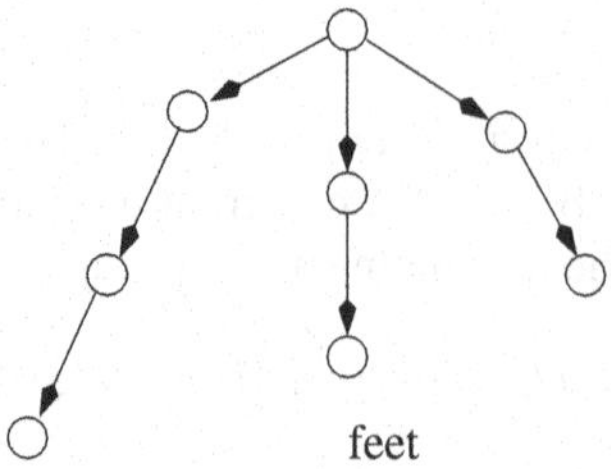

Figure 6.5: A spider.

Moreover, when s belongs to a minimum strongly connected dominating set,

$$\begin{aligned}|C_s| &\leq \alpha\big(opt_{\mathrm{B}}(G,s) + opt_{\mathrm{B}}(G^R,s)\big) + 1 \\ &\leq \alpha\big(opt_{\mathrm{S}}(G) - 1 + opt_{\mathrm{S}}(G^R) - 1)\big) + 1 \leq 2\alpha \cdot opt_{\mathrm{S}}(G),\end{aligned}$$

since a minimum strongly connected dominating set for G is also a minimum strongly connected dominating set for G^R. Thus, to ensure that C_s is of size at most $2\alpha \cdot opt_{\mathrm{S}}(G)$, we only need to find a node s in C^*. Choose an arbitrary node u and let $N(u) = \{u\} \cup \{x \in V \mid (x,u) \in E\}$. It is clear that $N(u) \cap C^* \neq \emptyset$. Therefore, we can just find, for each $s \in N(u)$, a connected dominating set C_s and use the smallest one among them as the approximation to C^*. □

Next, we describe a polynomial-time approximation for BT-MIN. First, we introduce some new terminologies. Assume that each node of the input digraph G has been assigned a unique ID. Let $H = (V_1, E_1)$ be a subgraph of G and s a source node. An *orphan* of H is a strongly connected component of H satisfying the following properties: (i) It does not contain s, and (ii) there is no edge in E that starts at a vertex in $V_1 \setminus C$ and ends at a vertex in C. For each orphan C of H, the node in C with the smallest ID is called the *head* of C. Note that if a subgraph H contains all nodes of G and has no orphans, then it must contain a broadcasting tree.

We call a subgraph S of G a *spider* if S consists of a body node and several disjoint directed paths from the body node to its foot nodes (see Figure 6.5). The general idea of our algorithm is to use a greedy strategy to build a broadcasting tree T by adding spiders to it one by one. More precisely, we start with the subgraph H that consists of all nodes in G and no edge (so that every node in H other than s is an orphan head). Then, we select, at each iteration, a spider S in G and add it to H until H has no more orphan heads. The selection of the spider is based on the greedy strategy that minimizes the number of internal nodes in S relative to the number of orphan heads (with respect to the current H) in S.

One of the problems with the above idea is that, at each iteration, there may be an exponential number of spiders to consider. To make the algorithm running in polynomial time, we need to limit our choices to some special spiders. We say a spider S is *legal* (with respect to H) if it satisfies the following three conditions:

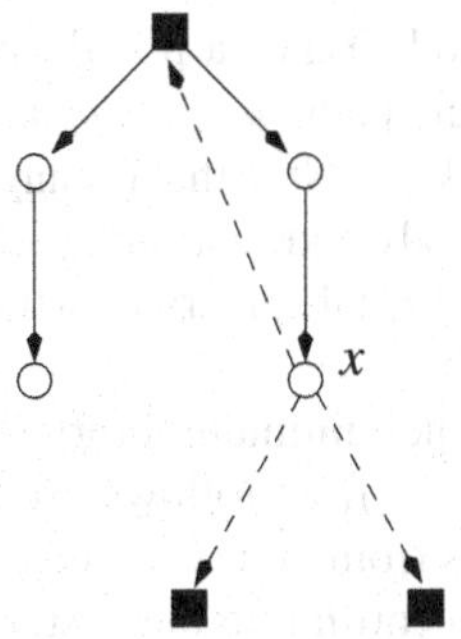

Figure 6.6: A new orphan is introduced when we add a spider.

(a) All feet of S are heads of some orphans of H,

(b) An orphan head can only occur in S at a foot or at the body node, and

(c) S contains at least two orphan heads of H, unless the body node of S is the source s.

In the above, conditions (a) and (b) allow us to decompose a broadcasting tree into the union of legal spiders at orphan heads. Condition (c) is required to make sure that the number of orphan heads in H decreases after each iteration. Note that when we add a legal spider S to H, the orphan heads in S at the feet of S are no longer orphan heads. In the meantime, a new orphan may emerge, which contains the body node of S. Thus, by condition (c), we reduce the number of orphan heads by at least one. Figure 6.6 shows such a case, where a dark square denotes an orphan head in H and the dashed edges denote a spider S. After spider S is added to H, an orphan is introduced that includes the body node x of S.

For a subgraph H and a legal spider S with respect to H, let $h_H(S)$ be the number of orphan heads in S and $cost_H(S)$ the number of internal nodes in S other than the internal nodes in H and the source s. When the subgraph H is clear, we write $h(S)$ for $h_H(S)$ and $cost(S)$ for $cost_H(S)$. Define $quot_H(S)$, or simply $quot(S)$, to be the ratio

$$quot(S) = \frac{cost(S)}{h(S)}.$$

Our intention is to use $quot(S)$ as the potential function and to add, in each iteration of our algorithm, the spider S with the minimum $quot(S)$ to H. However, even with the restriction to legal spiders, the number of possible choices of spiders is still too large and the spider with the minimum $quot(S)$ is still hard to find. To resolve this problem, we generalize the notion of spiders to pseudospiders.

Let u be a node in G. Suppose $p_1, p_2, \ldots, p_k$, for some $k \geq 2$ (or $k = 1$ and $u = s$), are k shortest paths from u to k different orphan heads such that none of the internal nodes of the paths $p_1, \ldots, p_k$ are orphan heads. Then we say the subtree $S = p_1 \cup p_2 \cup \cdots \cup p_k$ is a *legal pseudospider* (note that the paths $p_1, \ldots, p_k$ may share some common internal nodes). For a pseudospider $S = p_1 \cup \cdots \cup p_k$, we define $h(S)$ and $cost(S)$ as if the paths $p_1, \ldots, p_k$ are disjoint; that is, $cost(S) = length(p_1) + \cdots + length(p_k) - k + 1$. Note that for any legal spider S, there is a legal pseudospider S' with the same body node and same feet as S and having $cost(S') \le cost(S)$. Thus, when we consider legal spiders with the minimum $quot(S)$, we need only consider legal pseudospiders.

Moreover, we can compute the minimum $quot(S)$ over all legal pseudospiders S rooted at node u [called $quot(u)$] as follows: Suppose H has k orphan heads and $p_1, \ldots, p_k$ are shortest paths from node u to them, without passing through any orphan heads. Order the paths according to the cost: $cost(p_1) \le cost(p_2) \le \cdots \le cost(p_k)$. Then, it is easy to see that, for $u \ne s$,

$$quot(u) = \min_{2 \le i \le k} quot(p_1 \cup \cdots \cup p_i);$$

and for $u = s$,

$$quot(u) = \min_{1 \le i \le k} quot(p_1 \cup \cdots \cup p_i).$$

Now, we are ready to describe the algorithm.

Algorithm 6.F (*Greedy Approximation Algorithm for* BT-MIN)

Input: A strongly connected digraph $G = (V, E)$, a source node $s \in V$, and a unique ID for each node in V.

(1) $H \leftarrow V$; $A \leftarrow V \setminus \{s\}$.

(2) **For** each $v \in V$ **do** calculate $quot(v)$.

(3) **While** $A \ne \emptyset$ **do**

(3.1) Choose a node $u \in V$ with the minimum $quot(u)$;

(3.2) Let $S(u)$ be the legal pseudospider at u with $quot(S) = quot(u)$;

(3.3) $A \leftarrow A \setminus \{v \mid v \text{ is a head in } S(u)\}$;

(3.4) $H \leftarrow H \cup S(u)$;

(3.5) **If** the strongly connected component C_u of H that contains u becomes an orphan **then** $A \leftarrow A \cup \{\text{head of } C_u\}$;

(3.6) **For** each $v \in V$ **do** recalculate $quot(v)$.

(4) Let T be a broadcasting tree of H; **Return** T. ∎

We now analyze the performance of this algorithm.

Lemma 6.22 *For any subgraph H of G with q orphans, there exists a node u with*

$$quot(u) \le \frac{opt_{\mathrm{B}}(G, s)}{q}.$$

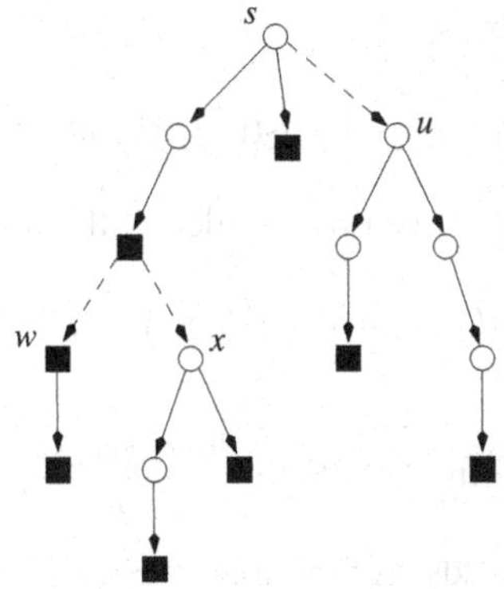

Figure 6.7: Spider decomposition.

Proof. Let T^* be an optimal broadcasting tree. We first prune T^* to obtain a subtree T such that every leaf of T is an orphan head of H. That is, we repeatedly remove the leaves in the tree that are not orphan heads of H until there are no more such leaves in H.

Next, we show that tree T can be decomposed to a sequence of legal spiders. For any leaf x of T, let $anc(x)$ be the lowest ancestor of x that is either a head of H or has out-degree greater than 1. Let $Anc(T) = \{anc(x) \mid x \text{ is a leaf of } T\}$. We remove legal spiders from T as follows:

Case 1. There exists a leaf x of T whose $anc(x)$ is a head and has out-degree 1. Note that the path from $anc(x)$ to x in T has no other branches and is a legal spider with respect to H. Therefore, we can remove this spider from T. The remaining part of T is still a tree, and we prune it to make all its leaves orphan heads.

Case 2. Not Case 1. Let y be a lowest node in $Anc(T)$. Assume that $y = anc(x)$. Then the subtree T_y rooted at y must have at least two leaves, since $anc(x)$ has out-degree at least 2. Furthermore, we know that all leaves w of T_y have $anc(w) = y$, for otherwise $anc(w)$ would be a proper descendant of y and so is a lower ancestor node in $Anc(T)$. Thus, all internal nodes other than y in T_y are not heads, and so T_y is a legal spider with respect to H. We can remove T_y from T, and again we prune T if necessary so that all its leaves are heads.

We perform the above procedure until T is a single node s. Then we get a sequence of legal spiders $S_1, S_2, \ldots, S_\ell$ such that

(i) Each spider S_i, $1 \le i \le \ell$, is a subtree of T;

(ii) The spiders $S_1, \ldots, S_\ell$ are mutually disjoint; and

(iii) Each orphan head of H is in one of the spiders $S_1, \ldots, S_\ell$.

For instance, the tree T in Figure 6.7 can be decomposed into spiders S_w, S_x, S_u and S_s, where S_t denotes the spider with body node t. In the figure, the nodes with labels are the body nodes of the spiders, the dark squares indicate orphan heads, and the dashed edges are the edges pruned in this process.

We have, from (i) and (ii),

$$cost(S_1) + \cdots + cost(S_\ell) \le opt_{\mathrm{B}}(G, s).$$

(Note that each S_i, $1 \le i \le \ell$, is a *real* spider, with all its legs disjoint.) We also have, from (iii),

$$h(S_1) + \cdots + h(S_\ell) = q.$$

Thus,

$$\min_{1 \le i \le \ell} quot(S_i) \le \frac{opt_{\mathrm{B}}(G, s)}{q}.$$

This means that one of the heads u of the spiders $S_1, \ldots, S_\ell$ meets our requirement. □

Theorem 6.23 *The problem* BT-MIN *has a polynomial-time approximation with the performance ratio* $(1 + 2\ln(n-1))$.

Proof. Suppose Algorithm 6.F runs on an input digraph G and a source node s and halts in k iterations. For each $i = 0, 1, \ldots, k-1$, let n_i denote the number of orphans in H right after the ith iteration. Also, let S_i, for $i = 1, \ldots, k$, be the legal pseudospider chosen at the ith iteration, and h_i be the number of heads in S_i. Note that we initially have $n_0 = n - 1$ orphans and, at the last iteration, $n_{k-1} = h_k$ orphans. In the following, we write *opt* to denote $opt_{\mathrm{B}}(G, s)$.

In each iteration i, for $i = 1, 2, \ldots, k$, we reduce at least $h_i - 1$ heads from H. Therefore, we get

$$n_i \le n_{i-1} - \frac{h_i}{2},$$

for each $i = 1, 2, \ldots, k$ (when $h_i = 1$, we reduce exactly one head in the ith iteration). Moreover, by Lemma 6.22, for each $i = 1, \ldots, k$,

$$\frac{cost(S_i)}{h_i} \le \frac{opt}{n_{i-1}}.$$

Together, for each $i = 1, \ldots, k$,

$$\frac{n_i}{n_{i-1}} \le 1 - \frac{cost(S_i)}{2 \cdot opt}.$$

Repeatedly applying the above inequality, we get

$$\frac{n_{k-1}}{n_0} \le \prod_{i=1}^{k-1} \left(1 - \frac{cost(S_i)}{2 \cdot opt}\right).$$

Hence,

$$\ln\left(\frac{n_{k-1}}{n_0}\right) \le -\frac{\sum_{i=1}^{k-1} cost(S_i)}{2 \cdot opt}.$$

Or, equivalently,

$$\sum_{i=1}^{k-1} cost(S_i) \leq 2 \cdot opt \cdot \ln\left(\frac{n_0}{n_{k-1}}\right) \leq 2 \cdot opt \cdot \ln(n-1).$$

Since $cost(S_k)/h_k \leq opt/n_{k-1}$ and $h_k = n_{k-1}$, we have $cost(S_k) \leq opt$. Therefore,

$$\sum_{i=1}^{k} cost(S_i) \leq (1 + 2\ln(n-1))opt. \qquad \square$$

As a consequence, we have

Corollary 6.24 *The problem* SCDS *has a polynomial-time* $(2 + 4\ln(n-1))$-*approximation.*

6.5 Multicast Routing in Optical Networks

In this section, we study the multicast routing problem in optical networks with both splitting and nonsplitting nodes. An optical network is usually formulated as an edge-weighted graph, with the switches represented as vertices. In the graph, there are two types of vertices, nonsplitting and splitting. A splitting vertex can send an input signal to several output vertices, while a nonsplitting vertex can only send the input signal to one output. A multicast route in a graph G is a subtree of G in which each edge is assigned a direction so that only a splitting vertex can have a higher out-degree than its in-degree.

> MINIMUM-WEIGHT MULTICAST ROUTING (MIN-MR): Given a graph $G = (V, E)$ with an edge-weight function $w : E \to \mathbb{R}^+$ that satisfies the triangle inequality, a subset $A \subseteq V$ of splitting vertices, a source $s \in V$, and a subset $M \subseteq V$ of multicast members, find a multicast route that spans all members in M with the minimum total edge-weight.

We notice that if all vertices are nonsplitting and $M = V$, then MIN-MR can be reduced to the minimum-weight Hamiltonian path problem, which is as hard as the traveling salesman problem (TSP). If all vertices are splitting, then MIN-MR is just the network Steiner minimum tree problem (NSMT). Thus, when both nonsplitting and splitting vertices are allowed, the problem MIN-MR is at least as hard as TSP and NSMT.

A simple idea for this problem is to first relax the problem to NSMT, and then modify the solution to get a multicast route. In the following, we assume that we have a polynomial-time ρ-approximation algorithm A_{NSMT} for NSMT.

Algorithm 6.G (*Relaxation Algorithm for* MIN-MR)

Input: An edge-weighted graph $G = (V, E)$ with $A \subseteq V$ identified as splitting vertices, a source vertex $s \in V$, and a subset $M \subseteq V$ of multicast members.

Stage 1: Relax the input graph G to a new graph G' that has the same vertex and edge sets as G but every vertex in G' is a splitting vertex;

Apply algorithm A_{NSMT} on G' to get a Steiner tree T.

Stage 2: Starting from the source vertex s, perform a depth-first search on T, treating every vertex as a nonsplitting vertex;

Output the resulting route R. ■

Let *opt* be the weight of the minimum-weight multicast route of G. It is easy to see that the weight $w(T^*)$ of the SMT T^* of graph G' is no greater than *opt*. Therefore, weight $w(T)$ of the tree obtained in Stage 1 is at most $\rho \cdot opt$. In addition, the weight $w(R)$ of the output is at most twice as large as the weight $w(T)$. Therefore, the above algorithm is a (2ρ)-approximation for MIN-MR.

We note that the second stage of Algorithm 6.G is a straightforward modification of T. Can we improve it with some more sophisticated modification? The answer is *yes*. The following algorithm, similar to Christofides's algorithm for TSP, uses minimum matching in the second stage to get a better approximation.

Algorithm 6.H (*Improved Relaxation Algorithm for* MIN-MR)

Input: An edge-weighted graph $G = (V, E)$ with $A \subseteq V$ identified as splitting vertices, a source vertex $s \in V$, and a subset $M \subseteq V$ of multicast members.

Stage 1:

(1.1) Let G' be the complete graph on vertices in $\{s\} \cup M \cup A$;

(1.2) **For** each edge $\{u, v\}$ of G' **do**
$w(\{u, v\}) \leftarrow$ the total weight of the shortest path between u and v in the input graph G;

(1.3) Apply A_{NSMT} to G' with weight w to get a Steiner tree T, treating all vertices in $M \cup \{s\}$ as terminals and all other vertices as Steiner vertices.

Stage 2:

(2.1) Let F be the subgraph of T that consists of all edges in T that are incident on some Steiner node;

(2.2) **For** each connected component C of F **do**
treat C as a rooted tree, with root being the node closest to s in T, and let $p(C)$ be a path from the root to a leaf in C;

(2.3) Let K be the subgraph of T consisting of all edges in $T \setminus F$ plus all edges in $p(C)$ for all connected components C of F;

(2.4) Let D be the set of vertices with an odd degree in K, and let M be a minimum-weight perfect matching for D;

(2.5) Find a multicast route R in $T \cup M$, and output R. ■

We first show that the above algorithm is well defined. In Stage 1, we note that the weight function w defined on G' satisfies the triangle inequality. Therefore, the algorithm A_{NSMT} works on G' with weight w. We also note that in the forest K constructed in Stage 2, every Steiner vertex has an even degree. Therefore, the number of multicast members with odd degrees in K must be even, and the minimum-weight perfect matching M for D exists. Finally, the following lemma shows that the last step (2.5) is well defined.

Lemma 6.25 *In* Algorithm 6.H, *the set* $T \cup M$ *contains a multicast route using each edge at most once.*

Proof. We note that since K is a forest, $K \cup M$ is a disjoint union of cycles; each cycle is a connected component of $K \cup M$. One of these cycles contains the source node s. We can construct a multicast route R in $T \cup M$ as follows:

(1) Initially, R contains a single vertex s.

(2) **While** there is a cycle Q of $K \cup M$ such that R contains a vertex x in Q but not all vertices in Q **do**

  (2.1) Traverse the cycle Q, starting from x, until all vertices in Q are visited; add these edges to R.

  (2.2) **While** there exists a Steiner vertex y in R whose neighbors in T are not all in R **do**
    Split R at y to include edges from y to all its neighbors that are not in R yet.

It is clear that this multicast route R uses each edge at most once. To see that the route R covers every multicast member, we note that the connected components of $K \cup M$ are connected by the Steiner vertices in T. So we can see, by a simple induction, that every cycle Q of $K \cup M$ will be visited by route R. □

We next estimate the total weight of $T \cup M$. To this end, it suffices to study the weight of M since the total weight of T is within a factor of ρ from the weight of a Steiner minimum tree, and hence is at most $\rho \cdot opt$, where *opt* is the minimum-weight of a multicast route.

Lemma 6.26 *The total weight of matching* M *found in* Algorithm 6.H *is at most opt.*

Proof. Let T^* be a minimum multicast tree in the input optical network. Starting from the source s, perform a depth-first search of tree T^*. Then we obtain a tour Q of the graph G whose total weight is at most $2 \cdot opt$. Note that the source node and all multicast members are in the cycle Q.

Recall that the set D consists of all vertices in K with odd degrees. We connect vertices in D along the cycle Q to get a cycle Q' over D. The total weight of cycle Q' is at most $2 \cdot opt$, since the edge-weight in G satisfies the triangle inequality. Since D contains an even number of vertices, the cycle Q' can be decomposed into two disjoint perfect matchings for D. One of them must have the total weight $\leq \rho$. Therefore, the total weight of M is at most *opt*. □

Theorem 6.27 *Assume that* NSMT *has a polynomial-time* ρ*-approximation. Then there is a polynomial-time* $(1+\rho)$*-approximation for* MIN-MR.

6.6 A Remark on Relaxation versus Restriction

In this and previous chapters, we have studied the restriction and relaxation techniques for approximation. It is useful, however, to point out that these techniques are only general ideas. When they are applied to specific problems, we often need to combine them with other techniques, such as greedy strategy and two-stage approximation to make them work. Moreover, these two techniques are not mutually exclusive. Indeed, an approximation can actually be derived from both techniques of relaxation and restriction. Let us look at a simple example.

> MULTIWAY CUT (MWC): Given a graph $G = (V, E)$ with a nonnegative edge-weight function $w : E \to \mathbb{N}$, and k terminals $x_1, \ldots, x_k \in V$, find a minimum total-weight subset of edges that, when removed, separate all k terminals from each other.

To get some idea of the approximation for this problem, let us examine an optimal solution C^* for MWC. Without loss of generality, we may assume that C^* has the minimum number of edges among all optimal solutions. Then removal of edges from C^* leaves the graph G with exactly k connected components $G_1, \ldots, G_k$, containing k terminals $x_1, \ldots, x_k$, respectively. Moreover, each edge $e \in C^*$ is between two different components G_i and G_j. For each $i = 1, 2, \ldots, k$, let

$$C_i = \{\{u, v\} \in C^* \mid u \in G_i, v \in G_j \text{ for some } j \neq i\}.$$

Then each C_i, with $1 \leq i \leq k$, is a cut separating x_i from other terminals, and each edge $\{u, v\} \in C^*$ appears in exactly two C_i's.

Motivated by the above fact, we can design the following approximation algorithm for MWC:

Algorithm 6.I (*Approximation Algorithm for* MWC)

Input: A graph $G = (V, E)$ with an edge-weight function $w : E \to \mathbb{Z}^+$ and k terminals $x_1, \ldots, x_k \in V$.

(1) **For** $i \leftarrow 1$ **to** k **do**
  compute a minimum weight cut D_i separating x_i from other terminals

(2) Output $C \leftarrow \bigcup_{i=1}^{k} D_i$. ■

It is well known that the minimum cut separating a terminal from some other terminals can be found in polynomial time. In addition, it is easy to see that this algorithm has a performance ratio 2.

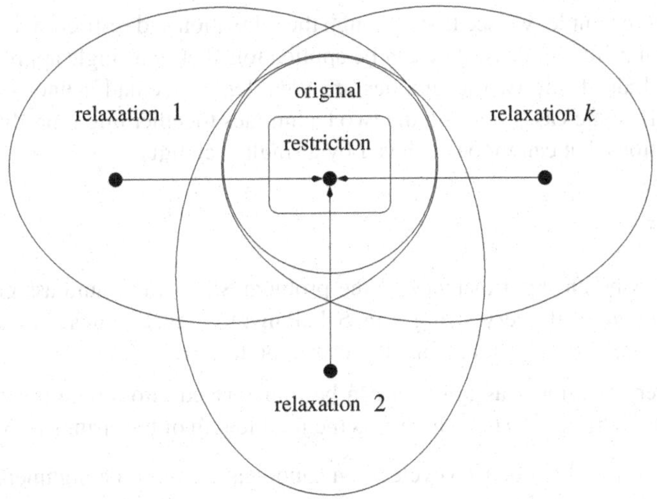

Figure 6.8: Relaxation and restriction.

Theorem 6.28 Algorithm 6.I *is a polynomial-time* 2*-approximation for the problem* MWC.

Proof. Since each D_i, for $1 \le i \le k$, is the minimum cut separating x_i from other terminals, we have $w(D_i) \le w(C_i)$ for every $i = 1, 2, \ldots, k$. It follows that

$$w(C) \le \sum_{i=1}^{k} w(D_i) \le \sum_{i=1}^{k} w(C_i) = 2w(C^*). \qquad \square$$

Now, let us examine the techniques used in the design of the above 2-approximation. First, we may view it as a two-stage relaxation algorithm. That is, we first relax the requirement of a multiway cut for all k terminals to a simpler requirement of cutting one terminal from the other $k-1$ terminals. This relaxation generates k new relaxed problems. Then, in the second stage, we combine the k optimal solutions for the k relaxed problems to an approximate solution for the original problem. Indeed, this type of two-stage approximation by relaxation is quite popular.

On the other hand, we may also consider the design of Algorithm 6.I as a restriction method. Namely, we restrict the feasible solutions to be the union of k solutions D_i each a minimum solution for separating one terminal from all other $k-1$ terminals. As illustrated in Chapters 3 and 4, this type of restriction of the feasible solutions to the unions of solutions of subproblems is also very popular.

The ideas behind these two viewpoints can be seen more clearly in Figure 6.8. In particular, when we relax the original problem into k new relaxed subproblems, the combined solution of the solutions from these relaxed subproblems is a *restricted* solution to the original problem. In addition, the restriction we imposed on the problem requires us to first solve k relaxed problems.

From this example, we see that although the relaxation and restriction techniques are based on different ideas, they can be applied together in a single approximation algorithm. Indeed, the two techniques are complementary and cannot be strictly separated. In some cases, mixing the two techniques together might produce better approximations that cannot be achieved by a single technique.

Exercises

6.1 Let S be an input instance of the problem SS. A minimum assignment in $F(S)$ is *canonical* if every string s in S belongs to a cycle whose weight is the smallest among all cycles that embed s. Prove the following:

(a) Every minimum assignment can be transformed into one in the canonical form in time $O(nL)$, where L is the total length of the strings in S.

(b) Let C_1 and C_2 be two cycles in a canonical minimum assignment and s_1, s_2 two strings belonging to C_1, C_2, respectively. Then

$$|ov(s_1, s_2)| + |ov(s_2, s_1)| < \max\{|s_1|, |s_2|\} + \min\{d(C_1), d(C_2)\}.$$

6.2 Show that two disks with radius 1 and center distance at most 1 can cover at most eight points that are apart from each other with distance bigger than 1. Use this fact to show that, in any unit disk graph G, the size of a maximal independent set is upper-bounded by $(3.8)opt + 1.2$, where *opt* is the size of the minimum connected dominating set of G.

6.3 Give an example to show that two disks with radius 1 and center distance at most 1 can cover nine points that are apart from each other with distance at least 1.

6.4 Consider a unit disk graph G. For any vertex subset C of G, define $f(C)$ to be the number of connected components of the subgraph of G induced by C. The following is a greedy algorithm to connect a maximal independent set D of G into a connected dominating set C.

(1) $C \leftarrow D$.

(2) **While** $f(C) \geq 2$ **do**
  choose a vertex x to maximize $-\Delta_x f(C)$;
  $C \leftarrow C \cup \{x\}$.

(3) Return C.

Show that this algorithm returns a connected dominating set of G of size at most $(6 + \ln 4)opt$ if $|D| \leq 4 \cdot opt + 1$, where *opt* is the size of a minimum connected dominating set of G.

6.5 Given four unit disks with one of them containing the centers of the other three, how many points can be covered by these four disks such that the distance between any two of the points is greater than $1/2$?

6.6 A *unit ball* is a ball in the three-dimensional space with radius 1/2. Prove the following:

(a) A unit ball can touch at most 12 unit balls without incurring any interior intersection point between any two balls.

(b) A unit ball can contain up to 12 points that are apart from each other with distance greater than $1/2$.

6.7 A graph is called a *unit ball graph* if each vertex is associated with a unit ball in the three-dimensional Euclidean space such that an edge $\{u, v\}$ exists if and only if the two unit balls associated with u and v have a nonempty intersection. Show that, in a unit ball graph, every maximal independent set has size at most $11 \cdot opt + 1$, where *opt* is the size of a minimum connected dominating set.

6.8 Let D be a maximal independent set in a unit disk graph G, of which every proper subset $D' \subseteq D$ is within distance 2 from $D \setminus D'$. Show that the following algorithm uses at most $3 \cdot opt$ vertices to connect D into a connected dominating set, where *opt* is the size of a minimum connected dominating set of G.

(1) Color all vertices in D in black and all other vertices in gray.

(2) **While** there exists a gray vertex x adjacent to at least three black components **do**
  Change the color of x to black.

(3) **While** there exists a gray vertex x adjacent to at least two black components **do**
  Change the color of x to black.

(4) **Return** all black vertices.

6.9 Show that Algorithm 6.E is a $(6\frac{7}{18})$-approximation for the problem CDS-UDG.

6.10 Design a polynomial-time greedy approximation for the minimum connected dominating set in a unit ball graph that produces a solution of size at most $(13 + \ln 10)opt + 1$, where *opt* is the size of the minimum connected dominating set.

6.11 Show that there exists a polynomial-time algorithm that, on a given connected graph G, finds a connected dominating set of G with the minimum diameter.

6.12 Show that there exists a polynomial-time algorithm that, on a given connected graph G, finds a connected dominating set C of G with the properties of

$$|C| \leq \alpha \cdot opt_S$$

and

$$diameter(C) \le \beta \cdot opt_{\mathrm{D}},$$

for some constants $\alpha, \beta > 1$, where opt_{S} is the size of a minimum connected dominating set of G, and opt_{D} is the diameter of a minimum-diameter connected dominating set of G.

6.13 Let D be a maximal independent set in a unit disk graph G. Consider the following algorithm to connect set D into a connected dominating set:

> **While** there exist $u, v \in D$ with distance 3 **do**
>
> connect u and v by adding all vertices on the shortest path between u and v to D.
>
> Return D.

Show that the size of the connected dominating set obtained by this algorithm is at most $192 \cdot opt + 48$, where *opt* is the size of a minimum connected dominating set of G.

6.14 A wireless network with different transmission ranges can be formulated as the following disk graph: Each vertex u is associated with a disk centered at u having radius equal to its transmission range. An edge exists between two vertices u and v if and only if the disk u covers the vertex v and the disk v covers the vertex u. Let G be such a disk graph. Prove the following:

(a) Every maximal independent set of G has size at most $K \cdot opt$, where

$$K = \begin{cases} 5, & \text{if } r_{\max}/r_{\min} = 1, \\ 10\left\lfloor \dfrac{\ln(r_{\max}/r_{\min})}{\ln(2\cos(\pi/5))} \right\rfloor, & \text{otherwise,} \end{cases}$$

$r_{\max}$ (and $r_{\min}$) is the maximum (and, respectively, minimum) radius of disks in G, and *opt* is the size of a minimum connected dominating set of G.

(b) There is a polynomial-time $(2+\ln K)$-approximation for the minimum connected dominating set of G.

6.15 Consider the following problem:

> Given a vertex-weighted graph $G = (V, E)$ and a vertex subset $A \subseteq V$, find a Steiner tree interconnecting the vertices in A with the minimum total vertex weight.

Show that this problem has a polynomial-time $(2 \ln n)$-approximation, where $n = |V|$.

6.16 Consider the following problem:

> Given a vertex-weighted strongly connected digraph $G = (V, E)$, find a strongly connected dominating set of G with the minimum total vertex weight.

Show that this problem has a polynomial-time $(2 \ln n)$-approximation, where $n = |V|$.

6.17 Consider the following problem:

> Given a vertex-weighted connected graph $G = (V, E)$, find a connected dominating set of G with the minimum total vertex weight.

Show that this problem has a polynomial-time $(\frac{3}{2} \ln n)$-approximation, where $n = |V|$.

6.18 Show that the problem SCDS has a polynomial-time $(3 \ln n)$-approximation, where $n = |V|$.

6.19 Show that the following algorithm is a 3-approximation for MIN-MR:

(1) Construct a graph G' and edge-weight w from the input network as described in Stage 1 of Algorithm 6.H.

(2) Construct a traveling salesman tour Q in G' with Christofides's approximation (Algorithm 1.H).

(3) Traverse along the tour Q, starting from the source vertex, to all multicast members. Convert this path in G' into a route in the original optical network.

Historical Notes

The 3-approximation for SS was given by Blum et al. [1991]. The performance ratio 3 has been improved subsequently to 2.889 by Teng and Yao [1997], to 2.833 by Czumaj et al. [1994], to 2.793 by Kosaraju et al. [1994], and to $2\frac{2}{3}$ by Armen and Stein [1996].

Connected dominating sets have important applications in multicast routing in wireless sensor networks (called *virtual backbone* in the literature of wireless networks). Much effort has been made to find approximations for the minimum connected dominating sets; see Das and Bhaghavan [1997], Sivakumar et al. [1998], Stojmenovic et al. [2002], Wu and Li [1999], Wan et al. [2002], Chen and Liestman [2002], and Alzoubi et al. [2002].

Guha and Khuller [1998a] showed a two-stage greedy $(\ln \Delta + 3)$-approximation for the minimum connected dominating sets in general graphs where Δ is the maximum degree in the graph. They also gave a lower bound $(\ln \Delta + 1)$ for any polynomial-time approximation for the minimum connected dominating set, provided $\mathbf{NP} \not\subseteq \mathbf{DTIME}(n^{\log \log n})$. Ruan et al. [2004] found a one-stage greedy $(\ln \Delta + 2)$-approximation.

Cheng et al. [2003] showed the existence of a PTAS for the minimum connected dominating sets in unit disk graphs. However, its high running time makes it hard to implement in practice. The two-stage approximation is a popular idea to construct connected dominating sets in unit disk graphs; see Wan et al. [2002], Alzoubi et al. [2002], Wan et al. [2008], Li et al. [2005], Cadei et al. [2002], Funke et al. [2006], and Min et al. [2006]. Among these approximations, the best performance ratio is $6\frac{7}{18}$ of Wan et al. [2008].

The $(4 \ln n)$-approximation for SCDS of Section 6.4 was given by Li, Du et al. [2008]. It has been improved to a $(3 \ln n)$-approximation by Li et al. [2009]. The spider decomposition technique was first used by Klein and Ravi [1995] in their analysis of an algorithm for vertex-weighted Steiner trees (Exercise 6.15). Guha and Khuller [1998b] applied this technique to get an improvement to the weighted dominating set problem.

For the problem MIN-MR, Yan et al. [2003] gave the first polynomial-time approximation (Algorithm 6.G). Suppose ρ is the performance ratio of the best polynomial-time approximation for the Steiner minimum tree known today, then the performance ratio of Algorithm 6.G is $2\rho \approx 3.1$. Du et al. [2005] improved it to a 3-approximation (see Exercise 6.19). Guo et al. [2005] further improved it by giving a $(1 + \rho)$-approximation (Algorithm 6.H). The polynomial-time 2-approximation for MWC (Algorithm 6.I) is from Dahlhaus et al. [1994].

7
Linear Programming

People take the longest possible paths, digress to numerous dead ends, and make all kinds of mistakes. Then historians come along and write summaries of this messy, nonlinear process and make it appear like a simple, straight line.

— Dean Kamen

A widely used relaxation technique for approximation algorithms is to convert an optimization problem into an integer linear program and then relax the constraints on the solutions allowing them to assume real, noninteger values. As the optimal solution to a (real-valued) linear program can be found in polynomial time, we can then solve the linear program and round the solutions to integers as the solutions for the original problem. In this chapter, we give a brief introduction to the theory of linear programming and discuss various rounding techniques.

7.1 Basic Properties of Linear Programming

Recall that an optimization problem is usually of the following form:

$$\begin{array}{ll} \text{minimize (or, maximize)} & c(x_1, x_2, \ldots, x_n) \\ \text{subject to} & (x_1, x_2, \ldots, x_n) \in \Omega, \end{array}$$

where c is a real-valued objective function and $\Omega \subseteq \mathbb{R}^n$ is the feasible region of the problem. An optimization problem is called a *linear program* (LP) if its objective function c is a linear function and its feasible region is constrained by linear equations and/or linear inequalities. Moreover, if its variables are required to be integers,

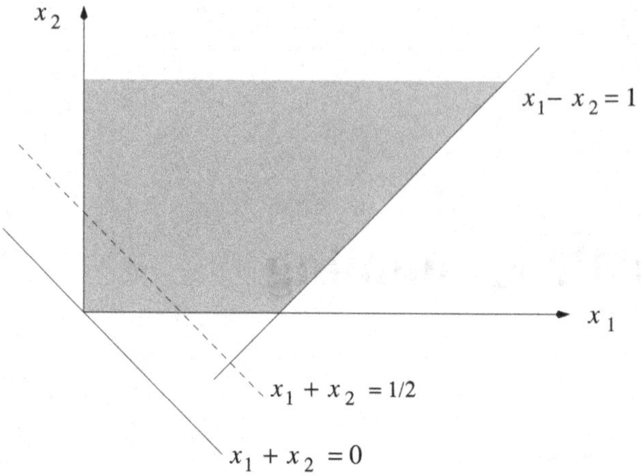

Figure 7.1: Feasible region of a linear program.

then it is called an *integer linear program* (ILP). For instance, the following is a linear program:

$$\begin{aligned} &\text{minimize} && x_1 + x_2 \\ &\text{subject to} && x_1 - x_2 \le 1, \\ & && x_1 \ge 0, \;\; x_2 \ge 0. \end{aligned} \tag{7.1}$$

In this example, the feasible region is constrained by three linear inequalities and can be easily seen to be a two-dimensional polyhedron as shown in Figure 7.1. Its objective function $x_1 + x_2$ reaches the minimum at point $(0, 0)$, which is a vertex of the polyhedron.

As another example, consider the problem KNAPSACK studied in Section 1.1. If we replace the constraints "$x_i \in \{0, 1\}$" by "$0 \le x_i \le 1$," for all $i = 1, 2, \ldots, n$, then we obtain the following linear program:

$$\begin{aligned} &\text{maximize} && c_1x_1 + c_2x_2 + \cdots + c_nx_n \\ &\text{subject to} && s_1x_1 + s_2x_2 + \cdots + s_nx_n \le S, \\ & && 0 \le x_i \le 1, \qquad \text{for } i = 1, 2, \ldots, n, \end{aligned}$$

where $c_1, \ldots, c_n, s_1, \ldots, s_n, S$ are nonnegative real numbers. If $S \ge \sum_{i=1}^{n} s_i$, then $(x_1, x_2, \ldots, x_n) = (1, 1, \ldots, 1)$ is a trivial optimal solution. Otherwise, the optimal solution can be computed as follows: First, sort all c_i/s_i in nonincreasing order. Without loss of generality, assume $c_1/s_1 \ge c_2/s_2 \ge \cdots \ge c_n/s_n$. Let k satisfy

$$\sum_{i=1}^{k} s_i \le S < \sum_{i=1}^{k+1} s_i.$$

Then the following is the optimal solution:

$$x_i = \begin{cases} 1, & \text{if } 1 \le i \le k, \\ (S - \sum_{i=1}^{k} s_i)/s_{k+1}, & \text{if } i = k+1, \\ 0, & \text{if } k+2 \le i \le n. \end{cases}$$

In fact, replacing the constraints "$x_i \in \{0,1\}$" by "$0 \le x_i \le 1$," for $1 \le i \le n$, is equivalent to allowing each item to be cut into smaller pieces of arbitrary size. Therefore, the best strategy for Ali Baba in this situation is to fill the knapsack with items in the decreasing order of the density c_i/s_i. We note that this optimal solution has at most one nonintegral component; that is, at most one item is to be cut into smaller pieces. Thus, if we give up this item, then we get an approximate solution to the original KNAPSACK problem within the difference of $\max\{c_i \mid 1 \le i \le n\}$ from the optimal solution. This is the essential idea of the greedy algorithm for KNAPSACK in Section 1.1.

Now, let us extend this idea to study a resource management problem with more than one type of resources:

$$\begin{aligned} \text{maximize} \quad & c_1x_1 + c_2x_2 + \cdots + c_nx_n \\ \text{subject to} \quad & a_{11}x_1 + a_{12}x_2 + \cdots + a_{1n}x_n \le b_1, \\ & a_{21}x_1 + a_{22}x_2 + \cdots + a_{2n}x_n \le b_2, \\ & \cdots \\ & a_{m1}x_1 + a_{m2}x_2 + \cdots + a_{mn}x_n \le b_m, \\ & x_i \in \{0,1\}, \qquad \text{for } i = 1, 2, \ldots, n, \end{aligned} \tag{7.2}$$

where $a_{ij}, b_i, c_j \in \mathbb{R}$ for all $i = 1, \ldots, m$ and $j = 1, \ldots, n$. Following the example of KNAPSACK, we may wish to convert this integer program into a linear program by relaxing the constraints "$x_i \in \{0,1\}$" to "$0 \le x_i \le 1$," for $1 \le i \le n$. Because of the complexity of this problem, however, we need to explore the theory of linear programming a little more before we can attack this problem.

Linear programs have a standard form as follows:

$$\begin{aligned} \text{minimize} \quad & \boldsymbol{c}\boldsymbol{x} \\ \text{subject to} \quad & \boldsymbol{A}\boldsymbol{x} = \boldsymbol{b}, \\ & \boldsymbol{x} \ge \boldsymbol{0}, \end{aligned} \tag{7.3}$$

where $\boldsymbol{A}$ is an $m \times n$ matrix over reals, with $m \le n$, $\boldsymbol{x}$ is an n-dimensional column vector over reals, $\boldsymbol{c}$ is an n-dimensional row vector over reals, and $\boldsymbol{b}$ is an m-dimensional column vector over reals. (For a vector $\boldsymbol{x} = (x_1, \ldots, x_n) \in \mathbb{R}^n$, we write $\boldsymbol{x} \ge \boldsymbol{0}$ to denote that $x_i \ge 0$ for all $i = 1, 2, \ldots, n$.) Every linear program can be transformed into an equivalent one in the standard form. In fact, if a variable x_i is not nonnegative, then we can use two nonnegative variables to replace it; that is, set $x_i = u_i - v_i$, $u_i \ge 0$, $v_i \ge 0$. Furthermore, an inequality can also be transformed to an equivalent equality by introducing a new nonnegative variable. For example, the linear program (7.1) can be transformed into an equivalent one in the standard form as follows:

$$\begin{array}{ll} \text{minimize} & x_1 + x_2 \\ \text{subject to} & x_1 - x_2 + w = 1, \\ & x_1 \geq 0,\ x_2 \geq 0,\ w \geq 0. \end{array} \tag{7.4}$$

In the standard form of linear programming, we usually assume that $rank(\boldsymbol{A}) = m$. In fact, if the feasible domain is not empty, then the property $rank(\boldsymbol{A}) < m$ means that there exist some useless constraints, and these useless constraints can be deleted to make the rank of the coefficient matrix equal to the number of rows.

It can be seen easily in Figure 7.1 that the optimal solution to (7.1) occurs at a vertex of the feasible region. Indeed, this is a very important general property of linear programs.

What is a vertex of the feasible region? Note that the feasible region of every linear program is a polyhedron. A point $\boldsymbol{x}$ in a polyhedron Ω is called a *vertex* or an *extreme point* if it has the following property:

$$\text{If } \boldsymbol{x} = (\boldsymbol{y} + \boldsymbol{z})/2 \text{ for some } \boldsymbol{y}, \boldsymbol{z} \in \Omega, \text{ then } \boldsymbol{x} = \boldsymbol{y} = \boldsymbol{z}.$$

With this definition, let us first give a formal proof for our observation.

Lemma 7.1 *Let* $\Omega = \{\boldsymbol{x} \mid \boldsymbol{A}\boldsymbol{x} = \boldsymbol{b}, \boldsymbol{x} \geq \boldsymbol{0}\}$. *If* $\min_{\boldsymbol{x}\in\Omega}(\boldsymbol{c}\boldsymbol{x})$ *has an optimal solution, then it can be found at one of its vertices.*

Proof. Consider an optimal solution $\boldsymbol{x}^*$ with the maximum number of zero components among all optimal solutions. We will show that $\boldsymbol{x}^*$ is a vertex of Ω. By way of contradiction, suppose x^* is not a vertex; that is, suppose there exist $\boldsymbol{y}, \boldsymbol{z} \in \Omega$ such that $\boldsymbol{x}^* = (\boldsymbol{y}+\boldsymbol{z})/2$ but $\boldsymbol{x}^*$, $\boldsymbol{y}$, and $\boldsymbol{z}$ are distinct (note that if two of them are equal, then they are all equal). Since $\boldsymbol{c}\boldsymbol{x}^* \leq \boldsymbol{c}\boldsymbol{y}$, $\boldsymbol{c}\boldsymbol{x}^* \leq \boldsymbol{c}\boldsymbol{z}$, and $\boldsymbol{c}\boldsymbol{x}^* = (\boldsymbol{c}\boldsymbol{y} + \boldsymbol{c}\boldsymbol{z})/2$, we must have $\boldsymbol{c}\boldsymbol{x}^* = \boldsymbol{c}\boldsymbol{y} = \boldsymbol{c}\boldsymbol{z}$. This means that $\boldsymbol{y}$ and $\boldsymbol{z}$ are also optimal solutions. It follows that all feasible points on the line $\boldsymbol{x}^* + \alpha(\boldsymbol{y} - \boldsymbol{x}^*)$, $\alpha \in \mathbb{R}$, are optimal solutions. However, by the constraints $x_i \geq 0$ for $i = 1, 2, \ldots, n$, Ω does not contain a whole line. Thus, the line $\boldsymbol{x}^* + \alpha(\boldsymbol{y} - \boldsymbol{x}^*)$ must have a point $\boldsymbol{x}'$ not in Ω; that is, $\boldsymbol{x}'$ violates at least one constraint.

Note that for any α, $\boldsymbol{A}(\boldsymbol{x}^* + \alpha(\boldsymbol{y} - \boldsymbol{x}^*)) = \boldsymbol{b}$. Thus, $\boldsymbol{x}'$ cannot violate constraint $\boldsymbol{A}\boldsymbol{x} = \boldsymbol{b}$. Moreover, suppose that $x_i^* = 0$ for some i, $1 \leq i \leq n$. Since $x_i^* = (y_i + z_i)/2$ and $y_i, z_i \geq 0$, we must have $z_i = y_i = x_i^* = 0$. Therefore, the ith component of $\boldsymbol{x}^* + \alpha(\boldsymbol{y} - \boldsymbol{x}^*)$ is equal to 0 for any α. This means that $\boldsymbol{x}'$ cannot violate any constraint $x_i \geq 0$ with $x_i^* = 0$. Hence, $\boldsymbol{x}'$ must violate a constraint $x_j \geq 0$ for some j with $x_j^* > 0$. We claim that there must exist some β, $0 < \beta < 1$, such that $\boldsymbol{x}'' = \beta\boldsymbol{x}^* + (1 - \beta)\boldsymbol{x}'$ is an optimal solution in Ω but has one more zero component than $\boldsymbol{x}^*$, contradicting the assumption that $\boldsymbol{x}^*$ has the maximum number of zero components among optimal solutions.

To prove the claim, let $J = \{j \mid 1 \leq j \leq n, x_j' < 0\}$ and, for each $j \in J$, define

$$\beta_j = \frac{-x_j'}{x_j^* - x_j'}.$$

Note that $0 < \beta_j < 1$, for all $j \in J$. Choose $j_0 \in J$ such that β_{j_0} is the maximum among all β_j's. Then we can see that $\boldsymbol{x}'' = \beta_{j_0}\boldsymbol{x}^* + (1 - \beta_{j_0})\boldsymbol{x}'$ has the properties $x''_{j_0} = 0$ and $x''_j \geq 0$ for all $j \in \{1, 2, \ldots, n\} - \{j_0\}$. So, x'' is an optimal solution in Ω. In addition, when $x^*_j = 0$, we must have $x'_j = 0$ and, hence, $x''_j = 0$. Also, since $j_0 \in J$, $x^*_{j_0} > 0$. Thus, $\boldsymbol{x}''$ has at least one more zero component than $\boldsymbol{x}^*$. □

Since the optimal solutions occur at the vertices of the feasible region, it is useful to give a necessary and sufficient condition for a feasible point to be a vertex.

Lemma 7.2 *Consider the linear program (7.3) in the standard form. Let $\boldsymbol{a}_j$, for $1 \leq j \leq n$, denote the jth column of $\boldsymbol{A}$. Then a feasible point $\boldsymbol{x} \in \Omega$ is a vertex if and only if the vectors in $\{\boldsymbol{a}_j \mid 1 \leq j \leq n,\ x_j \neq 0\}$ are linearly independent.*

Proof. Assume $\{j \mid 1 \leq j \leq n, x_j \neq 0\} = \{j_1, j_2, \ldots, j_k\}$.

For the "if" part, suppose $\boldsymbol{x} = (\boldsymbol{y}+\boldsymbol{z})/2$ and $\boldsymbol{y}, \boldsymbol{z} \in \Omega$. Note that $x_j = 0$ implies $y_j = z_j = 0$. Thus, $(x_{j_1}, x_{j_2}, \ldots, x_{j_k})$, $(y_{j_1}, y_{j_2}, \ldots, y_{j_k})$, and $(z_{j_1}, z_{j_2}, \ldots, z_{j_k})$ are all solutions to the following system of linear equations (over variables u_{j_1}, $u_{j_2}, \ldots, u_{j_k}$):

$$\boldsymbol{a}_{j_1}u_{j_1} + \boldsymbol{a}_{j_2}u_{j_2} + \cdots + \boldsymbol{a}_{j_k}u_{j_k} = \boldsymbol{b}. \tag{7.5}$$

Since $\boldsymbol{a}_{j_1}, \boldsymbol{a}_{j_2}, \ldots, \boldsymbol{a}_{j_k}$ are linearly independent, this system of linear equations has a unique solution. Thus, $(x_{j_1}, x_{j_2}, \ldots, x_{j_k}) = (y_{j_1}, y_{j_2}, \ldots, y_{j_k}) = (z_{j_1}, z_{j_2}, \ldots, z_{j_k})$. Hence, $\boldsymbol{x} = \boldsymbol{y} = \boldsymbol{z}$. This means that x is a vertex.

For the "only if" part, suppose that $\boldsymbol{x}$ is a vertex. We claim that the system of linear equations (7.5) has a unique solution. Suppose otherwise that (7.5) has a second solution $(x'_{j_1}, x'_{j_2}, \ldots, x'_{j_k}) \neq (x_{j_1}, x_{j_2}, \ldots, x_{j_k})$. Set $x'_j = 0$ for $j \in \{1, \ldots, n\} \setminus \{j_1, j_2, \ldots, j_k\}$. Then $\boldsymbol{A}\boldsymbol{x}' = \boldsymbol{b}$. In addition, for sufficiently small $\alpha > 0$, we have $\boldsymbol{x} + \alpha(\boldsymbol{x}' - \boldsymbol{x}) \geq \boldsymbol{0}$ and $\boldsymbol{x} - \alpha(\boldsymbol{x}' - \boldsymbol{x}) \geq \boldsymbol{0}$. Fix such an α and set $\boldsymbol{y} = \boldsymbol{x} + \alpha(\boldsymbol{x}' - \boldsymbol{x})$ and $\boldsymbol{z} = \boldsymbol{x} - \alpha(\boldsymbol{x}' - \boldsymbol{x})$. Then $\boldsymbol{y}, \boldsymbol{z} \in \Omega$, $\boldsymbol{x} = (\boldsymbol{y} + \boldsymbol{z})/2$, and $\boldsymbol{x} \neq \boldsymbol{y}$, contradicting the fact that $\boldsymbol{x}$ is a vertex. Thus, the claim is proven. It follows that $\boldsymbol{a}_{j_1}, \boldsymbol{a}_{j_2}, \ldots, \boldsymbol{a}_{j_k}$ are linearly independent. □

Recall that we may assume $rank(\boldsymbol{A}) = m$. Thus, by Lemma 7.2, a vertex $\boldsymbol{x}$ has at most m nonzero components. In the case of $\boldsymbol{x}$ having fewer than m nonzero components, we can add more columns to form a maximum independent subset of columns of $\boldsymbol{A}$. This means that a feasible point $\boldsymbol{x}$ is a vertex if and only if there exists a set $J = \{j_1, \ldots, j_m\}$ of m integers between 1 and n such that columns $\boldsymbol{a}_{j_1}, \boldsymbol{a}_{j_2}, \ldots, \boldsymbol{a}_{j_m}$ of $\boldsymbol{A}$ are linearly independent and $x_j = 0$ for $j \notin J$.

A vertex is also called a *basic feasible solution*. The index subset $J = \{j_1, j_2, \ldots, j_m\}$ associated with a basic feasible solution as described above is called a *feasible basis*. For any index subset $J = \{j_1, j_2, \ldots, j_m\}$, denote $\boldsymbol{A}_J = (\boldsymbol{a}_{j_1}, \boldsymbol{a}_{j_2}, \ldots, \boldsymbol{a}_{j_m})$ and $\boldsymbol{x}_J = (x_{j_1}, x_{j_2}, \ldots, x_{j_m})^T$. Then an index subset J is a feasible basis if and only if $rank(\boldsymbol{A}_J) = m = |J|$ and $\boldsymbol{A}_J^{-1}\boldsymbol{b} \geq \boldsymbol{0}$. Given a feasible basis J, we can determine the vertex x associated with J as follows:

$$\boldsymbol{x}_J = \boldsymbol{A}_J^{-1}\boldsymbol{b},$$
$$\boldsymbol{x}_{\bar{J}} = \boldsymbol{0},$$

where $\bar{J} = \{1, 2, \ldots, n\} - J$. Note that if the number of nonzero components of x is smaller than m, then x may correspond to more than one feasible basis.

A linear program is said to satisfy the *nondegeneracy assumption* if the number of nonzero components of every basic feasible solution is exactly m, or, equivalently, for every feasible basis J, $\boldsymbol{A}_J^{-1}\boldsymbol{b} > \boldsymbol{0}$. For a nondegenerate linear program, the above relationship between basic feasible solutions and feasible bases is a one-to-one correspondence.

Now, let us go back to the resource management problem (7.2). After relaxation, we obtain the following linear program:

$$\begin{array}{ll} \text{maximize} & c_1x_1 + c_2x_2 + \cdots + c_nx_n \\ \text{subject to} & a_{11}x_1 + a_{12}x_2 + \cdots + a_{1n}x_n \le b_1, \\ & a_{21}x_1 + a_{22}x_2 + \cdots + a_{2n}x_n \le b_2, \\ & \cdots \\ & a_{m1}x_1 + a_{m2}x_2 + \cdots + a_{mn}x_n \le b_m, \\ & 0 \le x_1, x_2, \ldots, x_n \le 1. \end{array}$$

That is,

$$\begin{array}{ll} \text{maximize} & \boldsymbol{cx} \\ \text{subject to} & \boldsymbol{Ax} \le \boldsymbol{b}, \\ & \boldsymbol{0} \le \boldsymbol{x} \le \boldsymbol{1}. \end{array} \tag{7.6}$$

This linear program can be transformed into the following one in the standard form:

$$\begin{array}{ll} \text{maximize} & \boldsymbol{cx} \\ \text{subject to} & \boldsymbol{Ax} + \boldsymbol{y} = \boldsymbol{b}, \\ & \boldsymbol{x} + \boldsymbol{z} = \boldsymbol{1}, \\ & \boldsymbol{x} \ge \boldsymbol{0}, \boldsymbol{y} \ge \boldsymbol{0}, \boldsymbol{z} \ge \boldsymbol{0}. \end{array} \tag{7.7}$$

It is easy to show that every vertex of the feasible region of (7.6) is transformed into a vertex of the feasible region of (7.7) and vice versa (see Exercise 7.2).

Now, we are going to study the basic feasible solutions of (7.7). We can write (7.7) in the matrix form as

$$\begin{pmatrix} \boldsymbol{A} & \boldsymbol{I}_m & \boldsymbol{0} \\ \boldsymbol{I}_n & \boldsymbol{0} & \boldsymbol{I}_n \end{pmatrix} \begin{pmatrix} \boldsymbol{x} \\ \boldsymbol{y} \\ \boldsymbol{z} \end{pmatrix} = \begin{pmatrix} \boldsymbol{b} \\ \boldsymbol{1}_n \end{pmatrix},$$

where $\boldsymbol{I}_n$ is the identity matrix of order n, and $\boldsymbol{1}_n = \underbrace{(1, 1, \ldots, 1)}_{n}{}^T$. Note that

$$rank \begin{pmatrix} \boldsymbol{A} & \boldsymbol{I}_m & \boldsymbol{0} \\ \boldsymbol{I}_n & \boldsymbol{0} & \boldsymbol{I}_n \end{pmatrix} = m + n.$$

Thus, every feasible basis contains $m + n$ column indices.

Lemma 7.3 *Every basic feasible solution to* (7.7) *[or* (7.6)*] contains at most* m *nonintegral components in* $\boldsymbol{x}$.

Proof. Consider a basic feasible solution $(\boldsymbol{x}, \boldsymbol{y}, \boldsymbol{z})$ determined by a feasible basis J. Observe the following facts:

(a) If J contains an index j, $1 \leq j \leq n$, but not $n + m + j$, then we must have $z_j = 0$ and hence $x_j = 1$.

(b) If J does not contain an index j, $1 \leq j \leq n$, but contains $n + m + j$, then we must have $x_j = 0$.

Thus, if $0 < x_j < 1$, then J must contain both indices j and $n + m + j$. Subtracting the $(m + n + j)$th column from the jth column, we obtain a column vector of the form $(\boldsymbol{a}_j^T, \boldsymbol{0})^T$, where $\boldsymbol{a}_j$ is the jth column of $\boldsymbol{A}$. We note that all these columns are still linearly independent. Since $rank(\boldsymbol{A}) \leq m$, we can have at most m such linearly independent columns. It follows that there exist at most m indices in $J \cap \{1, 2, \ldots, n\}$ such that $0 < x_j < 1$. □

This lemma suggests that if m is a fixed integer, then we can generalize the greedy algorithms for KNAPSACK (Algorithms 1.B and 1.C) to the resource management problem with arbitrarily small errors. (See Exercises 7.3 and 7.5.) Of course, these generalized algorithms must contain a subroutine for solving linear programming problems.

There are three important families of algorithms for linear programming: the *simplex method*, the *ellipsoid method*, and the *interior-point method*. The simplex method searches for the optimal solutions from a vertex to another vertex. It requires, in the worst case, more than polynomial time, but it runs in polynomial time in the average case and has been used widely in practice. The ellipsoid method is the first polynomial-time algorithm found for linear programming, but it is not efficient in practice. The best-known running time for the ellipsoid method is $O(n^6)$. The interior-point method runs efficiently both theoretically and practically. In the interior-point method, a nonlinear potential function is introduced, and it searches for the optimal solutions from points in the interior of the feasible region. Since it uses a nonlinear potential function, nonlinear programming techniques can be applied in this method. The best-known running time for an interior-point algorithm is $O(n^3)$.

Since the interior-point method involves nonlinear programming techniques, we will not present it in this book. We include a concise presentation of the simplex method in the next section, and a very brief discussion of the application of the ellipsoid method in Section 7.5.

From the (worst-case) polynomial running time of the ellipsoid and interior-point methods for linear programming, we see that, similar to the knapsack problem, it is not hard to design a linear programming-based PTAS for the resource management problem, when the number m of resources is fixed.

Theorem 7.4 *When the number* m *of resources is fixed, the resource management problem* (7.2) *has a PTAS.*

7.2 Simplex Method

The simplex method is motivated by the important observation made in Lemma 7.1: If an optimal solution exists for a linear program, then it can be found from a vertex of the feasible region.

Based on this observation, the simplex method starts from a vertex and, at each iteration, moves from one vertex to another, at which the value $\boldsymbol{cx}$ of the objective function decreases.

To describe it in detail, suppose $\boldsymbol{x}$ is a basic feasible solution associated with the feasible basis J. Let us explain how to determine whether $\boldsymbol{x}$ is an optimal solution and, if $\boldsymbol{x}$ is not optimal, how to find another feasible basic solution $\boldsymbol{x}^+$ with feasible basis J^+ such that $\boldsymbol{cx}^+ < \boldsymbol{cx}$. Let $\boldsymbol{y}$ be a feasible solution in Ω and $\boldsymbol{y}_J$ the vector composed of components y_j for $j \in J$. From $\boldsymbol{Ay} = \boldsymbol{b}$, we know that $\boldsymbol{A}_J\boldsymbol{y}_J + \boldsymbol{A}_{\bar{J}}\boldsymbol{y}_{\bar{J}} = \boldsymbol{b}$; or, equivalently, $\boldsymbol{y}_J = \boldsymbol{A}_J^{-1}(\boldsymbol{b} - \boldsymbol{A}_{\bar{J}}\boldsymbol{y}_{\bar{J}})$. Thus,

$$\boldsymbol{cy} = \boldsymbol{c}_J\boldsymbol{y}_J + \boldsymbol{c}_{\bar{J}}\boldsymbol{y}_{\bar{J}} = \boldsymbol{c}_J\boldsymbol{A}_J^{-1}\boldsymbol{b} + (\boldsymbol{c}_{\bar{J}} - \boldsymbol{c}_J\boldsymbol{A}_J^{-1}\boldsymbol{A}_{\bar{J}})\boldsymbol{y}_{\bar{J}}. \tag{7.8}$$

If $\boldsymbol{c}_{\bar{J}} - \boldsymbol{c}_J\boldsymbol{A}_J^{-1}\boldsymbol{A}_{\bar{J}} \geq 0$, then $\boldsymbol{cy} \geq \boldsymbol{c}_J\boldsymbol{A}_J^{-1}\boldsymbol{b}$ for all feasible solutions $\boldsymbol{y}$. In particular, if $\boldsymbol{y} = \boldsymbol{x}$, then we have $\boldsymbol{y}_{\bar{J}} = 0$, and so $\boldsymbol{cy}$ reaches the minimum value $\boldsymbol{c}_J\boldsymbol{A}_J^{-1}\boldsymbol{b}$. It follows that $\boldsymbol{x}$ is an optimal solution and we cannot improve over it. On the other hand, if $\boldsymbol{c}'_{\bar{J}} = \boldsymbol{c}_{\bar{J}} - \boldsymbol{c}_J\boldsymbol{A}_J^{-1}\boldsymbol{A}_{\bar{J}}$ has a negative component, say $c'_\ell < 0$ for some $\ell \in \bar{J}$, then increasing the value of x_ℓ may decrease the value of $\boldsymbol{cx}$. In other words, using ℓ to replace an index in J may result in a better feasible basis. How do we change the feasible basis? We next study this problem.

Denote $(a'_{ij}) = \boldsymbol{A}' = \boldsymbol{A}_J^{-1}\boldsymbol{A}$ and $\boldsymbol{b}' = \boldsymbol{A}_J^{-1}\boldsymbol{b}$. We note that to find the basic feasible solution $\boldsymbol{x}$ associated with the feasible basis $J = \{j_1, j_2, \ldots, j_m\}$, we transform the equation $\boldsymbol{Ax} = \boldsymbol{b}$ into $\boldsymbol{A}'\boldsymbol{x} = \boldsymbol{b}'$, and let $\boldsymbol{x}_J = \boldsymbol{b}'_J$. Now, suppose we want to replace the ith index j_i in J by ℓ to get a new basis $J^+ = (J - \{j_i\}) \cup \{\ell\}$. We need to perform a linear transformation to change the equation

$$\boldsymbol{A}'\boldsymbol{x} = (\boldsymbol{A}_J^{-1}\boldsymbol{A})\boldsymbol{x} = \boldsymbol{A}_J^{-1}\boldsymbol{b} = \boldsymbol{b}'$$

to

$$\boldsymbol{A}''\boldsymbol{x} = (\boldsymbol{A}_{J^+}^{-1}\boldsymbol{A})\boldsymbol{x} = \boldsymbol{A}_{J^+}^{-1}\boldsymbol{b} = \boldsymbol{b}''.$$

Note that the ℓth column of $\boldsymbol{A}_{J^+}^{-1}\boldsymbol{A}$ is a unit vector with value 1 in the ith component and value 0 in other components. It means that we need to perform the following operations to obtain $(\boldsymbol{A}'', \boldsymbol{b}'')$ from $(\boldsymbol{A}', \boldsymbol{b}')$:

(1) Divide the ith row of $(\boldsymbol{A}', \boldsymbol{b}')$ by $a'_{i\ell}$.

(2) For each k, $1 \leq k \leq m$, $k \neq i$, subtract $a'_{k\ell}$ times the ith row from the kth row of $(\boldsymbol{A}', \boldsymbol{b}')$.

In particular,

$$\begin{aligned} b''_i &= \frac{b'_i}{a'_{i\ell}}, \\ b''_k &= b'_k - a'_{k\ell}\frac{b'_i}{a'_{i\ell}} \qquad \text{if } k \neq i. \end{aligned} \tag{7.9}$$

In order to make $J^+ = (J - \{j_i\}) \cup \{\ell\}$ a feasible basis, we must have $b''_j \geq 0$ for all $j = 1, 2, \ldots, m$. First, we must have $a'_{i\ell} > 0$ in order to have $b''_i = b'_i / a'_{i\ell} \geq 0$.

Next, in order to have $b''_k \geq 0$ for indices $k \neq i$, we must have $b'_k \geq a'_{k\ell} b'_i / a'_{i\ell}$. For indices k with $a'_{k\ell} \leq 0$, this is clearly true. For indices k with $a'_{k\ell} > 0$, this amounts to a new requirement: We must have $b'_i / a'_{i\ell} \leq b'_k / a'_{k\ell}$. That is, we must choose index i by the condition

$$\frac{b'_i}{a'_{i\ell}} = \min \left\{ \frac{b'_k}{a'_{k\ell}} \;\middle|\; 1 \leq k \leq m, a'_{k\ell} > 0 \right\}. \tag{7.10}$$

Note that if there is an index i with a positive $a'_{i\ell}$, then an index i satisfying (7.10) always exists. On the other hand, if $a'_{i\ell} \leq 0$ for all $i = 1, \ldots, m$, then we can see that the linear program (7.3) has no optimal solution: Suppose that we set $x_j = 0$ for all $j \in \bar{J} - \{\ell\}$, x_ℓ to be an arbitrary large real number, and $\boldsymbol{x}_J = \boldsymbol{b}' - (a'_{1\ell} x_\ell, \ldots, a'_{m\ell} x_\ell)^T$. Then this is a feasible solution, and the objective function value on this feasible solution is equal to $\boldsymbol{c}_J \boldsymbol{A}_J^{-1} \boldsymbol{b} + c'_\ell x_\ell$. Since $c'_\ell < 0$, this value tends to $-\infty$ as x_ℓ goes to ∞.

In summary, for any feasible basis J, we have the following possibilities:

(1) If $\boldsymbol{c}'_{\bar{J}} \geq \boldsymbol{0}$, then the associated basic feasible solution $\boldsymbol{x}$ is an optimal solution to the linear program.

(2) If $\boldsymbol{c}'_{\bar{J}}$ has a negative component $c'_\ell < 0$, but $a'_{i\ell} \leq 0$ for all $i = 1, 2, \ldots, m$, then the linear program has no optimal solution.

(3) If $\boldsymbol{c}'_{\bar{J}}$ has a negative component $c'_\ell < 0$, and $a'_{i\ell} > 0$ for some $i = 1, 2, \ldots, m$, then we can choose index i by (7.10) and move our attention to a new feasible basis $J^+ = (J - \{j_i\}) \cup \{\ell\}$.

The transformation from basis J to basis J^+ is called a *pivot*. The simplex method begins with an initial feasible basis, and then perform a sequence of pivots until it finds an optimal solution or determines that the linear program has no optimal solution.

Algorithm 7.A (*Simplex Method* for LINEAR PROGRAMMING)

Input: A linear program in the standard form (7.3).

(1) Find an initial feasible basis J.

(2) **Repeat** the following:

(2.1) Let $\boldsymbol{c}' \leftarrow \boldsymbol{c} - \boldsymbol{c}_J \boldsymbol{A}_J^{-1} \boldsymbol{A}$; $\boldsymbol{A}' \leftarrow \boldsymbol{A}_J^{-1} \boldsymbol{A}$; $\boldsymbol{b}' \leftarrow \boldsymbol{A}_J^{-1} \boldsymbol{b}$.

(2.2) **If** $\boldsymbol{c}' \geq \boldsymbol{0}$, **then** stop and output the current basic feasible solution $((\boldsymbol{b}'_J)^T, \boldsymbol{0})^T$ associated with J.

(2.3) **If** $\boldsymbol{c}'$ has a component $c'_\ell < 0$ **then do**
  if $a'_{i\ell} \leq 0$ for all $1 \leq i \leq m$
    then stop and output "no optimal solution"
    else find an index i satisfying (7.10), and perform a pivot at $a'_{i\ell}$ to get a new feasible basis J^+;
      let $J \leftarrow J^+$. ■

Let us first demonstrate by a numerical example how the simplex method works.

Example 7.5 Consider the following linear program:

$$\begin{array}{ll} \text{minimize} & z = x_6 + x_7 \\ \text{subject to} & x_1 + 2x_2 + x_5 = 8, \\ & x_1 + x_2 - x_3 + x_6 = 3, \\ & -x_1 + x_2 - x_4 + x_7 = 1, \\ & x_1, x_2, \dots, x_7 \geq 0. \end{array}$$

To implement the simplex method, we introduce a simplex table to store all information of the program with respect to a feasible basis J:

$-z$	$\boldsymbol{c} - \boldsymbol{c}_J \boldsymbol{A}_J^{-1} \boldsymbol{A}$	
$\boldsymbol{A}_J^{-1} \boldsymbol{b}$	$\boldsymbol{A}_J^{-1} \boldsymbol{A}$	$\boldsymbol{x}_J$

Before we compute the initial feasible basis, the simplex table is as follows:

0	0	0	0	0	0	1	1
8	1	2	0	0	1	0	0
3	1	1	−1	0	0	1	0
1	−1	1	0	−1	0	0	1

Assume that we select $J = \{5, 6, 7\}$ as the initial feasible basis. Then the associated basic feasible solution is $(0, 0, 0, 0, 8, 3, 1)^T$, $\boldsymbol{c}_J = (c_5, c_6, c_7) = (0, 1, 1)$, and $\boldsymbol{A}_J = \boldsymbol{I}_3$. From $\boldsymbol{c}_J$ and $\boldsymbol{A}_J$, we obtain the following simplex table:

−4	0	−2	1	1	0	0	0	
8	1	2	0	0	1	0	0	x_5
3	1	1	−1	0	0	1	0	x_6
1	−1	[1]	0	−1	0	0	1	x_7

In the above, the vector $\boldsymbol{c}' = \boldsymbol{c} - \boldsymbol{c}_J \boldsymbol{A}_J^{-1} \boldsymbol{A}$ has only one negative component $c_2' = -2$. In addition, $b_3'/a_{3,2}'$ is the minimum among three values of b_k'/a_{k2}'. So we select $a_{3,2}'$ as the pivot element. (The pivot element is shown with a square around it.) That is, our new feasible basis is $J = \{5, 6, 2\}$, and $\boldsymbol{c}_J = (c_5, c_6, c_2) = (0, 1, 0)$. Let $\boldsymbol{a}_3'$ denote the bottom row of the above simplex table. To perform the pivot at $a_{3,2}'$, we subtract $c_2' \boldsymbol{a}_3'$ from the top row (or, the 0th row), and subtract $a_{i,2}' \boldsymbol{a}_3'$ from the ith row, for $i = 1, 2$, and we obtain

−2	−2	0	1	−1	0	0	2	
6	3	0	0	2	1	0	−2	x_5
2	$\boxed{2}$	0	−1	1	0	1	−1	x_6
1	−1	1	0	−1	0	0	1	x_2

The new $\boldsymbol{c}'$ has two negative components: $c'_1 = -2$ and $c'_4 = -1$. We arbitrarily let $\ell = 1$ and, from (7.10), select $a'_{2,1}$ as the new pivot element. After the second pivot, we obtain

0	0	0	0	0	0	1	1	
3	0	0	$\frac{3}{2}$	$\frac{1}{2}$	1	$-\frac{3}{2}$	$-\frac{1}{2}$	x_5
1	1	0	$-\frac{1}{2}$	$\frac{1}{2}$	0	$\frac{1}{2}$	$-\frac{1}{2}$	x_1
2	0	1	$-\frac{1}{2}$	$-\frac{1}{2}$	0	$\frac{1}{2}$	$\frac{1}{2}$	x_2

Now, the components in the top row $\boldsymbol{c} - \boldsymbol{c}_J \boldsymbol{A}_J^{-1} \boldsymbol{A}$ are all nonnegative. It means that $J^* = \{5, 1, 2\}$ is an optimal feasible basis. Its corresponding optimal solution is $(1, 2, 0, 0, 3, 0, 0)^T$. □

In the above example, the value of the objective function decreases after each pivot. Is this true for all linear programs? In other words, does the simplex method always halt after a finite number of pivots? From (7.8) and (7.9), we can see that when we change the feasible basis from J to $J^+ = (J - \{j_i\}) \cup \{\ell\}$, the value of the objective function on the new basic feasible solution $\boldsymbol{x}^+$ becomes $\boldsymbol{c}_J \boldsymbol{A}_J^{-1} \boldsymbol{b} + c'_\ell b'_i / a'_{i\ell}$. This value is less than the previous value $\boldsymbol{c}_J \boldsymbol{A}_J^{-1} \boldsymbol{b}$ as long as $b'_i > 0$. Therefore, if the linear program (7.3) satisfies the nondegeneracy assumption (and so $\boldsymbol{A}_J^{-1} \boldsymbol{b} > 0$ for all feasible bases J), then the value of the objective function decreases after each pivot. It follows that the algorithm will terminate after a finite number of pivots, since it must reach a new feasible basis after each pivot and the number of the feasible basis is finite.

Theorem 7.6 *Under the nondegeneracy assumption, the simplex method halts after a finite number of pivots. It either finds an optimal solution to the linear program* (7.3) *or outputs the fact that the linear program* (7.3) *has no optimal solution.*

What will happen if the given linear program does not satisfy the nondegeneracy assumption? In this case, the simplex method may fall into a cycle. We demonstrate this situation in the following example.

Example 7.7 Consider the following linear program:

$$\begin{aligned}
\text{minimize} \quad & z = -\tfrac{3}{4}x_4 + 20x_5 - \tfrac{1}{2}x_6 + 6x_7 \\
\text{subject to} \quad & x_1 + \tfrac{1}{4}x_4 - 8x_5 - x_6 + 9x_7 = 0, \\
& x_2 + \tfrac{1}{2}x_4 - 12x_5 - \tfrac{1}{2}x_6 + 3x_7 = 0, \\
& x_3 + x_6 = 1, \\
& x_1, x_2, \ldots, x_7 \geq 0.
\end{aligned}$$

The following are seven simplex tables that form a cycle.

0	0	0	0	$-\frac{3}{4}$	20	$-\frac{1}{2}$	6	
0	1	0	0	$\boxed{\frac{1}{4}}$	−8	−1	9	x_1
0	0	1	0	$\frac{1}{2}$	−12	$-\frac{1}{2}$	3	x_2
1	0	0	1	0	0	1	0	x_3

0	3	0	0	0	−4	$-\frac{7}{2}$	33	
0	4	0	0	1	−32	−4	36	x_4
0	−2	1	0	0	$\boxed{4}$	$\frac{3}{2}$	−15	x_2
1	0	0	1	0	0	1	0	x_3

0	1	1	0	0	0	−2	18	
0	−12	8	0	1	0	$\boxed{8}$	−84	x_4
0	$-\frac{1}{2}$	$\frac{1}{4}$	0	0	1	$\frac{3}{8}$	$-\frac{15}{4}$	x_5
1	0	0	1	0	0	1	0	x_3

0	−2	3	0	$\frac{1}{4}$	0	0	−3	
0	$-\frac{3}{2}$	1	0	$\frac{1}{8}$	0	1	$-\frac{21}{2}$	x_6
0	$\frac{1}{16}$	$-\frac{1}{8}$	0	$-\frac{3}{64}$	1	0	$\boxed{\frac{3}{16}}$	x_5
1	$\frac{3}{2}$	−1	1	$-\frac{1}{8}$	0	0	$\frac{21}{2}$	x_3

0	−1	1	0	$-\frac{1}{2}$	16	0	0	
0	$\boxed{2}$	−6	0	$-\frac{5}{2}$	56	1	0	x_6
0	$\frac{1}{3}$	$-\frac{2}{3}$	0	$-\frac{1}{4}$	$\frac{16}{3}$	0	1	x_7
1	−2	6	1	$\frac{5}{2}$	−56	0	0	x_3

$$
\begin{array}{c|ccccccc|l}
0 & 0 & -2 & 0 & -\frac{7}{4} & 44 & \frac{1}{2} & 0 & \\
\hline
0 & 1 & -3 & 0 & -\frac{5}{4} & 28 & \frac{1}{2} & 0 & x_1 \\
0 & 0 & \boxed{\frac{1}{3}} & 0 & \frac{1}{6} & -4 & -\frac{1}{6} & 1 & x_7 \\
1 & 0 & 0 & 1 & 0 & 0 & 1 & 0 & x_3
\end{array}
$$

$$
\begin{array}{c|ccccccc|l}
0 & 0 & 0 & 0 & -\frac{3}{4} & 20 & -\frac{1}{2} & 6 & \\
\hline
0 & 1 & 0 & 0 & \boxed{\frac{1}{4}} & -8 & -1 & 9 & x_1 \\
0 & 0 & 1 & 0 & \frac{1}{2} & -12 & -\frac{1}{2} & 3 & x_2 \\
1 & 0 & 0 & 1 & 0 & 0 & 1 & 0 & x_3
\end{array}
$$

□

In order to prevent the algorithm from falling into a cycle, we need to employ additional rules for the choice of the pivot element $a'_{i\ell}$. One such rule is the *lexicographical ordering method.* In the following, we discuss how this rule works.

First, let us explain what the lexicographical ordering $<_L$ is. Consider two vectors $\boldsymbol{x} = (x_1, x_2, \ldots, x_n)$ and $\boldsymbol{y} = (y_1, y_2, \ldots, y_n)$. The vector $\boldsymbol{x}$ is said to be *lexicographically less* than the vector $\boldsymbol{y}$, written as $\boldsymbol{x} <_L \boldsymbol{y}$, if $x_1 = y_1, \ldots, x_{i-1} = y_{i-1}$, and $x_i < y_i$, for some $1 \le i \le n$. A vector $\boldsymbol{x}$ is said to be *lexicographically positive* if $\boldsymbol{x} >_L \boldsymbol{0}$.

The lexicographical ordering method makes the following modifications on the simplex method:

(1) In step (1) of Algorithm 7.A, after the initial feasible basis J is found, rearrange the ordering of n columns such that the initial feasible basis J is placed at the first m columns. This ensures that every row in the initial simplex table, except the top row (i.e., $\boldsymbol{c} - \boldsymbol{c}_J \boldsymbol{A}_J^{-1} \boldsymbol{A}$), is lexicographically positive.

(2) In the "else" clause of step (2.2), instead of using (7.10) to choose the index i, we choose i by the following new rule:

$$\left(\frac{b'_i}{a'_{i\ell}}, \frac{a'_{i1}}{a'_{i\ell}}, \ldots, \frac{a'_{in}}{a'_{i\ell}}\right) = \min_L \left\{ \left(\frac{b'_k}{a'_{k\ell}}, \frac{a'_{k1}}{a'_{k\ell}}, \ldots, \frac{a'_{kn}}{a'_{k\ell}}\right) \,\Big|\, 1 \le k \le m, a'_{k\ell} > 0 \right\},$$

where $\min_L$ denotes the minimum element under the lexicographical ordering (i.e., for every row k with $a'_{k\ell} > 0$, divide it by $a'_{k\ell} > 0$, and then choose the lexicographically smallest row i among these rows).

The above new rule (2) guarantees that the lexicographical positiveness of all rows other than the top row is preserved under pivoting. For instance, suppose that we select $a'_{i\ell}$, with $i > 1$, as the pivot element under the new rule. Also, suppose that, for some $k \ne i$, $a'_{k\ell}$, a'_{k1}, and a'_{i1} are all positive, and $b'_k/a'_{k\ell} = b'_i/a'_{i\ell}$, $a'_{k1}/a'_{k\ell} > a'_{i1}/a'_{i\ell}$. Then after the pivoting, we get $b''_k = b'_k - a'_{k\ell} b'_i / a'_{i\ell} = 0$, and $a''_{k1} = a'_{k1} - a'_{k\ell} a'_{i1} / a'_{i\ell} > 0$, and row k is still lexicographically positive.

Now, we note that all rows other than the top row are lexicographically positive. In addition, since $c'_\ell < 0$ and $a'_{i\ell} > 0$, each pivot adds to the top row with a positive constant times one of the nontop rows. As a result, each pivot makes the top row increase strictly in the lexicographical ordering. Therefore, the modified simplex algorithm visits each feasible basis at most once and the objective function value is nonincreasing. It follows that it must halt after a finite number of pivots.

Theorem 7.8 *The simplex method with the additional lexicographical ordering rule always halts in a finite number of pivots, and it either finds an optimal solution to the linear program* (7.3) *or outputs the fact that the linear program* (7.3) *has no optimal solutions.*

Example 7.9 We observe that in the initial simplex table of Example 7.7, there are two choices of the pivot element: $a_{1,4} = 1/4$ and $a_{2,4} = 1/2$, because $b_1/a_{1,4} = b_2/a_{2,4} = 0$. In Example 7.7, we arbitrarily chose the element $a_{1,4}$ as the pivot point, and ended up in a cycle. If we apply the lexicographical ordering rule to this table, we can break the tie between $b_1/a_{1,4}$ and $b_2/a_{2,4}$ by comparing

$$\frac{a_{1,1}}{a_{1,4}} = 4 > 0 = \frac{a_{2,1}}{a_{2,4}},$$

and choosing instead $a_{2,4}$ as the pivot element. With respect to this pivot element, our new simplex table is as follows:

0	0	$\frac{3}{2}$	0	0	2	$-\frac{5}{4}$	$\frac{21}{2}$	
0	1	$-\frac{1}{2}$	0	0	-2	$-\frac{3}{4}$	$\frac{15}{2}$	x_1
0	0	2	0	1	-24	-1	6	x_4
1	0	0	1	0	0	$\boxed{1}$	0	x_3

From this table, $a'_{3,6}$ is the unique choice as the new pivot element, and the new simplex table becomes

$\frac{5}{4}$	0	$\frac{3}{2}$	$\frac{5}{4}$	0	2	0	$\frac{21}{2}$	
$\frac{3}{4}$	1	$-\frac{1}{2}$	$\frac{3}{4}$	0	-2	0	$\frac{15}{2}$	x_1
1	0	2	1	1	-24	0	6	x_4
1	0	0	1	0	0	1	0	x_6

Since the top row is all nonnegative, we see that $\boldsymbol{x} = (3/4, 0, 0, 1, 0, 1, 0)^T$ is an optimal solution, with the objective function value equal to $-5/4$. □

We summarize the relationship between the feasible basis and the optimal solution of a linear program as follows:

Theorem 7.10 *If a linear program* (7.3) *has an optimal solution, then it has an optimal basic feasible solution that is associated with a feasible basis J satisfying $\mathbf{c} - \mathbf{c}_J \boldsymbol{A}_J^{-1} \boldsymbol{A} \geq \mathbf{0}$. Moreover, if a feasible basis J satisfies $\mathbf{c} - \mathbf{c}_J \boldsymbol{A}_J^{-1} \boldsymbol{A} \geq \mathbf{0}$, then the basic feasible solution associated with J is optimal.*

7.3 Combinatorial Rounding

Many combinatorial optimization problems can be transformed into integer linear programs. By extending the feasible domain to allow real, noninteger numbers, we can relax an integer linear program to a linear program. Rounding the optimal solution of the resulting linear program to a feasible solution of the original combinatorial problem produces an approximation. This is a general approach to finding an approximation for a wide range of combinatorial optimization problems. In this section, we study some simple examples using this approach.

In Section 2.4, we studied the weighted vertex cover problem MIN-WVC and showed that the greedy algorithm for MIN-WVC has an $H(\delta)$-approximation, where δ is the maximum degree of the input graph. On the other hand, the unweighted version MIN-VC of this problem has a simple 2-approximation based on matching (see Exercise 1.10). It is therefore natural to ask whether this algorithm can be extended to the weighted version MIN-WVC with a better performance ratio than that of the greedy algorithm. To answer this question, we show, in the following, how to apply the linear programming approach to this problem to get a 2-approximation.

First, we transform the problem MIN-WVC into a 0–1 integer linear program. Suppose $V = \{v_1, v_2, \ldots, v_n\}$. We represent a subset $C \subseteq V$ by n variables $x_1, x_2, \ldots, x_n$, with $x_i = 1$ if $v_i \in C$ and $x_i = 0$ otherwise, for $i = 1, 2, \ldots, n$. Let w_i be the weight of vertex v_i. Then every vertex cover C corresponds to a feasible solution x in the following integer program, and the minimum-weight vertex cover corresponds to the optimal solution of this integer program:

$$\begin{array}{lll} \text{minimize} & w_1x_1 + w_2x_2 + \cdots + w_nx_n & \\ \text{subject to} & x_i + x_j \geq 1, & \{v_i, v_j\} \in E, \\ & x_i = 0 \text{ or } 1, & i = 1, 2, \ldots, n. \end{array} \tag{7.11}$$

By relaxing the constraints of $x_i = 0$ or 1, for $1 \leq i \leq n$, to the constraints of $0 \leq x_i \leq 1$ on real numbers x_i, for $1 \leq i \leq n$, this integer program is turned into a linear program:

$$\begin{array}{lll} \text{minimize} & w_1x_1 + w_2x_2 + \cdots + w_nx_n & \\ \text{subject to} & x_i + x_j \geq 1, & \{v_i, v_j\} \in E, \\ & 0 \leq x_i \leq 1, & i = 1, 2, \ldots, n. \end{array} \tag{7.12}$$

By solving this linear program (7.12) and rounding its optimal solution to the nearest integers, we obtain an approximation for MIN-WVC:

Algorithm 7.B (*Linear Programming Approximation for* MIN-WVC)
Input: A graph $G = (V, E)$ and a function $w : V \to \mathbb{N}$.

(1) Convert the input into a 0–1 integer program (7.11), and construct the corresponding linear program (7.12).

(2) Find an optimal solution $\boldsymbol{x}^*$ to the linear program (7.12).

(3) **For** $i \leftarrow 1, 2, \ldots, n$ **do**

$$\text{set } x_i^A = \begin{cases} 1, & \text{if } x_i^* \geq 1/2, \\ 0, & \text{otherwise.} \end{cases}$$

(4) Output $\boldsymbol{x}^A$. ■

For each $\{v_i, v_j\} \in E$, since $x_i^* + x_j^* \geq 1$, at least one of x_i^* or x_j^* must be greater than or equal to $1/2$. Therefore, at least one of x_i^A or x_j^A is equal to 1. This guarantees that $\boldsymbol{x}^A$ is a feasible solution to (7.11). In addition, it is clear that

$$\sum_{i=1}^{n} w_i x_i^A \leq 2 \sum_{i=1}^{n} w_i x_i^*$$

and that the optimal value of the objective function of (7.11) is no smaller than $\sum_{i=1}^{n} w_i x_i^*$. Therefore, the following theorem is proven.

Theorem 7.11 Algorithm 7.B *is a polynomial-time* 2*-approximation for* MIN-WVC.

In the above algorithm, the method we used to construct the approximate solution $\boldsymbol{x}^A$ from an optimal solution $\boldsymbol{x}^*$ to the linear program (7.12) is called *threshold rounding*. Next, we present another example of using the threshold rounding technique. Recall that a Boolean formula F is in conjunctive normal form (CNF) if it is a product of a finite number of clauses. If, in addition, each clause in a CNF formula F contains exactly two literals, then we say F is in 2-CNF.

> MINIMUM 2-SATISFIABILITY (MIN-2SAT): Given a Boolean formula in 2-CNF, determine whether it is satisfiable and, if it is, find a satisfying assignment that contains a minimum number of true variables.

MIN-2SAT can be seen as a generalization of the vertex cover problem MIN-VC. In fact, for each graph $G = (V, E)$, we can construct a 2-CNF $F(G)$ as follows: For each vertex $v_i \in V$, define a Boolean variable x_i, and for each edge $\{v_i, v_j\} \in E$, define a clause $(x_i \vee x_j)$. Then each vertex cover of G corresponds to a satisfying assignment of $F(G)$. Furthermore, the graph G has a vertex cover of size k if and only if $F(G)$ has a satisfying assignment with k true variables.

Similar to the problem MIN-VC, the problem MIN-2SAT can be transformed into a 0–1 integer program. Consider a 2-CNF formula F. Suppose that F has n Boolean variables $x_1, x_2, \ldots, x_n$. We will use the same symbols $x_1, \ldots, x_n$ to denote the corresponding 0–1 integer variables. Then the problem MIN-2SAT is equivalent to the following integer program:

$$\begin{aligned}
\text{minimize} \quad & x_1 + x_2 + \cdots + x_n \\
\text{subject to} \quad & x_i + x_j \geq 1, \quad \text{for each clause } (x_i \vee x_j) \text{ in } F, \\
& (1 - x_i) + x_j \geq 1, \quad \text{for each clause } (\bar{x}_i \vee x_j) \text{ in } F, \\
& (1 - x_i) + (1 - x_j) \geq 1, \quad \text{for each clause } (\bar{x}_i \vee \bar{x}_j) \text{ in } F, \\
& x_i = 0 \text{ or } 1, \quad i = 1, 2, \ldots, n.
\end{aligned} \tag{7.13}$$

Relaxing the constraints of $x_i = 0$ or 1 for $i = 1, 2, \ldots, n$, to the constraints of $0 \leq x_i \leq 1$ for $i = 1, 2, \ldots, n$, we obtain the following linear program:

$$\begin{aligned}
\text{minimize} \quad & x_1 + x_2 + \cdots + x_n \\
\text{subject to} \quad & x_i + x_j \geq 1, \quad \text{for each clause } (x_i \vee x_j) \text{ in } F, \\
& (1 - x_i) + x_j \geq 1, \quad \text{for each clause } (\bar{x}_i \vee x_j) \text{ in } F, \\
& (1 - x_i) + (1 - x_j) \geq 1, \quad \text{for each clause } (\bar{x}_i \vee \bar{x}_j) \text{ in } F, \\
& 0 \leq x_i \leq 1, \quad i = 1, 2, \ldots, n.
\end{aligned} \tag{7.14}$$

Suppose $\boldsymbol{x}^*$ is an optimal solution to (7.14). We may try to apply threshold rounding to $\boldsymbol{x}^*$ to get an approximate solution $\boldsymbol{x}^A$ to (7.13). For instance, we can set $x_i^A = 1$ if $x_i^* > 1/2$ and $x_i^A = 0$ if $x_i^* < 1/2$. This will satisfy all inequalities in which at least one variable x_i has $x_i^* \neq 1/2$. However, it is not clear how to determine the value of x_i^A when $x_i^* = 1/2$. For instance, if F contains both clauses $(x_i \vee x_j)$ and $(\bar{x}_i \vee \bar{x}_j)$ and if $x_i^* = x_j^* = 1/2$, then neither $x_i = x_j = 0$ nor $x_i = x_j = 1$ can satisfy both clauses.

What should we do in this case? We first note that since this problem is a generalization of MIN-WVC, we expect that the approximation algorithm based on the linear program (7.14) has a performance ratio at least 2. Now, let F_1 be the set of all clauses in F both of whose two variables have $\boldsymbol{x}^*$ value equal to $1/2$. We observe that for variables in F_1, the rounding of their values to either 1 or 0 keeps the performance ratio within constant 2. Thus, all we have to do is to find *any* satisfying assignment for F_1, without having to minimize the number of true variables in F_1. Based on this idea, we have the following approximation algorithm for MIN-2SAT.

Algorithm 7.C (*Linear Programming Approximation for* MIN-2SAT)

Input: A 2-CNF formula F over variables $x_1, x_2, \ldots, x_n$.

(1) Convert formula F into a linear program (7.14) and find an optimal solution $\boldsymbol{x}^*$ for it.

(2) **For** $i \leftarrow 1$ to n **do**
 if $x_i^* > 1/2$ **then** $x_i^A \leftarrow 1$
 else if $x_i^* < 1/2$ **then** $x_i^A \leftarrow 0$.

(3) Let F_1 be the collection of all clauses both of whose two variables have $\boldsymbol{x}^*$ value equal to $1/2$, and let $J \leftarrow \{j \mid 1 \leq j \leq n, x_j \text{ is in } F_1\}$.

(4) **For** $i \leftarrow 1$ to n **do**
 if $x_i^* = 1/2$ and $i \notin J$ **then** $x_i^A \leftarrow 0$.

(5) **If** F_1 is satisfiable
  then let x_j^A be a satisfying assignment for F_1 and output x^A
  else output "F is not satisfiable." ∎

It is easy to see that if F is satisfiable, then the solution x^A generated by Algorithm 7.C is a feasible solution to (7.13). First, by step (5), we know that every clause in F_1 is satisfied by x^A. For a clause $(x_i \vee x_j)$ not in F_1, we must have either $x_i^* > 1/2$ or $x_j^* > 1/2$ since $x_i^* + x_j^* \geq 1$. Thus, by step (2), either $x_i^A = 1$ or $x_j^A = 1$, and so x^A satisfies the clause $(x_i \vee x_j)$. A similar argument applies to other types of clauses, such as $(x_i \vee \bar{x}_j)$ or $(\bar{x}_i \vee \bar{x}_j)$.

In addition, we note that $x_i^A \leq 2x_i^*$ for each $i = 1, 2, \ldots, n$. Therefore, x^A is an approximation of performance ratio ≤ 2.

It remains to prove that Algorithm 7.C runs in polynomial time. To see this, we only need to demonstrate a polynomial-time algorithm for the following simpler problem:

> 2-SAT: For a given 2-CNF formula F_1, determine whether F_1 is satisfiable or not, and if F_1 is satisfiable, find a satisfying assignment for F_1.

In the following, we present an algorithm that converts the problem 2-SAT into a graph problem and solve it in polynomial time.

Algorithm 7.D (*Polynomial-Time Algorithm for* 2-SAT)
Input: A 2-CNF formula F_1 over variables $x_1, x_2, \ldots, x_n$.

(1) Construct a digraph $G(F_1) = (V, E)$ as follows:
  $V \leftarrow \{x_i, \bar{x}_i \mid 1 \leq i \leq n\}$,
  $E \leftarrow \{(\bar{y}_i, y_j), (\bar{y}_j, y_i) \mid (y_i \vee y_j)$ is a clause in $F_1\}$,
   where y_i denotes a literal x_i or $\bar{x}_i$.

(2) **For** $i \leftarrow 1$ to n **do**
  if vertices x_i and $\bar{x}_i$ are strongly connected
   then output "F_1 is not satisfiable" and halt.

(3) **For** $i \leftarrow 1$ to n **do**
  if there is a path from x_i to $\bar{x}_i$
   then for each literal y_j that is reachable from $\bar{x}_i$, set $\tau(y_j) \leftarrow 1$;[1]
  if there is a path from $\bar{x}_i$ to x_i
   then for each literal y_j that is reachable from x_i, set $\tau(y_j) \leftarrow 1$.

(4) **For** $i \leftarrow 1$ to n **do**
  if $\tau(x_i)$ is undefined
   then for each literal y_j that is reachable from x_i, set $\tau(y_j) \leftarrow 1$.

(5) Output τ. ∎

[1]This means that if $y_j = x_k$ for some variable x_k, then we set $\tau(x_k) \leftarrow 1$, and if $y_j = \bar{x}_k$, then we set $\tau(x_k) \leftarrow 0$.

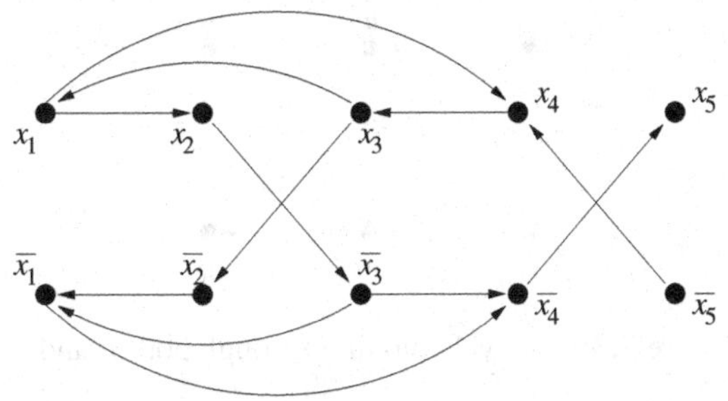

Figure 7.2: Digraph $G(F_1)$.

Example 7.12 We consider the formula

$$F_1 = (\bar{x}_1 \vee x_2) \wedge (\bar{x}_2 \vee \bar{x}_3) \wedge (\bar{x}_1 \vee x_3) \wedge (x_3 \vee \bar{x}_4) \wedge (x_4 \vee x_5) \wedge (x_1 \vee \bar{x}_4).$$

The corresponding graph $G(F_1)$ is shown in Figure 7.2.

Since there is a path from x_1 to $\bar{x}_1$, we set $x_1 = 0$ and consequently assign 1 to $\bar{x}_4$ and x_5. Now, for the remaining variables x_2 and x_3, we arbitrarily set $x_2 = 1$, and consequently $\bar{x}_3 = 1$. This gives us a satisfying assignment: $\tau(x_1) = 0$, $\tau(x_2) = 1$, $\tau(x_3) = 0$, $\tau(x_4) = 0$, $\tau(x_5) = 1$. □

Theorem 7.13 Algorithm 7.D *solves the problem* 2-SAT *correctly in polynomial time.*

Proof. To see that Algorithm 7.D works correctly, we first observe that the edge (y, z) in E indicates that, for any satisfying assignment τ for F_1, we must have $[\tau(y) = 1 \Rightarrow \tau(z) = 1]$. This property also extends to all pairs y and z for which there is a path from y to z. Thus, if some variable x_i and its negation $\bar{x}_i$ are strongly connected, then F_1 is unsatisfiable. This means that step (2) of Algorithm 7.E is correct.

Next, we consider step (3) of Algorithm 7.E. We observe another important property of the digraph $G(F_1)$: If there is a path from a vertex y to a vertex z, then there is a path from $\bar{z}$ to $\bar{y}$. From this property, we can prove that the assignment τ in step (3) is consistent; that is, it is not possible to assign, in step (3), both values 0 and 1 to a variable x_i.

To see this, suppose that a variable w is assigned with both values 0 and 1. Then, from the assignment $\tau(w) = 1$, we know that there must be a path from a literal $\bar{u}$ to u and then from u to w. From the assignment $\tau(\bar{w}) = 1$, we know that there must be a path from a literal $\bar{v}$ to v and then from v to $\bar{w}$. However, from the above property, we must also have a path from $\bar{w}$ to $\bar{u}$, and a path from w to $\bar{v}$. Together, they form a cycle that passes through both vertices w and $\bar{w}$ (see Figure 7.3), and Algorithm 7.E must have declared that F_1 is unsatisfiable and terminated in step (2).

The above property also extends to step (4). That is, in step (4), if x_i is unassigned and if y_j is reachable from x_i, then y_j either is unassigned or is assigned with value 1, for, otherwise, $\tau(\bar{y}_j)$ must have the value 1 and, hence, $\bar{x}_i$, which is reachable

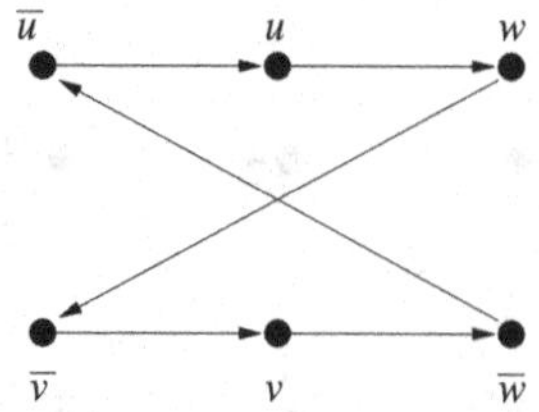

Figure 7.3: A cycle passing through both u and $\bar{u}$.

from $\bar{y}_j$, would have also been assigned value 1. Furthermore, we can see that if x_i is unassigned and if y_j is reachable from x_i, then $\bar{y}_j$ is not reachable from x_i, for otherwise there would be a path from y_j to $\bar{x}_i$, and hence a path from x_i to $\bar{x}_i$, which means $\bar{x}_i$ should have been assigned in step (3). Therefore, the assignment of τ in step (4) is also consistent.

Finally, we check that each clause $(y_i \vee y_j)$ in F_1 generates two edges $(\bar{y}_i, y_j)$ and $(\bar{y}_j, y_i)$ in E. From steps (3) and (4), we see that it is not possible to assign $\tau(y_i) = \tau(y_j) = 0$, and τ must be a satisfying assignment. □

From the above analysis, we conclude:

Theorem 7.14 Algorithm 7.C *is a polynomial-time* 2*-approximation to* MIN-2SAT.

In the above example, we used a polynomial-time algorithm for 2-SAT to find a rounding strategy. In the next example, we use the polynomial-time algorithm for matching to find a rounding strategy for a scheduling problem on unrelated parallel machines.

> SCHEDULING ON UNRELATED PARALLEL MACHINES (SCHEDULE-UPM): Given n jobs, m machines and, for each $1 \le i \le m$ and each $1 \le j \le n$, the amount of time t_{ij} required for the ith machine to process the jth job, find the schedule for all n jobs on these m machines that minimizes the *makespan*, i.e., the maximum processing time over all machines.

For each pair (i, j), with $1 \le i \le m$ and $1 \le j \le m$, let x_{ij} be the indicator for the ith machine to process the jth job; that is, $x_{ij} = 1$ if the jth job is processed on the ith machine, and $x_{ij} = 0$ otherwise. Then the problem SCHEDULE-UPM can be formulated as the following ILP:

$$\begin{array}{lll}
\text{minimize} & t & \\
\text{subject to} & \displaystyle\sum_{i=1}^{m} x_{ij} = 1, & 1 \le j \le n, \\
& \displaystyle\sum_{j=1}^{n} x_{ij} t_{ij} \le t, & 1 \le i \le m, \\
& x_{ij} \in \{0, 1\}, & 1 \le i \le m,\ 1 \le j \le n.
\end{array}$$

A natural relaxation of this ILP to an LP is as follows:

$$\begin{aligned}
\text{minimize} \quad & t \\
\text{subject to} \quad & \sum_{i=1}^{m} x_{ij} = 1, && 1 \le j \le n, \\
& \sum_{j=1}^{n} x_{ij} t_{ij} \le t, && 1 \le i \le m, \\
& 0 \le x_{ij} \le 1, && 1 \le i \le m,\ 1 \le j \le n.
\end{aligned} \tag{7.15}$$

Consider an optimal extreme point $\boldsymbol{x}^*$ to this LP. In order to devise a feasible rounding strategy, let us study the combinatorial properties of $\boldsymbol{x}^*$. Let $J = \{j \mid (\exists i)\, 0 < x^*_{ij} < 1\}$ and $M = \{1, \ldots, m\}$. Define a bipartite graph $H = (M, J, E)$ with $E = \{(i, j) \mid 0 < x^*_{ij} < 1\}$; that is, there is an edge (i, j) connecting j to i if and only if the jth job is partially assigned to the ith machine.

Lemma 7.15 *The bipartite graph H contains a matching covering J.*

Proof. It suffices to show that each connected component of H contains a matching covering all jobs in the connected component. Consider a connected component $H' = (M', J', E')$ of H. For each variable x_{ij} with $i \notin M'$ or $j \notin J'$, let us fix its value in LP (7.15) by $x_{ij} = x^*_{ij}$. Then we get a new LP over variables x_{ij}, for $i \in M'$ and $j \in J'$. It is easy to verify that $\boldsymbol{x}' = (x^*_{ij})_{i \in M', j \in J'}$ is an extreme point of this new LP. In fact, suppose $\boldsymbol{x}' = (\boldsymbol{y}' + \boldsymbol{z}')/2$ for some points $\boldsymbol{y}', \boldsymbol{z}'$ in the feasible region of the new LP. Define $\boldsymbol{y}$ to have $y_{ij} = x^*_{ij}$ for $i \notin M'$ or $j \notin J'$, and have $y_{ij} = y'_{ij}$ for $i \in M'$ and $j \in J'$; also, define $\boldsymbol{z}$ to have $z_{ij} = x^*_{ij}$ for $i \notin M'$ or $j \notin J'$, and have $z_{ij} = z'_{ij}$ for $i \in M'$ and $j \in J'$. Then we have $\boldsymbol{x}^* = (\boldsymbol{y} + \boldsymbol{z})/2$. It follows that $\boldsymbol{y} = \boldsymbol{z} = \boldsymbol{x}^*$ and, hence, $\boldsymbol{y}' = \boldsymbol{z}' = \boldsymbol{x}'$.

Let $\boldsymbol{a}_k$ be the kth row of the constraint matrix of the LP (7.15). We say an inequality constraint $\boldsymbol{a}_k \boldsymbol{x} \ge b_k$ is *active* at a point $\boldsymbol{x}^*$ if $\boldsymbol{a}_k \boldsymbol{x}^* = b_k$. Note that an extreme point $\boldsymbol{x}'$ of the new LP has $|M'| \cdot |J'|$ components, and hence must be determined by $|M'| \cdot |J'|$ active constraints. However, for each active constraint of the form $x_{ij} \ge 0$ or $x_{ij} \le 1$, the corresponding component x'_{ij} must be an integer. Note that there are only $|M'| + |J'|$ constraints not of such a form. Thus, $\boldsymbol{x}'$ can have at most $|M'| + |J'|$ nonintegral components. In other words, graph H' has at most $|M'| + |J'|$ edges. Since H' is connected, H' is either a tree or a tree plus an edge.

Case 1. H' is a tree. Fix any vertex $r \in J'$ as the root. Then H becomes a rooted tree. Note an important fact of this tree: A vertex $j \in J'$ cannot be a leaf. To see this, we note that for integer $j \in J'$, the constraint $\sum_{i \in M'} x_{ij} = 1$ on $\boldsymbol{x}'$ implies that $0 < x'_{ij} < 1$ for at least two different $i \in M'$. This means that there are at least two edges incident upon j, and so j is not a leaf. From this property, we have a simple way to find a matching covering J': For each $j \in J'$, match it to a child of j in the tree.

Case 2. H' is a tree plus an edge. This edge introduces a cycle, and H' is a cycle plus some trees growing out from the cycle (see Figure 7.4, in which a circle

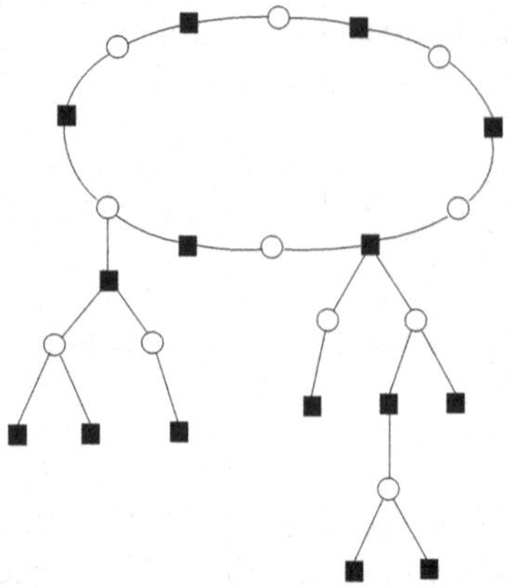

Figure 7.4: H' contains only one cycle.

○ denotes a job and a dark square ■ denotes a machine). Since H' is bipartite, the cycle has an even number of vertices and thus contains a matching covering all vertices on the cycle. Contracting the cycle of H' into a root point results in a rooted tree over other vertices. Again, a nonroot vertex $j \in J'$ in this tree cannot be a leaf. Thus, for each internal, nonroot vertex $j \in J'$ in the tree, we can match it to one of its children. Together with the matching of the cycle, we obtain a matching of H' covering every vertex in J'. □

Lemma 7.15 means that the partially assigned jobs in J can be assigned to machines in such a way that each machine receives at most one such job. It suggests a simple rounding strategy: First, for each job j with $x^*_{ij} = 1$ for some $i \in M$, we assign it to the ith machine. For the remaining jobs in J, we define the bipartite graph H and find a matching A of H that covers J, and for each $j \in J$, assign it to the ith machine if $(i, j) \in A$. This rounding strategy gives us an approximation with the makespan at most

$$opt + \max_{1 \le i \le m, 1 \le j \le n} t_{ij},$$

where *opt* is the minimum makespan. Thus, if we can bound the maximum t_{ij} by $c \cdot opt$ for some constant $c > 0$, then the above rounding strategy yields a constant-ratio approximation to SCHEDULE-UPM. Is such a bound possible? Unfortunately, the answer is no, as t_{ij} could be, in general, much greater than *opt*.

On the other hand, we observe that if a value t_{ij} is greater than *opt*, then the optimal solution must not assign the jth job to the ith machine. Therefore, we can prune the variable x_{ij} from the LP (7.15), and expect to get the same solution. This observation suggests that we set an upper bound T on t_{ij} and prune the variable x_{ij} if $t_{ij} > T$. How do we find the best value of the bound T? Since we do not know the value of *opt*, we cannot just set it to *opt*. Instead, we can search for the minimum T for which the following LP has a feasible solution:

$$
\begin{array}{llll}
\text{minimize} & t & & \\
\text{subject to} & \sum\limits_{1\le i\le m,\, t_{ij}\le T} x_{ij} = 1, & 1 \le j \le n, & \\
& \sum\limits_{1\le j\le n,\, t_{ij}\le T} x_{ij}t_{ij} \le t, & 1 \le i \le m, & \\
& 0 \le x_{ij} \le 1, & 1 \le i \le m,\ 1 \le j \le n. &
\end{array} \tag{7.16}
$$

Since the above LP (7.16) can be solved in polynomial time, we can use bisecting to find the minimum T for which (7.16) has a feasible solution. Denote this T as T^* and let $\boldsymbol{x}^*$ be an optimal extreme point of (7.16). Then $T^* \le opt$ and $t_{ij} \le opt$ for all $x^*_{ij} > 0$. Therefore, by the rounding based on Lemma 7.15, we obtain a polynomial-time 2-approximation to SCHEDULE-UPM.

Theorem 7.16 *The problem* SCHEDULE-UPM *has a polynomial-time approximation with performance ratio* 2.

7.4 Pipage Rounding

In this section, we introduce the idea of *pipage rounding*. Let us first look at an example.

> MAXIMUM-WEIGHT HITTING (MAX-WH): Given a collection $\mathcal{C}$ of subsets of a finite set E with a nonnegative weight function w on $\mathcal{C}$ and a positive integer p, find a subset A of E with $|A| = p$ that maximizes the total weight of subsets in $\mathcal{C}$ hit by A.

Assume $E = \{1, 2, \ldots, n\}$ and $\mathcal{C} = \{S_1, S_2, \ldots, S_m\}$. Denote $w_i = w(S_i)$ for $1 \le i \le m$. Let x_i be a 0–1 variable that indicates whether element i is in subset A. Then the problem MAX-WH can be formulated into the following integer program, which, as we will see later, can be relaxed to a linear program:

$$
\begin{array}{lll}
\text{maximize} & L(\boldsymbol{x}) = \sum\limits_{j=1}^{m} w_j \cdot \min\left\{1, \sum\limits_{i\in S_j} x_i\right\} & \\
\text{subject to} & \sum\limits_{i=1}^{n} x_i = p, & \\
& x_i \in \{0, 1\}, \qquad i = 1, 2, \ldots, n. &
\end{array} \tag{7.17}
$$

The following equivalent formulation of MAX-WH (as a nonlinear program) will be useful in the rounding algorithm:

$$\begin{aligned} \text{maximize} \quad & F(\boldsymbol{x}) = \sum_{j=1}^{m} w_j \cdot \left(1 - \prod_{i \in S_j} (1 - x_i)\right) \\ \text{subject to} \quad & \sum_{i=1}^{n} x_i = p, \\ & x_i \in \{0, 1\}, \qquad i = 1, 2, \ldots, n. \end{aligned} \tag{7.18}$$

The functions $L(\boldsymbol{x})$ and $F(\boldsymbol{x})$ have the same value when each x_i takes value 0 or 1. However, when the constraints $x_i \in \{0, 1\}$ are relaxed to the constraints $0 \le x_i \le 1$, they may have different values. Nevertheless, they satisfy the following relationship.

Lemma 7.17 *For the relaxed versions of* (7.17) *and* (7.18), *we must have* $F(\boldsymbol{x}) \ge (1 - 1/e)L(\boldsymbol{x})$.

Proof. Consider a fixed set S_j for some $j = 1, 2, \ldots, m$. Assume that $|S_j| = k$. Then, by the arithmetic mean–geometric mean inequality, we have

$$1 - \prod_{i \in S_j} (1 - x_i) \ge 1 - \left(\frac{\sum_{i \in S_j} (1 - x_i)}{k}\right)^k = 1 - \left(1 - \frac{\sum_{i \in S_j} x_i}{k}\right)^k.$$

Let $f(z) = 1 - (1 - z/k)^k$. Then, for $0 \le z \le k$, we have $f'(z) = (1 - z/k)^{k-1} \ge 0$ and $f''(z) = -((k-1)/k)(1 - z/k)^{k-2} \le 0$. Therefore, $f(z)$ is monotone increasing and concave in the interval $[0, k]$. Moreover, $f(0) = 0$. It follows that $f(z) \ge z \cdot f(1)$ for $z \in [0, 1]$, and so $f(z) \ge f(1) \cdot \min\{1, z\}$ for $z \in [0, k]$. Note that

$$f(1) = 1 - \left(1 - \frac{1}{k}\right)^k \ge 1 - \frac{1}{e}.$$

Thus,

$$1 - \prod_{i \in S_j} (1 - x_i) \ge \left(1 - \frac{1}{e}\right) \cdot \min\left\{1, \sum_{i \in S_j} x_i\right\},$$

and the lemma is proven. □

The relaxation of the integer program (7.17) is as follows:

$$\begin{aligned} \text{maximize} \quad & L(\boldsymbol{x}) = \sum_{j=1}^{m} w_j \cdot \min\left\{1, \sum_{i \in S_j} x_i\right\} \\ \text{subject to} \quad & \sum_{i=1}^{n} x_i = p, \\ & 0 \le x_i \le 1, \qquad i = 1, 2, \ldots, n. \end{aligned} \tag{7.19}$$

We can introduce m new variables to get an equivalent LP as follows:

$$\begin{aligned}
\text{maximize} \quad & \sum_{j=1}^{m} w_j z_j \\
\text{subject to} \quad & \sum_{i \in S_j} x_i \geq z_j, \qquad j = 1, \ldots, m, \\
& \sum_{i=1}^{n} x_i = p, \\
& 0 \leq x_i \leq 1, \qquad i = 1, 2, \ldots, n, \\
& 0 \leq z_j \leq 1, \qquad j = 1, 2, \ldots, m.
\end{aligned}$$

The optimal solution to this LP can be found in polynomial time. We will use function $F(\boldsymbol{x})$ to round the optimal solution $\boldsymbol{x}^*$ of (7.19) to get an integer solution $\boldsymbol{x}^A$ for (7.17). More precisely, we round, at each step, one or two nonintegral components of $\boldsymbol{x}^*$ to integers, with the criterion that the rounding does not decrease the value of $F(\boldsymbol{x})$.

Algorithm 7.E (*Pipage Rounding Algorithm for* MAX-WH)

Input: A set $E = \{1, 2, \ldots, n\}$, a collection $\mathcal{C}$ of subsets of E, a nonnegative weight function $w : \mathcal{C} \to \mathbb{N}$, and an integer $p > 0$.

(1) Construct the linear program (7.19) from the input, and find an optimal solution $\boldsymbol{x}^*$ to it.

(2) $\boldsymbol{x} \leftarrow \boldsymbol{x}^*$.

(3) **While** $\boldsymbol{x}$ has an nonintegral component **do**

(3.1) Choose $0 < x_k < 1$ and $0 < x_j < 1$ (with $k \neq j$);

(3.2) Define $\boldsymbol{x}(\varepsilon)$ by

$$x_i(\varepsilon) \leftarrow \begin{cases} x_i, & \text{if } i \neq k, j, \\ x_j + \varepsilon, & \text{if } i = j, \\ x_k - \varepsilon, & \text{if } i = k; \end{cases}$$

(3.3) Let $\varepsilon_1 \leftarrow \min\{x_j, 1 - x_k\}$;
$\varepsilon_2 \leftarrow \min\{1 - x_j, x_k\}$;

(3.4) **If** $F(\boldsymbol{x}(-\varepsilon_1)) \geq F(\boldsymbol{x}(\varepsilon_2))$ **then** $\boldsymbol{x} \leftarrow \boldsymbol{x}(-\varepsilon_1)$ **else** $\boldsymbol{x} \leftarrow \boldsymbol{x}(\varepsilon_2)$.

(4) Output $\boldsymbol{x}^A \leftarrow \boldsymbol{x}$. ■

We remark that, at step (3.4), we replace $\boldsymbol{x}$ by either $\boldsymbol{x}(-\varepsilon_1)$ or $\boldsymbol{x}(\varepsilon_2)$. In either case, the sum $\sum_{i=1}^{n} x_i$ remains an integer. Therefore, at step (3.1) of the next iteration, $\boldsymbol{x}$ must have at least two distinct nonintegral components. Thus, Algorithm 7.E is well defined.

The following is an important property of $F(\boldsymbol{x}(\varepsilon))$.

Lemma 7.18 *$F(\boldsymbol{x}(\varepsilon))$ is convex with respect to ε.*

Proof. We consider

$$F(\boldsymbol{x}(\varepsilon)) = \sum_{\ell=1}^{m} w_\ell \cdot \Big(1 - \prod_{i \in S_\ell} (1 - x_i(\varepsilon))\Big)$$

as a function of ε, with respect to a fixed $\boldsymbol{x}$ and fixed elements $j, k \in \{1, 2, \ldots, n\}$. Then, for each $\ell = 1, 2, \ldots, m$, we consider three cases.

Case 1. S_ℓ contains neither j nor k. Then the ℓth term of $F(\boldsymbol{x}(\varepsilon))$, $w_\ell \cdot (1 - \prod_{i \in S_\ell}(1 - x_i(\varepsilon)))$, is a constant with respect to ε, and so is convex.

Case 2. S_ℓ contains one of j or k. Then the ℓth term of $F(\boldsymbol{x}(\varepsilon))$ is linear with respect to ε, and so is convex.

Case 3. S_ℓ contains both k and j. Then the ℓth term of $F(\boldsymbol{x}(\varepsilon))$ is of the form

$$g(\varepsilon) = w_\ell - a(b + \varepsilon)(c - \varepsilon)$$

for some nonnegative constants a, b, and c. If $a = 0$, then this term is a constant w_ℓ and hence convex. If $a > 0$, then $g''(\varepsilon) = 2a > 0$, and so $g(\varepsilon)$ is convex.

Thus, each term of $F(\boldsymbol{x}(\varepsilon))$ is a convex function. Now, the lemma follows from the fact that the sum of a finite number of convex functions is still convex. □

By Lemma 7.18, $\max\{F(\boldsymbol{x}(-\varepsilon_1)), F(\boldsymbol{x}(\epsilon_2))\} \geq F(\boldsymbol{x})$, since $\varepsilon_1, \varepsilon_2 > 0$. Thus, the value of $F(\boldsymbol{x})$ is nondecreasing during step (3) (called the *Pipage Rounding* process) of Algorithm 7.E. Therefore, $F(\boldsymbol{x}^A) \geq F(\boldsymbol{x}^*)$.

Theorem 7.19 Algorithm 7.E *is a polynomial-time approximation to* MAX-WH *with performance ratio*$(e/(e-1))$.

Proof. First, we note that $\boldsymbol{x}^A$ has only integer components, and so $F(\boldsymbol{x}^A) = L(\boldsymbol{x}^A)$. It follows that

$$L(\boldsymbol{x}^A) = F(\boldsymbol{x}^A) \geq F(\boldsymbol{x}^*) \geq \Big(1 - \frac{1}{e}\Big) L(\boldsymbol{x}^*),$$

where $\boldsymbol{x}^*$ is the optimal solution to (7.19). □

The above example is a typical application of the pipage rounding technique. We can extend it to the following general setting: Consider a bipartite graph $G = (U, V, E)$ and an integer program with 0–1 variables x_e, each associated with an edge $e \in E$, and with constraints in the form

$$\sum_{e \in \delta(v)} x_e \leq p_v, \quad \text{or} \quad \sum_{e \in \delta(v)} x_e = p_v, \quad \text{or} \quad \sum_{e \in \delta(v)} x_e \geq p_v,$$

for some $v \in U \cup V$, where $\delta(v)$ is the set of all edges incident to $v \in U \cup V$ and p_v is a nonnegative integer. For instance, consider the following integer program:

$$\begin{array}{lll} \text{maximize} & L(\boldsymbol{x}) & \\ \text{subject to} & \displaystyle\sum_{e \in \delta(v)} x_e \leq p_v, & v \in U \cup V, \\ & x_e \in \{0, 1\}, & e \in E. \end{array} \tag{7.20}$$

(Intuitively, the above integer program asks for a subgraph $G_1 = (U, V, E_1)$ of G, with each vertex v having degree at most p_v, that maximizes $L(E_1)$.)

Suppose $L(\boldsymbol{x})$ has a companion function $F(\boldsymbol{x})$ such that

(A1) $L(\boldsymbol{x}) = F(\boldsymbol{x})$ when $x_e \in \{0, 1\}$ for all $e \in E$, and

(A2) $L(\boldsymbol{x}) \leq c \cdot F(\boldsymbol{x})$, for some constant $c > 0$, when $0 \leq x_e \leq 1$ for all $e \in E$.

Further assume that

(A3) The relaxation of the integer program (7.20) is equivalent to an LP:

$$
\begin{aligned}
\text{maximize} \quad & L(\boldsymbol{x}) & & \\
\text{subject to} \quad & \sum_{e \in \delta(v)} x_e \leq p_v, & & v \in U \cup V, \\
& 0 \leq x_e \leq 1, & & e \in E.
\end{aligned} \tag{7.21}
$$

Then we can apply the pipage rounding technique to the optimal solution $\boldsymbol{x}^*$ of (7.21) to obtain an integer solution $\boldsymbol{x}^A$ as follows:

Pipage Rounding

(1) Initially, set $\boldsymbol{x} \leftarrow \boldsymbol{x}^*$.

(2) **While** $\boldsymbol{x}$ is not an integer solution **do**

(2.1) Let $H_{\boldsymbol{x}}$ be the subgraph of G induced by all edges $e \in E$ with $0 < x_e < 1$. Let R be a cycle or a maximal path of $H_{\boldsymbol{x}}$. Then R can be decomposed into two matchings M_1 and M_2.

(2.2) Define $\boldsymbol{x}(\varepsilon)$ by

$$
x_e(\varepsilon) = \begin{cases} x_e, & \text{if } e \notin R, \\ x_e + \varepsilon, & \text{if } e \in M_1, \\ x_e - \varepsilon, & \text{if } e \in M_2. \end{cases}
$$

(2.3) Let $\varepsilon_1 \leftarrow \min\Big\{ \min_{e \in M_1} x_e, \min_{e \in M_2} (1 - x_e) \Big\}$,

$$\varepsilon_2 \leftarrow \min\Big\{ \min_{e \in M_1} (1 - x_e), \min_{e \in M_2} x_e \Big\}.$$

(2.4) **If** $F(\boldsymbol{x}(-\varepsilon_1)) \geq F(\boldsymbol{x}(\varepsilon_2))$ **then** $\boldsymbol{x} \leftarrow \boldsymbol{x}(-\varepsilon_1)$ **else** $\boldsymbol{x} \leftarrow \boldsymbol{x}(\varepsilon_2)$. ■

Lemma 7.20 *For $\varepsilon \in [-\varepsilon_1, \varepsilon_2]$, $\boldsymbol{x}(\varepsilon)$ is a feasible solution for* (7.21).

Proof. First, suppose R is a cycle. Then, for each vertex v in R, there is an edge in $\delta(v) \cap M_1$ and an edge in $\delta(v) \cap M_2$, and so $\sum_{e \in \delta(v)} x_e(\varepsilon) = \sum_{e \in \delta(v)} x_e$. Therefore, $\boldsymbol{x}(\varepsilon)$ is feasible.

Next, suppose R is a maximal path. Then, by a similar argument, we know that for each intermediate vertex v of R, $\sum_{e \in \delta(v)} x_e(\varepsilon) = \sum_{e \in \delta(v)} x_e$. That is,

$\sum_{e\in\delta(v)} x_e(\varepsilon) \neq \sum_{e\in\delta(v)} x_e$ only if v is an endpoint of R. Let v be an endpoint of R and $e' \in \delta(v) \cap R$. By the definitions of $\boldsymbol{x}(\varepsilon)$, ε_1, and ε_2, we know that, for $\varepsilon \in [-\varepsilon_1, \varepsilon_2]$, $0 \leq x_{e'}(\varepsilon) \leq 1$. In addition, we observe that for each $e \in \delta(v) \setminus \{e'\}$, x_e is an integer, since R is a maximal path in $H_{\boldsymbol{x}}$. Therefore, we have $p_v - \sum_{e\in\delta(v)\setminus\{e'\}} x_e \geq 1$. It follows that

$$p_v - \sum_{e\in\delta(v)} x_e(\varepsilon) = p_v - \sum_{e\in\delta(v)\setminus\{e'\}} x_e - x_{e'}(\varepsilon) \geq 1 - x_{e'}(\varepsilon) \geq 0.$$

Again, $\boldsymbol{x}(\varepsilon)$ is feasible. □

Finally, assume

(A4) For any R, $F(\boldsymbol{x}(\varepsilon))$ is convex with respect to ε.

Then, the above Pipage Rounding procedure results in an integer solution $\boldsymbol{x}^A$ such that $F(\boldsymbol{x}^A) \geq F(\boldsymbol{x}^*)$. Therefore,

$$L(\boldsymbol{x}^A) = F(\boldsymbol{x}^A) \geq F(\boldsymbol{x}^*) \geq c \cdot L(\boldsymbol{x}^*) \geq c \cdot opt.$$

For the problem MAX-WH, we can formulate it into a star bipartite graph $G = (U, V, E)$, with $U = \{u\}$, $V = \{v_1, v_2, \dots, v_n\}$, and $E = \{(u, v_1), (u, v_2), \dots, (u, v_2)\}$. Each variable x_i corresponds to an edge (u, v_i), and the constraint $\sum_{i=1}^n x_i = p$ becomes $\sum_{e\in\delta(u)} x_e = p$. Under this setting, the set R in step (2.1) of the Pipage Rounding procedure is always a maximal path consisting of two edges, which correspond to two nonintegral components x_j and x_k in step (3.1) of Algorithm 7.E.

7.5 Iterated Rounding

Recall the threshold rounding technique introduced in Section 7.3. We observe that it worked for the problem MIN-WVC, because the optimal fractional solution always has at least one variable in each clause taking value greater than or equal to $1/2$. Therefore, rounding these values to 1 yields a feasible solution that is a 2-approximation to the optimal integer solution. Suppose, however, that we are given some additional constraints of the form $\sum_{i\in A} x_i \geq k$. Then it is possible that there are not enough variables taking values at least $1/2$ in the optimal fractional solution to satisfy these constraints. Thus, the solution obtained by rounding these variables to 1 may not be feasible. What should we do in this situation? An idea is to perform a partial rounding, that is, to round those values greater than or equal to $1/2$ to 1, and then deal with the residual linear program. In the case that the fractional optimal solution of the residual linear program always contains a component of value greater than or equal to $1/2$, we can continue this rounding process and eventually obtain a feasible integer solution that is still a 2-approximation. This is the basic idea of iterated rounding. Now, let us apply this idea to a specific problem.

GENERALIZED SPANNING NETWORK (GSN): Given a graph $G = (V, E)$ with a nonnegative cost function $c : E \to \mathbb{R}^+$ on edges, and an integer $k > 0$, find a k-edge-connected subgraph with the minimum total edge cost.

The fact that a subgraph F is k-edge-connected may be verified as follows: For each partition $(S, V - S)$ of the vertex set V of G, there are at least k edges in F between S and $V - S$. Based on this concept, the problem GSN can be formulated as the following ILP:

$$\begin{array}{lll} \text{minimize} & \sum_{e \in E} c_e x_e & \\ \text{subject to} & \sum_{e \in \delta_G(S)} x_e \geq k, & \emptyset \neq S \subset V, \\ & x_e \in \{0, 1\}, & e \in E, \end{array}$$

where $\delta_G(S)$ denotes the set of edges with exactly one endpoint in S. Its LP relaxation is as follows:

$$\begin{array}{lll} \text{minimize} & \sum_{e \in E} c_e x_e & \\ \text{subject to} & \sum_{e \in \delta_G(S)} x_e \geq k, & \emptyset \neq S \subset V, \\ & 0 \leq x_e \leq 1, & e \in E. \end{array} \tag{7.22}$$

First, we need to point out that this LP, though having more than $2^{|V|}$ constraints, can be solved in polynomial time in $|V|$. This fact is somewhat surprising because it would take time $2^{|V|}$ even to write down all constraints explicitly. In the following, we present a brief description of how an algorithm based on the *ellipsoid method* can solve this LP in polynomial time in $|V|$.

The critical idea here is that we do not need to write down all constraints explicitly when we employ the ellipsoid method to solve the LP (7.22). What we need is, instead, an algorithm to find, for any infeasible solution $\boldsymbol{x}$, an unsatisfied constraint in polynomial time in $|V|$. This is called a *separation oracle*.

More precisely, solving a linear program can be reduced to solving a system of linear inequalities. For a system of linear inequalities, the algorithm based on the ellipsoid method maintains an ellipsoid (initially, a ball) that contains a feasible region of a certain volume if the system of linear inequalities has a solution. In each iteration, it checks whether or not the center of the ellipsoid is a solution of the system of linear inequalities. If not, it finds an unsatisfied constraint to cut the ellipsoid into two halves and uses a new ellipsoid to cover the half that satisfies the constraint. Moreover, the volume of the ellipsoid shrinks, in each iteration, by a fixed ratio $r < 1$ (which may depend on the input size n). Thus, if none of the centers of the ellipsoids is a solution, then the volume of the ellipsoid becomes, after a polynomial number of iterations, smaller than the volume of the possible feasible region and the algorithm terminates, reporting that the system of linear inequalities has no solutions.

The solution obtained by the ellipsoid method may not be a basic feasible solution. However, from the proof of Lemma 7.1, we can easily construct a polynomial-time algorithm to compute an optimal basic feasible solution from an optimal solution.

Thus, for a linear program with an exponential number of constraints, we can still solve it in polynomial time as long as we can construct separation oracles in polynomial time. In our case here, the separation oracles for the LP (7.22) can be constructed based on the *maximum-flow minimum-cut theorem* as follows: We first convert the constraints into a network flow problem. That is, for a potential solution x to (7.22), we assign, to each edge e, a capacity x_e. Then x is feasible if and only if, for every two nodes s, t of graph G, the maximum flow from s to t is at least k. Next, we compute the maximum flow for each pair (s, t) of nodes of graph G. When a pair (s, t) is found with the maximum flow from s to t less than k, we know that x is infeasible. In addition, by the maximum-flow minimum-cut theorem, there is a cut $(S, V - S)$ with the total capacity less than k. The constraint corresponding to this cut S is an unsatisfied constraint we are looking for; that is, $\sum_{e\in\delta_G(S)} x_e < k$. Note that here the minimum cut $(S, V - S)$ can be found in polynomial time in $|V|$, since the input to the minimum-cut problem is just the graph G.

Next, we note that if we make a partial assignment to the variables of the LP (7.22), the residual LP is still polynomial-time solvable with respect to $|V|$. Indeed, suppose for $e \in F \subset E$, x_e is already assigned value u_e. Now, suppose an assignment $(x_e)_{e\in E-F}$ is not feasible for the residual LP. Then this assignment $(x_e)_{e\in E-F}$, together with the partial assignment $(u_e)_{e\in F}$, forms an infeasible assignment for the original LP. Therefore, in polynomial time with respect to $|V|$, we can find an unsatisfied constraint

$$\sum_{e\in\delta_G(S)} x_e \geq k$$

in the original LP. The corresponding constraint

$$\sum_{e\in\delta_G(S)\setminus F} x_e \geq k - \sum_{e\in F} u_e$$

of the residual LP is then an unsatisfied constraint for $(x_e)_{e\in E-F}$. In other words, the separation oracles for the residual LP can also be constructed in polynomial time in $|V|$.

Now, let us study how iterated rounding works. First, we extend the notion of supmodular functions, which has been studied in Chapter 2, to weakly supmodular functions. A function $f : 2^V \to \mathbb{Z}$ is *weakly supmodular* if

(a) $f(V) = 0$, and

(b) For any two subsets $A, B \subseteq V$, either

$$f(A) + f(B) \leq f(A \setminus B) + f(B \setminus A)$$

or

$$f(A) + f(B) \leq f(A \cap B) + f(A \cup B).$$

The following is a key lemma in the application of iterated rounding to the problem GSN. Its proof is quite involved and is postponed to the end of this section.

Lemma 7.21 *Suppose $f : 2^V \to \mathbb{Z}$ is a weakly supmodular function. Then, for the following LP,*

$$\begin{aligned} \text{minimize} \quad & \sum_{e\in E} c_e x_e \\ \text{subject to} \quad & \sum_{e\in\delta_G(S)} x_e \geq f(S), \qquad S \subseteq V, \\ & 0 \leq x_e \leq 1, \qquad e \in E, \end{aligned} \tag{7.23}$$

every basic feasible solution $\boldsymbol{x}$ contains at least one component $x_e \geq 1/3$.

Note that the function

$$f(S) = \begin{cases} 0, & \text{if } S = \emptyset \text{ or } V, \\ k, & \text{otherwise} \end{cases} \tag{7.24}$$

is weakly supmodular. By Lemma 7.21, every basic feasible solution of (7.12) contains at least one component $x_e \geq 1/3$. We round such variables x_e to 1 and study the residual LP. After setting $x_e = 1$ for edges $e \in F$ for some subset $F \subseteq E$, the residual LP can be represented as follows:

$$\begin{aligned} \text{minimize} \quad & \sum_{e\in E-F} c_e x_e \\ \text{subject to} \quad & \sum_{e\in\delta_{G-F}(S)} x_e \geq f(S) - |\delta_F(S)|, \qquad S \subseteq V, \\ & 0 \leq x_e \leq 1, \qquad e \in E, \end{aligned} \tag{7.25}$$

where F also represents the subgraph of G with edge set F and vertex set V. It is not hard to verify that $f(S) - |\delta_F(S)|$ is still weakly supmodular (see Exercise 7.15). By Lemma 7.21, every basic feasible solution of (7.25) must contain a component $x_e \geq 1/3$, which can be rounded to 1. From the above analysis, we can now present the iterated rounding algorithm for GSN as follows.

Algorithm 7.F (*Iterated Rounding Algorithm for* GSN)

Input: A graph $G = (V, E)$ with an edge-cost function $c : E \to \mathbb{Q}^+$, and an integer $k > 0$.

(1) Construct an LP (7.25) with $f(S)$ of (7.24) and $F = \emptyset$.

(2) **While** F is not k-edge-connected **do**

(2.1) Find an optimal basic feasible solution $\boldsymbol{x}^*$ of (7.25);

(2.2) $F \leftarrow F \cup \{e \mid x_e^* \geq 1/3\}$.

(3) Output F. ■

Theorem 7.22 Algorithm 7.F *produces a* 3*-approximation for the problem* GSN.

Proof. Suppose F is the output obtained from Algorithm 7.F through t iterations. For $i = 1, 2, \ldots, t$, let F_i be the set of edges added to F in the first i iterations; thus, $F = F_t$. Also, denote, for $i = 1, 2, \ldots, t$, $\overline{F}_i = E - F_i$. Let $\boldsymbol{x}^i$ denote the optimal fractional solution of (7.25) with respect to $F = F_i$. Thus, under the condition that $x_e = 1$ for $e \in F_i$, $\boldsymbol{x}^i$ is a better solution to (7.25) than any other solution, including $\boldsymbol{x}^{i-1}$. It follows that

$$\begin{aligned}
\sum_{e \in F} c_e &\leq \sum_{e \in F_{t-1}} c_e + 3 \sum_{e \in \overline{F}_{t-1}} c_e x_e^{t-1} \\
&\leq \sum_{e \in F_{t-1}} c_e + 3 \sum_{e \in \overline{F}_{t-1}} c_e x_e^{t-2} \\
&\leq \sum_{e \in F_{t-2}} c_e + 3 \sum_{e \in \overline{F}_{t-2}} c_e x_e^{t-2} \\
&\leq \quad \cdots \quad \leq 3 \sum_{e \in E} c_e x_e^0 \leq 3 \cdot opt,
\end{aligned}$$

where *opt* is the value of optimal integer solution of (7.25) for $F = \emptyset$. □

The rest of this section is devoted to the proof of Lemma 7.21.

We first prove an important property of the basic feasible solutions of (7.23). Let $\boldsymbol{a}_S$ denote the row of the constraint matrix of (7.23) corresponding to a set $S \subseteq V$; that is, each nonzero component of $\boldsymbol{a}_S$ has value 1 and corresponds to an edge in $\delta_G(S)$. So, we have $\boldsymbol{a}_S \boldsymbol{x} = \sum_{e \in \delta_G(S)} x_e$. Recall that an inequality constraint $\boldsymbol{a}_S \boldsymbol{x} \geq f(S)$ is *active* for a basic feasible solution $\boldsymbol{x}$ if the constraint holds as an equality; that is, $\boldsymbol{a}_S \boldsymbol{x} = f(S)$. We say a set $S \subseteq V$ is *active* for $\boldsymbol{x}$ if its corresponding constraint $\boldsymbol{a}_S$ is active for $\boldsymbol{x}$. We note that for a basic feasible solution $\boldsymbol{x}$ with k fractional components (i.e., $0 < x_e < 1$ for k edges e), there are at least k active constraints. Furthermore, the corresponding rows $\boldsymbol{a}_S$ of these active constraints have rank equal to k.

We say a set $A \subseteq V$ *crosses* another set $B \subseteq V$ if $A \setminus B \neq \emptyset$, $B \setminus A \neq \emptyset$, and $A \cap B \neq \emptyset$. A family $\mathcal{F}$ of sets is called a *laminar family* if no member of $\mathcal{F}$ crosses another member. In the following lemma, we will show that each basic feasible solution $\boldsymbol{x}$ of (7.23) is determined by a laminar family of active sets for $\boldsymbol{x}$. In the following, we assume, without loss of generality, that $0 < x_e < 1$ for all $e \in E$. Indeed, if $x_e = 0$ for some edges e, we may delete these edges from G and the proof works for the resulting graph; and if $x_e = 1$, then Lemma 7.21 holds trivially.

Lemma 7.23 *Let $\boldsymbol{x}$ be a basic feasible solution of* (7.23)*, with* $0 < x_e < 1$ *for all* $e \in E$*. Then there is a laminar family $\mathcal{F}$ of active sets in G such that* *(a)* $|\mathcal{F}| = |E|$,

(b) The set of vectors $\boldsymbol{a}_S$, *over all* $S \in \mathcal{F}$, *is linearly independent, and*

(c) $f(S) \geq 1$, *for all* $S \in \mathcal{F}$.

Proof. It suffices to show that for every maximal laminar family $\mathcal{L}$ of active sets, $\{\boldsymbol{a}_S \mid S \in \mathcal{L}\}$ has rank $|E|$. In fact, if this is true, then we can simply choose a subfamily $\mathcal{F}$ of a maximal laminar family $\mathcal{L}$ such that $\{\boldsymbol{a}_S \mid S \in \mathcal{F}\}$ forms a basis of $\{\boldsymbol{a}_S \mid S \in \mathcal{L}\}$. It is clear that this laminar family $\mathcal{F}$ satisfies conditions (a) and (b). For condition (c), we note that for an active set S, $f(S)$ must be nonnegative. In addition, if $f(S) = 0$, then $\boldsymbol{a}_S$ would be equal to $\mathbf{0}$ because $x_e > 0$ for all edges $e \in E$, contradicting condition (b). Thus, condition (c) also holds.

For the sake of contradiction, suppose that $\mathcal{L}$ is a maximal laminar family of active sets such that the rank of $\{\boldsymbol{a}_S \mid S \in \mathcal{L}\}$ is less than $|E|$. Let $Span(\mathcal{L})$ denote the set of all linear combinations of all $\boldsymbol{a}_S$ with $S \in \mathcal{L}$. Since the set of all active constraints has rank equal to $|E|$, there exists an active set A such that $\boldsymbol{a}_A \notin Span(\mathcal{L})$. Since $\mathcal{L}$ is maximal, A must cross a set in $\mathcal{L}$. We choose A to be the active set that crosses the *minimum* number of sets in $\mathcal{L}$, among all active sets S whose corresponding constraint $\boldsymbol{a}_S$ is not in $Span(\mathcal{L})$.

Let $B \subseteq V$ be a set in $\mathcal{L}$ that crosses A. Note that f is weakly supmodular. Thus, we have either

$$f(A) + f(B) \leq f(A \setminus B) + f(B \setminus A)$$

or

$$f(A) + f(B) \leq f(A \cup B) + f(A \cap B).$$

First, we assume that

$$f(A) + f(B) \leq f(A \setminus B) + f(B \setminus A). \tag{7.26}$$

For two disjoint sets $C, D \subseteq V$, let $E(C, D)$ denote the set of all edges in E with one endpoint in C and the other in D. Also, denote $S_1 = A \setminus B$, $S_2 = A \cap B$, $S_3 = B \setminus A$, and $S_4 = V - (A \cup B)$. For $1 \leq i, j \leq 4$, let

$$m_{i,j} = \sum_{e \in E(S_i, S_j)} x_e.$$

Since A and B are both active, we have

$$f(A) = m_{1,3} + m_{1,4} + m_{2,3} + m_{2,4},$$
$$f(B) = m_{1,2} + m_{1,3} + m_{2,4} + m_{3,4}.$$

Moreover, for constraints S_1 and S_3, we have

$$f(S_1) \leq m_{1,2} + m_{1,3} + m_{1,4},$$
$$f(S_3) \leq m_{1,3} + m_{2,3} + m_{3,4}.$$

Thus,

$$f(S_1) + f(S_3) + 2m_{2,4} \le f(A) + f(B).$$

However, by (7.26), we know that $f(A) + f(B) \le f(S_1) + f(S_3)$. Therefore, $m_{2,4}$ must be equal to 0, and

$$f(A) + f(B) = f(S_1) + f(S_3).$$

It means that S_1 and S_3 are active. In addition, $m_{2,4} = 0$ implies $E(S_2, S_4) = \emptyset$, since $x_e > 0$ for all $e \in E$. It follows that

$$\boldsymbol{a}_A + \boldsymbol{a}_B = \boldsymbol{a}_{S_1} + \boldsymbol{a}_{S_3}.$$

Since $\boldsymbol{a}_A \notin Span(\mathcal{L})$ and $B \in \mathcal{L}$, either $\boldsymbol{a}_{S_1}$ or $\boldsymbol{a}_{S_3}$ is not in $Span(\mathcal{L})$.

Case 1. $\boldsymbol{a}_{S_1} \notin Span(\mathcal{L})$. We claim that every set $C \in \mathcal{L}$ crossing set S_1 must also cross set A. To see this, suppose that $C \in \mathcal{L}$ crosses S_1. Note that A is a superset of S_1. Therefore, $S_1 \cap C \neq \emptyset$ implies $A \cap C \neq \emptyset$, and $S_1 \setminus C \neq \emptyset$ implies $A \setminus C \neq \emptyset$.

Furthermore, $S_1 \cap C \neq \emptyset$ also implies that $C \setminus B \neq \emptyset$. Since B and C are both in $\mathcal{L}$, we have either $B \subset C$ or $B \cap C = \emptyset$. In either case, we must have $C \setminus A \neq \emptyset$: If $B \subset C$, then $(C \setminus A) \supseteq (B \setminus A) \neq \emptyset$, and if $B \cap C = \emptyset$, then $(C \setminus A) = (C \setminus S_1) \neq \emptyset$. It follows that C crosses A, and the claim is proven.

Now we observe that set B crosses A but does not cross S_1. Together with the above claim, we see that the number of sets in $\mathcal{L}$ crossing S_1 is strictly less than the number of sets in $\mathcal{L}$ crossing A. This is a contradiction to our choice of A.

Case 2. $\boldsymbol{a}_{S_3} \notin Span(\mathcal{L})$. Then, similar to Case 1, we claim that every set C in $\mathcal{L}$ crossing S_3 must also cross A. To prove this claim, suppose that $C \in \mathcal{L}$ crosses S_3. Then $S_3 \setminus C \neq \emptyset$ implies $B \setminus C \neq \emptyset$, and $S_3 \cap C \neq \emptyset$ implies $B \cap C \neq \emptyset$. Since B and C are both in $\mathcal{L}$, we must have $C \subset B$. It follows that $\emptyset \neq (C \setminus S_3) \subseteq (A \cap C)$. Moreover, $(A \setminus C) \supset (A \setminus B) \neq \emptyset$ and $(C \setminus A) = (C \cap S_3) \neq \emptyset$. Therefore, C crosses A.

Now, we observe that set B crosses A, but not S_3, and this, together with the claim, leads to a contradiction to our choice of A.

Finally, we note that for the case

$$f(A) + f(B) \le f(A \cup B) + f(A \cap B),$$

a contradiction can be derived by a similar argument. □

Next, we use a counting argument to show a nice property of the laminar family $\mathcal{F}$ given by Lemma 7.23.

Lemma 7.24 *Suppose x_e is fractional for every $e \in E$. Then the laminar family $\mathcal{F}$ of* Lemma 7.23 *contains a set S with $|\delta_G(S)| \le 3$.*

Proof. Suppose to the contrary that for every $S \in \mathcal{F}$, $|\delta_G(S)| \ge 4$. We construct a forest T over set $\mathcal{F}$ such that (A, B) is an edge in T if and only if $A \supset B$ and there is no other set C such that $A \supset C \supset B$. (Note that if $A \subset B$ and $A \subset C$, then

$B \cap C \neq \emptyset$, and hence either $B \subset C$ or $C \subset B$, since $\mathcal{F}$ is a laminar family. Thus, T is a forest.) Next, we will count the number of *endpoints* in T. For each vertex $u \in V$, we count it as an endpoint for *each* edge incident on u. To be more precise, let $E' = \{(u, e) \mid u \text{ is an endpoint of } e\}$, and we call each $(u, e) \in E'$ an endpoint.

We assign an endpoint $(u, e) \in E'$ to a set $S \in \mathcal{F}$, and write $(u, e) \in P(S)$, if $u \in S$ and $u \notin S'$ for any proper subset S' of S in $\mathcal{F}$. For a subtree T' of T, we define $P(T')$ to be the set of endpoints (u, e) that are in $P(S)$ for some node S of T'. Note that each leaf S of T has $|\delta_G(S)| \geq 4$, and hence $P(S)$ has at least four endpoints. We claim that for any subtree T' of T, $|P(T')| \geq 2|V(T')| + 2$, where $V(T')$ is the set of nodes in T'.

If T' contains only a single leaf S, then the claim holds trivially, as, by the above observation, $|P(S)| \geq 4 = 2|V(T')| + 2$. In general, suppose T' contains at least two nodes.

Assume that R is the root of T'. Suppose R has $k \geq 2$ children which are the roots of k subtrees $T_1, T_2, \ldots, T_k$. By the induction hypothesis, the number of endpoints in $P(T')$ is at least

$$2|V(T_1)| + 2 + 2|V(T_2)| + 2 + \cdots + 2|V(T_k)| + 2 \geq 2|V(T')| + 2.$$

Suppose R has only one child S. Let T_1 be the subtree rooted at S. By the induction hypothesis, the number of endpoints in $P(T_1)$ is at least $2|V(T_1)| + 2$. If there are at least two endpoints in $P(R)$, then the number of endpoints in $P(T')$ is at least $2|V(T_1)|+2+2 = 2|V(T')|+2$. Otherwise, if there is at most one endpoint in $P(R)$, then $\delta_G(R)$ and $\delta_G(S)$ must differ in exactly one edge. Indeed, since $\boldsymbol{a}_R$ and $\boldsymbol{a}_S$ are linearly independent, $\delta_G(R)$ and $\delta_G(S)$ must be different. If there is an edge $e = \{u, v\} \in \delta_G(R) \setminus \delta_G(S)$, with $u \in R$ and $v \notin R$, then $(u, e) \in P(R)$. In addition, if $e = \{u, v\} \in \delta_G(S) \setminus \delta_G(R)$, with $u \in S$ and $v \notin S$, then v must be in R and so $(v, e) \in P(R)$. Therefore, $\delta_G(S)$ and $\delta_G(R)$ can differ in at most one edge. Let e be the edge in $\delta_G(R)\Delta\delta_G(S)$. Then $x_e = |f(R) - f(S)|$ must be an integer, contradicting the assumption that all components x_e are fractional. This completes the proof of our claim.

The above claim implies that there are totally at least $2|\mathcal{F}| + 2 = 2|E| + 2$ endpoints. However, since each edge can generate only two endpoints, there are only $2|E|$ endpoints in E', and we have reached a contradiction. □

To finish the proof of Lemma 7.21, we note that if $x_e = 1$ for some edge e, then Lemma 7.21 holds. Otherwise, let S be an active set in the laminar family $\mathcal{F}$ of Lemma 7.24 with $|\delta_G(S)| \leq 3$. Then, by condition (c) of Lemma 7.23, we have

$$\sum_{e \in \delta_G(S)} x_e = f(S) \geq 1,$$

and at least one of the edges $e \in \delta_G(S)$ has $x_e \geq 1/3$.

By exploring more properties of the laminar families, people have found ways to further improve the result of Lemma 7.21. The reader is referred to Jain [2001], Gabow and Gallagher [2008], and Gabow et al. [2009] for these results.

7.6 Random Rounding

A general idea in rounding is to round a fractional optimal solution point randomly to an integer point. With a natural probability distribution, such a random rounding scheme often gets a reasonably good expected performance ratio. Moreover, for some types of simple random rounding schemes, derandomization techniques may be applied to get a deterministic approximation algorithm with the same performance ratio. The following is a simple example.

> MAXIMUM SATISFIABILITY (MAX-SAT): Given a CNF Boolean formula F, find a Boolean assignment to maximize the number of satisfied clauses.

Suppose F contains m clauses $C_1, \ldots, C_m$ over n variables $x_1, \ldots, x_n$. Then the problem MAX SAT on input F can be formulated as the following integer linear program:

$$\begin{array}{lll} \text{maximize} & z_1 + z_2 + \cdots + z_m & \\ \text{subject to} & \sum_{x_i \in C_j} y_i + \sum_{\overline{x}_i \in C_j} (1 - y_i) \geq z_j, & j = 1, 2, \ldots, m, \\ & y_i \in \{0, 1\}, & i = 1, 2, \ldots, n, \\ & z_j \in \{0, 1\}, & j = 1, 2, \ldots, m, \end{array}$$

in which the value of the integer variable y_i, $1 \leq i \leq n$, corresponds to the value assigned to the Boolean variable x_i.

After relaxing the integer variables y_i's and z_j's to real number variables, we get the following linear program:

$$\begin{array}{lll} \text{maximize} & z_1 + z_2 + \cdots + z_m & \\ \text{subject to} & \sum_{x_i \in C_j} y_i + \sum_{\overline{x}_i \in C_j} (1 - y_i) \geq z_j, & j = 1, 2, \ldots, m, \\ & 0 \leq y_i \leq 1, & i = 1, 2, \ldots, n, \\ & 0 \leq z_j \leq 1, & j = 1, 2, \ldots, m. \end{array} \tag{7.27}$$

Let $(\boldsymbol{y}^*, \boldsymbol{z}^*)$ be an optimal solution of the above LP, and let opt_{LP} be its corresponding optimal objective function value; that is, $opt_{\mathrm{LP}} = z_1^* + z_2^* + \cdots + z_m^*$. Now, to get an integer solution to F, we randomly round each y_i^* to 1 or 0 independently as follows:

Algorithm 7.G (*Independent Random Rounding Algorithm for* MAX-SAT)

Input: A CNF Boolean formula F of clauses $C_1, C_2, \ldots, C_m$ over variables $x_1, x_2, \ldots, x_n$.

(1) Construct LP (7.27) and find an optimal solution $(\boldsymbol{y}^*, \boldsymbol{z}^*)$.

(2) **For** $i \leftarrow 1$ **to** n **do**
Set $x_i \leftarrow 1$ with probability y_i^*. ■

To analyze the performance of this independent random rounding, let Z_j be the indicator random variable for the event that clause C_j is satisfied.

Lemma 7.25 *For any clause* C_j, $1 \le j \le m$, $E[Z_j] \ge z_j^*(1 - 1/e)$.

Proof. We note that Z_j is an indicator random variable, and so

$$\begin{aligned} E[Z_j] &= \Pr[Z_j = 1] = 1 - \Pr[Z_j = 0] \\ &= 1 - \prod_{x_i \in C_j} (1 - y_i^*) \cdot \prod_{\overline{x}_i \in C_j} y_i^*. \end{aligned} \tag{7.28}$$

By an argument similar to that of Lemma 7.17, we can prove that $E[Z_j] \ge z_j^*(1 - 1/e)$. We omit the detail. □

Denote $Z_F = Z_1 + Z_2 + \cdots + Z_m$, and let *opt* be the optimal objective function value of MAX-SAT. By Lemma 7.25, we have

$$\begin{aligned} E[Z_F] &= E[Z_1] + E[Z_2] + \cdots + E[Z_m] \\ &\ge \left(1 - \frac{1}{e}\right)(z_1^* + z_2^* + \cdots + z_m^*) \\ &\ge opt_{\mathrm{LP}} \cdot \left(1 - \frac{1}{e}\right) \ge opt \cdot \left(1 - \frac{1}{e}\right), \end{aligned}$$

and we get a performance ratio $e/(e-1)$ for Algorithm 7.G:

Theorem 7.26 *The expected output value of* Algorithm 7.G *is an* $(e/(e-1))$-*approximation to* MAX-SAT.

The random rounding of Algorithm 7.G rounds each variable x_i independently. For such a simple random rounding, we can derandomize it by the method of conditional probability. Namely, we note that

$$E[Z_F] = E[Z_F \mid x_1 = 1] \cdot y_1^* + E[Z_F \mid x_1 = 0] \cdot (1 - y_1^*).$$

Therefore, we have either

$$E\big[Z_{F|_{x_1=1}}\big] = E\big[Z_F \mid x_1 = 1\big] \ge opt_{\mathrm{LP}} \cdot \left(1 - \frac{1}{e}\right)$$

or

$$E\big[Z_{F|_{x_1=0}}\big] = E\big[Z_F \mid x_1 = 0\big] \ge opt_{\mathrm{LP}} \cdot \left(1 - \frac{1}{e}\right),$$

where $F|_{x_1=b}$, $b \in \{0, 1\}$, denotes the Boolean formula obtained from F with the partial assignment $x_1 = b$. Moreover, as shown in (7.28), each $E[Z_j]$, and hence $E[Z_F]$, can be computed in polynomial time. This also applies to $E[Z_F \mid x_1 =$

0] and $E[Z_F \mid x_1 = 1]$. Therefore, we can find out, in polynomial time, which of the two assignments $x_1 = 0$ or $x_i = 1$ has a better expected output value. This observation suggests the following derandomization of Algorithm 7.G.

Algorithm 7.H (*Derandomization of* Algorithm 7.G *for* MAX-SAT)

Input: A CNF Boolean formula F of clauses $C_1, C_2, \ldots, C_m$ over variables $x_1, x_2, \ldots, x_n$.

(1) Construct LP (7.27) and find an optimal solution $(\boldsymbol{y}^*, \boldsymbol{z}^*)$.

(2) **For** $i \leftarrow 1$ **to** n **do**
 if $E[Z_F \mid x_i = 1] \ge E[Z_F \mid x_i = 0]$
  then $x_i \leftarrow 1;\ F \leftarrow F|_{x_i=1}$
  else $x_i \leftarrow 0;\ F \leftarrow F|_{x_i=0}$. ■

Theorem 7.27 MAX SAT *has a polynomial-time* $e/(e-1)$*-approximation.*

Proof. We observe that, at each iteration,

$$\max\{E[Z_F \mid x_i = 0], E[Z_F \mid x_i = 1]\} \ge E[Z_F].$$

Thus, we can prove, by a simple induction, that the formula F at the end of each iteration must satisfy $E[Z_F] \ge (1 - 1/e)\, opt_{\rm LP}$. Note that, at the end of the nth iteration, F contains no variable, and so

$$Z_F = E[Z_F] \ge \Big(1 - \frac{1}{e}\Big)\, opt_{\rm LP} \ge \Big(1 - \frac{1}{e}\Big)\, opt. \qquad \square$$

In the above example, each variable is rounded to an integer independently. Next, we introduce some general random rounding techniques in which the roundings for different variables are not independent.

Recall the pipage rounding technique introduced in Section 7.4, where the rounding at each stage is determined by a companion function which is closely related to the objective function. Within the setting of pipage rounding, we can apply random rounding to avoid the use of the companion function. This technique of combining random rounding with pipage rounding has many applications.

We first study the general framework of random pipage rounding. Consider a bipartite graph $G = (U, V, E)$ and variables x_e, for $e \in E$. Let $\boldsymbol{x}^*$ be an optimal solution to an LP of the form (7.21).

Random Pipage Rounding

(1) Initially, set $\boldsymbol{x} \leftarrow \boldsymbol{x}^*$.

(2) **While** $\boldsymbol{x}$ is not an integer solution **do**

 (2.1) Let $H_{\boldsymbol{x}}$ be the subgraph of G induced by all edges $e \in E$ with $0 < x_e < 1$. Let R be a cycle or a maximal path of $H_{\boldsymbol{x}}$. Then R can be decomposed into two matchings M_1 and M_2.

(2.2) Define $\boldsymbol{x}(\varepsilon)$ by

$$x_e(\varepsilon) = \begin{cases} x_e & \text{if } e \notin R, \\ x_e + \varepsilon & \text{if } e \in M_1, \\ x_e - \varepsilon & \text{if } e \in M_2. \end{cases}$$

(2.3) Let $\varepsilon_1 \leftarrow \min\Big\{ \min_{e \in M_1} x_e, \min_{e \in M_2} (1 - x_e) \Big\}$,

$$\varepsilon_2 \leftarrow \min\Big\{ \min_{e \in M_1} (1 - x_e), \min_{e \in M_2} x_e \Big\}.$$

(2.4) Set

$$\boldsymbol{x} \leftarrow \begin{cases} \boldsymbol{x}(\varepsilon_2), & \text{with probability } \varepsilon_1/(\varepsilon_1 + \varepsilon_2), \\ \boldsymbol{x}(-\varepsilon_1), & \text{with probability } \varepsilon_2/(\varepsilon_1 + \varepsilon_2). \end{cases}$$

■

Lemma 7.28 *For each edge e, let X_e be the random variable denoting the value of x_e output by the* Random Pipage Rounding *procedure. Then the following properties hold for X_e:*

(P1) (Marginal Distribution) *For every edge e,* $\Pr[X_e = 1] = x_e^*$.

(P2) (Degree Preservation) *For any vertex $v \in U \cup V$,*

$$\Pr\big[D_v \in \{\lfloor d_v \rfloor, \lceil d_v \rceil\}\big] = 1,$$

where $D_v = \sum_{e \in \delta(v)} X_e$ and $d_v = \sum_{e \in \delta(v)} x_e^*$.

(P3) (Negative Correlation) *For any $v \in U \cup V$, $S \subseteq \delta(v)$, and $b \in \{0, 1\}$,*

$$\Pr\Big[\bigwedge_{e \in S} (X_e = b)\Big] \le \prod_{e \in S} \Pr\big[X_e = b\big].$$

Proof. For property (P1), we prove it by induction on the number k of edges e with nonintegral x_e^*. For $k = 0$, it is trivial. Now, we consider the case $k \ge 1$. Let x'_e be the random variable for the value of x_e at the end of the first iteration, and write $\boldsymbol{x}' = (x'_e)_{e \in E}$. Note that within steps (2.1)–(2.3) of the first iteration, we have $\boldsymbol{x}' = \boldsymbol{x}^*$. Then, after step (2.4), the number of nonintegral components of $\boldsymbol{x}'$ is at most $k - 1$. Therefore, by the induction hypothesis, we have

$$\Pr\big[X_e = 1 \,\big|\, \boldsymbol{x}' = \boldsymbol{x}(-\varepsilon_1)\big] = x'_e = x_e(-\varepsilon_1) = x_e^*(-\varepsilon_1),$$
$$\Pr\big[X_e = 1 \,\big|\, \boldsymbol{x}' = \boldsymbol{x}(\varepsilon_2)\big] = x'_e = x_e(\varepsilon_2) = x_e^*(\varepsilon_2).$$

It follows that

$$\begin{aligned} \Pr[X_e = 1] &= \Pr\big[X_e = 1 \,\big|\, \boldsymbol{x}' = \boldsymbol{x}(-\varepsilon_1)\big] \cdot \Pr[\boldsymbol{x}' = \boldsymbol{x}(-\varepsilon_1)] \\ &\quad + \Pr\big[X_e = 1 \,\big|\, \boldsymbol{x}' = \boldsymbol{x}(\varepsilon_2)\big] \cdot \Pr[\boldsymbol{x}' = \boldsymbol{x}(\varepsilon_2)] \\ &= x_e^*(-\varepsilon_1) \cdot \frac{\varepsilon_2}{\varepsilon_1 + \varepsilon_2} + x_e^*(\varepsilon_2) \cdot \frac{\varepsilon_1}{\varepsilon_1 + \varepsilon_2}. \end{aligned}$$

Now, if $e \notin R$, then $x_e^*(-\varepsilon_1) = x_e^*(\varepsilon_2) = x_e^*$ and, hence, $\Pr[X_e = 1] = x_e^*$. If $e \in M_1$, then $x_e^*(-\varepsilon_1) = x_e^* - \varepsilon_1$ and $x_e^*(\varepsilon_2) = x_e^* + \varepsilon_2$. Hence,

$$\Pr[X_e = 1] = (x_e^* - \varepsilon_1) \cdot \frac{\varepsilon_2}{\varepsilon_1 + \varepsilon_2} + (x_e^* + \varepsilon_2) \cdot \frac{\varepsilon_1}{\varepsilon_1 + \varepsilon_2} = x_e^*.$$

If $e \in M_2$, then $x_e^*(-\varepsilon_1) = x_e^* + \varepsilon_1$ and $x_e^*(\varepsilon_2) = x_e^* - \varepsilon_2$. Hence,

$$\Pr[X_e = 1] = (x_e^* + \varepsilon_1) \cdot \frac{\varepsilon_2}{\varepsilon_1 - \varepsilon_2} + (x_e^* - \varepsilon_2) \cdot \frac{\varepsilon_1}{\varepsilon_1 + \varepsilon_2} = x_e^*.$$

For property (P2), we consider three cases.

Case 1. For all edges $e \in \delta(v)$, x_e^* is an integer. Then $X_e = x_e^*$ for all $e \in \delta(v)$ and so $D_v = d_v$.

Case 2. There exists exactly one edge $e \in \delta(v)$ such that x_e^* is nonintegral. Then, $D_v = \lfloor d_v \rfloor$ if $X_e = 0$ and $D_v = \lceil d_v \rceil$ if $X_e = 1$. So, (P2) holds in this case.

Case 3. There exists more than one edge $e \in \delta(v)$ such that x_e^* is nonintegral. Then, at the beginning of an iteration, if there is more than one edge $e \in \delta(v)$ with nonintegral x_e, then, by the argument in the proof of Lemma 7.20, the value $\sum_{e \in \delta(v)} x_e$ does not change after this iteration and so is still equal to d_v. If, at the end of an iteration, the number of nonintegral components x_e, for $e \in \delta(v)$, drops below two, then either case 1 or case 2 applies, and so $D_v = \lfloor d_v \rfloor$ or $\lceil d_v \rceil$. This shows that (P2) also holds for this case.

For property (P3), we will also prove it by induction on the number k of edges e with nonintegral x_e^*. For $k = 0$, (P3) holds trivially with equality. Now, we consider the case of $k \geq 1$. Let x_e' be the random variable for the value of x_e at the end of the first iteration and let $\boldsymbol{x}' = (x_e')_{e \in \delta(v)}$. So, by the induction hypothesis

$$\Pr\Big[\bigwedge_{e \in S} (X_e = b) \;\Big|\; \boldsymbol{x}' = \boldsymbol{x}(-\varepsilon_1)\Big] \leq \prod_{e \in S} \Pr\big[X_e = b \mid \boldsymbol{x}' = \boldsymbol{x}(-\varepsilon_1)\big]$$

and

$$\Pr\Big[\bigwedge_{e \in S} (X_e = b) \;\Big|\; \boldsymbol{x}' = \boldsymbol{x}(\varepsilon_2)\Big] \leq \prod_{e \in S} \Pr\big[X_e = b \mid \boldsymbol{x}' = \boldsymbol{x}(\varepsilon_2)\big].$$

Note that $S \subseteq \delta(v)$ may have at most two edges in R. We consider the following three cases.

Case 1. No edge in S belongs to R. Then, by property (P1), for any $e \in S$,

$$\begin{aligned}\Pr\big[X_e = 1 \mid \boldsymbol{x}' = \boldsymbol{x}(-\varepsilon_1)\big] &= \Pr\big[X_e = 1 \mid \boldsymbol{x}' = \boldsymbol{x}(\varepsilon_2)\big] \\ &= x_e' = x_e^* = \Pr[X_e = 1]\end{aligned}$$

and

$$\begin{aligned}\Pr\big[X_e = 0 \mid \boldsymbol{x}' = \boldsymbol{x}(-\varepsilon_1)\big] &= \Pr\big[X_e = 0 \mid \boldsymbol{x}' = \boldsymbol{x}(\varepsilon_2)\big] \\ &= 1 - x_e^* = \Pr[X_e = 0].\end{aligned}$$

Therefore, we have

$$\begin{aligned}\Pr\Big[\bigwedge_{e\in S}(X_e=b)\Big] &= \Pr\Big[\bigwedge_{e\in S}(X_e=b)\,\Big|\,\boldsymbol{x}'=\boldsymbol{x}(-\varepsilon_1)\Big]\cdot\Pr[\boldsymbol{x}'=\boldsymbol{x}(-\varepsilon_1)]\\ &\quad+\Pr\Big[\bigwedge_{e\in S}(X_e=b)\,\Big|\,\boldsymbol{x}'=\boldsymbol{x}(\varepsilon_2)\Big]\cdot\Pr[\boldsymbol{x}'=\boldsymbol{x}(\varepsilon_2)]\\ &\le \prod_{e\in S}\Pr[X_e=b]\cdot\frac{\varepsilon_2}{\varepsilon_1+\varepsilon_2}+\prod_{e\in S}\Pr[X_e=b]\cdot\frac{\varepsilon_1}{\varepsilon_1+\varepsilon_2}\\ &= \prod_{e\in S}\Pr[X_e=b],\end{aligned}$$

and so (P3) holds for case 1.

Case 2. S contains only one edge e' in R. Without loss of generality, assume that $e'\in M_1$. Then, at the end of the first iteration, $x_{e'}(-\varepsilon_1)=x^*_{e'}-\varepsilon_1$ and $x_{e'}(\varepsilon_2)=x^*_{e'}+\varepsilon_2$. So, by (P1),

$$\begin{aligned}\Pr\big[X_{e'}=1\,\big|\,\boldsymbol{x}'=\boldsymbol{x}(-\varepsilon_1)\big]&=x_{e'}(-\varepsilon_1)=x^*_{e'}-\varepsilon_1,\\ \Pr\big[X_{e'}=1\,\big|\,\boldsymbol{x}'=\boldsymbol{x}(\varepsilon_2)\big]&=x_{e'}(\varepsilon_2)=x^*_{e'}+\varepsilon_2.\end{aligned}$$

Therefore,

$$\begin{aligned}&\Pr\Big[\bigwedge_{e\in S}(X_e=1)\Big]\\ &\le\Big[(x^*_{e'}-\varepsilon_1)\cdot\frac{\varepsilon_2}{\varepsilon_1+\varepsilon_2}+(x^*_{e'}+\varepsilon_2)\cdot\frac{\varepsilon_1}{\varepsilon_1+\varepsilon_2}\Big]\cdot\prod_{e\in S-\{e'\}}\Pr[X_e=1]\\ &=x^*_{e'}\prod_{e\in S-\{e'\}}\Pr[X_e=1]=\prod_{e\in S}\Pr[X_e=1].\end{aligned}$$

Similarly,

$$\begin{aligned}&\Pr\Big[\bigwedge_{e\in S}(X_e=0)\Big]\\ &\le\Big[(1-x^*_{e'}+\varepsilon_1)\cdot\frac{\varepsilon_2}{\varepsilon_1+\varepsilon_2}+(1-x^*_{e'}-\varepsilon_2)\cdot\frac{\varepsilon_1}{\varepsilon_1+\varepsilon_2}\Big]\cdot\prod_{e\in S-\{e'\}}\Pr[X_e=0]\\ &=(1-x^*_{e'})\prod_{e\in S-\{e'\}}\Pr[X_e=0]=\prod_{e\in S}\Pr[X_e=0].\end{aligned}$$

This shows that (P3) holds for case 2.

Case 3. S contains two edges e' and e'' in R. Then we must have $e'\in M_1$ and $e''\in M_2$. By (P1), we know that

$$\begin{aligned}
\Pr\big[X_{e'} = 1 \;\big|\; \boldsymbol{x}' = \boldsymbol{x}(-\varepsilon_1)\big] &= x^*_{e'} - \varepsilon_1,\\
\Pr\big[X_{e'} = 1 \;\big|\; \boldsymbol{x}' = \boldsymbol{x}(\varepsilon_2)\;\;\big] &= x^*_{e'} + \varepsilon_2,\\
\Pr\big[X_{e''} = 1 \,\big|\, \boldsymbol{x}' = \boldsymbol{x}(-\varepsilon_1)\big] &= x^*_{e''} + \varepsilon_1,\\
\Pr\big[X_{e''} = 1 \,\big|\, \boldsymbol{x}' = \boldsymbol{x}(\varepsilon_2)\;\;\big] &= x^*_{e''} - \varepsilon_2.
\end{aligned}$$

Therefore,

$$\begin{aligned}
\Pr\Big[\bigwedge_{e\in S}(X_e = 1)\Big] \le{}& \Big[(x^*_{e'} - \varepsilon_1)(x^*_{e''} + \varepsilon_1)\cdot\frac{\varepsilon_2}{\varepsilon_1+\varepsilon_2}\\
&+ (x^*_{e'} + \varepsilon_2)(x^*_{e''} - \varepsilon_2)\cdot\frac{\varepsilon_1}{\varepsilon_1+\varepsilon_2}\Big]\cdot\prod_{e\in S-\{e',e''\}}\Pr[X_e = 1]\\
={}& (x^*_{e'}x^*_{e''} - \varepsilon_1\varepsilon_2)\cdot\prod_{e\in S-\{e',e''\}}\Pr[X_e = 1]\\
\le{}& \prod_{e\in S}\Pr[X_e = 1].
\end{aligned}$$

For the case of $b = 0$, the proof is similar. □

For a simple application of the above properties of the Random Pipage Rounding procedure, consider the problem MAX-WH again. Let $\boldsymbol{x}^* = (x^*_i)_{1\le i\le n}$ be an optimal (fractional) solution for the LP-relaxation (7.19) of MAX-WH. Applying the Random Pipage Rounding procedure to $\boldsymbol{x}^*$, we round each variable x_i, $1 \le i \le n$, to a random variable $X_i \in \{0,1\}$. Let $L_j(\boldsymbol{X}) = \min\{1, \sum_{i\in S_j} X_i\}$ and $L(\boldsymbol{X}) = \sum_{j=1}^m w_j L_j(\boldsymbol{X})$; that is, $L(\boldsymbol{X})$ is the objective function value of the random pipage rounding. The following theorem shows that the expected value of $L(\boldsymbol{X})$ is as good as the approximate solution produced by the deterministic pipage rounding of Algorithm 7.E. In the following, *opt* denotes the optimal objective function value of the problem MAX-WH.

Theorem 7.29 $E\big[L(\boldsymbol{X})\big] \ge \Big(1 - \dfrac{1}{e}\Big)\, opt.$

Proof. Note that for each $j = 1, 2, \ldots, n$,

$$\begin{aligned}
\Pr\big[L_j(\boldsymbol{X}) = 1\big] &= 1 - \Pr\big[L_j(\boldsymbol{X}) = 0\big]\\
&= 1 - \Pr\Big[\bigwedge_{i\in S_j}(X_i = 0)\Big]\\
&\ge 1 - \prod_{i\in S_j}\Pr\big[X_i = 0\big] && \text{(by negative correlation)}\\
&= 1 - \prod_{i\in S_j}(1 - x^*_i) && \text{(by marginal distribution)}\\
&\ge \Big(1 - \frac{1}{e}\Big)\cdot\min\Big\{1, \sum_{i\in S_j} x^*_i\Big\},
\end{aligned}$$

where the last inequality follows from the proof of Lemma 7.17. Thus, we have

$$\begin{aligned} E\big[L(\boldsymbol{X})\big] &= \sum_{j=1}^{m} w_j \cdot E\big[L_j(\boldsymbol{X})\big] = \sum_{j=1}^{m} w_j \cdot \Pr\big[L_j(\boldsymbol{X}) = 1\big] \\ &\geq \Big(1 - \frac{1}{e}\Big) \sum_{j=1}^{m} w_j \cdot \min\Big\{1, \sum_{i \in S_j} x_i^*\Big\} \geq \Big(1 - \frac{1}{e}\Big)\, opt. \qquad \Box \end{aligned}$$

Next, we study a random rounding technique based on the geometric structure of the feasible region. Consider an n-dimensional polytope P with integer vertices. Then, every point $\boldsymbol{x}$ in P can be expressed as a convex combination of at most $n+1$ vertices. The following is a simple rounding scheme based on this property.

Vector Rounding

Input: An n-dimensional polytope P with integer vertices, and a noninteger solution $\boldsymbol{x}$ in P.

(1) Write $\boldsymbol{x} = \sum_{i=1}^{n+1} \alpha_i \boldsymbol{v}_i$, where $\boldsymbol{v}_1, \ldots, \boldsymbol{v}_{n+1}$ are vertices of P, $\alpha_i \geq 0$, and $\sum_{i=1}^{n+1} \alpha_i = 1$.

(2) Round $\boldsymbol{x}$ to vertex $\boldsymbol{v}_i$ with probability α_i, $1 \leq i \leq n+1$. ■

The above vector rounding can be extended to the following more general geometric rounding scheme. In the following, we write $\langle \boldsymbol{v}_1, \ldots, \boldsymbol{v}_n, \boldsymbol{v}_{n+1} \rangle$ to denote the n-dimensional simplex generated by points $\boldsymbol{v}_1, \ldots, \boldsymbol{v}_{n+1}$.

Geometric Rounding

Input: A simplex $P = \langle \boldsymbol{v}_1, \ldots, \boldsymbol{v}_n, \boldsymbol{v}_{n+1} \rangle$ and a point $\boldsymbol{x}$ in the simplex.

(1) **For** $i \leftarrow 1$ **to** $n+1$ **do**

Select a random number β_i from $(0, 1]$.

(2) Let $\boldsymbol{u} \leftarrow \dfrac{\sum_{i=1}^{n+1} \beta_i\, \boldsymbol{v}_i}{\sum_{i=1}^{n+1} \beta_i}$.

(3) Round $\boldsymbol{x}$ to $\boldsymbol{v}_i$ if $\boldsymbol{u}$ lies in the simplex $\langle \boldsymbol{v}_1, \ldots, \boldsymbol{v}_{i-1}, \boldsymbol{x}, \boldsymbol{v}_{i+1}, \ldots, \boldsymbol{v}_{n+1} \rangle$ (see Figure 7.5). ■

Indeed, if each β_i, $1 \leq i \leq n+1$, is chosen randomly based on the unit-exponential distribution, then the corresponding geometric rounding is equivalent to vector rounding. This relationship can be seen from the following two lemmas. Let $P = \langle \boldsymbol{v}_1, \ldots, \boldsymbol{v}_{n+1} \rangle$ be a nondegenerate simplex, and $\boldsymbol{x} = \sum_{i=1}^{n+1} \alpha_i \boldsymbol{v}_i$, where $\alpha_i \geq 0$ for each $1 \leq i \leq n+1$, and $\sum_{i=1}^{n+1} \alpha_i = 1$. Also, let $\boldsymbol{u}$ be the point defined in the Geometric Rounding procedure about the simplex P and point $\boldsymbol{x}$.

Lemma 7.30 *The point $\boldsymbol{u}$ lies in the simplex $\langle \boldsymbol{v}_1, \ldots, \boldsymbol{v}_{i-1}, \boldsymbol{x}, \boldsymbol{v}_{i+1}, \ldots, \boldsymbol{v}_{n+1} \rangle$ if and only if*

$$\frac{\beta_i}{\alpha_i} = \min_{1 \leq k \leq n+1} \frac{\beta_k}{\alpha_k}.$$

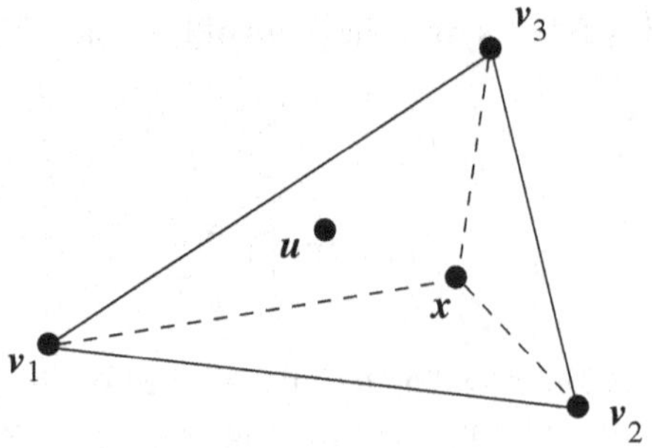

Figure 7.5: Geometric rounding: rounding $\boldsymbol{x}$ to $\boldsymbol{v}_2$.

Proof. It suffices to consider the case $i = 1$. Suppose $\boldsymbol{u} \in \langle \boldsymbol{x}, \boldsymbol{v}_2, \ldots, \boldsymbol{v}_{n+1}\rangle$. Then $\boldsymbol{u}$ can be written as a convex combination of $\boldsymbol{x}, \boldsymbol{v}_2, \ldots, \boldsymbol{v}_{n+1}$; that is, $\boldsymbol{u} = \lambda_1 \boldsymbol{x} + \lambda_2 \boldsymbol{v}_2 + \cdots + \lambda_{n+1}\boldsymbol{v}_{n+1}$, with $\lambda_i \geq 0$, for each $1 \leq i \leq n+1$, and $\sum_{i=1}^{n+1} \lambda_i = 1$. Substituting $\sum_{i=1}^{n+1} \alpha_i \boldsymbol{v}_i$ for $\boldsymbol{x}$, we obtain

$$\boldsymbol{u} = \lambda_1 \alpha_1 \boldsymbol{v}_1 + \sum_{i=2}^{n+1} (\lambda_1 \alpha_i + \lambda_i)\boldsymbol{v}_i.$$

Since $\langle \boldsymbol{v}_1, \ldots, \boldsymbol{v}_{n+1}\rangle$ is a nondegenerate simplex, the convex combination for $\boldsymbol{u}$, in terms of $\boldsymbol{v}_i$, $1 \leq i \leq n+1$, is unique. Hence,

$$\frac{\beta_1}{\beta} = \lambda_1 \alpha_1,$$

$$\frac{\beta_i}{\beta} = \lambda_1 \alpha_i + \lambda_i, \quad 2 \leq i \leq n+1,$$

where $\beta = \sum_{i=1}^{n+1} \beta_i$. Thus, for $2 \leq i \leq n+1$,

$$\frac{\beta_1}{\alpha_1} = \beta\lambda_1 = \frac{\beta_i - \lambda_i}{\alpha_i} \leq \frac{\beta_i}{\alpha_i}.$$

Conversely, assume that $\beta_k/\alpha_k = \min_{1 \leq i \leq n+1} \beta_i/\alpha_i$ and yet $\boldsymbol{u} \notin \langle \boldsymbol{v}_1, \ldots, \boldsymbol{v}_{k-1}, \boldsymbol{x}, \boldsymbol{v}_{k+1}, \ldots, \boldsymbol{v}_{n+1}\rangle$. Without loss of generality, assume that $k \neq 1$ and $\boldsymbol{u} \in \langle \boldsymbol{x}, \boldsymbol{v}_2, \ldots, \boldsymbol{v}_{n+1}\rangle$. So we can write $\boldsymbol{u} = \lambda_1 \boldsymbol{x} + \sum_{i=2}^{n+1} \lambda_i \boldsymbol{v}_i$. Then, as shown above, we must have

$$\frac{\beta_1}{\alpha_1} = \min_{1 \leq i \leq n+1} \frac{\beta_i}{\alpha_i} = \frac{\beta_k}{\alpha_k}.$$

Furthermore, from the above proof, we know that $\lambda_k = 0$. In other words, $\boldsymbol{u}$ can be written as a convex combination of $\boldsymbol{x}, \boldsymbol{v}_2, \ldots, \boldsymbol{v}_{k-1}, \boldsymbol{v}_{k+1}, \ldots, \boldsymbol{v}_{n+1}$. However, this means that $\boldsymbol{u} \in \langle \boldsymbol{v}_1, \ldots, \boldsymbol{v}_{k-1}, \boldsymbol{x}, \boldsymbol{v}_{k+1}, \ldots, \boldsymbol{v}_{n+1}\rangle$, which leads to a contradiction. □

In the following, we write $X \sim \exp(\lambda)$ to denote that the random variable X has the exponential distribution with rate parameter λ; that is, its probability density function is $f(x) = \lambda e^{-\lambda x}$.

Lemma 7.31 *In the* Geometric Rounding *procedure, if* $\beta_1, \ldots, \beta_{n+1}$ *are chosen independently with the distribution* $\beta_i \sim \exp(1)$ *for* $1 \le i \le n+1$*, then*

$$\Pr[\boldsymbol{x} \text{ is rounded to } \boldsymbol{v}_i] = \alpha_i.$$

Proof. It suffices to prove the case that $\boldsymbol{x}$ is rounded to $\boldsymbol{v}_1$. Since, for each $1 \le i \le n+1$, $\beta_i \sim \exp(1)$, we have $\beta_i/\alpha_i \sim \exp(\alpha_i)$. By Lemma 7.30, $\boldsymbol{x}$ is rounded to $\boldsymbol{v}_1$ by the Geometric Rounding procedure if and only if $\beta_1/\alpha_1 = \min_{1\le i\le n+1} \beta_i/\alpha_i$. Note that the cumulative distribution of $X \sim \exp(\lambda)$ is $F(x) = 1 - e^{-\lambda x}$, and

$$\int_a^\infty \lambda e^{-\lambda x} dx = 1 - e^{-\lambda x}\Big|_{x=a}^{\infty} = e^{-a\lambda}.$$

Therefore, we have

$$\Pr[\boldsymbol{x} \text{ is rounded to } \boldsymbol{v}_1] = \Pr\left[\frac{\beta_1}{\alpha_1} = \min_{1\le i\le n+1} \frac{\beta_i}{\alpha_i}\right]$$

$$= \int_0^\infty \alpha_1 e^{-\alpha_1 x_1} \Big(\int_{x_1}^\infty \alpha_2 e^{-\alpha_2 x_2} dx_2 \cdots \int_{x_1}^\infty \alpha_{n+1} e^{-\alpha_{n+1} x_{n+1}} dx_{n+1} \Big) dx_1$$

$$= \int_0^\infty \alpha_1 e^{-\alpha_1 x_1} e^{-\alpha_2 x_1} \cdots e^{-\alpha_{n+1} x_1} dx_1$$

$$= \int_0^\infty \alpha_1 e^{-x_1} dx_1 = \alpha_1.$$

□

Applications of geometric rounding techniques can be found in Ge, Ye, and Zhang [2010] and Ge, He et al. [2010].

Exercises

7.1 Give an example to show that a linear program not satisfying the nondegeneracy assumption may still have a one-to-one and onto correspondence between basic feasible solutions and feasible bases.

7.2 Show that vertices of the feasible region are invariant under the transformation from a linear program to its standard form.

7.3 (a) Generalize the greedy algorithm for KNAPSACK in Section 1.1 to obtain a polynomial-time $(m+1)$-approximation for the resource management problem (7.2).

(b) Generalize the generalized greedy algorithm for KNAPSACK in Section 1.1 to obtain a PTAS for the resource management problem (7.2).

7.4 Recall that a subset A of the vertex set V of a graph $G = (V, E)$ is an *independent set* if no two vertices in A form an edge in E. The maximum independent set problem (MAX-IS) asks one, for a given graph G, to find an independent set of G of the maximum cardinality. Formulate the problem MAX-IS as a resource management problem with an unlimited number of resources.

7.5 Discuss whether the simplex method can avoid cycles for a linear program if it is known that this linear program has a one-to-one and onto correspondence between its basic feasible solutions and feasible bases.

7.6 Use the optimal feasible basis obtained in Example 7.5 to solve the following linear program with the simplex method:

$$\begin{array}{llrcl}
\text{minimize} & z = & -2x_1 - x_2 & & \\
\text{subject to} & & x_1 + 2x_2 + x_5 & = & 8, \\
& & x_1 + x_2 - x_3 + x_6 & = & 3, \\
& & -x_1 + x_2 - x_4 + x_7 & = & 1, \\
& & x_1, x_2, \ldots, x_7 & \geq & 0.
\end{array}$$

7.7 Design an $(e/(e-1))$-approximation for MAX-SAT with the pipage rounding technique.

7.8 Consider the following problem.

MAXIMUM k-CUT IN A HYPERGRAPH (MAX-k-CUT-HYPER): Given a hypergraph $H = (V, E)$ with nonnegative edge weight $w : E \to \mathbb{N}$, and k positive integers $p_1, \ldots, p_k$, with $p_1 + \cdots + p_k = |V|$, partition V into k parts $X_1, \ldots, X_k$ such that $|X_t| = p_t$, $1 \leq t \leq k$, to maximize the total weight of edges not lying entirely in any part.

This problem can be formulated as the following ILP:

$$\begin{array}{lll}
\text{maximize} & \displaystyle\sum_{S \in E} w_S z_S & \\
\text{subject to} & z_S \leq |S| - \displaystyle\sum_{i \in S} x_{it}, & S \in E,\ t = 1, 2, \ldots, k, \\
& \displaystyle\sum_{t=1}^{k} x_{it} = 1, & i \in V, \\
& \displaystyle\sum_{i=1}^{n} x_{it} = p_t, & t = 1, 2, \ldots, k, \\
& x_{it}, z_S \in \{0, 1\}, & S \in E,\ i \in V,\ t = 1, 2, \ldots, k.
\end{array}$$

Define

$$F(\boldsymbol{x}) = \sum_{S \in E} w_S \left(1 - \sum_{t=1}^{k} \prod_{i \in S} x_{it}\right)$$

and

$$L(\boldsymbol{x}) = \sum_{S \in E} w_S \cdot \min\left\{1, \min_{1 \le t \le k} \left(|S| - \sum_{i \in S} x_{it}\right)\right\}.$$

Show that

$$F(\boldsymbol{x}) \ge \rho \cdot L(\boldsymbol{x}),$$

where $\rho = \min\{\lambda_{|S|} \mid S \in E\}$ and $\lambda_r = 1 - (1 - 1/r)^r - (1/r)^r$. Use the pipage rounding scheme to design a $(1/\rho)$-approximation for MAX-k-CUT-HYPER.

7.9 Show that for $r \ge 3$, $\lambda_r = 1 - (1 - 1/r)^r - (1/r)^r \ge 1 - e^{-1}$.

7.10 Consider the following problem:

> MAXIMUM COVERAGE WITH KNAPSACK CONSTRAINTS (MAX-COVER-KC): Given a set $I = \{1, 2, \ldots, n\}$ with weights $c_i \ge 0$ for $i \in I$, a family $\mathcal{F} = \{S_1, S_2, \ldots, S_m\}$ of subsets of I with weights $w_j \ge 0$ for $j \in \{1, \ldots, m\}$, and a positive integer B, find a subset $X \subseteq I$ with $\sum_{i \in X} c_i \le B$ that maximizes the total weight of the sets in $\mathcal{F}$ having nonempty intersections with X.

This problem can be formulated as the following ILP:

$$\begin{aligned}
\text{maximize} \quad & \sum_{j=1}^{m} w_j z_j \\
\text{subject to} \quad & \sum_{i \in S_j} x_i \ge z_j, \qquad 1 \le j \le m, \\
& \sum_{i=1}^{n} c_i x_i \le B, \\
& x_i, z_j \in \{0, 1\}, \quad 1 \le j \le m,\ 1 \le i \le n.
\end{aligned}$$

Define

$$F(\boldsymbol{x}) = \sum_{j=1}^{m} w_j \left(1 - \prod_{i \in S_j} (1 - x_i)\right).$$

Use the pipage rounding scheme with function $F(\boldsymbol{x})$ to design a ρ-approximation for MAX-COVER-KC, where $\rho = 1/(1 - (1 - 1/k)^k)$ and $k = \max\{|S_j| \mid 1 \le j \le m\}$.

7.11 Let X be a finite set. We say a function $f : [0, 1]^X \to \mathbb{R}^+$ is *submodular* if

$$f(\boldsymbol{x} \vee \boldsymbol{y}) + f(\boldsymbol{x} \wedge \boldsymbol{y}) \le f(\boldsymbol{x}) + f(\boldsymbol{y}),$$

where, for $\boldsymbol{x}, \boldsymbol{y} \in [0, 1]^X$, $\boldsymbol{x} \vee \boldsymbol{y}$ and $\boldsymbol{x} \wedge \boldsymbol{y}$ are members of $[0, 1]^X$ defined by $(\boldsymbol{x} \vee \boldsymbol{y})_i = \max\{x_i, y_i\}$ and $(\boldsymbol{x} \wedge \boldsymbol{y})_i = \min\{x_i, y_i\}$. We say f is *monotone increasing* if $f(\boldsymbol{x}) \le f(\boldsymbol{y})$ as long as $\boldsymbol{x} \le \boldsymbol{y}$ coordinatewise. Show that if f is smooth, that is, if its second partial derivatives exist everywhere in $[0, 1]^X$, then

(a) f is monotone increasing if and only if $\partial f / \partial y_j \geq 0$ for each $j \in X$;

(b) f is submodular if and only if $\partial^2 f / (\partial y_i \partial y_j) \leq 0$ for any $i, j \in X$.

7.12 Let $f : 2^X \to \mathbb{R}^+$ be a monotone increasing, submodular function.

(a) Show that f can be extended to a smooth, monotone increasing, submodular function $f : [0,1]^X \to \mathbb{R}^+$ as follows:

$$f(\boldsymbol{y}) = \sum_{R \subseteq X} f(R) \prod_{i \in R} y_i \prod_{i \notin R} (1 - y_i).$$

(b) Show that for any $\boldsymbol{y} \in [0,1]^X$ and $i, j \in X$, $f(\boldsymbol{y} + t(\boldsymbol{e}_i - \boldsymbol{e}_j))$ is convex with respect to t, where $\boldsymbol{e}_i$ is the unit vector with the ith component equal to 1.

(c) Design a pipage rounding algorithm with the potential function f over the feasible domain

$$\left\{ \boldsymbol{y} \in [0,1]^X \;\middle|\; \boldsymbol{y} \geq 0, \sum_{j \in S} y_j \leq r(S), \text{ for all } S \subseteq X \right\},$$

where $r : 2^X \to \mathbb{N}$ is a polymatroid function with $r(\{i\}) = 1$ for all $i \in X$.

7.13 Consider a graph $G = (V, E)$ and a function $f : 2^V \to \mathbb{N}$. Show that f is weakly supmodular if f satisfies the following conditions:

(i) $f(V) = 0$;

(ii) For every subset S of V, $f(S) = f(V - S)$; and

(iii) For any two disjoint subsets A and B of V, $f(A \cup B) \leq \max\{f(A), f(B)\}$.

7.14 We say a function $f : 2^V \to \mathbb{Z}$ is *strongly submodular* if $f(V) = 0$ and for every two subsets A and B of V,

$$f(A) + f(B) \geq f(A \setminus B) + f(B \setminus A)$$

and

$$f(A) + f(B) \geq f(A \cup B) + f(A \cap B).$$

For any subset $S \subseteq V$, let $\delta_G(S)$ denote the set of edges with exactly one endpoint in S. Show that for any graph $G = (V, E)$, the function $f(S) = |\delta_G(S)|$ is strongly submodular.

7.15 Consider a graph $G = (V, E)$ with a weakly supmodular function $f : 2^V \to \mathbb{Z}$. Show that for any subgraph F of G, $f(S) - |\delta_F(S)|$ is still weakly supmodular.

7.16 Consider a graph $G = (V, E)$ and the following LP:

$$\begin{array}{lll} \text{minimize} & \sum_{e \in E} c_e x_e & \\ \text{subject to} & \sum_{e \in \delta_G(S)} x_e \geq k, & S \subseteq V, \\ & 0 \leq x_e \leq 1, & e \in E. \end{array}$$

Show that the constraint matrix of this LP has rank $|E|$.

7.17 Consider the following problem:

> GENERALIZED STEINER NETWORK: Given a graph $G = (V, E)$ with a positive edge cost function $c : E \to \mathbb{Z}^+$, and a subset P of V, find a minimum-cost k-edge-connected subgraph containing P.

Use the iterated rounding technique to construct a 3-approximation for this problem.

7.18 Improve Lemma 7.21 by showing that every basic feasible solution of the linear program (7.23) has a component whose value is at least 1/2.

7.19 Show that the following algorithm is a 2-approximation for MAX-SAT:

Input: A CNF formula F over variables $x_1, \ldots, x_n$.
(1) **For** $i \leftarrow 1$ to n **do**
  assign $x_i \leftarrow 1$ with probability $1/2$.
 {Let Z_F be the number of clauses satisfied by this assignment.}
(2) **For** $i \leftarrow 1$ to n **do**
  if $E[Z_F \mid x_i = 1] \geq E[Z_F \mid x_i = 0]$ **then** $x_i \leftarrow 1$; $F \leftarrow F|_{x_i=1}$
  else $x_i \leftarrow 0$; $F \leftarrow F|_{x_i=0}$.

7.20 Show that the following algorithm is a $(4/3)$-approximation for MAX-SAT:

Input: A CNF formula F over variables $x_1, \ldots, x_n$.
(1) Construct the LP-relaxation of F, and find its optimal solution y^*.
 {Let opt_{LP} denote the optimal objective function value of this LP.}
(2) **For** $i \leftarrow 1$ to n **do**
  assign $x_i \leftarrow 1$ with probability $1/4 + y^*/2$.
 {Let Z_F be the number of clauses satisfied by this assignment.}
(3) **For** $i \leftarrow 1$ **to** n **do**
  if $E[Z_F \mid x_i = 1] \geq opt_{LP} \cdot (1 - 1/e)$ **then** $x_i \leftarrow 1$; $F \leftarrow F|_{x_i=1}$
  else $x_i \leftarrow 0$; $F \leftarrow F|_{x_i=0}$.

7.21 Extend Algorithm 7.H to get an $(e/(e-1))$-approximation for the following problem:

MAXIMUM-WEIGHT SATISFIABILITY (MAX-WSAT): Given a CNF formula with nonnegative weight on clauses, find a Boolean assignment to its variables that maximizes the total weight of true clauses.

7.22 Show that if two random variables $X \sim \exp(\mu)$ and $Y \sim \exp(\lambda)$ are independent, then for $0 \le \alpha \le \beta$,

$$\Pr[\alpha X < Y < \beta X] = \mu\Big(\frac{1}{\mu + \lambda\alpha} - \frac{1}{\mu + \lambda\beta}\Big).$$

7.23 Suppose that in the Geometric Rounding scheme, we choose $\boldsymbol{u}$ uniformly from the simplex $P = \langle v_1, \ldots, v_{n+1} \rangle$, and round $\boldsymbol{x}$ to $\widehat{\boldsymbol{x}}$ and $\boldsymbol{y}$ to $\widehat{\boldsymbol{y}}$ by this method. Show that

$$E[d(\widehat{\boldsymbol{x}}, \widehat{\boldsymbol{y}})] \le 2 \cdot d(\boldsymbol{x}, \boldsymbol{y}),$$

where $d(\boldsymbol{x}, \boldsymbol{y})$ is the Euclidean distance between $\boldsymbol{x}$ and $\boldsymbol{y}$.

7.24 Consider the following problem:

MINIMUM FEASIBLE CUT: Given a graph $G = (V, E)$ with edge weight $c : E \to \mathbb{R}^+$, a vertex $s \in V$, and a set M of pairs of vertices in G, find a subset of V, with the minimum-weight edge cut, that contains s but does not contain any pair in M.

This problem can be formulated as the following ILP:

$$\begin{array}{lll} \text{minimize} & \sum_{e \in E} c_e x_e & \\ \text{subject to} & x_e \ge y_u - y_v, & e = \{u, v\} \in E, \\ & x_e \ge y_v - y_u, & e = \{u, v\} \in E, \\ & y_u + y_v \le 1, & \{u, v\} \in M, \\ & y_s = 1, & \\ & y_u, x_e \in \{0, 1\}, & u \in V,\ e \in E. \end{array}$$

Let $(\boldsymbol{x}, \boldsymbol{y})$ be an optimal solution of the LP relaxation of the above ILP, and opt_{LP} the corresponding optimal objective function value. Choose a number U uniformly from $[0, 1]$, and round each y_u to 1 if $y_u \ge U$, and to 0 if $y_u < U$. Let $\widehat{\boldsymbol{y}}$ denote the resulting $\boldsymbol{y}$, and set $\widehat{x}_e = |\widehat{y}_u - \widehat{y}_v|$ if $e = \{u, v\}$. Show that

$$opt_{\text{LP}} \le E\Big[\sum_{e \in E} c_e \widehat{x}_e\Big] \le 2 \cdot opt_{\text{LP}}.$$

7.25 Consider the following problem:

MIN-SAT: Given a CNF Boolean formula F with weighted clauses $C_1, \ldots, C_m$ over variables $x_1, \ldots, x_n$, find an assignment to the variables that minimizes the total weight of satisfied clauses.

This problem can be formulated as the following ILP:

$$\begin{aligned} \text{minimize} \quad & \sum_{j=1}^{m} w_j z_j \\ \text{subject to} \quad & z_j \geq y_i, && x_i \in C_j, \\ & z_j \geq 1 - y_i, && \bar{x}_i \in C_j, \\ & y_i, z_j \in \{0,1\}, && 1 \leq i \leq n,\ 1 \leq j \leq m. \end{aligned}$$

Let $(\boldsymbol{y}, \boldsymbol{z})$ be an optimal solution of the LP-relaxation of the above ILP, and opt_{LP} the corresponding optimal objective function value. Split set $\{1, 2, \ldots, n\}$ into two sets A and B randomly, with probability 1/2 of assigning each i, $1 \leq i \leq n$, to set A. Choose U uniformly from $[0, 1]$. For each $i \in A$, set $\widehat{x}_i = 1$ if $y_i > U$, and 0 otherwise, and for each $y_i \in B$, set $\widehat{x}_i = 1$ if $y_i > 1 - U$, and 0 otherwise. For each $j = 1, 2, \ldots, m$, set

$$\widehat{z}_j = \max\Big\{ \max_{x_i \in C_j} \widehat{y}_i,\ \max_{\bar{x}_i \in C_j} (1 - \widehat{y}_i) \Big\}.$$

Show that

$$E\Big[\sum_{j=1}^{m} w_j \widehat{z}_j\Big] \leq 2\Big(1 - \frac{1}{2^k}\Big) \cdot opt_{\mathrm{LP}}.$$

Historical Notes

The simplex method for linear programming was first proposed by Dantzig in 1947 [Dantzig, 1951, 1963]. Charnes [1952] gave the first method, called the perturbation method, which is equivalent to the lexicographical ordering method, to deal with degeneracy in linear programming. Bland [1977] found another rule to overcome the degeneracy problem. Klee and Minty [1972] presented an example showing that the simplex method does not run in polynomial time in the worst case. Khachiyan [1979] found the first polynomial-time solution, the ellipsoid method, for linear programming, with the worst-case running time $O(n^6)$. Karmarkar [1984] discovered the interior-point method for linear programming, which runs in time $O(n^3)$.

The application of linear programming in combinatorial optimization began in the early 1950s. However, its application to approximation algorithms came only after 1970. The works of Lovász [1975], Chvátal [1979], and Wolsey [1980] were pioneering in this direction.

Bellare et al. [1995] showed that MIN-VC does not have a polynomial-time ρ-approximation for $\rho < 16/15$ unless **P** = **NP**. So far, no polynomial-time ρ-approximation, with $\rho < 2$, has been found for MIN-VC. A survey on MIN-VC and GC can be found in Hochbaum [1997a]. The 2-approximation for MIN-2SAT of Section 7.3 is due to Gusfield and Pitt [1992]. The 2-approximation for SCHEDULE-UPM of the same section was given by Lenstra et al. [1990].

The pipage rounding technique was proposed by Ageev and Sviridenko [2004]. Gandhi et al. [2006] applied this technique to dependent rounding. With pipage rounding, Calinescu et al. [2007] studied the maximization of monotone submodular functions subject to matroid constraints. Exercises 7.8–7.10 are from Ageev and Sviridenko [2004]. The iterated rounding scheme was proposed by Jain [2001] and was later improved by Gabow and Gallagher [2008] and Gabow et al. [2009]. It has found a lot of applications [Fleischer et al., 2001; Cheriyan et al., 2006; Chen, 2007; Melkonian and Tardos, 2004]. Exercise 7.11 is from Wolsey [1982b], Exercise 7.12 is from Calinescu et al. [2007], and Exercises 7.13 and 7.15 are from Goemans, Goldberg et al. [1994]. For the improvement over Lemma 7.21 (Exercise 7.18), see Jain [2001]. It is known that MAX-SAT has no PTAS unless **P** = **NP** (see Chapter 10). Its approximation has been studied extensively [Johnson, 1974; Yannakakis, 1994; Goemans and Williamson, 1994; Karloff and Zwick, 1997]. Exercise 7.19 is from Johnson [1974], and Exercise 7.20 is from Goemans and Williamson [1994]. The techniques of dependent randomized rounding were initiated by Bertsimas et al. [1999]. They also proposed the vector rounding scheme in an earlier version of the paper. Its generalization, the geometric rounding scheme, can be found in Ge, Ye, and Zhang [2010] and Ge, He et al. [2010]. Exercise 7.23 is from Ge, Ye, and Zhang [2010]. Exercises 7.24 and 7.25 are from Bertsimas et al. [1999].

8
Primal-Dual Schema and Local Ratio

We believe, in fact, that the one act of respect has little force
unless matched by the other—in balance with it.
The acting out of that dual respect I would
name as precisely the source of our power.

— Barbara Deming

Based on the duality theory of linear programming, a new approximation technique, called the *primal-dual* schema, has been developed. With this technique, we do not need to compute the optimal solution of the relaxed linear program in order to get an approximate solution of the integer program. Thus, we can reduce the running time of many linear programming–based approximation algorithms from $O(n^3)$ to at most $O(n^2)$. Moreover, this method can actually be formulated in an equivalent form, called the *local ratio* method, which does not require the knowledge of the theory of linear programming. In this chapter, we study these two techniques and their relationship.

8.1 Duality Theory and Primal-Dual Schema

One of the most important and intriguing elements of linear programming is the duality theory. Consider a linear program of the standard form

$$\begin{array}{ll} \text{minimize} & \boldsymbol{c}\boldsymbol{x} \\ \text{subject to} & \boldsymbol{A}\boldsymbol{x} = \boldsymbol{b}, \\ & \boldsymbol{x} \geq \boldsymbol{0}, \end{array} \tag{8.1}$$

where A is an $m \times n$ matrix over reals, $\boldsymbol{c}$ an n-dimensional row vector, $\boldsymbol{x}$ an n-dimensional column vector, and $\boldsymbol{b}$ an m-dimensional column vector. We can define a new linear program

$$\begin{array}{ll} \text{maximize} & \boldsymbol{y}\boldsymbol{b}, \\ \text{subject to} & \boldsymbol{y}\boldsymbol{A} \leq \boldsymbol{c}, \end{array} \tag{8.2}$$

where $\boldsymbol{y}$ is an m-dimensional row vector.[1] This linear program (8.2) is called the *dual* linear program of the *primal* linear program (8.1). These two linear programs have a very interesting relationship.

Theorem 8.1 *Suppose $\boldsymbol{x}$ and $\boldsymbol{y}$ are feasible solutions of* (8.1) *and* (8.2), *respectively. Then $\boldsymbol{c}\boldsymbol{x} \geq \boldsymbol{y}\boldsymbol{b}$.*

Proof. Since $\boldsymbol{x}$ and $\boldsymbol{y}$ satisfy the constraints of (8.1) and (8.2), respectively, we have $\boldsymbol{c}\boldsymbol{x} \geq (\boldsymbol{y}\boldsymbol{A})\boldsymbol{x} = \boldsymbol{y}\boldsymbol{b}$. □

Corollary 8.2 *The linear programs* (8.1) *and* (8.2) *satisfy one of the following conditions:*

(1) Neither (8.1) *nor* (8.2) *has a feasible solution.*

(2) The linear program (8.1) *has a feasible solution but has no optimal solutions, and the dual linear program* (8.2) *has no feasible solutions.*

(3) The linear program (8.1) *has no feasible solutions, and its dual linear program* (8.2) *has a feasible solution but has no optimal solutions.*

(4) Both the linear program (8.1) *and its dual linear program* (8.2) *have an optimal solution.*

Proof. From Theorem 8.1, if either (8.1) or (8.2) has unbounded solutions, then the other linear program cannot have a feasible solution. Thus, if none of cases (1), (2), or (3) is satisfied, then both (8.1) and (8.2) have bounded solutions and, hence, have optimal solutions. □

From the proof of Theorem 8.1, it is easy to see that, for two feasible solutions $\boldsymbol{x}$ and $\boldsymbol{y}$ of linear programs (8.1) and (8.2), respectively, $\boldsymbol{c}\boldsymbol{x} = \boldsymbol{y}\boldsymbol{b}$ if and only if

$$(\boldsymbol{c} - \boldsymbol{y}\boldsymbol{A})\boldsymbol{x} = \boldsymbol{0}.$$

The above equation is called the *complementary slackness condition*. This condition can be used to verify whether $\boldsymbol{x}$ and $\boldsymbol{y}$ are optimal solutions.

[1] Note that we write, for convenience, $\boldsymbol{b}$ and $\boldsymbol{x}$ as column vectors, while $\boldsymbol{c}$ and $\boldsymbol{y}$ are row vectors.

Theorem 8.3 *(a) Suppose $\boldsymbol{x}$ and $\boldsymbol{y}$ are feasible solutions of the primal and dual linear programs* (8.1) *and* (8.2)*, respectively. If $(\boldsymbol{c} - \boldsymbol{yA})\boldsymbol{x} = \boldsymbol{0}$, then $\boldsymbol{x}$ and $\boldsymbol{y}$ are optimal solutions of* (8.1) *and* (8.2)*, respectively.*

(b) Suppose $\boldsymbol{x}^*$ and $\boldsymbol{y}^*$ are optimal solutions of the primal and dual linear programs (8.1) *and* (8.2)*, respectively. Then $\boldsymbol{cx}^* = \boldsymbol{y}^*\boldsymbol{b}$.*

Proof. Part (a) follows immediately from Theorem 8.1.

For part (b), it suffices to show that if (8.1) and (8.2) have optimal solutions, then there exist feasible solutions $\boldsymbol{x}$ and $\boldsymbol{y}$ for (8.1) and (8.2), respectively, such that $(\boldsymbol{c}-\boldsymbol{yA})\boldsymbol{x} = \boldsymbol{0}$. From Theorem 7.10, we know that if (8.1) has an optimal solution, then it has a feasible basis J such that $\boldsymbol{c} - \boldsymbol{c}_J\boldsymbol{A}_J^{-1}\boldsymbol{A} \geq \boldsymbol{0}$. Suppose $\boldsymbol{x}$ is the basic feasible solution of (8.1) associated with basis J and $\boldsymbol{y} = \boldsymbol{c}_J\boldsymbol{A}_J^{-1}$. Then $\boldsymbol{c} \geq \boldsymbol{yA}$, and so $\boldsymbol{y}$ is a feasible solution of (8.2). In addition, we have

$$(\boldsymbol{c} - \boldsymbol{yA})\boldsymbol{x} = \boldsymbol{0}$$

since $\boldsymbol{c}_J - \boldsymbol{yA}_J = \boldsymbol{c}_J - \boldsymbol{c}_J\boldsymbol{A}_J^{-1}\boldsymbol{A}_J = \boldsymbol{0}$, and $\boldsymbol{x}_{\bar{J}} = \boldsymbol{0}$. □

We notice that the primal linear program (8.1) and its dual (8.2) are of different forms. In general, the primal linear program does not have to be in standard form. The following is such a pair of primal and dual linear programs of the symmetric form:

(primal LP)		*(dual LP)*		
minimize	$\boldsymbol{cx}$	maximize	$\boldsymbol{yb}$	(8.3)
subject to	$\boldsymbol{Ax} \geq \boldsymbol{b}$,	subject to	$\boldsymbol{yA} \leq \boldsymbol{c}$,	
	$\boldsymbol{x} \geq \boldsymbol{0}$,		$\boldsymbol{y} \geq \boldsymbol{0}$.	

For this pair of linear programs, Theorem 8.1 still holds, but the complementary slackness condition is changed to

$$(\boldsymbol{c} - \boldsymbol{yA})\boldsymbol{x} + \boldsymbol{y}(\boldsymbol{Ax} - \boldsymbol{b}) = 0;$$

or, equivalently,

$$(\boldsymbol{c} - \boldsymbol{yA})\boldsymbol{x} = 0 = \boldsymbol{y}(\boldsymbol{Ax} - \boldsymbol{b}).$$

In the above, $(\boldsymbol{c} - \boldsymbol{yA})\boldsymbol{x} = 0$ is called the *primal complementary slackness* condition, while $\boldsymbol{y}(\boldsymbol{Ax} - \boldsymbol{b}) = 0$ is called the *dual complementary slackness* condition.

The duality theory of linear programming provides us with a new tool to approach some approximation problems from a different direction. For instance, we mentioned, in Section 7.3, that the 2-approximation for MIN-VC, which is based on maximum matching, cannot be extended immediately to the weighted version MIN-WVC. Nevertheless, with the duality theory, we can look at this approximation from a different angle and get an extension.

Consider the unweighted case of the vertex cover problem MIN-VC. Assume that the input to MIN-VC is a graph $G = (V, E)$, where $V = \{v_1, v_2, \ldots, v_n\}$. We may formulate the problem as the following integer program:

$$\begin{array}{lll} \text{minimize} & x_1 + x_2 + \cdots + x_n & \\ \text{subject to} & x_i + x_j \ge 1, & \{v_i, v_j\} \in E, \\ & x_i \in \{0, 1\}, & i = 1, 2, \ldots, n. \end{array} \tag{8.4}$$

A natural relaxation of the above integer linear program is as follows:

$$\begin{array}{lll} \text{minimize} & x_1 + x_2 + \cdots + x_n & \\ \text{subject to} & x_i + x_j \ge 1, & \{v_i, v_j\} \in E, \\ & x_i \ge 0, & i = 1, 2, \ldots, n. \end{array}$$

Its dual linear program is as follows:

$$\begin{array}{lll} \text{maximize} & \sum_{\{v_i, v_j\} \in E} y_{ij} & \\ \text{subject to} & \sum_{j:\{v_i, v_j\} \in E} y_{ij} \le 1, & i = 1, 2, \ldots, n, \\ & y_{ij} \ge 0, & \{v_i, v_j\} \in E. \end{array} \tag{8.5}$$

Now, consider any 0–1 dual feasible solution $\boldsymbol{y}$ (i.e., a feasible solution to the dual linear program (8.5)). Note that for each vertex v_i, the constraint

$$\sum_{j:\{v_i, v_j\} \in E} y_{ij} \le 1$$

requires that, among all edges incident upon v_i, there is at most one edge $\{v_i, v_j\} \in E$ having $y_{ij} = 1$. This means that the set $Y = \big\{\{v_i, v_j\} \in E \mid y_{ij} = 1\big\}$ forms a matching of the graph G. When Y is a maximal matching, the following assignment is then a primal feasible solution for (8.4):

$$x_i = \begin{cases} 1, & \text{if } \sum_{j:(v_i, v_j) \in E} y_{ij} = 1, \\ 0, & \text{otherwise.} \end{cases}$$

Indeed, this is exactly the 2-approximation for MIN-VC based on maximum matching. Next, we show how to follow this approach to extend this 2-approximation algorithm to the weighted case.

We first formulate the weighted version MIN-WVC into an integer program:

$$\begin{array}{lll} \text{minimize} & c_1 x_1 + c_2 x_2 + \cdots + c_n x_n & \\ \text{subject to} & x_i + x_j \ge 1, & \{v_i, v_j\} \in E, \\ & x_i \in \{0, 1\}, & i = 1, 2, \ldots, n. \end{array}$$

Then we relax it to the following linear program:

$$\begin{array}{lll} \text{minimize} & c_1x_1 + c_2x_2 + \cdots + c_nx_n & \\ \text{subject to} & x_i + x_j \geq 1, & \{v_i, v_j\} \in E, \\ & x_i \geq 0, & i = 1, 2, \ldots, n. \end{array}$$

Its dual linear program is

$$\begin{array}{lll} \text{maximize} & \sum_{\{v_i,v_j\}\in E} y_{ij} & \\ \text{subject to} & \sum_{j:\{v_i,v_j\}\in E} y_{ij} \leq c_i, & i = 1, 2, \ldots, n, \\ & y_{ij} \geq 0, & \{v_i, v_j\} \in E. \end{array} \tag{8.6}$$

In terms of the graph G, this dual linear program may be viewed as a generalized maximum matching problem: Maximize the total value of y_{ij} over all edges $\{v_i, v_j\}$, under the constraint that the total value of all edges incident on a vertex v_i is bounded by c_i. A simple idea of the algorithm, thus, is to repeatedly select an edge $\{v_i, v_j\}$ into the generalized matching, with the value y_{ij} of the edge maximized within the bound $\max\{c_i, c_j\}$. This idea leads to the following 2-approximation algorithm for MIN-WVC.

Algorithm 8.A (*Primal-Dual Approximation Algorithm for* MIN-WVC)

Input: Graph $G = (\{v_1, \ldots, v_n\}, E)$, and vertex weights $\mathbf{c} = (c_1, \ldots, c_n)$.

(1) Construct the dual linear program (8.6) from G and $\mathbf{c}$.

(2) **For** each $\{v_i, v_j\} \in E$ **do** set $y_{ij} \leftarrow 0$.

(3) **While** there exists some $\{v_i, v_k\} \in E$ such that

$$\sum_{j:\{v_i,v_j\}\in E} y_{ij} < c_i \text{ and } \sum_{j:\{v_k,v_j\}\in E} y_{kj} < c_k \textbf{ do}$$

$$y_{ik} \leftarrow y_{ik} + \min\Big\{ c_i - \sum_{j:\{v_i,v_j\}\in E} y_{ij},\ c_k - \sum_{j:\{v_k,v_j\}\in E} y_{kj} \Big\}.$$

(4) **For** $i \leftarrow 1$ **to** n **do**

$$x_i \leftarrow \begin{cases} 1, & \text{if } \sum_{j:\{v_i,v_j\}\in E} y_{ij} = c_i, \\ 0, & \text{otherwise.} \end{cases}$$

■

Theorem 8.4 *Let x^A be the output of* Algorithm 8.A. *Then set $C = \{v_i \mid x_i^A = 1\}$ is a 2-approximation for* MIN-WVC.

Proof. Let *opt* denote the optimal objective value of the input $(G, \boldsymbol{c})$. For each edge $\{v_i, v_j\} \in E$, let $\widehat{y}_{ij}$ denote the final value of y_{ij} in Algorithm 8.A. From step (3), we see that for every edge $\{v_i, v_k\} \in E$, at least one of the endpoints v_i has $\sum_{j:(v_i,v_j)\in E} \widehat{y}_{ij} = c_i$. Hence, every edge $\{v_i, v_k\}$ in E is covered by set C.

To show $\sum_{i=1}^{n} c_i x_i^A \leq 2 \cdot opt$, we note that $\widehat{\boldsymbol{y}} = (\widehat{y}_{ij})_{\{v_i,v_j\}\in E}$ is a dual feasible solution to (8.6), and hence

$$\sum_{\{v_i,v_j\}\in E} \widehat{y}_{ij} \leq opt. \tag{8.7}$$

Note that for each $i = 1, 2, \ldots, n$, $x_i^A = 1$ if and only if $\sum_{j:\{v_i,v_j\}\in E} \widehat{y}_{ij} = c_i$. Thus,

$$\sum_{i=1}^{n} c_i x_i^A = \sum_{x_i^A=1} c_i = \sum_{x_i^A=1} \sum_{j:\{v_i,v_j\}\in E} \widehat{y}_{ij} \leq 2 \sum_{\{v_i,v_j\}\in E} \widehat{y}_{ij} \leq 2 \cdot opt. \qquad \square$$

Now, let us examine more carefully the relationship between $\boldsymbol{x}^A$ and $\widehat{\boldsymbol{y}}$ obtained from Algorithm 8.A. From step (4) we see that for each $i = 1, 2, \ldots, n$,

$$x_i^A\Big(c_i - \sum_{j:\{v_i,v_j\}\in E} \widehat{y}_{ij}\Big) = 0.$$

That is, the primal complementary slackness condition holds. On the other hand, we can see that the dual complementary slackness condition does not necessarily hold. More precisely, for some edges $\{v_i, v_j\} \in E$, we may not have the relationship

$$\widehat{y}_{ij}(x_i^A + x_j^A - 1) = 0.$$

Instead, we only have the following relaxed relationship:

$$\widehat{y}_{ij} > 0 \quad \Longrightarrow \quad 1 \leq x_i^A + x_j^A \leq 2,$$

which allows us to establish the performance ratio 2 for Algorithm 8.A. In other words, we do not actually need the full power of the dual complementary slackness condition to prove that the solution $\boldsymbol{x}^A$ is a good approximation to the original problem MIN-WVC. All we need is that $\widehat{\boldsymbol{y}} = (\widehat{y}_{ij})_{\{v_i,v_j\}\in E}$ is a dual feasible solution of (8.6). This property alone is, by the duality theory, sufficient to imply the bound (8.7), which in turn gives us the constant bound 2 for the performance ratio of Algorithm 8.A. This observation suggests a general idea of designing approximation algorithms based on the duality theory of linear programming. We elaborate in the following.

In the LP-based approximations, we first relax a minimization (or, a maximization) problem Π to a linear program $\Pi_{\rm LP}$. We then solve the linear program $\Pi_{\rm LP}$, and round its optimal solution $opt_{\rm LP}$ to a feasible solution for Π. Note that $opt_{\rm LP}$ is a lower bound (or, respectively, an upper bound) for the optimal solution *opt* of Π, and we often use the difference between $opt_{\rm LP}$ and *opt* to estimate the performance ratio of this approximation. Now, from the duality theory, we know that *every* dual feasible solution provides us a lower bound (or, respectively, an upper bound) for opt_{LP} of Π_{LP} and, hence, also a lower bound (or, respectively, an upper bound) for the optimal solution *opt* of Π. This means that a "reasonably good" dual feasible solution can also be used to establish the performance ratio of approximation. Thus, we do not need to compute the exact value opt_{LP} of the optimal primal solution of Π_{LP}. Instead, we may simply compute a reasonably good dual feasible solution and convert it to a feasible solution of problem Π, and then use the difference between them to estimate the performance ratio. This method is called the *primal-dual schema*.

The advantage of the primal-dual schema is that by avoiding the step of finding the optimal primal solution, we can speed up the computation a lot, as the running time of the software implementations for linear programming tends to be high. In particular, the best-known implementation of the interior point method for linear programming runs in time $O(n^{3.5})$ (even though the theoretical time bound for it is $O(n^3)$). Indeed, for applications to certain types of online problems, computing the optimal solution for the primal LP is impractical, and this speedup is necessary.

The following lemma gives a more precise mathematical interpretation of the above idea.

Lemma 8.5 *Let Π be a minimization integer program and Π_{LP} its LP-relaxation. Suppose a primal (integer) feasible solution $\boldsymbol{x}$ of Π and a dual feasible solution $\boldsymbol{y}$ of Π_{LP} satisfy the following conditions:*

(i) (Relaxed primal condition) $\dfrac{\boldsymbol{cx}}{r_1} \le \boldsymbol{yAx} \le \boldsymbol{cx}$; *and*

(ii) (Relaxed dual condition) $\boldsymbol{yb} \le \boldsymbol{yAx} \le r_2\boldsymbol{yb}$.

Then $\boldsymbol{cx} \le (r_1r_2)\boldsymbol{yb}$; *that is,* $\boldsymbol{x}$ *is an* (r_1r_2)*-approximation.*

Proof. $\boldsymbol{cx} \le r_1\boldsymbol{yAx} \le (r_1r_2)\boldsymbol{yb}$. □

For instance, for the problem MIN-WVC, the primal complementary slackness condition implies $r_1 = 1$, and the relation (8.7) gives us the bound $r_2 = 2$, and so Algorithm 8.A is a 2-approximation for MIN-WVC. In the next two sections, we study the application of the primal-dual schema to two specific problems.

8.2 General Cover

Recall the problem GENERAL COVER (GC) defined in Chapter 2, which can be formulated as the following integer linear program:

$$\begin{aligned} &\text{minimize} && \boldsymbol{cx} \\ &\text{subject to} && \boldsymbol{Ax} \ge \boldsymbol{b}, \\ &&& \boldsymbol{x} \in \{0,1\}^n, \end{aligned} \tag{8.8}$$

where $\boldsymbol{A}$ is an $m \times n$ matrix over $\mathbb{N}$, $\boldsymbol{c}$ is an n-dimensional row vector over $\mathbb{N}$, and $\boldsymbol{b}$ is an m-dimensional column vector over $\mathbb{N}$. In this section, we consider a subproblem of GC in which all the components of $\boldsymbol{b}$ are equal to 1:

$$\text{GC}_1: \quad \begin{array}{ll} \text{minimize} & \boldsymbol{cx} \\ \text{subject to} & \boldsymbol{Ax} \geq \mathbf{1}_m, \\ & \boldsymbol{x} \in \{0,1\}^n, \end{array} \tag{8.9}$$

where $\boldsymbol{A}$ is an $m \times n$ nonnegative integer matrix, $\boldsymbol{c}$ is a positive integer n-dimensional row vector, and $\mathbf{1}_m$ is the m-dimensional column vector with all of its components having value 1.[2]

Suppose $\boldsymbol{A} = (a_{ij})_{1 \leq i \leq m, 1 \leq j \leq n}$. Let f be the maximum of the row sum of matrix $\boldsymbol{A}$; that is, $f = \max_{1 \leq i \leq m} \sum_{j=1}^{n} a_{ij}$. We are going to apply the primal-dual schema to get an f-approximation algorithm for GC_1 that runs in time $O(n^2)$.

The following are the primal and dual linear programs of a natural LP-relaxation of the problem GC_1:

$$\begin{array}{llcll} & \textit{(primal LP)} & & & \textit{(dual LP)} \\ \text{minimize} & \boldsymbol{cx} & & \text{maximize} & \boldsymbol{y}\mathbf{1}_m \\ \text{subject to} & \boldsymbol{Ax} \geq \mathbf{1}_m, & & \text{subject to} & \boldsymbol{yA} \leq \boldsymbol{c}, \\ & \boldsymbol{x} \geq \mathbf{0}, & & & \boldsymbol{y} \geq \mathbf{0}, \end{array} \tag{8.10}$$

where $\boldsymbol{x}$ is an n-dimensional column vector, and $\boldsymbol{y}$ is an m-dimensional row vector.

An idea of approximation for GC_1 based on the dual LP above is, similar to that of Algorithm 8.A, to increase the values of y_i as much as possible, without violating the constraint $\boldsymbol{yA} \leq \boldsymbol{c}$. However, as this constraint $\boldsymbol{yA} \leq \boldsymbol{c}$ is more complicated than that in (8.6), it is not clear how we should increase the values of variables y_i in each stage. Let us study this issue more carefully through the complementary slackness condition.

The complementary slackness condition between the two linear programs for GC_1 is

$$(\boldsymbol{c} - \boldsymbol{yA})\boldsymbol{x} = 0 = \boldsymbol{y}(\boldsymbol{Ax} - \mathbf{1}_m).$$

Suppose that $\boldsymbol{x}$ is a primal feasible solution and $\boldsymbol{y}$ is a dual feasible solution. Then, by the constraints $\boldsymbol{yA} \leq \boldsymbol{c}$ and $\boldsymbol{Ax} \geq \mathbf{1}_m$, the above complementary slackness condition can be divided into the following subconditions:

(C_P) For each $j = 1, 2, \ldots, n$, if $\sum_{i=1}^{m} a_{ij} y_i < c_j$, then $x_j = 0$; and

(C_D) For each $i = 1, 2, \ldots, m$, if $\sum_{j=1}^{n} a_{ij} x_j > 1$, then $y_i = 0$.

---

[2]Note that the requirement of $\mathbf{c}$ being a positive vector is not too restrictive: If a component of $\mathbf{c}$, say c_j, is equal to 0, then we may set $x_j = 1$ and remove, for each i, the ith row of A if $a_{ij} \geq 1$ to get an equivalent LP with $\mathbf{c} \geq \mathbf{1}_n$.

Our goal is to keep the difference $\boldsymbol{cx} - \boldsymbol{y}\mathbf{1}_m$ between the objective function values of the two linear programs as small as possible. Note that

$$\boldsymbol{cx} - \boldsymbol{y}\mathbf{1}_m = (\boldsymbol{c} - \boldsymbol{yA})\boldsymbol{x} + \boldsymbol{y}(\boldsymbol{Ax} - \mathbf{1}_m). \tag{8.11}$$

Thus, the more conditions in ($C_{\rm P}$) and ($C_{\rm D}$) above are satisfied, the closer the values $\boldsymbol{cx}$ and $\boldsymbol{y}\mathbf{1}_m$ are. On the other hand, we cannot expect all subconditions to be satisfied when we round $\boldsymbol{x}$ and $\boldsymbol{y}$ to integer solutions.

For instance, we may, following the approach of Algorithm 8.A for problem MIN-WVC, try to satisfy all the primal subconditions in ($C_{\rm P}$), and simply define the approximate solution $\boldsymbol{x}^A$ from $\boldsymbol{y}$ as follows:

$$x_j^A = \begin{cases} 1, & \text{if } \sum_{i=1}^m a_{ij} y_i = c_j, \\ 0, & \text{if } \sum_{i=1}^m a_{ij} y_i < c_j. \end{cases} \tag{8.12}$$

The problem with this approach is that, while this assignment for $\boldsymbol{x}^A$ would satisfy the primal complementary slackness condition, it may not be primal feasible itself. In this case, we need to go back to modify $\boldsymbol{y}$ to make the corresponding $\boldsymbol{x}^A$ primal feasible. Thus, it suggests the following general structure of the algorithm: We start with an initial dual feasible $\boldsymbol{y}$ and iteratively modify it until the corresponding $\boldsymbol{x}^A$ (as defined by (8.12)) becomes primal feasible.

Now, under this framework, how do we proceed in each iteration? We observe that, in each iteration, we want to make $\boldsymbol{x}^A$ closer to a feasible solution for the primal problem. To do so, we need to increase the number of components of $\boldsymbol{x}^A$ that have value 1 (since $\boldsymbol{A}$ is nonnegative); or, equivalently, from (8.12), we need to modify $\boldsymbol{y}$ to increase the number of j's satisfying $\sum_{i=1}^m a_{ij} y_i = c_j$. This in turn amounts to increasing some values of y_i. For which indices i and for what amount should we increase the values of y_i? Let us examine the complementary slackness condition (8.11) again.

First, we note that if $\boldsymbol{x}^A$ does not satisfy $\boldsymbol{Ax}^A \geq \mathbf{1}_m$, then the set $I = \{i \mid 1 \leq i \leq m, \sum_{j=1}^n a_{ij} x_j^A = 0\}$ is nonempty. For an index $i \notin I$, increasing y_i could increase the second term $\boldsymbol{y}(\boldsymbol{Ax} - \mathbf{1}_m)$ of (8.11), and hence increase the gap between $\boldsymbol{cx}$ and $\boldsymbol{y}\mathbf{1}_m$. This means that we should not increase these y_i's. On the other hand, for an index $i \in I$, increasing y_i will actually decrease the gap between $\boldsymbol{cx}$ and $\boldsymbol{y}\mathbf{1}_m$. So we should try to increase y_i's only for those $i \in I$. In addition, we note that we need to keep the new $\boldsymbol{y}$ dual feasible. That is, the new values of y_i must still satisfy $\sum_{i=1}^m a_{ij} y_i \leq c_j$ for all j. This condition suggests that we should increase the values of y_i, for all $i \in I$, simultaneously, until one of the sum $\sum_{i=1}^m a_{ij} y_i$ reaches the value c_j.

The above analysis yields the following algorithm.

Algorithm 8.B (*Primal-Dual Schema for* GC_1)

Input: An $m \times n$ nonnegative integer matrix $\boldsymbol{A}$ and $\boldsymbol{c} \in (\mathbb{Z}^+)^n$.

(1) Set $\boldsymbol{x}^0 \leftarrow \mathbf{0}$; $\boldsymbol{y}^0 \leftarrow \mathbf{0}$; $k \leftarrow 0$.

(2) **While** $\boldsymbol{x}^k$ is not primal feasible **do**

$J_k \leftarrow \{j \mid 1 \le j \le n,\ x_j^k = 0\}$;

$I_k \leftarrow \{i \mid 1 \le i \le m,\ \sum_{j=1}^n a_{ij} x_j^k \le 0\}$;

Choose $r \in J_k$ such that

$$\frac{c_r - \sum_{i=1}^m a_{ir} y_i^k}{\sum_{i \in I_k} a_{ir}} = \alpha = \min_{j \in J_k} \frac{c_j - \sum_{i=1}^m a_{ij} y_i^k}{\sum_{i \in I_k} a_{ij}};$$

For $j \leftarrow 1$ **to** n **do**

if $j = r$ **then** $x_j^{k+1} \leftarrow 1$ **else** $x_j^{k+1} \leftarrow x_j^k$;

For $i \leftarrow 1$ **to** m **do**

if $i \in I_k$ **then** $y_i^{k+1} \leftarrow y_i^k + \alpha$ **else** $y_i^{k+1} \leftarrow y_i^k$;

$k \leftarrow k + 1$.

(3) Output $\boldsymbol{x}^A = \boldsymbol{x}^k$. ■

Algorithm 8.B runs in time $O(n(m+n))$ because the algorithm runs at most n iterations and each iteration takes time $O(m+n)$ (note that the value $c_j - \sum_{i=1}^m a_{ij} y_i^k$ can be updated from that of the $(k-1)$st iteration in time $O(1)$).

Next, we show that it has the performance ratio f.

Lemma 8.6 *During the execution of* Algorithm 8.B, *the following properties hold for all* $k \ge 0$*:*

(a) $\boldsymbol{y}^k$ *is dual feasible.*

(b) $(\boldsymbol{c} - \boldsymbol{y}^k A)\boldsymbol{x}^k = 0$.

(c) $\boldsymbol{y}^k A \boldsymbol{x}^k \le f \boldsymbol{y}^k \mathbf{1}_m$, *where* $f = \max_{1 \le i \le m} \sum_{j=1}^n a_{ij}$.

Proof. We prove properties (a) and (b) by induction on k. It is clear that conditions (a) and (b) are true with respect to the initial values $\boldsymbol{x}^0 = \boldsymbol{y}^0 = \mathbf{0}$. Next, suppose they hold true for some $k \ge 0$ and consider the case of $k+1$.

For condition (a), we note that, from condition (a) of the induction hypothesis, $\boldsymbol{y}^k$ is dual feasible, and so α must be nonnegative, and so $y_i^{k+1} \ge y_i^k \ge 0$ for all $i = 1, 2, \ldots, m$. First, consider the case of $j \notin J_k$. From condition (b) of the induction hypothesis, we know that if $j \notin J_k$, then $c_j - \sum_{i=1}^m a_{ij} y_i^k = 0$. Furthermore, for each $i \in I_k$, we have $\sum_{j=1}^n a_{ij} x_j^k = 0$, and so $a_{ij} = 0$ for each $j \notin J_k$. It follows that

$$c_j - \sum_{i=1}^m a_{ij} y_i^{k+1} = c_j - \sum_{i \notin I_k} a_{ij} y_i^{k+1} = c_j - \sum_{i \notin I_k} a_{ij} y_i^k = 0.$$

Next, for the case $j \in J_k$, we know, by the choice of α, that

$$\alpha \sum_{i \in I_k} a_{ij} \le c_j - \sum_{i=1}^m a_{ij} y_i^k.$$

Thus, for $j \in J_k$,

$$c_j - \sum_{i=1}^{m} a_{ij} y_i^{k+1} = c_j - \sum_{i=1}^{m} a_{ij} y_i^k - \alpha \sum_{i \in I_k} a_{ij} \geq 0.$$

So, $\boldsymbol{y}^{k+1}$ is dual feasible.

For condition (b), consider an index $j \in \{1, 2, \ldots, n\}$ with $\sum_{i=1}^{m} a_{ij} y_i^{k+1} < c_j$. Since $\boldsymbol{y}^k \leq \boldsymbol{y}^{k+1}$, we know that $\sum_{i=1}^{m} a_{ij} y_i^k < c_j$. By the induction hypothesis, $x_j^k = 0$. In addition, we have, from the choice of r,

$$\sum_{i=1}^{m} a_{ir} y_i^{k+1} = \sum_{i=1}^{m} a_{ir} y_i^k + \sum_{i \in I_k} a_{ir} \alpha = \sum_{i=1}^{m} a_{ir} y_i^k + c_r - \sum_{i=1}^{m} a_{ir} y_i^k = c_r.$$

Therefore, $j \neq r$, and we must have $x_j^{k+1} = x_j^k = 0$.

Finally, for condition (c), we note that

$$\boldsymbol{y}^k \boldsymbol{A} \boldsymbol{x}^k = \sum_{i=1}^{m} y_i^k \left(\sum_{j=1}^{n} a_{ij} x_j^k \right) \leq \sum_{i=1}^{m} y_i^k \left(\sum_{j=1}^{n} a_{ij} \right) \leq f \sum_{i=1}^{m} y_i^k = f \boldsymbol{y}^k \mathbf{1}_m,$$

and the lemma is proven. □

Theorem 8.7 *Let opt be the optimal value of the problem* GC_1. *The solution* $\boldsymbol{x}^A$ *produced by* Algorithm 8.B *satisfies*

$$\boldsymbol{c}\boldsymbol{x}^A \leq f \cdot opt,$$

where $f = \max_{1 \leq i \leq m} \sum_{j=1}^{n} a_{ij}$.

Proof. By Lemma 8.6 and Theorem 8.1, we have

$$\boldsymbol{c}\boldsymbol{x}^A = \boldsymbol{y}^k \boldsymbol{A} \boldsymbol{x}^k \leq f \cdot \boldsymbol{y}^k \mathbf{1}_m \leq f \cdot opt,$$

where k is the final value of the variable k in Algorithm 8.B. □

From the proof of Lemma 8.6, we see that property (c) of Lemma 8.6 holds for every dual feasible solution $\boldsymbol{y}^k$. Therefore, we have $\boldsymbol{c}\boldsymbol{x} \leq f \cdot opt$, as long as a primal feasible solution $\boldsymbol{x}$ and a dual feasible solution $\boldsymbol{y}$ satisfy the primal complementary slackness condition $(\boldsymbol{c} - \boldsymbol{y}\boldsymbol{A})\boldsymbol{x} = 0$. This observation shows that the following variation of Algorithm 8.B has the same performance ratio f as Algorithm 8.B.

Algorithm 8.C (*Second Primal-Dual Schema for* GC_1)

Input: An $m \times n$ nonnegative integer matrix $\boldsymbol{A}$ and $\boldsymbol{c} \in (\mathbb{Z}^+)^n$.

(1) Set $\boldsymbol{x}^0 \leftarrow \mathbf{0}$; $\boldsymbol{y}^0 \leftarrow \mathbf{0}$; $k \leftarrow 0$.

(2) **While** $\boldsymbol{x}^k$ is not primal feasible **do**

Select an index i' such that $\sum_{j=1}^{n} a_{i'j} x_j^k = 0$;

$J_k \leftarrow \{j \mid x_j^k = 0 \text{ and } a_{i'j} > 0\}$;

Choose $r \in J_k$ such that

$$\frac{c_r - \sum_{i=1}^{m} a_{ir} y_i^k}{a_{i'r}} = \alpha = \min_{j \in J_k} \frac{c_j - \sum_{i=1}^{m} a_{ij} y_i^k}{a_{i'j}};$$

For $j \leftarrow 1$ **to** n **do**

  if $j = r$ **then** $x_j^{k+1} \leftarrow 1$ **else** $x_j^{k+1} \leftarrow x_j^k$;

For $i \leftarrow 1$ **to** m **do**

  if $i \in I_k$ **then** $y_i^{k+1} \leftarrow y_i^k + \alpha$ **else** $y_i^{k+1} \leftarrow y_i^k$;

$k \leftarrow k + 1$.

(3) Output $\boldsymbol{x}^A = \boldsymbol{x}^k$. ∎

It is interesting to point out that neither Algorithm 8.B nor Algorithm 8.C requires solving a linear program. The theory of linear programming is used as an inspiration and as an analysis tool only. It is therefore natural to ask whether we can design such algorithms without the knowledge of linear programming at all. The answer is affirmative. We will introduce an equivalent local ratio method in later sections.

Finally, we remark that, for a single integer program, there are often more than one way to relax it to linear programs. For instance, in Algorithms 8.B and 8.C, we used the primal and dual linear programs obtained from GC_1 by relaxing the condition "$x_j \in \{0, 1\}$" to "$x_j \geq 0$." One might ask why we did not relax it to a stronger condition "$0 \leq x_j \leq 1$." As to be seen below, the reason is that the primal-dual algorithm obtained from the stronger relaxation is actually weaker than Algorithms 8.B and 8.C. To see this, let us consider this relaxation:

$$\begin{array}{ll} \text{minimize} & \boldsymbol{cx} \\ \text{subject to} & \boldsymbol{Ax} \geq \mathbf{1}_m, \\ & \mathbf{0} \leq \boldsymbol{x} \leq \mathbf{1}_n. \end{array}$$

To find a primal-dual algorithm based on this relaxation, we first write this linear program and its dual linear program in the symmetric form of (8.3):

(primal LP)		*(dual LP)*	
minimize	$\boldsymbol{cx}$	maximize	$\boldsymbol{y}\mathbf{1}_m - \boldsymbol{z}\mathbf{1}_n$
subject to	$\boldsymbol{Ax} \geq \mathbf{1}_m$,	subject to	$\boldsymbol{yA} - \boldsymbol{z} \leq \boldsymbol{c}$,
	$-\boldsymbol{x} \geq -\mathbf{1}_n$,		$\boldsymbol{y} \geq \mathbf{0}$, $\boldsymbol{z} \geq \mathbf{0}$,
	$\boldsymbol{x} \geq \mathbf{0}$,		

where $\boldsymbol{y} \in \mathbb{R}^m$, $\boldsymbol{z} \in \mathbb{R}^n$ are row vectors.

Following the analysis of the primal and dual linear programs of (8.10), we can express the difference between the two objective functions as

$$\boldsymbol{cx} - \boldsymbol{y1}_m + \boldsymbol{z1}_n = \boldsymbol{y}(\boldsymbol{Ax} - \mathbf{1}_m) + \boldsymbol{z}(\mathbf{1}_n - \boldsymbol{x}) + (\boldsymbol{c} - \boldsymbol{yA} + \boldsymbol{z})\boldsymbol{x}.$$

Correspondingly, the complementary slackness condition of the new pair of primal and dual linear programs above is

$$\boldsymbol{y}(\boldsymbol{Ax} - \mathbf{1}_m) + \boldsymbol{z}(\mathbf{1}_n - \boldsymbol{x}) + (\boldsymbol{c} - \boldsymbol{yA} + \boldsymbol{z})\boldsymbol{x} = 0.$$

Now, we can follow the approach of Algorithm 8.B to approximate GC_1. Namely, we want to increase the number of components of $\boldsymbol{x}$ to have value 1 and, in the meantime, keep $(\boldsymbol{y}, \boldsymbol{z})$ dual feasible. Notice that when we increase the value of x_j from 0 to 1, we need to change the values of the y_i's and z_j's to satisfy $\sum_{i=1}^{m} a_{ij} y_i - z_j = c_j$. Since increasing the values of the z_j's only means we need to increase more to the values of the y_i's, we can just focus on increasing the y_i's. Thus, the criteria for selecting the components of $\boldsymbol{y}$ to increase are the same as those for Algorithm 8.B. The only difference here is that we need to, if necessary, adjust the values of other z_k's to make sure that $\sum_{i=1}^{n} a_{ik} y_k - z_k$ is no greater than c_k. These observations lead to the following primal-dual algorithm for GC_1:

Algorithm 8.D (*Third Primal-Dual Schema* for GC_1)
Input: An $m \times n$ nonnegative integer matrix $\boldsymbol{A}$ and $\boldsymbol{c} \in (\mathbb{Z}^+)^n$.

(1) Set $\boldsymbol{x}^0 \leftarrow \mathbf{0}$; $\boldsymbol{y}^0 \leftarrow \mathbf{0}$;, $\boldsymbol{z}^0 \leftarrow \mathbf{0}$; $k \leftarrow 0$.

(2) **While** $\boldsymbol{x}^k$ is not prime feasible **do**

$J_k \leftarrow \{j \mid x_j^k = 0\}$;

$I_k \leftarrow \{i \mid \sum_{j=1}^{n} a_{ij} x_j^k = 0\}$;

Choose $r \in J_k$ such that

$$\frac{c_r - \sum_{i=1}^{m} a_{ir} y_i^k}{\sum_{i \in I_k} a_{ir}} = \alpha = \min_{j \in J_k} \frac{c_j - \sum_{i=1}^{m} a_{ij} y_i^k}{\sum_{i \in I_k} a_{ij}};$$

For $i \leftarrow 1$ **to** m **do**

if $i \in I_k$ **then** $y_i^{k+1} \leftarrow y_i^k + \alpha$ **else** $y_i^{k+1} \leftarrow y_i^k$;

For $j \leftarrow 1$ **to** n **do**

if $j = r$ **then** $x_j^{k+1} \leftarrow 1$ **else** $x_j^{k+1} \leftarrow x_j^k$;

$z_j^{k+1} \leftarrow \max\left\{ \sum_{i=1}^{m} a_{ij} y_i^{k+1} - c_j, 0 \right\}$;

$k \leftarrow k + 1$.

(3) Output $\boldsymbol{x}^A = \boldsymbol{x}^k$. ∎

Comparing Algorithm 8.D with Algorithm 8.B, we find that $\boldsymbol{z}$ is redundant. Indeed, in Algorithm 8.D, we did not use z^k in the computation of x^{k+1} and y^{k+1}. So we may as well remove the variables in $\boldsymbol{z}$ from the relaxed LP. Note that $\boldsymbol{z}$ was introduced by the extra constraints $\boldsymbol{x} \le \mathbf{1}_n$, and so removing the variables in $\boldsymbol{z}$ is equivalent to removing the constraints $\boldsymbol{x} \le \mathbf{1}_n$.

Another interesting observation about the removal of $\boldsymbol{z}$ is that after $\boldsymbol{z}$ is removed, the lower bound for the optimal solution of the original integer linear program is actually improved from $\boldsymbol{yb} - \boldsymbol{z}\mathbf{1}_n$ to $\boldsymbol{yb}$.

8.3 Network Design

For many subproblems of GENERAL COVER (called *covering-type* problems), we can often use the primal-dual method to obtain approximations with performance ratios better than f as shown in Theorem 8.7. For instance, consider the following subclass of covering-type problems:

> NETWORK DESIGN: Given a graph $G = (V, E)$ with nonnegative edge costs c_e, for $e \in E$, solve the integer program
>
> $$\begin{aligned} \text{minimize} \quad & \sum_{e \in E} c_e x_e \\ \text{subject to} \quad & \sum_{e \in \delta(S)} x_e \ge f(S), \qquad \emptyset \ne S \subset V, \qquad (8.13) \\ & x_e \in \{0, 1\}, \qquad e \in E, \end{aligned}$$
>
> where $\delta(S)$ is the set of edges between S and $V - S$ (i.e., the *cut* between S and $V - S$), and $f(S)$ is a 0–1 function over 2^V.

The following are two specific instances of the network design problem:

> TREE PARTITION: Given a graph $G = (V, E)$ with nonnegative edge costs c_e, for $e \in E$, and a positive integer k, find the minimum-cost subset of edges that partitions all vertices into trees of at least k vertices.

> STEINER FOREST: Given a graph $G = (V, E)$ with edge costs c_e, for $e \in E$, and m disjoint subsets $P_1, P_2, \ldots, P_m$ of vertices, find a minimum-cost forest F of G such that every set P_i is contained in a connected component of F.

The problem TREE PARTITION can be formulated as the integer program (8.13) with the following $f(S)$:

$$f(S) = \begin{cases} 1, & \text{if } 0 < |S| < k, \\ 0, & \text{otherwise.} \end{cases}$$

STEINER FOREST can be formulated as the integer program (8.13) with the following $f(S)$:

$$f(S) = \begin{cases} 1, & \text{if } (\exists P_i)\,[S \cap P_i \neq \emptyset \neq (V - S) \cap P_i], \\ 0, & \text{otherwise.} \end{cases}$$

In both instances above, the function $f(S)$ satisfies the following *maximality property*: For any two disjoint sets $A, B \subseteq V$,

$$f(A \cup B) \leq \max\{f(A), f(B)\}.$$

In the network design problem, if a vector $\boldsymbol{x} = (x_e)_{e \in E}$ is not a feasible solution, then there must be a nonempty vertex subset $S \subseteq V$ such that

$$\sum_{e \in \delta(S)} x_e < f(S).$$

We call such a set $S \subseteq V$ a *violated set* (with respect to $\boldsymbol{x}$). If, furthermore, no proper nonempty subset T of S satisfies

$$\sum_{e \in \delta(T)} x_e < f(T),$$

then we call S a *minimal* violated set. We denote by *Violate*($\boldsymbol{x}$) the collection of all minimal violated sets with respect to $\boldsymbol{x}$. When a network design problem has the maximality property, the minimal violated sets have a nice characterization.

Lemma 8.8 *Suppose $f(S)$ is a 0–1 function over 2^V with the maximality property. Then, for any $\boldsymbol{x}$, every minimal violated set S is a connected component of graph $G_{\boldsymbol{x}} = (V, \{e \mid x_e = 1\})$.*

Proof. Note that if S is a violated set, then we must have

$$0 = \sum_{e \in \delta(S)} x_e < f(S) = 1.$$

This means that for any edge $e \in \delta(S)$, $x_e = 0$. Thus, S is a union of connected components of the graph $G_{\boldsymbol{x}}$.

If S contains more than one connected component, then, by the maximality property, $f(T) = 1$ for some connected component T in S. Thus,

$$\sum_{e \in \delta(T)} x_e = 0 < f(T) = 1,$$

and T is a violated set. It follows that S is not a minimal violated set. □

The above lemma indicates that for each $\boldsymbol{x}$, the set of all minimal violated sets is easy to compute, and hence suggests the following simplified primal-dual algorithm.

Algorithm 8.E (*Primal-Dual Schema* for NETWORK DESIGN)

Input: A graph $G = (V, E)$ with edge costs c_e, for $e \in E$, and a function $f : 2^V \to \{0, 1\}$ (given implicitly).

(1) $\boldsymbol{x} \leftarrow \mathbf{0}$; **For** every $S \subseteq V$ **do** $y_S \leftarrow 0$.

(2) **While** *Violate*$(\boldsymbol{x}) \neq \emptyset$ **do**

{Increase the values of y_S simultaneously for all minimal violated sets S until some edge e becomes tight.}

Let e^* be the edge that reaches the minimum

$$\alpha = \min_{e \in E, x_e = 0} \frac{c_e - \sum_{S: e \in \delta(S)} y_S}{|\mathit{Violate}(\boldsymbol{x}) \cap \{S \mid e \in \delta(S)\}|};$$

For each $S \in \mathit{Violate}(\boldsymbol{x})$ **do** $y_S \leftarrow y_S + \alpha$;

$x_{e^*} \leftarrow 1$.

(3) **For** each $e \in E$ **do**

let $\boldsymbol{x}'$ be the vector $\boldsymbol{x}$ modified with $x'_e \leftarrow 0$;

if $\boldsymbol{x}'$ is primal feasible **then** $\boldsymbol{x} \leftarrow \boldsymbol{x}'$.

(4) Output $\boldsymbol{x}$. ■

Let us analyze the running time of Algorithm 8.E first. We note that, in general, the network design problem has an exponential number of constraints (with respect to the size of the input graph G). Thus, a straightforward implementation of Algorithm 8.E would take superpolynomial time. However, when the function $f(S)$ has the maximality property, Algorithm 8.E can be implemented to run in polynomial time. To see this, we note that if $f(S)$ has the maximality property, then, by Lemma 8.8, each set $S \in \mathit{Violate}(\boldsymbol{x})$ is a connected component of $G_{\boldsymbol{x}}$. So, in each iteration of Algorithm 8.E, there are only polynomially many minimal violated sets and we can compute them in polynomial time. Moreover, the value of y_S may become nonzero only if S is a minimal violated set. Therefore, in each iteration, there are only polynomially many nonzero terms in the sum $t_e = \sum_{S: e \in \delta(S)} y_S$. From this observation, we can implement steps (1) and (2) of Algorithm 8.E as follows to make it run in polynomial time:

(1) $\boldsymbol{x} \leftarrow \mathbf{0}$; **For** every $e \in E$ **do** $t_e \leftarrow 0$.

(2) **While** *Violate*$(\boldsymbol{x}) \neq \emptyset$ **do**

Let e^* be the edge that reaches the minimum

$$\alpha = \min_{e \in E, x_e = 0} \frac{c_e - t_e}{|\mathit{Violate}(\boldsymbol{x}) \cap \{S \mid e \in \delta(S)\}|};$$

For each $e \in E$ **do**

  for each $S \in \mathit{Violate}(\boldsymbol{x})$ **do**

    if $e \in \delta(S)$ then $t_e \leftarrow t_e + \alpha$;

$x_{e^*} \leftarrow 1$.

Next, we consider the performance ratio of Algorithm 8.E. A function f is *downward monotone* if

$$\emptyset \neq T \subset S \Rightarrow f(S) \leq f(T).$$

Clearly, downward monotonicity implies maximality. We note that the function f defining TREE PARTITION is downward monotone, while that for STEINER FOREST is not.

Theorem 8.9 *Suppose the input function* $f(S)$ *in* Algorithm 8.E *is downward monotone. Then* Algorithm 8.E *is a 2-approximation for the associated network design problem.*

Proof. For any primal value $\boldsymbol{x}$, let $F(\boldsymbol{x}) = \{e \in E \mid x_e = 1\}$, and let F^* denote the set $F(\boldsymbol{x})$ corresponding to the output $\boldsymbol{x}$ of Algorithm 8.E. Note that for each $e \in F^*$, $\sum_{S:e\in\delta(S)} y_S = c_e$. Therefore, we have

$$\sum_{e\in F^*} c_e = \sum_{e\in F^*} \sum_{S:e\in\delta(S)} y_S$$
$$= \sum_{S\subseteq V} \sum_{e\in\delta(S)\cap F^*} y_S = \sum_{S\subseteq V} \deg_{F^*}(S) \cdot y_S,$$

where $\deg_{F^*}(S) = |\delta(S) \cap F^*|$. Now, from Lemma 8.5, it suffices to prove

$$\sum_{S\subseteq V} \deg_{F^*}(S) \cdot y_S \leq 2 \sum_{S\subseteq V} y_S. \tag{8.14}$$

To get (8.14), we note that it is sufficient to show that at each iteration,

$$\sum_{S\in \mathit{Violate}(\boldsymbol{x})} \deg_{F^*}(S) \leq 2 \cdot |\mathit{Violate}(\boldsymbol{x})|. \tag{8.15}$$

To see this, let $\boldsymbol{x}^k$ denote the value of $\boldsymbol{x}$ at the beginning of the kth iteration, and let α_k be the minimum value α found in the kth iteration. Thus, in the kth iteration, we added α_k to y_S for each $S \in \mathit{Violate}(\boldsymbol{x}^k)$. So the right-hand side of (8.14) can be decomposed into

$$2\sum_{S\subseteq V} y_S = 2\sum_{k=1}^{K} \alpha_k \cdot |\mathit{Violate}(\boldsymbol{x}^k)|,$$

assuming Algorithm 8.E halts after K iterations. Moreover, the sum on the left-hand side of (8.14) can also be decomposed into

$$\sum_{S\subseteq V} \deg_{F^*}(S) \cdot y_S = \sum_{k=1}^{K} \sum_{S\in \mathit{Violate}(\boldsymbol{x}^k)} \deg_{F^*}(S) \cdot \alpha_k$$
$$= \sum_{k=1}^{K} \alpha_k \sum_{S\in \mathit{Violate}(\boldsymbol{x}^k)} \deg_{F^*}(S).$$

Thus, to get (8.14), it suffices to show that for each k,

$$\sum_{S\in \mathit{Violate}(\boldsymbol{x}^k)} \deg_{F^*}(S) \leq 2 \cdot |\mathit{Violate}(\boldsymbol{x}^k)|.$$

Now, in order to prove (8.15), construct a graph H with the vertex set $V(H)$ containing all connected components of the graph $G_{\boldsymbol{x}} = (V, F(\boldsymbol{x}))$ and the edge set $E(H) = F^* - F(\boldsymbol{x})$. From step (3) of Algorithm 8.E and the fact that $f(S) \in \{0,1\}$, we know that H is acyclic. Therefore, the number of edges in H equals the number of vertices minus the number of connected components in H. It follows that

$$\sum_{S \in \mathit{Violate}(\boldsymbol{x})} \deg_{F^*}(S) = 2|F^* - F(\boldsymbol{x})| \le 2(|V(H)| - c),$$

where c is the number of connected components in H.

To prove (8.15), we show that each connected component of H contains at most one vertex S such that $f(S) = 0$. For the sake of contradiction, suppose there exist two vertices S_1 and S_2 in a connected component C of H such that $f(S_1) = f(S_2) = 0$. Let e be an edge of H in the path between S_1 and S_2. Then $e \in F^*$ and, by step (3) of Algorithm 8.E, $F^* - \{e\}$ is not feasible. Thus, there exists a set $S \subset V$ such that $e \in \delta(S)$, $f(S) = 1$, and $(F^* - \{e\}) \cap \delta(S) = \emptyset$. Since H is acyclic, the removal of e splits the connected component C into two connected components A and B. Since $(F^* - \{e\}) \cap \delta(S) = \emptyset$, we must have either $A \subseteq S$ or $B \subseteq S$ and, consequently, either $S_1 \subseteq S$ or $S_2 \subseteq S$. However, by the downward monotone property of f, we would have either $f(S_1) = 1$ or $f(S_2) = 1$, which leads to a contradiction.

Since each connected component of H contains at most one vertex S with $f(S) = 0$, all but c many vertices S of H are in *Violate*$(\boldsymbol{x})$. We conclude that $|V(H)| - c \le |\mathit{Violate}(x)|$, and (8.15) is proven. □

Corollary 8.10 Algorithm 8.E *is a 2-approximation for* TREE PARTITION.

A function f over 2^V is said to be *symmetric* if $f(S) = f(V - S)$ for all $S \subset V$. The function f defining the problem STEINER FOREST is symmetric with the maximality property.

Lemma 8.11 *Let f be a 0–1 symmetric function on 2^V with the maximality property. Then $f(A) = f(B) = 0$ implies $f(A \setminus B) = 0$.*

Proof. By the symmetry property of f, $f(V - A) = 0$. So, by the maximality property, $f((V - A) \cup B) = 0$. Now the lemma follows from the fact of $V - (A \setminus B) = (V - A) \cup B$. □

Theorem 8.12 *Assume that f is a 0–1 symmetric function on 2^V with the maximality property. Then* Algorithm 8.E *is a 2-approximation for the associated network design problem.*

Proof. Following the proof of Theorem 8.9, we see that it is sufficient to show (8.15). Also, consider the graph H constructed in the same proof. We claim that for every leaf vertex S of H, $f(S) = 1$. For the sake of contradiction, suppose that S is a leaf of H with $f(S) = 0$. Let e be the unique edge in $E(H) = F^* - F(\boldsymbol{x})$ incident upon S, and let C be the connected component of graph (V, F^*) that contains S. Since F^* is feasible, we must have $f(C) = 0$ and so, by Lemma 8.11, $f(C - S)$

is also equal to 0. However, we note that $F^* - \{e\}$ is not feasible, which implies either $f(S) = 1$ or $f(C - S) = 1$ and gives us a contradiction.

The above claim implies that every vertex S of H that is not in *Violate*$(\boldsymbol{x})$ has degree at least 2. Therefore,

$$\begin{aligned}\sum_{S \in \mathit{Violate}(\boldsymbol{x})} \deg_{F^*}(S) &= \sum_{S \in V(H)} \deg_{F^*}(S) - \sum_{S \notin \mathit{Violate}(\boldsymbol{x})} \deg_{F^*}(S) \\ &\le 2(|V(H)| - 1) - 2(|V(H)| - |\mathit{Violate}(x)|) \\ &= 2|\mathit{Violate}(x)| - 2. \end{aligned}$$ □

Corollary 8.13 Algorithm 8.E *is a* 2*-approximation for* STEINER FOREST.

8.4 Local Ratio

Local ratio is a simple, yet powerful, technique for designing approximation algorithms with broad applications. It also has a close relationship with the primal-dual schemas in linear programming. In this section, we study some examples.

The main idea of the local ratio method comes from the following observation:

Theorem 8.14 (Local Ratio Theorem) *Assume that in a minimization problem*

$$\min\{c(\boldsymbol{x}) \mid \boldsymbol{x} \in \Omega\},$$

we can decompose the cost function c into $c = c_1 + c_2$. If $\boldsymbol{x} \in \Omega$ is an r-approximation with respect to both cost functions c_1 and c_2, then $\boldsymbol{x}$ is also an r-approximation with respect to the cost function c.

Proof. Suppose $\boldsymbol{x}_1^*$, $\boldsymbol{x}_2^*$, and $\boldsymbol{x}^*$ are optimal solutions with respect to cost functions c_1, c_2, and c, respectively. Then we have

$$\begin{aligned} c_1(\boldsymbol{x}) &\le rc_1(\boldsymbol{x}_1^*) \le rc_1(\boldsymbol{x}^*), \\ c_2(\boldsymbol{x}) &\le rc_2(\boldsymbol{x}_2^*) \le rc_2(\boldsymbol{x}^*). \end{aligned}$$

Therefore,

$$c(\boldsymbol{x}) = c_1(\boldsymbol{x}) + c_2(\boldsymbol{x}) \le rc_1(\boldsymbol{x}^*) + rc_2(\boldsymbol{x}^*) = rc(\boldsymbol{x}^*).$$ □

To see the applications of the local ratio theorem, let us first review the weighted vertex cover problem, MIN-WVC. Given a graph $G = (V, E)$ with nonnegative vertex weight c, we choose an edge $\{u, v\}$ with $c(u) > 0$ and $c(v) > 0$. (If such an edge does not exist, then all vertices with weight zero form an optimal solution.) Suppose $c(u) \le c(v)$. Define $c_1(u) = c_1(v) = c(u)$, and $c_1(x) = 0$ for $x \in V - \{u, v\}$. Then, any feasible solution is a 2-approximation with respect to c_1. So the problem is reduced to finding a 2-approximation for the problem with respect to the cost function $c_2 = c - c_1$. If all vertices x with $c_2(x) = 0$ form a vertex cover, then it is optimal with respect to c_2 and clearly also a 2-approximation solution with respect to c. Otherwise, we can continue the above process to decompose the weight function c_2 and to generate a new subproblem with more vertices having weight zero. This algorithm is summarized as follows.

Algorithm 8.F (*Local Ratio Algorithm for* MIN-WVC)

Input: A graph $G = (V, E)$ with a nonnegative vertex weight function $c : V \to \mathbb{N}$.

(1) **While** $\exists\{u, v\} \in E$ with $c(u) > 0$ and $c(v) > 0$ **do**
  $c_1 \leftarrow \min\{c(u), c(v)\}$;
  $c(u) \leftarrow c(u) - c_1$;
  $c(v) \leftarrow c(v) - c_1$.

(2) Output $\{v \mid c(v) = 0\}$. ■

It is inspiring to compare this algorithm with the Second Primal-Dual Schema (Algorithm 8.C). We rewrite Algorithm 8.C in the following for the problem MIN-WVC, in which we write, for a vertex $v \in V$, $E(v)$ to denote the set of all edges incident on v.

Algorithm 8.C (*Revisited, for* MIN-WVC)

Input: A graph $G = (V, E)$ with a nonnegative vertex weight function $c : V \to \mathbb{N}$.

(1) $\boldsymbol{x}^0 \leftarrow \boldsymbol{0}$; $\boldsymbol{y}^0 \leftarrow \boldsymbol{0}$; $k \leftarrow 0$;

(2) **While** $\boldsymbol{x}^k$ is not primal feasible (i.e., $\{j \mid x_j^k = 1\}$ is not a vertex cover) **do**

(2.1) Choose an uncovered edge $i' = \{u, v\}$;

(2.2) Choose $r \in \{u, v\}$ such that

$$\alpha = c_r - \sum_{i \in E(r)} y_i^k = \min_{j \in \{u,v\}} \left\{ c_j - \sum_{i \in E(j)} y_i^k \right\};$$

(2.3) **For** $j \leftarrow 1$ **to** n **do**
  if $j = r$ **then** $x_j^{k+1} \leftarrow 1$ **else** $x_j^{k+1} \leftarrow x_j^k$;

(2.4) **For** $i \leftarrow 1$ **to** m **do**
  if $i = i'$ **then** $y_i^{k+1} \leftarrow y_i^k + \alpha$ **else** $y_i^{k+1} \leftarrow y_i^k$;

(2.5) $k \leftarrow k + 1$.

(3) Output $\boldsymbol{x}^k$. ■

Note that if we update the cost function by setting

$$c_j \leftarrow c_j - \sum_{i \in E(j)} y_i^{k+1}$$

after line (2.4), and replace the definition of α of line (2.2) by

$$\alpha = c_r = \min_{j \in \{u,v\}} c_j,$$

then Algorithm 8.C is reduced to exactly Algorithm 8.F. In other words, these two algorithms are actually equivalent.

In general, it is easy to see that Algorithm 8.C is equivalent to the following local ratio algorithm for GC_1.

Algorithm 8.G (*Local Ratio Algorithm for* GC_1)

Input: An $m \times n$ nonnegative integer matrix $\boldsymbol{A}$ and $\mathbf{c} \in (\mathbb{Z}^+)^n$.

(1) Set $\boldsymbol{x} \leftarrow \mathbf{0}$.

(2) **While** $\boldsymbol{x}$ is not feasible **do**

Select an index i' such that $\sum_{j=1}^{n} a_{i'j} x_j = 0$;

Set $J \leftarrow \{j \mid x_j = 0 \text{ and } a_{i'j} > 0\}$;

Choose j' such that $\alpha = \dfrac{c_{j'}}{a_{i'j'}} = \min_{j \in J} \dfrac{c_j}{a_{i'j}}$;

$x_{j'} \leftarrow 1$;

For $j \leftarrow 1$ **to** n **do** $c_j \leftarrow c_j - a_{i'j}\alpha$.

(3) Output $\boldsymbol{x}$. ■

What about the First Primal-Dual Schema? Is there a local ratio algorithm equivalent to Algorithm 8.B? The answer is *yes*. The following is such an algorithm for the problem MIN-WVC. We leave the general local ratio algorithm for GC_1 as an exercise.

Algorithm 8.H (*Second Local Ratio Algorithm for* MIN-WVC)

Input: A graph $G = (V, E)$ with a nonnegative vertex weight function $c : V \to \mathbb{N}$.

(1) $C \leftarrow \emptyset$.

(2) **While** $G \neq \emptyset$ **do**

Choose $u \in V$ such that $\dfrac{c(u)}{\deg_G(u)} = \min_{v \in V} \dfrac{c(v)}{\deg_G(v)}$;

For every $\{u, v\} \in G$ **do** $c(v) \leftarrow c(v) - \dfrac{c(u)}{\deg_G(u)}$;

$C \leftarrow C \cup \{u\}$;
$V \leftarrow V - \{u\}$;
$G \leftarrow G|_V$.

(3) Output C. ■

In each iteration of the above algorithm, the cost function c is decomposed into two parts $c = c_1 + c_2$, where $c_1(u) = c(u)$ and $c_1(v) = c(u)/\deg_G(u)$ for each

$v \in G$ that is adjacent to u, and $c_1(v) = 0$ otherwise. Thus, any vertex cover for G is a 2-approximation with respect to c_1. So it provides us with another 2-approximation for MIN-WVC.

In general, a local ratio algorithm can be divided into the following two steps:

Step 1. Find a type of weight function c_1 with which an r-approximation can be constructed.

Step 2. Reduce the general weight c by a weight function c_1 of the above special type iteratively until a feasible solution can be found trivially.

In all of Algorithms 8.F, 8.G, and 8.H, step 1 is somewhat trivial, in the sense that the cost function c_1 found has the property that any feasible solution for the problem is a 2-approximation for c_1. In general, can we expect to always find such a trivial function c_1? The answer is *no*, as demonstrated by the following example.

> PARTIAL VERTEX COVER (PVC): Given a graph $G = (V, E)$ with nonnegative vertex weight $c : V \to \mathbb{N}$, and an integer $k > 0$, find a minimum-weight subset of vertices that covers at least k edges.

We note that in the general cases of this problem, no single vertex subset must contribute to all feasible solutions. Thus, it is hard to find a function c_1 with respect to which any feasible solution is trivially a 2-approximation. In such situations, we focus instead on *minimal feasible solutions*.

A feasible solution is said to be *minimal* if none of its proper subsets is feasible. The idea here is to find a cost function c_1 with respect to which every minimal feasible solution is a 2-approximation. To do so, we consider the minimum cost needed to cover *a single edge* in graph G. Suppose a feasible solution includes a vertex v, which has degree $\deg(v) \le k$. Then, vertex v covers $\deg(v)$ edges with cost $c(v)$, and so each edge incident on v incurs cost $c(v)/\deg(v)$. If $\deg(v) > k$, then each edge incurs cost $c(v)/k$ since we only need to cover k edges. This observation suggests we assign $c_1(v)$ as follows:

First, let α be the minimum cost to cover a single edge; that is,

$$\alpha = \min_{v \in V} \frac{c(v)}{\min\{k, \deg(v)\}}.$$

Next, for every $u \in V$, define $c_1(u)$ to be the cost of covering all edges (up to k many) incident on u:

$$c_1(u) = \alpha \cdot \min\{k, \deg(u)\}.$$

Lemma 8.15 *Every minimal feasible solution for G is a 2-approximation with respect to cost function c_1.*

Proof. From the definition of α, we know that covering any edge in G costs at least α. Therefore, $k\alpha$ is a lower bound for the optimal solution *opt*.

Now, consider a minimal feasible solution C for graph G. If C contains a vertex v such that $\deg(v) \ge k$, then $C = \{v\}$ with cost $k\alpha$. Therefore, we may assume that C contains at least two vertices, all with degree $< k$. In this case, the total cost of C is equal to $\alpha \cdot \sum_{v\in C} \deg(v)$. Since $k\alpha$ is a lower bound of the optimal cost, it suffices to show $\sum_{v\in C} \deg(v) \le 2k$.

For each vertex $v \in C$ and $i \in \{1,2\}$, let $d_i(v)$ denote the number of edges incident on v that have i endpoints in C. Then $\deg(v) = d_1(v)+d_2(v)$. Choose $v^* \in C$ with $d_1(v^*) = \min_{v\in C} d_1(v)$. Then, $d_1(v^*) \le \sum_{v\in C-\{v^*\}} d_1(v)$. Next, observe that the total number of edges covered by C is equal to $\sum_{v\in C}(d_1(v) + d_2(v)/2)$. Since C is minimal, we must have

$$\frac{1}{2}\sum_{v\in C} d_2(v) + \sum_{v\in C-\{v^*\}} d_1(v) < k,$$

for otherwise $C - \{v^*\}$ would be feasible, violating the minimality assumption about C. Therefore,

$$\begin{aligned}\sum_{v\in C} \deg(v) &= d_1(v^*) + \sum_{v\in C-\{v^*\}} d_1(v) + \sum_{v\in C} d_2(v)\\ &\le 2\left(\frac{1}{2}\sum_{v\in C} d_2(v) + \sum_{v\in C-\{v^*\}} d_1(v)\right) < 2k. \qquad \square\end{aligned}$$

Corollary 8.16 *The problem* PARTIAL VERTEX COVER *has a polynomial-time* 2*-approximation.*

Next, we consider the following problem. We say a subset F of vertices of a graph $G = (V, E)$ is a *feedback vertex set* if the removal of F results in an acyclic graph, that is, if $G|_{V-F}$ is acyclic.

> FEEDBACK VERTEX SET (FVS): Given a graph $G = (V, E)$ with non-negative vertex weight $w : V \to \mathbb{N}$, find a minimum-weight feedback vertex set of G.

A feedback vertex set F is said to be *minimal* if no proper subset of F is a feedback vertex set. To design a local ratio algorithm for this problem, we follow the same idea in the design of function c_1 for the problem PVC and define the following special weight function:

$$w_1(u) = \varepsilon \cdot \deg(u),$$

where ε is a positive constant.

Lemma 8.17 *Let G be a graph and w_1 a weight function defined above. Suppose each vertex in G has degree at least* 2*. Then every minimal feedback vertex set F is a* 2*-approximation for* FVS *with respect to weight w_1.*

Proof. Since F is minimal, for each $u \in F$, there exists a cycle C_u such that u is the only vertex in F contained in C_u. For each $u \in F$, fix the cycle C_u and let P_u be the path obtained from C_u by deleting u. Denote by G_1 the subgraph of G consisting of all connected components of $G|_{V-F}$ that contain such a path P_u. Let V_1 be the vertex set of G_1, and $V_2 = V - F - V_1$. For $i = 1, 2$, define $n_i = |V_i|$, and define m_i to be the number of edges in G incident on vertices in V_i. In addition, define m_F to be the number of edges in G between vertices in F and, for $i = 1, 2$, define m'_i to be the number of edges in G between a vertex in V_i and a vertex in F. Now, we observe the following relationships between these parameters:

(a) The total degree of vertices in F can be expressed as

$$\sum_{u \in F} \deg(u) = m'_1 + m'_2 + 2m_F.$$

(b) The total degree of vertices in V_2 is

$$\sum_{u \in V_2} \deg(u) = 2(m_2 - m'_2) + m'_2 = 2m_2 - m'_2.$$

Since each vertex in G has degree at least 2, we have $2n_2 \leq \sum_{u \in V_2} \deg(u)$ and, hence,

$$m'_2 \leq 2(m_2 - n_2).$$

(c) Let F^* be a minimum feedback vertex set with respect to weight w_1. We claim that

$$m'_1 \leq m_1 - n_1 + |F^*|.$$

To see this, we first note that each connected component of G_1 is a tree, and so $m'_1 = m_1 - n_1 + k$, where k is the number of connected components of G_1. Next, we note that each connected component of G_1 contains a P_u, and each C_u must contain a vertex in F^*. Thus, either $u \in F^*$ or P_u contains a vertex in $F^* \setminus F$. It follows that each connected component of G_1 contains either a vertex in $F^* \setminus F$ or a P_u with $u \in F^* \cap F$. This means that $k \leq |F^* \setminus F| + |F^* \cap F| = |F^*|$, and the claim is proven.

(d) Since each vertex in F has at least two edges going to vertices in V_1, we have

$$2|F| \leq m'_1.$$

From the above relationships, we get

$$\begin{aligned}\sum_{u \in F} \deg(u) &\leq m_1 - n_1 + |F^*| + 2m_2 - 2n_2 + 2m_F \\ &= 2(m_1 + m_2 + m_F) - 2(n_1 + n_2) - (m_1 - n_1 + |F^*|) + 2|F^*| \\ &\leq 2|E| - 2|V| + 2|F| - m'_1 + 2|F^*| \\ &\leq 2(|E| - |V| + |F^*|) \leq 2 \sum_{u \in F^*} \deg(u).\end{aligned}$$

The last inequality above is derived as follows: After removing F^*, the graph G has no cycles and, hence, has at most $|V| - |F^*| - 1$ edges left. This means that at least

$|E| - |V| + |F^*| + 1$ edges have been removed, a number that cannot exceed the total degree $\sum_{u \in F^*} \deg(u)$ of vertices in F^*. □

The above lemma suggests the following local ratio algorithm.

Algorithm 8.I (*Local Ratio Algorithm for* FVS)

Function FVS(G, w)

(1) **If** $G = \emptyset$ **then** return $\emptyset$.

(2) **If** $\exists u \in V(G)$ with $\deg(u) \leq 1$ **then** return FVS$(G - \{u\}, w)$.

(3) **If** $\exists u \in V(G)$ with $w(u) = 0$
  then $F \leftarrow$ FVS$(G - \{u\}, w)$;
    if F is a feedback set for G **then** return F
                else return $F \cup \{u\}$
  else set $\varepsilon \leftarrow \min_{u \in V(G)} \dfrac{w(u)}{\deg(u)}$;
    for all $u \in V(G)$ **do** $w_1(u) \leftarrow \varepsilon \cdot \deg(u)$;
    return FVS$(G, w - w_1)$. ■

Theorem 8.18 Algorithm 8.I *is a 2-approximation for* FVS.

Proof. Let $F^*(G, w)$ denote an optimal solution for FVS on input (G, w). Also, let F be the set returned by FVS(G, w). We show by induction that F is a minimal feedback vertex set of G and is a 2-approximation to $F^*(G, w)$. For $G = \emptyset$, this is trivially true. For general G, suppose u is the first vertex deleted from G in Algorithm 8.I. There are two cases.

Case 1. $\deg(u) \leq 1$. In this case, a vertex subset is a feedback vertex set of G if and only if it is a feedback vertex set of $G - \{u\}$. By the induction hypothesis, F is a minimal feedback vertex set of $G - \{u\}$ and is a 2-approximation to $F^*(G - \{u\}, w)$. It follows that F is also a minimal feedback vertex set of G and is a 2-approximation to $F^*(G, w) = F^*(G - \{u\}, w)$.

Case 2. $w(u) = 0$. In this case, every vertex v of G has $\deg(v) \geq 2$. Now consider two subcases:

Subcase 2.1. $u \notin F$. From line 3 of step (3), we know that F is a feedback vertex set of G. By the induction hypothesis, F is a minimal feedback vertex set of $G - \{u\}$ and hence is also minimal for G. In addition, F is a 2-approximation to $F^*(G - \{u\}, w)$, and so it is also a 2-approximation to $F^*(G, w)$.

Subcase 2.2. $u \in F$. By the induction hypothesis, $F - \{u\}$ is a minimal feedback vertex set of $G - \{u\}$ but not a feedback vertex set of G. Therefore, F must be a feedback vertex set of G and must also be minimal. Since $w(u) = 0$, F and $F - \{u\}$ have the same weight. Therefore, the induction hypothesis that $F - \{u\}$ is a 2-approximation to $F^*(G - \{u\}, w)$ implies that F is a 2-approximation to $F^*(G, w)$.

Finally, we notice that, before a vertex u with $w(u) = 0$ is deleted from G, the algorithm may have reduced the weight w to $w - w_1$. In such a case, the above argument in case 2 showed that F is a minimal feedback vertex set of G and is a 2-approximation to $F^*(G, w - w_1)$. By Lemma 8.17, F is also a 2-approximation to $F^*(G, w_1)$. Hence, by the local ratio theorem, F is also a 2-approximation to $F^*(G, w)$. □

Next, we study a maximization problem. Recall that a vertex subset $S \subseteq V$ of a graph $G = (V, E)$ is an *independent set* if no two vertices in S are connected by an edge in E.

> MAXIMUM-WEIGHT INDEPENDENT SET (MAX-WIS): Given a graph $G = (V, E)$ with a nonnegative vertex weight function $w : V \to \mathbb{N}$, find an independent set with the maximum total weight.

In the analysis of the local ratio algorithm for PVC (Lemma 8.15), we introduced a new analysis technique. Instead of comparing the approximate solution with the optimal solution *opt*, we compare it with a lower bound $k\alpha$ of *opt*. Here we will apply this technique again, in a more sophisticated way, by comparing the approximate solution of MAX-WIS with an *upper bound* of the optimal solution (as this is a *maximization* problem while PVC is a minimization problem).

To find an upper bound of the optimal solution, we can first formulate the problem as an integer linear program:

$$\begin{array}{lll} \text{maximize} & \sum_{u \in V} w(u) x_u & \\ \text{subject to} & x_u + x_v \le 1, & \{u, v\} \in E, \\ & x_u \in \{0, 1\}, & u \in V. \end{array}$$

Then we relax this ILP to the following LP by replacing the constraints $x_u \in \{0, 1\}$ with $0 \le x_u \le 1$:

$$\begin{array}{lll} \text{maximize} & \sum_{u \in V} w(u) x_u & \\ \text{subject to} & x_u + x_v \le 1, & \{u, v\} \in E, \\ & 0 \le x_u \le 1, & u \in V. \end{array} \tag{8.16}$$

Let $\boldsymbol{x}^*$ be an optimal solution of this LP. Then, $\sum_{u \in V} w(u) x_u^*$ is an upper bound for the optimal solution *opt* of the ILP. Now, instead of defining a weight function w_1 for which an r-approximation is easy to find, we only need to define a weight function w_1 for which a feasible solution $\boldsymbol{x}$ satisfying

$$\sum_{u \in V} w_1(u) x_u \ge \frac{1}{r} \sum_{u \in V} w_1(u) x_u^*$$

is easy to find.

Let $V_+ = \{u \in V \mid w(u) > 0\}$. For each $u \in V$, let $N(u)$ denote the set consisting of vertex u and its neighbors in G. Choose a vertex $v \in V_+$ to minimize $\sum_{u \in N(v) \cap V_+} x_u^*$. Let $\varepsilon = w(v)$, and define

$$w_1(u) = \begin{cases} \varepsilon, & \text{if } u \in N(v) \cap V_+, \\ 0, & \text{otherwise.} \end{cases}$$

Lemma 8.19 *For any independent subset I of V_+ with $I \cap N(v) \neq \emptyset$, we have*

$$\sum_{u \in V} w_1(u) x_u^* \leq \frac{\delta + 1}{2} \cdot w_1(I),$$

where δ is the maximum vertex degree of the input graph G.

Proof. From the definition of w_1, we see that

$$\sum_{u \in V} w_1(u) x_u^* = \varepsilon \cdot \sum_{u \in N(v) \cap V_+} x_u^*.$$

Since $I \cap (N(v) \cap V_+) \neq \emptyset$, we have $w_1(I) \geq \varepsilon$. This means that we only need to show

$$\sum_{u \in N(v) \cap V_+} x_u^* \leq \frac{\delta + 1}{2}.$$

By the choice of v, it suffices to show the existence of a vertex $s \in V_+$ with

$$\sum_{u \in N(s) \cap V_+} x_u^* \leq \frac{\delta + 1}{2}.$$

Choose $s = \arg\max_{u \in V_+} x_u^*$. Without loss of generality, we assume $|N(s)| \geq 2$. Now, if $x_s^* \leq 1/2$, then $x_u^* \leq 1/2$ for all $u \in N(s)$, and so

$$\sum_{u \in N(s) \cap V_+} x_u^* \leq \frac{\deg(s) + 1}{2} \leq \frac{\delta + 1}{2}.$$

On the other hand, if $x_s^* > 1/2$, then, by the constraint $x_s + x_u \leq 1$, we know that $x_u^* < 1/2$ for all $u \in N(s) - \{s\}$. Pick a neighbor t of s, and let $N'(s) = N(s) - \{s, t\}$; then we get

$$\sum_{u \in N(s) \cap V_+} x_u^* \leq (x_s^* + x_t^*) + \sum_{u \in N'(s) \cap V_+} x_u^* \leq 1 + \frac{\deg(s) - 1}{2} \leq \frac{\delta + 1}{2}. \quad \square$$

The following is the local ratio algorithm for MAX-WIS, which decomposes the input weight recursively to simpler weights of the form w_1.

Algorithm 8.J (*Local Ratio Algorithm for* MAX-WIS)
Input: A graph $G = (V, E)$, with a nonnegative vertex weight function $w : V \to \mathbb{N}$.

(1) Solve LP (8.16); let $\boldsymbol{x}^*$ be an optimal solution.

(2) Output WIS$(G, w, \boldsymbol{x}^*)$. ■

The function WIS$(G, w, \boldsymbol{x}^*)$ is defined as follows:

Function WIS$(G, w, \boldsymbol{x}^*)$.

(1) $V_+ \leftarrow \{u \mid w(u) > 0\}$.

(2) **If** V_+ is independent in G **then** return V_+.

(3) Choose $v \in V_+$ to minimize $\sum_{u \in N(v) \cap V_+} x^*_u$.

(4) $\varepsilon \leftarrow w(v)$.

(5) **For** all $u \in V$ **do** $w_1(u) \leftarrow \begin{cases} \varepsilon, & \text{if } u \in N(v) \cap V_+, \\ 0, & \text{otherwise.} \end{cases}$

(6) $S \leftarrow$ WIS$(G, w - w_1, \boldsymbol{x}^*)$.

(7) **If** $S \cup \{v\}$ is independent in G **then** return $S \cup \{v\}$
else return S. ■

Theorem 8.20 Algorithm 8.J *is a* $((\delta+1)/2)$*-approximation for* MAX-WIS, *where* δ *is the maximum degree of the input graph.*

Proof. Let I denote the set returned by the function WIS$(G, w, \boldsymbol{x}^*)$. We claim that I is an independent subset of V_+ and that

$$\sum_{u \in V} w(u) x^*_u \le \frac{\delta+1}{2} \cdot w(I).$$

We prove this claim by induction on the number of recursive calls made to get the output I. In the case that no recursive call is made, V_+ is independent. Clearly, our claim is true since $I = V_+$.

In general, we consider the first recursive call of the form WIS$(G, w - w_1, \boldsymbol{x}^*)$. Suppose this call returns set S. Denote $w_2 = w - w_1$. By the induction hypothesis, we have

$$\sum_{u \in V} w_2(u) x^*_u \le \frac{\delta+1}{2} \cdot w_2(S) \tag{8.17}$$

and S is an independent subset of $V'_+ = \{u \mid w_2(u) > 0\}$. Note that $V_+ = V'_+ \cup (N(v) \cap V_+)$. If $S \cup \{v\}$ is independent, then $I = S \cup \{v\}$, which is clearly an independent subset of V_+. If $S \cup \{v\}$ is not independent, then $I = S$, and it must contain a vertex in $N(v)$. Thus, in either case, I is an independent subset of V_+, with $I \cap N(v) \neq \emptyset$. We have, by Lemma 8.19,

$$\sum_{u \in V} w_1(u) x^*_u \le \frac{\delta+1}{2} \cdot w_1(I).$$

In addition, we note that $w_2(v) = 0$. Therefore, by (8.17), we have

$$\sum_{u \in V} w_2(u)x_u^* \le \frac{\delta+1}{2} \cdot w_2(S) = \frac{\delta+1}{2} \cdot w_2(I).$$

Together, we get

$$\sum_{u \in V} w(u)x_u^* \le \frac{\delta+1}{2} \cdot w(I),$$

and the claim is proven. □

We remark that the recursive Algorithm 8.J for MAX-WIS may be further improved. In each recursive call, we may compute a new point $\boldsymbol{x}^{**}$ corresponding to the weight $w_2 = w - w_1$, and call function WIS with parameters $(G, w_2, \boldsymbol{x}^{**})$ instead of $(G, w_2, \boldsymbol{x}^*)$. Then we can use the total weight at x^{**} as an upper bound for the optimal solution for MAX-WIS of G with respect to weight w_2. This way, we might get a better performance ratio. Indeed, the idea of this extension is exactly that of iterated rounding introduced in Section 7.5. In other words, the iterated rounding technique can also be seen as an application of the local ratio technique in LP-based approximations.

8.5 More on Equivalence

In the last section, we demonstrated the equivalence between the primal-dual schema and the local ratio method for the problems MIN-WVC and GC_1. In this section, we further discuss the relationship between these two techniques.

We first make two observations on the problems studied in this chapter with the primal-dual schema. The first observation is that all problems studied so far in this chapter are of the covering type; that is, they are the following special cases of the problem GENERAL COVER:

> Consider a base set X, a collection $\mathcal{C}$ of subsets of X, and a nonnegative cost function c on X. For each subset C of X, denote $c(C) = \sum_{x \in C} c(x)$. A minimization problem
>
> $$\min\{c(C) \mid C \in \mathcal{C}\}$$
>
> is said to be of the *covering type* if $A \subset B$ and $A \in \mathcal{C}$ imply $B \in \mathcal{C}$.

The second observation is that every primal-dual schema studied so far preserves the primal complementary slackness condition and relaxes the dual complementary slackness condition. To be more specific, let us consider the problem GC_1 and its dual:

(primal LP)		*(dual LP)*	
minimize	$\boldsymbol{c}\boldsymbol{x}$	maximize	$\boldsymbol{y}\mathbf{1}_m$
subject to	$A\boldsymbol{x} \ge \mathbf{1}_m,$	subject to	$\boldsymbol{y}A \le \boldsymbol{c},$
	$\boldsymbol{x} \ge \mathbf{0},$		$\boldsymbol{y} \ge \mathbf{0}.$

The primal complementary slackness condition is

$$(\boldsymbol{c} - \boldsymbol{y}A)\boldsymbol{x} = \boldsymbol{0}.$$

To keep this condition holding, we set $\boldsymbol{x}$ in the following way:

$$x_j = 1 \iff \sum_{i=1}^{m} a_{ij} y_i = c_j.$$

The condition $\sum_{i=1}^{m} a_{ij} y_i = c_j$ provides us with a decomposition of the cost function. Note that in a local ratio algorithm, we usually set $x_j \leftarrow 1$ when the weight c_j is reduced to 0. Therefore, there is a simple correspondence between the condition $\sum_{i=1}^{m} a_{ij} y_i = c_j$ in the primal-dual schema and the assignment $c_j \leftarrow 0$ in the local ratio algorithm. Suppose y_i^k is the value of y_i after the kth iteration in a primal-dual schema. Then

$$c'_j = \sum_{i=1}^{m} a_{ij}(y_i^{k+1} - y_i^k)$$

is the cost reduction in the $(k+1)$st iteration of the local ratio algorithm that corresponds to the primal-dual schema, and a translation between the primal-dual schema and the local ratio algorithm can be built upon this relationship. As an example, let us consider the problem NETWORK DESIGN. Its primal-dual schema, Algorithm 8.E, can be translated into the following equivalent local ratio algorithm:

Algorithm 8.K (*Local Ratio Algorithm for* NETWORK DESIGN)

Input: A graph $G = (V, E)$ with edge costs c_e, for $e \in E$, and a function $f : 2^V \to \{0, 1\}$ (given implicitly).

(1) $\boldsymbol{x} \leftarrow \boldsymbol{0}$.

(2) **While** $\boldsymbol{x}$ is not primal feasible **do**

Set $\alpha \leftarrow \min_{e \in E} \dfrac{c_e}{|\mathit{Violate}(\boldsymbol{x}) \cap \{S \mid e \in \delta(S)\}|}$;

For each $e \in E$ **do**

$c'_e \leftarrow \alpha \cdot |\mathit{Violate}(\boldsymbol{x}) \cap \{S \mid e \in \delta(S)\}|$;

$c_e \leftarrow c_e - c'_e$;

if $c_e = 0$ then $x_e \leftarrow 1$.

(3) **For** each $e \in F$ **do**

Let $\boldsymbol{x}'$ be the vector $\boldsymbol{x}$ modified with $x'_e \leftarrow 0$;

If $\boldsymbol{x}'$ is primal feasible **then** $\boldsymbol{x} \leftarrow \boldsymbol{x}'$.

(4) Output $\boldsymbol{x}$. ■

Now, let us look at how we analyze this local ratio algorithm.

Let $\boldsymbol{x}^*$ be the output of Algorithm 8.K, and let $F^* = \{e \mid x_e^* = 1\}$. Also, let $\boldsymbol{x}^k$ be the value of $\boldsymbol{x}$ at the beginning of the kth iteration, α_k the minimum value

α found in the kth iteration, and $c'_e(k)$ the value of c'_e at the kth iteration. That is, in the kth iteration, we decompose the cost function c_e into the sum of $c'_e(k)$ and $c_e - c'_e(k)$. By the local ratio theorem, all we need to prove is that solution $\boldsymbol{x}^*$, as a local solution to the problem with respect to the cost function $c'_e(k)$, is a 2-approximation. That is, we need to show

$$\sum_{e\in E} c'_e(k)x^*_e \leq 2 \cdot opt_k, \tag{8.18}$$

where opt_k is the cost value of the optimal solution with respect to the cost function $c'_e(k)$. Note that

$$\begin{aligned}\sum_{e\in E} c'_e(k)x^*_e &= \sum_{e\in F^*} c'_e(k) = \sum_{e\in F^*} \sum_{\substack{S\in \mathit{Violate}(\boldsymbol{x}^k)\\ e\in\delta(S)}} \alpha_k \\ &= \sum_{S\in \mathit{Violate}(\boldsymbol{x}^k)} \deg_{F^*}(S)\cdot \alpha_k\end{aligned}$$

and

$$opt_k \geq |\mathit{Violate}(\boldsymbol{x}^k)| \cdot \alpha_k.$$

The second inequality follows from the fact that for every $S \in \mathit{Violate}(\boldsymbol{x}^k)$, there must be an edge $e \in F^* \cap \delta(S)$. So, to show (8.18), it suffices to prove

$$\sum_{S\in \mathit{Violate}(\boldsymbol{x})} \deg_{F^*}(S) \leq 2\cdot |\mathit{Violate}(\boldsymbol{x})|.$$

This is exactly the inequality (8.15) that we encountered in the analysis of the primal-dual schema (see Theorem 8.9). Thus, not only does the cost decomposition in Algorithm 8.K follow from the primal-dual schema of Algorithm 8.E, but the analysis can also be done in a similar way.

From the above observations, we see that the equivalence between the primal-dual schema and the local ratio method is built on the covering-type problems and the preservation of the primal complementary slackness condition. A natural question arises: For a noncovering-type problem and a primal-dual schema that does not preserve the primal complementary slackness condition, can we still find an equivalent local ratio algorithm? This question is difficult to answer, because there are very few primal-dual schemas known that relax the primal complementary slackness condition. One of the proposed primal-dual schema of this type is about the following facility location problem.

Consider a set C of m cities and a set F of n possible locations for facilities, with two cost functions c_{ij}, for $i \in F$ and $j \in C$, and f_i, for $i \in F$. Intuitively, c_{ij} is the cost for city j to use facility at location i, and f_i is the cost of installing the facility at location i. We say the costs c_{ij} satisfy the *extended triangle inequality* if $c_{ij} \leq c_{i'j} + c_{i'j'} + c_{ij'}$, for any $i, i' \in F$ and $j, j' \in C$.

FACILITY LOCATION: Given sets C and F, costs c_{ij}, f_i, for $i \in F$ and $j \in C$, with c_{ij} satisfying the extended triangle inequality, find a subset

$S \subseteq F$ to install facilities such that the total cost of installingfacilities and the use of these facilities is minimized, under the condition that each city is assigned to exactly one facility.

This problem can be formulated into the following integer linear program, in which we use $x_{ij} = 1$ to indicate that city j is assigned to use facility at location i, and $y_i = 1$ to indicate a facility is installed at location i:

$$\begin{array}{lll} \text{minimize} & \sum_{i\in F, j\in C} c_{ij}x_{ij} + \sum_{i\in F} f_i y_i & \\ \text{subject to} & \sum_{i\in F} x_{ij} \geq 1, & j \in C, \\ & y_i - x_{ij} \geq 0, & i \in F,\ j \in C, \\ & x_{ij}, y_i \in \{0,1\}, & i \in F,\ j \in C. \end{array}$$

The following are a relaxation of this ILP and its corresponding dual LP:

$$\begin{array}{llll} \textit{(primal LP)} & \text{minimize} & \sum_{i\in F, j\in C} c_{ij}x_{ij} + \sum_{i\in F} f_i y_i & \\ & \text{subject to} & \sum_{i\in F} x_{ij} \geq 1, & j \in C, \\ & & y_i - x_{ij} \geq 0, & i \in F,\ j \in C, \\ & & x_{ij} \geq 0,\ y_i \geq 0, & i \in F,\ j \in C; \end{array}$$

$$\begin{array}{llll} \textit{(dual LP)} & \text{maximize} & \sum_{j\in C} \alpha_j & \\ & \text{subject to} & \alpha_j - \beta_{ij} \leq c_{ij}, & i \in F,\ j \in C, \\ & & \sum_{j\in C} \beta_{ij} \leq f_i, & i \in F, \\ & & \alpha_j \geq 0,\ \beta_{ij} \geq 0, & i \in F,\ j \in C. \end{array}$$

The intuitive meaning of the variables α_j and β_{ij} of the above dual LP is as follows: For each $i \in F$, city j pays β_{ij} toward the installation of the facility i. Also, each city j pays altogether α_j for the installation and the use of these facilities. The primal complementary slackness conditions of the above primal and dual LPs are

$$\begin{aligned} x_{ij}(c_{ij} - (\alpha_j - \beta_{ij})) &= 0, && \text{for } i \in F,\ j \in C, \\ y_i\Big(f_i - \sum_{j\in C} \beta_{ij}\Big) &= 0, && \text{for } i \in F, \end{aligned}$$

and the dual complementary slackness conditions are

$$\alpha_j\Big(\sum_{i\in F} x_{ij} - 1\Big) = 0, \qquad \text{for } j \in C,$$

$$\beta_{ij}(y_i - x_{ij}) = 0, \qquad \text{for } i \in F,\ j \in C.$$

As this is not a covering-type problem, and the objective function of the primal LP is complicated, there does not seem to be a simple primal-dual schema for it that preserves the primal complementary slackness condition. Instead, Jain and Vazirani [2001] proposed the following idea to get a primal-dual schema that preserves the dual complementary slackness condition but relaxes the primal complementary slackness condition.

(1) Keep the primal solutions x_{ij} and y_i, for $i \in F$ and $j \in C$, integral. Also, each city $j \in C$ is to be assigned to a unique facility $\phi(j)$.

(2) Cities in C are partitioned into two sets D and $C - D$. Only cities in D pay for the installation cost of the facilities; that is, $\beta_{ij} = 0$ if $j \notin D$ or if $i \neq \phi(j)$.

(3) For $j \in C - D$, the first primary complementary slackness condition is relaxed to

$$\frac{1}{3} c_{\phi(j)j} \le \alpha_j \le c_{\phi(j)j}.$$

(4) All other dual and primary complementary slackness conditions are to be satisfied. In particular, for $j \in D$,

$$\alpha_j - \beta_{\phi(j)j} = c_{\phi(j)j},$$

and, for each i with $y_i = 1$,

$$f_i = \sum_{j:\phi(j)=i} \beta_{ij}.$$

The above proposed method appears interesting. It is not clear, however, whether it can be implemented in such a way that the algorithm always outputs a feasible solution, as the details of the implementation were not presented in the paper (see Exercise 8.10). It is also not known whether there is an equivalent local ratio algorithm for FACILITY LOCATION, even if the above ideas can indeed be implemented in a polynomial-time approximation with a constant performance ratio.

Finally, we point out that weight decomposition is a well-known proof technique in discrete mathematics. Essentially, the local ratio method may be viewed as the extension of this old proof technique to the design of algorithms. In particular, we note that this proof technique has been used in the analysis of the greedy approximation for the problem MIN-SMC (see Theorem 2.29). As the local ratio algorithms we studied in this chapter can be converted to equivalent primal-dual schemas, we may ask whether the weight decomposition analysis can also be proved by certain primal-dual relationships. The answer is affirmative for some problems. For instance, for the analysis of the greedy approximation for MIN-SMC, we can employ the duality theory of linear programming as follows.

First, let us recall the problem MIN-SMC. Let $E = \{1, 2, \ldots, n\}$, $f : 2^E \to \mathbb{R}$ a polymatroid function, and $c : E \to \mathbb{R}^+$ a nonnegative cost function. The problem

MIN-SMC asks us to minimize $c(A) = \sum_{a\in A} c(a)$ for $A \in \Omega_f = \{A \mid f(A) = f(E)\}$.

This problem can be formulated as an integer linear program as follows:[3]

$$\begin{aligned} &\text{minimize} && \sum_{i\in E} c(i)v_i \\ &\text{subject to} && \sum_{i\in E-S} \Delta_i f(S)\, v_i \ge \Delta_{E-S} f(S), \quad && S \in 2^E, \\ & && v_i \in \{0,1\}, && i \in E. \end{aligned} \tag{8.19}$$

To see this, let $A \in \Omega_f$; that is, $f(A) = f(E)$. We claim that

$$v_i = \begin{cases} 1, & \text{if } i \in A, \\ 0, & \text{otherwise,} \end{cases}$$

is a feasible solution of LP (8.19). Indeed, for any $S \in 2^E$,

$$\begin{aligned} \sum_{i\in E-S} \Delta_i f(S)\, v_i &= \sum_{i\in A\setminus S} \Delta_i f(S) \ge \Delta_{A\setminus S} f(S) \\ &= f(A) - f(S) = f(E) - f(S) = \Delta_{E-S} f(S). \end{aligned}$$

Conversely, if $\boldsymbol{v}$ is a feasible solution of LP (8.19), then we can see that $A = \{i \mid v_i = 1\}$ satisfies $f(A) = f(E)$. In fact, considering the inequality constraint for $S = A$, we have

$$\sum_{i\in E-A} \Delta_i f(A)\, v_i \ge \Delta_{E-A} f(A);$$

that is,

$$0 \ge f(E) - f(A).$$

Since f is monotone increasing, we must have $f(E) = f(A)$. The above shows that the ILP (8.19) is equivalent to the problem MIN-SMC.

Now, we can relax this ILP to an LP and get its dual LP as follows:

$$\begin{aligned} &\text{maximize} && \sum_{S\in 2^E} \Delta_{E-S} f(S)\, y_S \\ &\text{subject to} && \sum_{S:i\notin S} \Delta_i f(S)\, y_S \le c(i), \quad && i \in E, \\ & && y_S \ge 0, && S \in 2^E. \end{aligned}$$

Next, we review the analysis of the greedy Algorithm 2.D on the functions f and c. Suppose $x_1, x_2, \ldots, x_k$ are the elements selected by the greedy Algorithm 2.D in

[3] We use v_i, instead of x_i, to denote a variable corresponding to element $i \in E$, to avoid confusion with the name x_i used in the analysis in Theorem 2.29.

the order of their selection into the approximate solution A. Denote $A_0 = \emptyset$ and, for $i = 1, \ldots, k$, $A_i = \{x_1, \ldots, x_i\}$. In the proof of Theorem 2.29, we decomposed the total weight $c(A)$ to $\sum_{i=1}^{k} w(x_i)$, where, for each $a \in E$,

$$w(a) = \sum_{j=1}^{k} (z_{a,j} - z_{a,j+1}) \frac{c(x_j)}{r_j},$$

$z_{a,j} = \Delta_a f(A_{j-1})$, and $r_j = \Delta_{x_j} f(A_{j-1})$. Also, recall that in the proof of Theorem 2.29, we established property (b), which states that for any $a \in E$,

$$w(a) = \frac{c(x_1)}{r_1} z_{a,1} + \sum_{j=2}^{k} \left(\frac{c(x_j)}{r_j} - \frac{c(x_{j-1})}{r_{j-1}} \right) z_{a,j} \le c(a) \cdot H(\gamma), \qquad (8.20)$$

where $\gamma = \max_{x \in E} f(\{x\})$.

Now, set

$$y_S = \begin{cases} \dfrac{1}{H(\gamma)} \cdot \dfrac{c(x_1)}{r_1}, & \text{if } S = A_0, \\[2mm] \dfrac{1}{H(\gamma)} \left(\dfrac{c(x_{i+1})}{r_{i+1}} - \dfrac{c(x_i)}{r_i} \right), & \text{if } S = A_i,\ 1 \le i \le k-1, \\[2mm] 0 & \text{otherwise.} \end{cases}$$

Then, from (8.20), we see that for any $a \in E$,

$$\begin{aligned} \sum_{S: a \notin S} \Delta_a f(S)\, y_S &= \sum_{j=0}^{k-1} \Delta_a f(A_j) y_{A_j} \\ &= \frac{1}{H(\gamma)} \left(\frac{c(x_1)}{r_1} z_{a,1} + \sum_{j=2}^{k} \left(\frac{c(x_j)}{r_j} - \frac{c(x_{j-1})}{r_{j-1}} \right) z_{a,j} \right) \\ &= \frac{1}{H(\gamma)} \cdot w(a) \le c(a), \end{aligned}$$

and, hence, y_S is feasible for the dual LP of MIN-SMC. In addition, we observe that

$$\begin{aligned} \sum_{S \in 2^E} \Delta_{E-S} f(S)\, y_S = \frac{1}{H(\gamma)} \Bigg(& \frac{c(x_1)}{r_1} (f(E) - f(A_0)) \\ & + \sum_{j=2}^{k} \left(\frac{c(x_j)}{r_j} - \frac{c(x_{j-1})}{r_{j-1}} \right) \cdot (f(E) - f(A_{j-1})) \Bigg). \end{aligned}$$

Thus, from $f(A_k) = f(E)$, we have

$$\begin{aligned}
c(A_k) &= \sum_{i=1}^{k} c(x_i) = \sum_{i=1}^{k} \frac{c(x_i)}{r_i}(f(A_i) - f(A_{i-1})) \\
&= \frac{c(x_1)}{r_1}(f(E) - f(A_0)) + \sum_{j=2}^{k}\Big(\frac{c(x_j)}{r_j} - \frac{c(x_{j-1})}{r_{j-1}}\Big)(f(E) - f(A_{j-1})) \\
&= H(\gamma) \sum_{S \in 2^E} \Delta_{E-S} f(S)\, y_S \le H(\gamma) \cdot opt,
\end{aligned}$$

where *opt* is the minimum value of the objective function of LP (8.19). So, we have obtained a new proof for Theorem 2.29 using the duality theory of linear programming.

Exercises

8.1 Consider the dual linear program (8.6) of the relaxation of MIN-WVC. A dual feasible solution $\boldsymbol{y}$ is *maximal* if no $\boldsymbol{y}'$ exists such that $\boldsymbol{y}' \geq \boldsymbol{y}$ and $\sum_{\{v_i,v_j\}\in E} y'_{ij} > \sum_{\{v_i,v_j\}\in E} y_{ij}$. Define

$$x_i = \begin{cases} 1, & \text{if } \displaystyle\sum_{j:\{v_i,v_j\}\in E} y_{ij} = c_i, \\ 0, & \text{otherwise.} \end{cases}$$

Show that if $\boldsymbol{y}$ is a maximal dual feasible solution, then $\{v_i \mid x_i = 1\}$ is a 2-approximation for the optimal weighted vertex cover.

8.2 Consider the following approximation algorithm for MIN-WVC:

(1) Set $C \leftarrow \emptyset$.

(2) **For** each $v_i \in V$ **do** $w'_i \leftarrow c_i$.

(3) **While** $E \neq \emptyset$ **do** $\{E$ denotes the set of uncovered edges$\}$
  Choose an edge $\{v_i, v_j\} \in E$;
  If $w'_i \leq w'_j$
   then $C \leftarrow C \cup \{v_i\}$;
    $E \leftarrow E - \{\{v_i, v_k\} \mid \{v_i, v_k\} \in E\}$;
    $w'_j \leftarrow w'_j - w'_i$
   else $C \leftarrow C \cup \{v_j\}$;
    $E \leftarrow E - \{\{v_j, v_k\} \mid \{v_j, v_k\} \in E\}$;
    $w'_i \leftarrow w'_i - w'_j$.

(4) Output C. ■

Now, compute a dual feasible solution $\boldsymbol{y}$ along with the above algorithm as follows:

(i) Initially, in step (1), set $\boldsymbol{y} \leftarrow \mathbf{0}$.

(ii) In step (3), when an edge $\{v_i, v_j\}$ is chosen from E, set $y_{ij} \leftarrow \min\{w'_i, w'_j\}$.

Show that $\boldsymbol{y}$ is a maximal dual feasible solution (see Exercise 8.1 for definition) and $v_i \in C$ implies $\sum_{j:\{v_i,v_j\}\in E} y_{ij} = c_i$. Furthermore, show that C is a 2-approximation for MIN-WVC, running in time $O(n)$.

8.3 Consider the following approximation algorithm for MIN-WVC:

(1) Set $C \leftarrow \emptyset$.

(2) **For** each $v_i \in V$ **do** $w'_i \leftarrow c_i$.

(3) **While** $E \neq \emptyset$ **do**

Choose $v_i \in V$ satisfying $\dfrac{w'_i}{d_E(v_i)} = \min\limits_{k\in V-C} \dfrac{w'_k}{d_E(v_k)}$;

$\{d_E(v_i)$ is the number of edges in E with endpoint v_i.$\}$

For each $v_k \in V$ with $\{v_i, v_k\} \in E$ **do** $w'_k \leftarrow w'_k - \dfrac{w'_i}{d_E(v_i)}$;

$C \leftarrow C \cup \{v_i\}$;

$E \leftarrow E - \{\{v_i, v_k\} \mid \{v_i, v_k\} \in E\}$.

(4) Output C. ■

Compute a dual feasible solution $\boldsymbol{y}$ along with the above algorithm as follows:

(i) Initially, in step (1), set $\boldsymbol{y} \leftarrow \mathbf{0}$.

(ii) In step (3), when a vertex v_i is chosen, set $y_{ik} \leftarrow w'_i/d_E(v_i)$ for each $v_k \in V$ such that $\{v_i, v_k\} \in E$.

Show that $\boldsymbol{y}$ is a maximal dual feasible solution (see Exercise 8.1 for definition), and $v_i \in C$ implies $\sum_{j:\{v_i,v_j\}\in E} y_{ij} = c_i$. Furthermore, show that C is a 2-approximation for MIN-WVC.

8.4 Consider the problem GC as defined in (8.8). The following is a modification of Algorithm 8.B for the general case of GC. Explain why this algorithm is not an approximation algorithm for GC.

(1) Set $\boldsymbol{x}^0 \leftarrow \mathbf{0}$; $\boldsymbol{y}^0 \leftarrow \mathbf{0}$; $k \leftarrow 0$.

(2) **While** $\boldsymbol{x}^k$ is not primal feasible **do**

$J_k \leftarrow \{j \mid 1 \le j \le n,\ x_j^k = 0\}$;

$I_k \leftarrow \{i \mid 1 \le i \le m,\ \sum_{j=1}^n a_{ij}x_j^k \le b_i - 1\}$;

Choose $r \in J_k$ such that

$$\frac{c_r - \sum_{i=1}^m a_{ir}y_i^k}{\sum_{i\in I_k} a_{ir}} = \alpha = \min_{j\in J_k} \frac{c_j - \sum_{i=1}^m a_{ij}y_i^k}{\sum_{i\in I_k} a_{ij}};$$

For $j \leftarrow 1$ **to** n **do**

if $j = r$ **then** $x_j^{k+1} \leftarrow 1$ **else** $x_j^{k+1} \leftarrow x_j^k$;

For $i \leftarrow 1$ **to** m **do**

if $i \in I_k$ **then** $y_i^{k+1} \leftarrow y_i^k + \alpha$ **else** $y_i^{k+1} \leftarrow y_i^k$;

$k \leftarrow k + 1$.

(3) Output $\boldsymbol{x}^A = \boldsymbol{x}^k$. ∎

8.5 Recall the weighted version of the set cover problem MIN-WSC defined in Section 2.4. The following is an LP-relaxation of MIN-WSC:

$$\begin{aligned}
\text{minimize} \quad & \sum_{j=1}^{n} w_j x_j \\
\text{subject to} \quad & \sum_{j=1}^{n} |S_j \cap T| x_j \geq |T|, && T \subseteq S \\
& x_j \geq 0, && j = 1, 2, \ldots, n,
\end{aligned}$$

where S is the given set and $\mathcal{C} = \{S_j \mid j = 1, 2, \ldots, n\}$ is the given family. Based on this formulation, design an approximation algorithm for MIN-WSC. Discuss the relationships between your algorithm and that of Exercise 8.3 for MIN-VC.

8.6 Design a primal-dual approximation algorithm for the problem MIN-WSC.

8.7 Consider the following problem:

PRIZE COLLECTING VERTEX COVER: Given a graph $G = (V, E)$ with vertex weight and edge weight $w : V \cup E \to \mathbb{N}$, find a vertex subset C to minimize

$$\sum_{u \in C} w(u) + \sum_{\{u,v\} \in E,\, u \notin C,\, v \notin C} w(\{u, v\}).$$

(a) Show that the following local ratio algorithm is a 2-approximation for this problem:

While $\exists \{u, v\} \in E$ with $\min\{w(u), w(v), w(\{u, v\})\} > 0$ **do**

Set $\varepsilon \leftarrow \min\{w(u), w(v), w(\{u, v\})\}$;

$w(u) \leftarrow w(u) - \varepsilon$;

$w(v) \leftarrow w(v) - \varepsilon$;

$w(\{u, v\}) \leftarrow w(\{u, v\}) - \varepsilon$.

Return $C = \{u \mid w(u) = 0\}$. ∎

(b) Design a primal-dual algorithm for this problem that is equivalent to the above algorithm.

8.8 Consider the network design problem given in Section 8.3. Prove the following properties to get an improvement over Theorem 8.9.

(a) Suppose f is a 0–1 downward monotone function. Then, for any $\boldsymbol{x}$, by Lemma 8.8, every minimal violated set S is a connected component of graph $G_{\boldsymbol{x}}$. However, not every connected component is a minimal violated set. Suppose $\boldsymbol{x}^*$ is a minimal primal feasible solution and $F^* = \{e \mid x_e^* = 1\}$. Let H^* be the graph obtained from $G_{\boldsymbol{x}}$ by adding edges in F^* to it. Show that each connected component of H^* contains at most one connected component of $G_{\boldsymbol{x}}$ which is not a minimal violated set.

(b) Show that if f is a 0–1 downward monotone function, then Algorithm 8.E is a 2-approximation for NETWORK DESIGN.

8.9 Consider the problem NETWORK DESIGN given in Section 8.3. Suppose f is a 0–1 downwards monotone function. Show that the following algorithm is a 2-approximation for it.

(1) $T \leftarrow MST(G)$. $\{MST(G)$ is the minimum spanning tree of $G.\}$

(2) Sort edges of T in the nonincreasing order of cost.
  {Without loss of generality, assume $c(e_1) \geq c(e_2) \geq \cdots \geq c(e_n)$.}

(3) **For** $j = 1$ **to** n **do**
  if $T - \{e_j\}$ is feasible **then** $T \leftarrow T - \{e_j\}$. ∎

8.10 Consider the problem FACILITY LOCATION.

(a) Design a primal-dual schema for FACILITY LOCATION based on the ideas presented in Section 8.5, and prove that if this algorithm outputs a primal feasible solution, then the solution is a 3-approximation to the optimal solution.

(b) Can you prove that the algorithm you designed above always produces a feasible solution?

8.11 Design a primal-dual approximation algorithm for the problem PVC with performance ratio 2.

8.12 A *tournament* is a directed graph $G = (V, E)$ without self-loops such that for any two vertices u and v, either $(u, v) \in E$ or $(v, u) \in E$, but not both.

(a) Show that a tournament contains a cycle if and only if it contains a triangle (a cycle of size 3).

(b) Use part (a) above to design a local ratio approximation for the problem FVS on tournaments with performance ratio 3.

(c) Design a primal-dual approximation for the problem FVS on tournaments with performance ratio 3.

8.13 A *t-interval system* is a collection $\{\mathcal{I}_1, \mathcal{I}_2, \ldots, \mathcal{I}_n\}$ of nonempty sets each of at most t disjoint real intervals. A *t-interval graph* $G = (V, E)$ is the intersection of a t-interval system $\{\mathcal{I}_1, \mathcal{I}_2, \ldots, \mathcal{I}_n\}$; i.e., $V = \{\mathcal{I}_1, \mathcal{I}_2, \ldots, \mathcal{I}_n\}$ and $\{\mathcal{I}_i, \mathcal{I}_j\} \in$

E if and only if $A \cap B \neq \emptyset$ for some intervals $A \in \mathcal{I}_i$ and $B \in \mathcal{I}_j$. Let R be the set of right endpoints of intervals in the system. Given a t-interval graph $G = (V, E)$ with nonnegative node weight $w : V \to \mathbb{N}$, we consider the problem MAX-WIS, i.e., the problem of finding a maximum-weight independent set in G. Let $\boldsymbol{x}^*$ be an optimal solution of the following linear program:

$$\begin{aligned} \text{maximize} \quad & \sum_{u \in V} w(u) x_u \\ \text{subject to} \quad & \sum_{u : p \in\in u} x_u \leq 1, \qquad p \in R, \\ & 0 \leq x_u \leq 1, \qquad u \in V, \end{aligned}$$

where $p \in\in u$ means p belongs to an interval $A \in u$.

(a) Recall that $V_+ = \{u \in V \mid w(u) > 0\}$ and, for each $v \in V$, $N(v)$ is the set consisting of v and all its neighbors. Choose $v \in V_+$ to minimize $\sum_{u \in N(v) \cap V_+} x_u^*$. Show that $\sum_{u \in N(v) \cap V_+} x_u^* \leq 2t$.

(b) Design a local ratio algorithm that is a $(2t)$-approximation for MAX-WIS on t-interval graphs.

8.14 For a vertex v in a graph $G = (V, E)$, let $\deg(v)$ denote the degree of the vertex v and $\delta(v)$ the set of neighbors of v in V. Consider the following problem:

> Given a simple graph $G = (V, E)$ and an integer $t \geq 0$, find the minimum subset $D \subseteq V$ such that $D_0 \cup D_1 \cup \cdots \cup D_t = V$, where $D_0 = D$ and $D_{i+1} = \{v \mid |(D_0 \cup \cdots \cup D_i) \cap \delta(v)| \geq \deg(v)/2\}$.

(a) Find an integer linear programming formulation for this problem.

(b) Construct a greedy approximation for this problem with performance ratio $O(\log(t\delta))$, where δ is the maximum vertex degree of the input graph G.

Historical Notes

The primal-dual method for linear programming was proposed by Dantzig, Ford, and Fulkerson [1956]. The primal-dual approximation as a modified version of this method was first used by Bar-Yehuda and Even [1981] for the weighted set cover problem. Since then, the primal-dual schema has become a major technique for the design of approximations for covering-type problems, including many network design problems [Agrawal et al., 1995; Goemans and Williamson, 1995a, 1997; Ravi and Klein, 1993; Williamson et al., 1995; Bertsimas and Teo, 1998]. Exercises 8.8 and 8.9 are from Goemans and Williamson [1997].

The initial idea of primal-dual approximation is to enforce the primal complementary slackness condition and relax the dual complementary slackness conditions. Jain and Vazirani [2001] presented ideas of primal-dual schemas to enforce the dual complementary slackness condition and relax the primal complementary slackness condition for the noncovering-type problems FACILITY LOCATION and k-MEDIAN. It is, however, not clear how to implement the ideas. For the special case of METRIC FACILITY LOCATION, the currently best-known lower bound for the approximation ratio is 1.463 [Guha and Khuller, 1998c], and the best-known upper bound is 1.5 [Mahdian et al., 2002; Byrka, 2007].

The primal complementary slackness condition is the root of the equivalence of the primal-dual schema and local ratio method. The local ratio method was first proposed by Bar-Yehuda and Even [1985]. Later, this method has been used to design approximation algorithms for the feedback vertex set problem [Bafna et al., 1999], the node deletion problem [Fujito, 1998], resource allocation and scheduling problems [Bar-Noy et al., 2001], the minimum s-t cut problem, the assignment problems [Bar-Yehuda and Rawitz, 2004], and MAX-WIS on t-interval graphs (Exercise 8.13) [Bar-Yehuda et al. 2004]. Bar-Yehuda and Rawitz [2005a] gave a framework for describing the equivalence between the primal-dual schema and local ratio method for the covering-type problems. Other interesting issues on the primal-dual schema and the local ratio method can be found in Bar-Yehuda and Rawitz [2004, 2005b], Freund and Rawitz [2003], and Jain et al. [2003].

Wolsey [1982] was the first to analyze the greedy approximation for MIN-SMC with the primal-dual method. This method has been extended to more general problems [Fujito, 1999; Fujito and Yabuta, 2004; Chvátal, 1979]. Exercise 8.14 is from Wang et al. [2009].

9
Semidefinite Programming

A set definite objective must be established
if we are to accomplish anything in a big way.
— John McDonald

Semidefinite programming studies optimization problems with a linear objective function over semidefinite constraints. It shares many interesting properties with linear programming. In particular, a semidefinite program can be solved in polynomial time. Moreover, an integer quadratic program can be transformed into a semidefinite program through relaxation. Therefore, if a combinatorial optimization problem can be formulated as an integer quadratic program, then we can approximate it using the semidefinite programming relaxation and other related techniques such as the primal-dual schema. As the semidefinite programming relaxation is a higher-order relaxation, it often produces better results than the linear programming relaxation, even if the underlying problem can be formulated as an integer linear program. In this chapter, we introduce the fundamental concepts of semidefinite programming, and demonstrate its application to the approximation of **NP**-hard combinatorial optimization problems, with various rounding techniques.

9.1 Spectrahedra

Let $\mathcal{S}_n$ be the family of symmetric matrices of order n over real numbers. Recall that if a square matrix $\boldsymbol{A}$ over real numbers is symmetric, then all of its eigenvalues are real. If, in addition, all the eigenvalues of $\boldsymbol{A}$ are nonnegative, then $\boldsymbol{A}$ is called a *positive semidefinite matrix*. Also, if all eigenvalues are positive, then it is called a *positive definite matrix*. Consider any two matrices $\boldsymbol{A} = (a_{ij})_{n\times n}$, $\boldsymbol{B} = (b_{ij})_{n\times n}$ in $\mathcal{S}_n$. The *Frobenius inner product* of $\boldsymbol{A}$ and $\boldsymbol{B}$ is defined to be

$$\boldsymbol{A} \bullet \boldsymbol{B} = \text{Tr}(\boldsymbol{A}^T\boldsymbol{B}) = \sum_{i=1}^{n}\sum_{j=1}^{n} a_{ij}b_{ij}.$$

That is, if we treat each of $\boldsymbol{A}$ and $\boldsymbol{B}$ as an n^2-dimensional vector, then the Frobenius inner product is just the inner product of two vectors. If $\boldsymbol{A} - \boldsymbol{B}$ is positive semidefinite, then we write $\boldsymbol{A} \succeq \boldsymbol{B}$. If $\boldsymbol{A} - \boldsymbol{B}$ is positive definite, then we write $\boldsymbol{A} \succ \boldsymbol{B}$.

Positive semidefinite matrices have a number of useful characterizations. We list some of them below.

Proposition 9.1 *Let $\boldsymbol{A}$ be a matrix in $\mathcal{S}_n$. Then the following are equivalent:*

(i) $\boldsymbol{A}$ is positive semidefinite.

(ii) For any $\boldsymbol{x} \in \mathbb{R}^n$, $\boldsymbol{x}^T\boldsymbol{A}\boldsymbol{x} \geq 0$.

(iii) $\boldsymbol{A} = \boldsymbol{V}^T\boldsymbol{V}$ for some matrix $\boldsymbol{V}$.

It is useful to consider the geometric meaning of a semidefinite inequality. For given matrices $\boldsymbol{Q}_0, \boldsymbol{Q}_1, \ldots, \boldsymbol{Q}_m$, the solution set of a semidefinite inequality

$$\mathcal{S} = \left\{ \boldsymbol{x} \;\middle|\; \sum_{i=1}^{n} x_i\boldsymbol{Q}_i \preceq \boldsymbol{Q}_0 \right\}$$

is a closed convex set and is called a *spectrahedron*. This spectrahedron may be viewed as a generalization of the polyhedron defined by a system of linear inequalities:

$$\mathcal{P} = \{\boldsymbol{x} \mid \boldsymbol{A}\boldsymbol{x} \leq \boldsymbol{b}\},$$

where $\boldsymbol{A}$ is an $m \times n$ matrix and $\boldsymbol{b}$ is an m-dimensional vector. In fact, suppose $\boldsymbol{A} = (\boldsymbol{a}_1, \boldsymbol{a}_2, \ldots, \boldsymbol{a}_n)$, where each $\boldsymbol{a}_i$ is an m-dimensional vector. Then $\mathcal{P}$ may be represented as the spectrahedron of the following form:

$$\left\{ \boldsymbol{x} \;\middle|\; \sum_{i=1}^{n} x_i \cdot \text{Diag}(\boldsymbol{a}_i) \preceq \text{Diag}(\boldsymbol{b}) \right\},$$

where

$$\text{Diag}(\boldsymbol{b}) = \begin{pmatrix} b_1 & 0 & \cdots & 0 \\ 0 & b_2 & \cdots & 0 \\ \vdots & \vdots & \ddots & \vdots \\ 0 & 0 & \cdots & b_m \end{pmatrix}.$$

Spectrahedra share many properties with polyhedra. The following is an example.

Proposition 9.2 *The intersection of two spectrahedra is still a spectrahedron.*

Proof. Consider two spectrahedra

$$\mathcal{G} = \left\{ \boldsymbol{x} \;\middle|\; \sum_{i=1}^{m} x_i \boldsymbol{G}_i \preceq \boldsymbol{G}_0 \right\}, \qquad \mathcal{H} = \left\{ \boldsymbol{x} \;\middle|\; \sum_{i=1}^{m} x_i \boldsymbol{H}_i \preceq \boldsymbol{H}_0 \right\}.$$

Define

$$\boldsymbol{Q}_i = \begin{pmatrix} \boldsymbol{G}_i & \\ & \boldsymbol{H}_i \end{pmatrix}.$$

Note that two symmetric matrices $\boldsymbol{A}$ and $\boldsymbol{B}$ are both positive semidefinite if and only if the matrix

$$\begin{pmatrix} \boldsymbol{A} & \\ & \boldsymbol{B} \end{pmatrix}$$

is positive semidefinite. Now, we observe that

$$\mathcal{G} \cap \mathcal{H} = \left\{ \boldsymbol{x} \;\middle|\; \sum_{i=1}^{m} x_i \boldsymbol{Q}_i \preceq \boldsymbol{Q}_0 \right\},$$

and so it is a spectrahedron. □

An immediate consequence of this proposition is that, for any matrices $\boldsymbol{Q}_1$, $\boldsymbol{Q}_2, \ldots, \boldsymbol{Q}_m$ and real numbers $c_1, c_2, \ldots, c_m$, the set

$$\Omega = \{ \boldsymbol{U} \mid \boldsymbol{Q}_i \bullet \boldsymbol{U} = c_i,\ i = 1, 2, \ldots, m;\ \boldsymbol{U} \succeq \boldsymbol{0} \}$$

is a spectrahedron because Ω is the intersection of a polyhedron

$$\{ \boldsymbol{U} \mid \boldsymbol{U} \bullet \boldsymbol{Q}_i = c_i, 1 \le i \le m \}$$

with a spectrahedron

$$\{ \boldsymbol{U} \mid \boldsymbol{U} \succeq \boldsymbol{0} \}.$$

9.2 Semidefinite Programming

A *semidefinite program* is a maximization or minimization problem with a linear objective function whose feasible domain is a spectrahedron. It shares many properties with a linear program.

A standard form of the semidefinite program is as follows:

$$\begin{aligned} &\text{minimize} && \boldsymbol{U} \bullet \boldsymbol{Q}_0 \\ &\text{subject to} && \boldsymbol{U} \bullet \boldsymbol{Q}_i = c_i, \qquad i = 1, 2, \ldots, m, \\ & && \boldsymbol{U} \succeq \boldsymbol{0}, \end{aligned} \tag{9.1}$$

where $\boldsymbol{Q}_0, \boldsymbol{Q}_1, \ldots, \boldsymbol{Q}_m$ are given linearly independent symmetric matrices of order n, and $c_1, \ldots, c_m$ are given constants. As we pointed out in the last section, its feasible domain

$$\Omega = \{\boldsymbol{U} \mid \boldsymbol{U} \bullet \boldsymbol{Q}_i = c_i,\ 1 \le i \le m;\ \boldsymbol{U} \succeq \mathbf{0}\}$$

is a spectrahedron.

The semidefinite program (9.1) has a dual program

$$\begin{array}{ll} \text{maximize} & \mathbf{c}^T \boldsymbol{x} \\ \text{subject to} & \sum_{i=1}^{m} x_i \boldsymbol{Q}_i \preceq \boldsymbol{Q}_0, \end{array} \tag{9.2}$$

where $\mathbf{c} = (c_1, c_2, \ldots, c_m)^T$.

The primal program (9.1) and the dual program (9.2) have the following relations:

Lemma 9.3 *Suppose $\boldsymbol{U}$ is a primal feasible solution of* (9.1) *and $\boldsymbol{x}$ a dual feasible solution of* (9.2)*. Then $\mathbf{c}^T\boldsymbol{x} \le \boldsymbol{U} \bullet \boldsymbol{Q}_0$. In addition, if $\mathbf{c}^T\boldsymbol{x} = \boldsymbol{U} \bullet \boldsymbol{Q}_0$, then $\boldsymbol{U}$ and $\boldsymbol{x}$ are, respectively, the optimal primal and dual solutions.*

Proof. We observe that

$$\mathbf{c}^T \boldsymbol{x} = \sum_{i=1}^{m} c_i x_i = \sum_{i=1}^{m} (\boldsymbol{U} \bullet \boldsymbol{Q}_i) x_i = \boldsymbol{U} \bullet \left(\sum_{i=1}^{m} x_i \boldsymbol{Q}_i \right).$$

Now, we note that the trace of the product of two positive semidefinite matrices must be nonnegative [see Exercise 9.1(b)]. Thus, we have

$$\begin{aligned} \boldsymbol{U} \bullet \boldsymbol{Q}_0 - \mathbf{c}^T \boldsymbol{x} &= \boldsymbol{U} \bullet \left(\boldsymbol{Q}_0 - \sum_{i=1}^{m} x_i \boldsymbol{Q}_i \right) \\ &= \mathrm{Tr}\left(\boldsymbol{U} \left(\boldsymbol{Q}_0 - \sum_{i=1}^{m} x_i \boldsymbol{Q}_i \right) \right) \ge 0. \end{aligned}$$

Clearly, if $\boldsymbol{U} \bullet \boldsymbol{Q}_0 = \mathbf{c}^T \boldsymbol{x}$, then $\boldsymbol{U}$ must be an optimal primal solution to (9.1) and $\boldsymbol{x}$ an optimal dual solution to (9.2). □

Semidefinite programs have an equivalent form called *vector programs*. A vector program is an optimization problem on vector variables, with a linear objective function and linear constraints with respect to inner products between the vector variables. The following is an example of a vector program on n vector variables $\boldsymbol{v}_1, \boldsymbol{v}_2, \ldots, \boldsymbol{v}_n$:

$$\begin{array}{lll} \text{maximize} & \frac{1}{4} \sum_{1 \le i,j \le n} w_{ij}(1 - \boldsymbol{v}_i \cdot \boldsymbol{v}_j) & \\ \text{subject to} & \sum_{i=1}^{n} \sum_{j=1}^{n} \boldsymbol{v}_i \cdot \boldsymbol{v}_j = 0, & \\ & \boldsymbol{v}_i \cdot \boldsymbol{v}_i = 1, & i = 1, 2, \ldots, n. \end{array} \tag{9.3}$$

To see the relations between semidefinite programs and vector programs, we note, from Proposition 9.1, that every positive semidefinite matrix U can be expressed as $U = V^T V$ for some matrix V. Thus, we can convert a semidefinite program (9.1) into a vector program as follows: Let $V = (v_1, v_2, \ldots, v_n)$. Substituting $U = V^T V$ into the semidefinite program (9.1), we obtain the following equivalent vector program:

$$\begin{array}{lll} \text{minimize} & Q_0 \bullet V^T V & \\ \text{subject to} & Q_i \bullet V^T V = c_i, & \text{for } i = 1, 2, \ldots, m. \end{array}$$

Conversely, for each vector program, we can obtain an equivalent semidefinite program by replacing $v_i \cdot v_j$ with variable u_{ij}. For instance, the above vector program (9.3) can be converted into the following equivalent semidefinite program:

$$\begin{array}{lll} \text{maximize} & \frac{1}{4} W \bullet (J - U) & \\ \text{subject to} & J \bullet U = 0, & \\ & u_{ii} = 1, & i = 1, 2, \ldots, n, \\ & U \succeq \mathbf{0}. & \end{array} \tag{9.4}$$

where $W = (w_{ij})$, $U = (u_{ij})$, and J is the $n \times n$ matrix with all entries having value 1.

Thus, for a given vector program such as (9.3), we can solve it as follows: We first convert it into a semidefinite program (9.4). Then we solve (9.4) to get a positive semidefinite matrix solution U. Finally, we compute matrix V such that $U = V^T V$. The computation of the last step is called the *Cholesky factorization*. In the following, we show that it can be done in time $O(n^3)$.

We first show a simple lemma about submatrices of a positive semidefinite matrix.

Lemma 9.4 *Let U be a positive semidefinite matrix of order n. Assume that*

$$U = \begin{pmatrix} a & b^T \\ b & N \end{pmatrix},$$

where $a \in \mathbb{R}$ and $b \in \mathbb{R}^{n-1}$.

(a) If $a > 0$, then $N - \frac{1}{a} bb^T \succeq \mathbf{0}$.

(b) If $a = 0$, then $b = \mathbf{0}$.

Proof. (a) We prove this result by the characterization (ii) of Proposition 9.1. For any $x \in \mathbb{R}^{n-1}$,

$$x^T \Big(N - \frac{1}{a} bb^T \Big) x = \Big(-\frac{1}{a} b^T x, x^T \Big) U \begin{pmatrix} -\frac{1}{a} b^T x \\ x \end{pmatrix} \geq 0.$$

Hence, $N - \frac{1}{a} bb^T \succeq \mathbf{0}$.

(b) For the sake of contradiction, suppose $b \neq \mathbf{0}$. Note that N is also positive semidefinite. Choose $c > b^T N b / (2\|b\|^2)$. Then

$$(-c,\ b^T) \begin{pmatrix} 0 & b^T \\ b & N \end{pmatrix} \begin{pmatrix} -c \\ b \end{pmatrix} = -2c\|b\|^2 + b^T N b < 0,$$

contradicting the assumption that U is positive semidefinite. □

Now, we are ready to present the $O(n^3)$-time algorithm for Cholesky factorization.

Theorem 9.5 *Given a positive semidefinite matrix U, we can compute a matrix V satisfying $U = V^T V$ in $O(n^3)$ time.*

Proof. We prove the theorem by induction on n. For $n = 1$, suppose $U = (a)$. Then $V = (\sqrt{a})$.

For $n \geq 2$, suppose

$$U = \begin{pmatrix} a & b^T \\ b & N \end{pmatrix} \succeq \mathbf{0},$$

where $a \in \mathbb{R}$ and $b \in \mathbb{R}^{n-1}$. Then a is nonnegative.

If $a > 0$, then we can express U as

$$U = \begin{pmatrix} \sqrt{a} & \mathbf{0}^T \\ \frac{1}{\sqrt{a}} b & I_{n-1} \end{pmatrix} \begin{pmatrix} 1 & \mathbf{0}^T \\ \mathbf{0} & N - \frac{1}{a} bb^T \end{pmatrix} \begin{pmatrix} \sqrt{a} & \frac{1}{\sqrt{a}} b^T \\ \mathbf{0} & I_{n-1} \end{pmatrix}.$$

By Lemma 9.4(a), $N - \frac{1}{a} bb^T \succeq 0$. Thus, we can compute its Cholesky factorization

$$N - \frac{1}{a} bb^T = M^T M$$

recursively, and get

$$U = \begin{pmatrix} \sqrt{a} & \frac{1}{\sqrt{a}} b^T \\ \mathbf{0} & M \end{pmatrix}^T \begin{pmatrix} \sqrt{a} & \frac{1}{\sqrt{a}} b^T \\ \mathbf{0} & M \end{pmatrix}.$$

If $a = 0$, then by Lemma 9.4(b),

$$U = \begin{pmatrix} 0 & \mathbf{0}^T \\ \mathbf{0} & N \end{pmatrix}$$

and $N \succeq \mathbf{0}$. Compute the Cholesky factorization

$$N = M^T M,$$

and we obtain

$$U = \begin{pmatrix} 0 & \mathbf{0}^T \\ \mathbf{0} & M \end{pmatrix}^T \begin{pmatrix} 0 & \mathbf{0}^T \\ \mathbf{0} & M \end{pmatrix}.$$

Since there are only $O(n)$ recursive steps, and since each step needs at most time $O(n^2)$ to compute $N - \frac{1}{a}\boldsymbol{b}\boldsymbol{b}^T$, the total computation time is $O(n^3)$. □

The most important property of semidefinite programs is their polynomial-time solvability.

Theorem 9.6 *Semidefinite programs can be solved within a factor* $1 + \varepsilon$ *from the optimal solutions in time polynomial in* $n + 1/\varepsilon$, *where* n *is the input size of the semidefinite program and* ε *is an arbitrary positive number.*

As the emphasis of this book is on the application, rather than the theory, of semidefinite programming, we omit the proof of the polynomial-time algorithm for semidefinite programming. The reader is referred to Alizadeh [1991] and de Klerk [2002] for details.

9.3 Hyperplane Rounding

In the remainder of this chapter, we present some applications of semidefinite programming in the design of approximation algorithms, together with various rounding techniques.

We first consider the following problem.

> MAX-CUT: Given a graph $G = (V, E)$, where $V = \{1, 2, \ldots, n\}$, and a nonnegative edge weight w_{ij} for each edge $\{i, j\} \in E$, find a cut $(S, V - S)$ of G that maximizes the total weight of the cut $\sum\{w_{ij} \mid \{i, j\} \in E, i \in S, j \in V - S\}$.

First, let us extend the weight w_{ij} to arbitrary pairs $(i, j) \in V \times V$, with $w_{ij} = 0$ if $\{i, j\} \notin E$. Then the problem MAX-CUT can be formulated as an integer linear program as follows:

$$\begin{array}{lll} \text{maximize} & \displaystyle\sum_{1 \le i < j \le n} w_{ij} x_{ij} & \\ \text{subject to} & x_{ij} \le 1 - \dfrac{y_i + y_j}{2}, & 1 \le i < j \le n, \\ & x_{ij} \le 1 + \dfrac{y_i + y_j}{2}, & 1 \le i < j \le n, \\ & y_i \in \{-1, 1\}, & 1 \le i \le n, \\ & x_{ij} \in \{0, 1\}, & 1 \le i < j \le n. \end{array}$$

If we relax the constraints $y_i \in \{-1, 1\}$ to $-1 \le y_i \le 1$, then it is easy to see that the optimal solution would be reached at $y_i = 0$, for all $i = 1, 2, \ldots, n$, and $x_{ij} = 1$, for all edges $\{i, j\} \in E$. This optimal solution nevertheless does not offer any help in finding an approximation for the original problem of MAX-CUT, as the feasible domain of the relaxed linear program is too big. In such a case, a general idea is to add some additional constraints to get a relaxed linear program with a smaller feasible domain. With this approach, we can obtain a linear programming-based 2-approximation for MAX-CUT.

On the other hand, as we will see below, the semidefinite programming relaxation on the following quadratic programming formulation will give us a better approximation:

$$\begin{array}{lll} \text{maximize} & \displaystyle\sum_{1 \le i < j \le n} w_{ij} \cdot \frac{1}{2}(1 - x_i x_j) & \\ \text{subject to} & x_i^2 = 1, & i = 1, 2, \ldots, n. \end{array}$$

First, we change this quadratic program to a vector program by substituting an n-dimensional vector $\boldsymbol{v}_i$ for the variable x_i. The constraint $x_i^2 = 1$ is thus replaced by the constraint $\boldsymbol{v}_i \in S_1$, where $S_1 = \{(1, 0, \ldots, 0)^T, (-1, 0, \ldots, 0)^T\}$.

$$\begin{array}{lll} \text{maximize} & \displaystyle\sum_{1 \le i < j \le n} \frac{1}{2} w_{ij}(1 - \boldsymbol{v}_i \cdot \boldsymbol{v}_j) & \\ \text{subject to} & \boldsymbol{v}_i \in S_1, & i = 1, 2, \ldots, n. \end{array}$$

Next, we further relax the constraint $\boldsymbol{v} \in S_1$ to $\boldsymbol{v} \in S_n$, where S_n is the n-dimensional unit sphere $S_n = \{\boldsymbol{y} \mid \|y\| = 1\}$, and arrive at the following vector program:

$$\begin{array}{lll} \text{maximize} & \displaystyle\sum_{1 \le i < j \le n} \frac{1}{2} w_{ij}(1 - \boldsymbol{v}_i \cdot \boldsymbol{v}_j) & \\ \text{subject to} & \boldsymbol{v}_i \cdot \boldsymbol{v}_i = 1, & i = 1, 2, \ldots, n. \end{array} \tag{9.5}$$

Finally, following the idea explained in the last section, we can convert this vector program into an equivalent semidefinite program as follows:

$$\begin{array}{lll} \text{maximize} & \displaystyle\frac{1}{4} \boldsymbol{W} \bullet (\boldsymbol{J} - \boldsymbol{U}) & \\ \text{subject to} & u_{ii} = 1, & i = 1, 2, \ldots, n, \\ & \boldsymbol{U} \succeq \boldsymbol{0}, & \end{array} \tag{9.6}$$

where $\boldsymbol{W} = (w_{ij})$, $\boldsymbol{U} = (u_{ij})$, and $\boldsymbol{J}$ is the $n \times n$ matrix with all entries having value 1.

Now, we can solve this semidefinite program and obtain, through the Cholesky factorization, the optimum solution $(\boldsymbol{v}_1, \boldsymbol{v}_2, \ldots, \boldsymbol{v}_n)$ for the vector program (9.5).

Note that the endpoints of these n vectors are all located on the unit sphere S_n. These points correspond to n vertices in the graph G. That is, the solution to the above semidefinite programming relaxation is an embedding of the graph G on the unit sphere S_n.

To obtain an approximation to the original instance of MAX-CUT, we need to partition these vertices into two parts and maintain as much weight between the two parts as possible. In other words, we need to *round* the solution and move each vertex to either $(1, 0, \ldots, 0)^T$ or $(-1, 0, \ldots, 0)^T$. A simple idea of this rounding is to select a hyperplane that passes through the origin to cut the unit sphere into two parts and move the vertices in one part to $(1, 0, \ldots, 0)^T$ and the vertices in the other part to $(-1, 0, \ldots, 0)^T$. As it appears hard to find such a hyperplane by a deterministic method that maintains near-optimal weight between the two parts, we resort to a simple random method. That is, we simply select a random hyperplane uniformly and show that the expected weight between the two parts is high. Algorithmically, this is equivalent to first selecting a random normal vector $\boldsymbol{a}$ of a hyperplane uniformly on the unit sphere, and then setting $x_i = 1$ or $x_i = -1$ depending on whether $\boldsymbol{a}^T \boldsymbol{v}_i \geq 0$ or $\boldsymbol{a}^T \boldsymbol{v}_i < 0$, respectively. This method is called *hyperplane rounding*. We summarize it as follows.

Algorithm 9.A (*Semidefinite Programming Approximation for* MAX-CUT)
Input: A graph $G = (V, E)$ and nonnegative edge weights w_{ij}, for $i, j \in V$.

(1) Construct the semidefinite program (9.6).

(2) Solve the semidefinite program (9.6);
Compute $\boldsymbol{v}_1, \boldsymbol{v}_2, \ldots, \boldsymbol{v}_n$ by Cholesky factorization.

(3) Choose a random vector $\boldsymbol{a}$ uniformly from S_n;
For $i \leftarrow 1$ **to** n **do**
  if $\boldsymbol{a}^T \boldsymbol{v}_i \geq 0$ **then** $x_i \leftarrow 1$ **else** $x_i \leftarrow -1$.

(4) Output the cut $(S, V - S)$, where $S = \{i \mid x_i = 1\}$. ■

To evaluate the performance of this approximation, we first show the following two lemmas.

Lemma 9.7 *Assume that x_i and x_j are defined from vectors $\boldsymbol{v}_i$ and $\boldsymbol{v}_j$ as in step* (3) *of* Algorithm 9.A. *Then we have*

$$\Pr[x_i x_j = -1] = \frac{\arccos \boldsymbol{v}_i^T \boldsymbol{v}_j}{\pi}.$$

Proof. Let P be the two-dimensional plane spanned by vectors $\boldsymbol{v}_i$ and $\boldsymbol{v}_j$. The hyperplane with normal vector $\boldsymbol{a}$ separates $\boldsymbol{v}_i$ and $\boldsymbol{v}_j$ if and only if the projection of $\boldsymbol{a}$ onto plane P lies in the two dark regions shown in Figure 9.1. Each region is a fan-shaped area with the angle equal to the angle formed by the two vectors $\boldsymbol{v}_i$ and $\boldsymbol{v}_j$, whose size is $\arccos \boldsymbol{v}_i^T \boldsymbol{v}_j$. The lemma follows from this observation. □

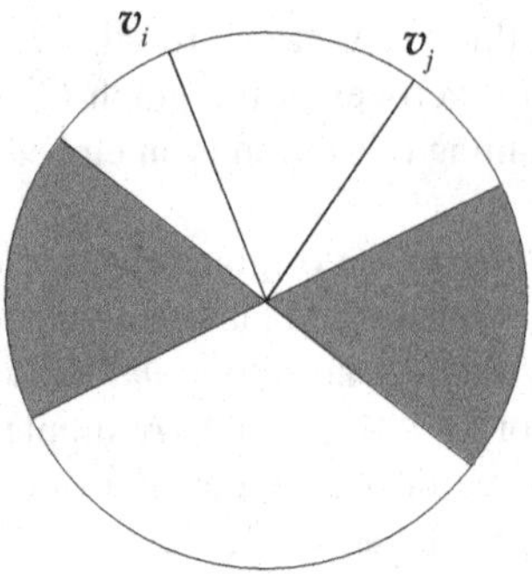

Figure 9.1: The area of the normal vectors $\boldsymbol{a}$ that separate $\boldsymbol{v}_i$ from $\boldsymbol{v}_j$.

Lemma 9.8 *For* $0 \le \theta \le \pi$,

$$\frac{\theta}{\pi} \ge \alpha \cdot \frac{1-\cos\theta}{2},$$

where $\alpha = 0.878567$.

Proof. First, we note that

$$\frac{\theta}{\pi} = \frac{1-\cos\theta}{2},$$

for $\theta = 0, \pi/2$, or π. Moreover,

$$\left(\frac{1-\cos\theta}{2}\right)'' = \frac{\cos\theta}{2} \ge 0,$$

for $0 \le \theta \le \pi/2$; that is, $(1-\cos\theta)/2$ is convex on $[0, \pi/2]$. Therefore, we have

$$\frac{\theta}{\pi} \ge \frac{1-\cos\theta}{2},$$

for $\theta \in [0, \pi/2]$ (cf. Figure 9.2).

Next, we consider the case of $\theta \in [\pi/2, \pi]$. Define

$$f(\theta) = \frac{\theta}{\pi} - \alpha \cdot \frac{1-\cos\theta}{2},$$

where $\alpha = 0.878567$. Then,

$$f'(\theta) = \frac{1}{\pi} - \alpha \cdot \frac{\sin\theta}{2} \quad \text{and} \quad f''(\theta) = -\frac{\alpha}{2} \cdot \cos\theta.$$

Note that $f''(\theta) \ge 0$ for $\theta \in [\pi/2, \pi]$. Hence, $f(\theta)$ is convex on $[\pi/2, \pi]$. Also, note that $f'(\pi/2) = 1/\pi - \alpha/2 < 0$ and $f'(\pi) = 1/\pi > 0$. Thus, $f(\theta)$ is not monotone on $[\pi/2, \pi]$, and it reaches its minimum at the point $\theta^* = \pi - \arcsin(2/(\pi\alpha))$, where $f'(\theta^*) = 0$. The proof of the lemma is completed by verifying that

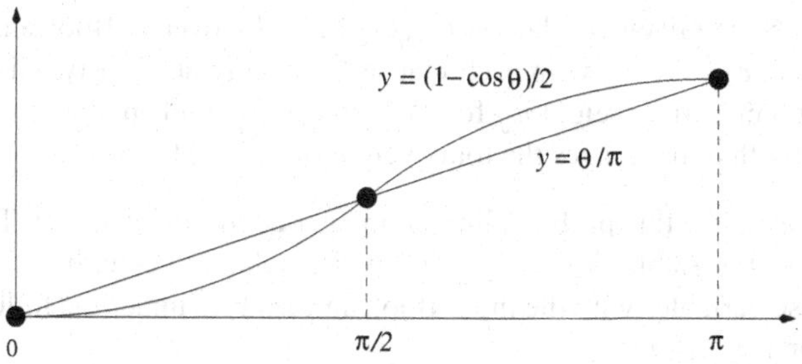

Figure 9.2: Function $(1 - \cos\theta)/2$ versus θ/π.

$$f(\theta^*) = \frac{\pi - \arcsin\frac{2}{\pi\alpha}}{\pi} - \alpha \cdot \frac{1 + \sqrt{1 - \left(\frac{2}{\pi\alpha}\right)^2}}{2} \geq 0. \qquad \square$$

From these two lemmas, we get the following performance ratio for Algorithm 9.A.

Theorem 9.9 *Let* opt_{CUT} *denote the objective function value of the optimum solution to* MAX-CUT. *We have*

$$E\left[\sum_{1\leq i<j\leq n} w_{ij} \cdot \frac{1}{2}(1 - x_i x_j)\right] \geq \alpha \cdot opt_{\text{CUT}},$$

where $\alpha = 0.878567$.[1]

Proof. The inequality can be derived as follows:

$$\begin{aligned} &E\left[\sum_{1\leq i<j\leq n} w_{ij} \cdot \frac{1}{2}(1 - x_i x_j)\right] \\ &= \sum_{1\leq i<j\leq n} w_{ij} \cdot E\left[\frac{1}{2}(1 - x_i x_j)\right] = \sum_{1\leq i<j\leq n} w_{ij} \cdot \frac{\arccos \boldsymbol{v}_i^T \boldsymbol{v}_j}{\pi} \\ &\geq \alpha \sum_{1\leq i<j\leq n} w_{ij} \cdot \frac{1 - \boldsymbol{v}_i^T \boldsymbol{v}_j}{2} \geq \alpha \cdot opt_{\text{CUT}}. \end{aligned} \qquad \square$$

Finally, we remark that the above random rounding can be derandomized by a standard, but nontrivial derandomization technique. The reader is referred to Mahajan and Ramesh [1999] for details.

Next, we apply the hyperplane rounding technique to the following problem.

---

[1]In this chapter, we follow the literature in semidefinite programming–based approximation using $\inf_I \mathcal{A}(I)/opt(I)$, where I ranges over all input instances, to measure the performance of an approximation algorithm $\mathcal{A}$ on a maximization problem. For deterministic algorithms $\mathcal{A}$, this is the reciprocal of the performance ratio defined in Section 1.6.

MAX-2SAT: Given m clauses $C_1, C_2, \ldots, C_m$ over n Boolean variables $x_1, x_2, \ldots, x_n$, with each clause C_j having at most two literals, and a nonnegative weight w_j for each clause C_j, find an assignment to variables that maximizes the total weight of satisfied clauses.

We first formulate this problem into an integer quadratic program. To do so, we introduce $n+1$ variables $y_0, y_1, \ldots, y_n$, which take values either -1 or 1, and associate these variables with the input Boolean variables under the following interpretation: For $1 \le i \le n$,

$$x_i = \begin{cases} \text{TRUE}, & \text{if } y_i \neq y_0, \\ \text{FALSE}, & \text{if } y_i = y_0. \end{cases} \tag{9.7}$$

For convenience, we define $n+1$ additional variables $y_{n+1}, \ldots, y_{2n+1}$, and use the quadratic constraints $y_0 y_{2n+1} = 1$ and $y_i y_{n+i} = -1$, for $i = 1, 2, \ldots, n$, to make $y_{2n+1} = y_0$ and $y_{n+i} = -y_i$, for $i = 1, 2, \ldots, n$.

Under this setting, we can now encode each clause C_j, $1 \le j \le m$, by some quadratic inequalities over these integer variables. We first define, for each j, $1 \le j \le m$, two integers j_1 and j_2 as follows:

(1) If C_j contains only one literal x_i (or, $\bar{x}_i$), then let $j_1 = i$ (or, respectively, $j_1 = n + i$) and $j_2 = 2n + 1$.

(2) If $C_j = x_i \vee x_{i'}$, then let $j_1 = i$ and $j_2 = i'$.

(3) If $C_j = x_i \vee \bar{x}_{i'}$, then let $j_1 = i$ and $j_2 = n + i'$.

(4) If $C_j = \bar{x}_i \vee \bar{x}_{i'}$, then let $j_1 = n + i$ and $j_2 = n + i'$.

With these choices of j_1 and j_2 and the interpretation (9.7), we get the following relationship between clause C_j and the three variables y_0, y_{j_1}, and y_{j_2}:

$$C_j = \text{FALSE} \iff y_0 = y_{j_1} = y_{j_2}.$$

Or, equivalently,

$$C_j = \text{TRUE} \iff \frac{3 - y_0 y_{j_1} - y_0 y_{j_2} - y_{j_1} y_{j_2}}{4} = 1,$$

$$C_j = \text{FALSE} \iff \frac{3 - y_0 y_{j_1} - y_0 y_{j_2} - y_{j_1} y_{j_2}}{4} = 0.$$

From this, we obtain the following integer quadratic program for MAX-2SAT:

$$\begin{aligned} \text{maximize} \quad & \sum_{j=1}^{m} w_j \cdot \frac{3 - y_0 y_{j_1} - y_0 y_{j_2} - y_{j_1} y_{j_2}}{4} \\ \text{subject to} \quad & y_0 y_{2n+1} = 1, \\ & y_i y_{n+i} = -1, \qquad 1 \le i \le n, \\ & y_i^2 = 1, \qquad 0 \le i \le 2n+1. \end{aligned} \tag{9.8}$$

By a semidefinite programming relaxation similar to the one used for the problem MAX-CUT, we get the following semidefinite program:

$$\begin{aligned}
\text{maximize} \quad & \sum_{j=1}^{m} w_j \cdot \frac{3 - u_{0,j_1} - u_{0,j_2} - u_{j_1,j_2}}{4} \\
\text{subject to} \quad & u_{0,2n+1} = 1, \\
& u_{i,n+i} = -1, \qquad 1 \le i \le n, \\
& u_{ii} = 1, \qquad 0 \le i \le 2n+1, \\
& \boldsymbol{U} \succeq \boldsymbol{0},
\end{aligned} \tag{9.9}$$

where $\boldsymbol{U} = (u_{ij})_{0 \le i,j \le 2n+1}$.

We can now solve this semidefinite program and apply hyperplane rounding to get an approximation for MAX-2SAT.

Algorithm 9.B (*Semidefinite Programming Approximation for* MAX-2SAT)

Input: A CNF formula with clauses $C_1, \ldots, C_m$, each with at most two literals, and weights $w_1, \ldots, w_m$.

(1) Formulate the semidefinite program (9.9) as above.

(2) Solve the semidefinite program (9.9) to obtain $\boldsymbol{U}^*$;
Compute $\boldsymbol{v}_0, \boldsymbol{v}_1, \ldots, \boldsymbol{v}_{2n+1}$ by the Cholesky factorization.

(3) Choose a random vector $\boldsymbol{a}$ uniformly from S_{2n+2};
For $i \leftarrow 0$ **to** n **do**
  if $\boldsymbol{a}^T \boldsymbol{v}_i \ge 0$ **then** $y_i \leftarrow 1$ **else** $y_i \leftarrow -1$.

(4) **For** $i \leftarrow 1$ **to** n **do**
  if $y_i \ne y_0$ **then** $x_i \leftarrow$ TRUE **else** $x_i \leftarrow$ FALSE.

(5) Output $\boldsymbol{x}$. ∎

The following analysis shows that this algorithm has the same performance ratio as Algorithm 9.A.

Theorem 9.10 *Let* opt_{2SAT} *denote the objective function value of the optimal solution to the problem* MAX-2SAT. *Then we have*

$$E\left[\sum_{j=1}^{m} w_j \cdot \frac{3 - y_0 y_{j_1} - y_0 y_{j_2} - y_{j_1} y_{j_2}}{4}\right] \ge \alpha \cdot opt_{2SAT},$$

where $\alpha = 0.878567$.

Proof. Denote $\theta_{ij} = \arccos \boldsymbol{v}_i^T \boldsymbol{v}_j$. Then we have, from Lemmas 9.7 and 9.8,

$$\begin{aligned}
\mathrm{E}\Big[\sum_{j=1}^{m} w_j \cdot \frac{3 - y_0 y_{j_1} - y_0 y_{j_2} - y_{j_1} y_{j_2}}{4}\Big]
&= \sum_{j=1}^{m} w_j \cdot \Big(\mathrm{E}\Big[\frac{1 - y_0 y_{j_1}}{4}\Big] + \mathrm{E}\Big[\frac{1 - y_0 y_{j_2}}{4}\Big] + \mathrm{E}\Big[\frac{1 - y_{j_1} y_{j_2}}{4}\Big]\Big) \\
&= \sum_{j=1}^{m} w_j \cdot \Big(\frac{\theta_{0,j_1}}{2\pi} + \frac{\theta_{0,j_2}}{2\pi} + \frac{\theta_{j_1,j_2}}{2\pi}\Big) \\
&\geq \alpha \cdot \sum_{j=1}^{m} w_j \cdot \Big(\frac{1 - u_{0,j_1}}{4} + \frac{1 - u_{0,j_2}}{4} + \frac{1 - u_{j_1,j_2}}{4}\Big) \\
&= \alpha \cdot \sum_{j=1}^{m} w_j \cdot \frac{3 - u_{0,j_1} - u_{0,j_2} - u_{j_1,j_2}}{4} \geq \alpha \cdot opt_{\text{2SAT}}. \qquad \square
\end{aligned}$$

9.4 Rotation of Vectors

The hyperplane rounding technique studied in the last section works in three steps. First, we apply semidefinite programming relaxation to the input instance to get a semidefinite program. Next, we solve the semidefinite program and get a mapping of the input variables to vectors in S_n. Finally, we select a hyperplane to cut the unit sphere S_n into two parts and round the vectors to the one-dimensional unit sphere S_1. We observe, from the two examples of the last section, that the performance of such an algorithm often depends on the angles $\theta_{ij} = \arccos \boldsymbol{v}_i^T \boldsymbol{v}_j$ between the vectors on S_n. This observation suggests the following idea to improve the performance of the hyperplane rounding–based approximation algorithms: Before the third step of hyperplane rounding, shift the vectors on S_n so that the angles between these vectors are changed to effect a better rounding result. In this section, we explore this idea on some examples.

First, let's look at the problem MAX-2SAT again. In the analysis of the performance of Algorithm 9.B, we notice that the expected total weight is equal to

$$\sum_{j=1}^{m} w_j \cdot \frac{\theta_{0,j_1} + \theta_{0,j_2} + \theta_{j_1,j_2}}{2\pi}. \tag{9.10}$$

Thus, we would like to find a way of changing the angles θ_{ij} between the vectors to get a larger value for the above sum. To do so, we observe that, among the variables in the integer quadratic program (9.8), y_0 is a special one, as it is involved in every term of the summation (9.10) above. Therefore, we may focus on changing the angles $\theta_{0,i}$ between vector $\boldsymbol{v}_0$ and other vectors $\boldsymbol{v}_i$. That is, we want to *rotate* the vectors $\boldsymbol{v}_i$ toward or away from the vector $\boldsymbol{v}_0$ to increase the sum (9.10). More precisely, let $f(\theta)$ be a function defined on $\theta \in [0, \pi]$. Then, we can define a rotation operation on vectors $\boldsymbol{v}_i$, for $i \neq 0$, as follows: For each vector $\boldsymbol{v}_i$, $i \neq 0$, we map $\boldsymbol{v}_i$ to a new vector $\boldsymbol{v}'_i$ located in the plane spanned by vectors $\boldsymbol{v}_0$ and $\boldsymbol{v}_i$ such that $\boldsymbol{v}'_i$ lies on the same side of $\boldsymbol{v}_0$ as $\boldsymbol{v}_i$ and forms an angle $f(\theta_{0,i})$ with vector $\boldsymbol{v}_0$.

Let $\theta'_{i,j}$ denote the new angle between v'_i and v'_j after the rotation. [Thus, $\theta'_{0,i} = f(\theta_{0,i})$ for all $i = 1, \ldots, 2n$.] How do we choose the rotation function f to maximize the sum

$$\sum_{j=1}^{m} w_j \cdot \frac{\theta'_{0,j_1} + \theta'_{0,j_2} + \theta'_{j_1,j_2}}{2\pi}?$$

First, as a general rule, a rotation function f is usually required to satisfy the property

$$f(\pi - \theta) = \pi - f(\theta),$$

so that the vectors v_i move toward or away from the line passing through v_0 in a symmetric way. Next, for any fixed rotation function f, we need to calculate $\theta'_{i,j}$ to estimate the effect of the rotation on the sum (9.10),

Motivated by the analysis in the proof of Theorem 9.10, let us consider the following family of rotation functions:

$$f_\lambda(\theta) = (1 - \lambda)\theta + \lambda \cdot \frac{\pi}{2}(1 - \cos\theta),$$

where λ is a parameter between 0 and 1. The angle θ'_{j_1,j_2} under this rotation function f_λ can be computed as follows: First, from spherical trigonometry, we have

$$\cos\theta_{j_1,j_2} = \cos\theta_{0,j_1}\cos\theta_{0,j_2} + \cos\beta\sin\theta_{0,j_1}\sin\theta_{0,j_2},$$
$$\cos\theta'_{j_1,j_2} = \cos\theta'_{0,j_1}\cos\theta'_{0,j_2} + \cos\beta\sin\theta'_{0,j_1}\sin\theta'_{0,j_2},$$

where β is the angle between the plane spanned by vectors v_0 and v_{j_1} and the plane spanned by vectors v_0 and v_{j_2}. From these equations we obtain the following formula for θ'_{j_1,j_2}:

$$\theta'_{j_1,j_2} = \arccos\Big[\cos\theta'_{0,j_1}\cos\theta'_{0,j_2} + \big(\cos\theta_{j_1,j_2} - \cos\theta_{0,j_1}\cos\theta_{0,j_2}\big)\cdot\frac{\sin\theta'_{0,j_1}\sin\theta'_{0,j_2}}{\sin\theta_{0,j_1}\sin\theta_{0,j_2}}\Big].$$

We note that for a fixed λ, θ'_{j_1,j_2} is a function of variables θ_{0,j_1}, θ_{0,j_2}, and θ_{j_1,j_2}. Let us denote it by $g_\lambda(\theta_{0,j_1}, \theta_{0,j_2}, \theta_{j_1,j_2})$. Then, from the proof of Theorem 9.10, we see that the effect of the rotation f_λ is, for each clause C_j, to use

$$\frac{f_\lambda(\theta_{0,j_1}) + f_\lambda(\theta_{0,j_2}) + g_\lambda(\theta_{0,j_1}, \theta_{0,j_2}, \theta_{j_1,j_2})}{2\pi}$$

to approximate

$$\frac{3 - \cos\theta_{0,j_1} - \cos\theta_{0,j_2} - \cos\theta_{j_1,j_2}}{4}.$$

Therefore, the reciprocal of the performance ratio of the new algorithm is at least

$$\rho_\lambda = \min_{(\theta_1,\theta_2,\theta_3)\in\Omega} \frac{2}{\pi} \cdot \frac{f_\lambda(\theta_1) + f_\lambda(\theta_2) + g_\lambda(\theta_1,\theta_2,\theta_3)}{3 - \cos\theta_1 - \cos\theta_2 - \cos\theta_3},$$

where Ω is the area bounded by the following constraints:

$$\begin{aligned} 0 \le \theta_i \le \pi, & \qquad i = 1, 2, 3, \\ \theta_1 + \theta_2 + \theta_3 \le 2\pi. & \end{aligned}$$

By selecting the best λ, we obtain

$$\sum_{j=1}^{m} w_j \cdot \frac{\theta'_{0,j_1} + \theta'_{0,j_2} + \theta'_{j_1,j_2}}{2\pi} \ge \rho \cdot opt_{2\mathrm{SAT}},$$

where $\rho = \max_{0\le\lambda\le1} \rho_\lambda$.

Unfortunately, it can be verified, through numerical evaluation, that this new ratio ρ is actually very close to the ratio $\alpha = 0.878567$ obtained without the rotation. How do we get a more significant improvement over ρ? We notice that the estimate of ρ_λ is made over the feasible domain Ω and may have been too loose. It is easy to see that when the feasible domain shrinks, the minimum value increases. This observation suggests that we should try to add some constraints to shrink the feasible domain Ω and get a greater ρ. We note that for any $y_i, y_j, y_k \in \{1, -1\}$, they must satisfy

$$\begin{aligned} y_iy_j + y_jy_k + y_ky_i &\ge -1, \\ y_iy_j - y_jy_k - y_ky_i &\ge -1, \\ -y_iy_j + y_jy_k - y_ky_i &\ge -1, \\ -y_iy_j - y_jy_k + y_ky_i &\ge -1. \end{aligned}$$

This means that we can add constraints

$$\begin{aligned} u_{ij} + u_{jk} + u_{ki} &\ge -1, \\ u_{ij} - u_{jk} - u_{ki} &\ge -1, \\ -u_{ij} + u_{jk} - u_{ki} &\ge -1, \\ -u_{ij} - u_{jk} + u_{ki} &\ge -1 \end{aligned}$$

to the semidefinite program (9.9) about MAX-2SAT. This means that Ω can be constrained by

$$\begin{aligned} \cos\theta_1 + \cos\theta_2 + \cos\theta_3 &\ge -1, \\ \cos\theta_1 - \cos\theta_2 - \cos\theta_3 &\ge -1, \\ -\cos\theta_1 + \cos\theta_2 - \cos\theta_3 &\ge -1, \\ -\cos\theta_1 - \cos\theta_2 + \cos\theta_3 &\ge -1, \\ 0 \le \theta_i \le \pi, & \qquad i = 1, 2, 3, \\ \theta_1 + \theta_2 + \theta_3 \le 2\pi. & \end{aligned} \tag{9.11}$$

With these constraints, we get a smaller Ω and a greater ρ. To be more precise, let Ω_1 denote the area bounded by the constraints of (9.11). Also, for $0 \le \lambda \le 1$, let

$$\rho_\lambda = \min_{(\theta_1,\theta_2,\theta_3)\in\Omega_1} \frac{2}{\pi} \cdot \frac{f_\lambda(\theta_1) + f_\lambda(\theta_2) + g_\lambda(\theta_1,\theta_2,\theta_3)}{3 - \cos\theta_1 - \cos\theta_2 - \cos\theta_3},$$

and $\rho = \max_{0\le\lambda\le1} \rho_\lambda$. Based on this setting, Feige and Goemans [1995] and Zwick [2000] have computed that $\rho_\lambda \ge 0.93109$ for $\lambda = 0.806765$.

We summarize the above discussion in the following approximation algorithm for MAX-2SAT.

Algorithm 9.C (*Second Semidefinite Programming Approximation for* MAX-2SAT)

Input: A CNF formula with clauses $C_1, \ldots, C_m$, each with at most two literals, weights $w_1, \ldots, w_m$, and a real number $0 \le \lambda \le 1$.

(1) Formulate the following semidefinite program:

$$\text{maximize} \quad \sum_{j=1}^{m} w_j \cdot \frac{3 - u_{0,j_1} - u_{0,j_2} - u_{j_1,j_2}}{4}$$

$$\begin{aligned}
\text{subject to} \quad & u_{0,2n+1} = 1, && \\
& u_{i,n+i} = -1, && 1 \le i \le n, \\
& u_{ii} = 1, && 0 \le i \le 2n+1, \\
& u_{0i} + u_{0j} + u_{ij} \ge -1, && 1 \le i < j \le 2n+1, \\
& u_{0i} - u_{0j} - u_{ij} \ge -1, && 1 \le i < j \le 2n+1, \\
& -u_{0i} + u_{0j} - u_{ij} \ge -1, && 1 \le i < j \le 2n+1, \\
& -u_{0i} - u_{0j} + u_{ij} \ge -1, && 1 \le i < j \le 2n+1, \\
& \boldsymbol{U} \succeq \boldsymbol{0}. &&
\end{aligned}$$

(2) Solve the above semidefinite program to obtain $\boldsymbol{U}^*$;
Compute $\boldsymbol{v}_0, \boldsymbol{v}_1, \ldots, \boldsymbol{v}_{2n+1}$ by the Cholesky factorization;
Compute $\boldsymbol{v}'_1, \boldsymbol{v}'_2, \ldots, \boldsymbol{v}'_n$ from $\boldsymbol{v}_0, \boldsymbol{v}_1, \ldots, \boldsymbol{v}_n$, where each $\boldsymbol{v}'_i$ is obtained by rotating $\boldsymbol{v}_i$ on the plane spanned by vectors $\boldsymbol{v}_0$ and $\boldsymbol{v}_i$ to form an angle $\theta'_{0,i} = f_\lambda(\theta_{0,i})$ with $\boldsymbol{v}_0$.

(3) Choose a random vector $\boldsymbol{a}$ uniformly from S_{2n+2};
For $i \leftarrow 0$ **to** n **do**
  if $\boldsymbol{a}^T\boldsymbol{v}'_i \ge 0$ **then** $y_i \leftarrow 1$ **else** $y_i \leftarrow -1$.

(4) **For** $i \leftarrow 1$ **to** n **do**
  if $y_i \ne y_0$ **then** $x_i \leftarrow$ TRUE **else** $x_i \leftarrow$ FALSE.

(5) Output $\boldsymbol{x}$. ■

Theorem 9.11 *The expected total weight of satisfied clauses obtained by* Algorithm 9.C *is at least* $\rho_\lambda \cdot opt_{2\text{SAT}}$, *and* $\rho_\lambda \ge 0.93109$ *when* $\lambda = 0.806765$.

In the above, we used $f_\lambda(\theta)$ to rotate vectors. An alternative way to perform the rotation of the vectors is to calculate the new vectors directly from Cholesky factorization.

Algorithm 9.D (*Third Semidefinite Programming Approximation for* MAX-2SAT)

Input: Same input as Algorithm 9.C.

(1) Same as step (1) of Algorithm 9.C.

(2) Solve the above semidefinite program to obtain $\boldsymbol{U}^*$;
Compute vectors $\boldsymbol{v}'_0, \boldsymbol{v}'_1, \ldots, \boldsymbol{v}'_{2n+1}$ through Cholesky factorization of $\lambda \boldsymbol{U}^* + (1-\lambda)\boldsymbol{I}$, where $\boldsymbol{I}$ is the identity matrix of order $2n+2$.

(3)–(5) Same as steps (3)–(5) of Algorithm 9.C. ■

For this approximation, it can be verified that the expected total weight of satisfied clauses is at least

$$\zeta_\lambda \cdot opt_{2\mathrm{SAT}},$$

where

$$\zeta_\lambda = \min_{(\theta_1,\theta_2,\theta_3)\in\Omega_1} \frac{2}{\pi} \cdot \frac{\arccos(\lambda\cos\theta_1) + \arccos(\lambda\cos\theta_2) + \arccos(\lambda\cos\theta_3)}{3 - \cos\theta_1 - \cos\theta_2 - \cos\theta_3},$$

and Ω_1 is the region defined by (9.11). As ζ_λ has a simpler expression than ρ_λ, this second way of rotation has been used more often in the literature. However, no solid results of comparison between these two ways of rotation have been obtained regarding which one will give us a better performance ratio.

Most approximation problems that have been studied with the method of semidefinite programming relaxation are maximization problems. In the following we consider a minimization problem.

> SCHEDULING ON PARALLEL MACHINES (SCHEDULE-PM): Given n jobs $J = \{1, 2, \ldots, n\}$, m machines $M = \{1, 2, \ldots, m\}$, and the processing time p_{ij} for job $j \in J$ on machine $i \in M$, schedule all jobs to m machines to minimize the total weighted completion time $\sum_{j=1}^{n} w_j C_j$, where C_j is the completion time of job j (i.e., the total processing time of the first k jobs on machine i if job j is assigned as the kth job on machine i).

For the case of $m = 1$, we can find the best scheduling by a simple greedy algorithm. For $i \in M$ and $j, k \in J$, define $j \prec_i k$ if $[w_j/p_{ij} > w_k/p_{ik}]$ or $[w_j/p_{ij} = w_k/p_{ik}$ and $j < k]$.

Lemma 9.12 *For the problem* SCHEDULE-PM *with* $m = 1$, *an optimal solution is to schedule all jobs in ordering* $\prec_1$.

Proof. Suppose job j is scheduled right after job k, but $j \prec_1 k$. Exchanging job j and job k, we reduce the objective function value by

$$w_j p_{1k} - w_k p_{1j} \geq 0. \qquad \square$$

From the above lemma, we know that if jobs $j_1, j_2, \ldots, j_k$ are assigned to machine i, then the scheduling of these jobs on machine i is fixed according to $\prec_i$. Therefore, the problem SCHEDULE-PM can be formulated as the following integer quadratic program, where $x_{ij} \in \{0, 1\}$ is the variable indicating whether job j is assigned to machine i:

$$\begin{array}{lll} \text{minimize} & \displaystyle\sum_{j=1}^{n} w_j \sum_{i=1}^{m} x_{ij}\Bigg(p_{ij} + \sum_{k \prec_i j} x_{ik} p_{ik}\Bigg) & \\ \text{subject to} & \displaystyle\sum_{i=1}^{m} x_{ij} = 1, & j = 1, 2, \ldots, n, \\ & x_{ij} \in \{0, 1\}. & \end{array} \tag{9.12}$$

In the rest of this section, we consider only the case of $m = 2$. We introduce $n+2$ variables $y_1, y_2, \ldots, y_n, y_{n+1}, y_{n+2} \in \{-1, 1\}$ satisfying the following constraints:

(a) $y_{n+1} y_{n+2} = -1$.

(b) $x_{1j} = 1$ if and only if $y_j = y_{n+1}$ (and, hence, $x_{2j} = 1$ if and only if $y_j = y_{n+2}$).

From these constraints, we have, for $i \in \{1, 2\}$ and $j, k \in J$,

$$x_{ij} = \frac{1 + y_{n+i} y_j}{2},$$

$$x_{ij} x_{ik} = \frac{1 + y_j y_k + y_{n+i} y_j + y_{n+i} y_k}{4}.$$

Substituting these formulas into (9.12), we obtain a new integer quadratic program:

$$\begin{array}{lll} \text{minimize} & \displaystyle\sum_{j=1}^{n} w_j \sum_{i=n+1}^{n+2} \left(\frac{1 + y_i y_j}{2} \cdot p_{ij} + \sum_{k \prec_i j} \frac{1 + y_j y_k + y_i y_j + y_i y_k}{4} \cdot p_{ik}\right) & \\ \text{subject to} & y_j^2 = 1, & j = 1, 2, \ldots, n+2, \\ & y_{n+1} y_{n+2} = -1. & \end{array}$$

By the semidefinite programming relaxation, we get the following semidefinite program:

$$\begin{array}{lll} \text{minimize} & \displaystyle\sum_{j=1}^{n} w_j \sum_{i=n+1}^{n+2} \left(\frac{1 + u_{ij}}{2} \cdot p_{ij} + \sum_{k \prec_i j} \frac{1 + u_{jk} + u_{ij} + u_{ik}}{4} \cdot p_{ik}\right) & \\ \text{subject to} & u_{jj} = 1, & j = 1, 2, \ldots, n+2, \\ & u_{n+1,n+2} = -1, & \\ & U \succeq \mathbf{0}. & \end{array}$$

We can now apply the hyperplane rounding technique with the rotation of vectors to design an approximation algorithm for SCHEDULE-PM. Here, variables y_{n+1} and y_{n+2} are the special vectors. We can rotate each vector of $\boldsymbol{v}_1, \boldsymbol{v}_2, \ldots, \boldsymbol{v}_n$ toward or away from $\boldsymbol{v}_{n+1}$ to find a better performance ratio. The details are left to the reader as an exercise.

Finally, let us make some remarks about the rotation. Given a semidefinite programming relaxation, how do we rotate vectors to get a better rounding? A general method is called the *outward rotation*. Let θ_{ij} denote the angle between vectors $\boldsymbol{v}_i$ and $\boldsymbol{v}_j$ before the rotation, and θ'_{ij} the angle between them after the rotation. In an outward rotation, we rotate vectors so that $\pi/2 > \theta'_{ij} > \theta_{ij}$ if $0 < \theta_{ij} < \pi/2$, and $\pi/2 < \theta'_{ij} < \theta_{ij}$ if $\pi/2 < \theta_{ij} < \pi$, for all vectors $\boldsymbol{v}_i$ and $\boldsymbol{v}_j$. This can be achieved by embedding the original vector space to a larger vector space and then rotating the vectors *out of* the original space. We note that if the objective function of some problem attains its maximum value at some configurations in which many angles are less than $\pi/2$, then the outward rotation is potentially helpful. This is indeed the case for many maximization problems. On the other hand, for minimization problems, the "inward rotation," in which $0 < \theta'_{ij} < \theta_{ij}$ for $0 < \theta_{ij} < \pi/2$ and $\theta_{ij} < \theta'_{ij} < \pi$ for $\pi/2 < \theta_{ij} < \pi$, seems to be more helpful. The reader is referred to, for instance, Zwick [1999] for more details.

9.5 Multivariate Normal Rounding

There is another rounding technique, called *multivariate normal rounding*, in the semidefinite programming-based approximation. We demonstrate the idea on the problem MAX-CUT.

Algorithm 9.E (*Multivariate Normal Rounding for* MAX-CUT)

Input: A graph $G = (V, E)$ and nonnegative edge weights w_{ij}, for $i, j \in V$.

(1) Construct the semidefinite program (9.6).

(2) Find the optimal solution $\boldsymbol{U}^*$ of the semidefinite program (9.6).

(3) Generate a random vector $\boldsymbol{y}$ from a multivariate normal distribution with mean $\boldsymbol{0}$ and covariance matrix $\boldsymbol{U}^*$; that is, $\boldsymbol{y} \in N(\boldsymbol{0}, \boldsymbol{U}^*)$.

(4) **For** $i \leftarrow 1$ **to** n **do**
 if $y_i \geq 0$ **then** $x_i \leftarrow 1$ **else** $x_i \leftarrow -1$.

(5) Output the cut $S = \{i \mid x_i = 1\}$. ■

It should be pointed out first that although we did not state it explicitly, we actually need to apply the Cholesky factorization to implement step (3). More precisely, step (3) can be implemented as follows:

(3.1) Compute $\boldsymbol{V}$ with $\boldsymbol{V}^T\boldsymbol{V} = \boldsymbol{U}^*$ by Cholesky factorization.

(3.2) Choose $\boldsymbol{a} \in N(\boldsymbol{0}, \boldsymbol{I})$;
Set $\boldsymbol{y} \leftarrow \boldsymbol{V}\boldsymbol{a}$.

Next, let us show that Algorithm 9.E has the same performance ratio as Algorithm 9.A. To see this, we only need the following property of the random vector $\boldsymbol{y}$, which plays a similar role to Lemma 9.7 for hyperplane rounding.

Define, for any real number x,

$$sgn(x) = \begin{cases} 1, & \text{if } x \geq 0, \\ -1, & \text{if } x < 0. \end{cases}$$

Lemma 9.13 *For* $\boldsymbol{y} \in N(\mathbf{0}, \boldsymbol{U}^*)$,

$$\mathrm{E}[sgn(y_i) \cdot sgn(y_j)] = \frac{2}{\pi} \arcsin u_{ij}^*.$$

Proof. Let $\boldsymbol{y} \in N(\mathbf{0}, \boldsymbol{U}^*)$. It can be found from Johnson and Kotz [1972] that

$$\Pr[y_i \geq 0, y_j \geq 0] = \Pr[y_i < 0, y_j < 0] = \frac{1}{4} + \frac{1}{2\pi} \arcsin u_{ij}^*,$$

$$\Pr[y_i \geq 0, y_j < 0] = \Pr[y_i < 0, y_j \geq 0] = \frac{1}{4} - \frac{1}{2\pi} \arcsin u_{ij}^*.$$

Hence, we get

$$\begin{aligned} \mathrm{E}[sgn(y_i) \cdot sgn(y_j)] &= \Pr[y_i \geq 0, y_j \geq 0] + \Pr[y_i < 0, y_j < 0] \\ &\quad - \Pr[y_i \geq 0, y_j < 0] - \Pr[y_i < 0, y_j \geq 0] \\ &= \frac{2}{\pi} \arcsin u_{ij}^*. \end{aligned}$$

□

Theorem 9.14 *The expected value of the total weight of the output of* Algorithm 9.E *is at least* $\alpha \cdot opt_{\mathrm{CUT}}$, *where* $\alpha = 0.878567$.

Proof. The proof is essentially identical to that of Theorem 9.9 by noting that $\arcsin x = \pi/2 - \arccos x$. □

For another example of the application, consider the following problem.

> MAXIMUM BISECTION (MAX-BISEC): Given a graph $G = (V, E)$, where $V = \{1, 2, \ldots, n\}$, and a nonnegative weight w_{ij} for each edge $\{i, j\}$ in E, find a partition (V_1, V_2) of the vertex set V to maximize the total weight of edges between V_1 and V_2 under the condition that $|V_1| = |V_2|$.

This problem can be formulated as

$$\begin{aligned} \text{maximize} \quad & \frac{1}{4} \sum_{1 \leq i,j \leq n} w_{ij}(1 - x_i x_j) \\ \text{subject to} \quad & \sum_{i=1}^{n} x_i = 0, \\ & x_i^2 = 1, \qquad\qquad i = 1, 2, \ldots, n. \end{aligned} \tag{9.13}$$

For each variable x_i, introduce a vector variable $\boldsymbol{v}_i = (x_i, 0, \ldots, 0)^T$. Then $x_i x_j = \boldsymbol{v}_i \cdot \boldsymbol{v}_j$. Note that $\sum_{i=1}^n x_i = 0$ is equivalent to $\sum_{1 \le i,j \le n} x_i x_j = 0$. Therefore, the quadratic program (9.13) is equivalent to the following:

$$
\begin{array}{lll}
\text{maximize} & \displaystyle\frac{1}{4} \sum_{1 \le i,j \le n} w_{ij}(1 - \boldsymbol{v}_i \cdot \boldsymbol{v}_j) & \\
\text{subject to} & \displaystyle\sum_{1 \le i,j \le n} \boldsymbol{v}_i \cdot \boldsymbol{v}_j = 0, & \\
& \boldsymbol{v}_i \cdot \boldsymbol{v}_i = 1, & i = 1, 2, \ldots, n, \\
& \boldsymbol{v}_i \in S_1, & i = 1, 2, \ldots, n,
\end{array}
\tag{9.14}
$$

where $S_1 = \{(1, 0, \ldots, 0), (-1, 0, \ldots, 0)\}$ is the one-dimensional unit sphere. Now, we relax S_1 to the n-dimensional unit sphere $S_n = \{\boldsymbol{v} \mid \|\boldsymbol{v}\| = 1\}$. Then the above formulation becomes a vector program equivalent to the following semidefinite program:

$$
\begin{array}{lll}
\text{maximize} & \frac{1}{4} \boldsymbol{W} \bullet (\boldsymbol{J} - \boldsymbol{U}) & \\
\text{subject to} & \boldsymbol{J} \bullet \boldsymbol{U} = 0, & \\
& u_{ii} = 1, & i = 1, 2, \ldots, n, \\
& \boldsymbol{U} \succeq \boldsymbol{0},
\end{array}
\tag{9.15}
$$

where $\boldsymbol{W} = (w_{ij})$, $\boldsymbol{U} = (u_{ij})$, and $\boldsymbol{J}$ is the $n \times n$ matrix with every entry having value 1.

Suppose $\boldsymbol{U}^*$ is an optimal solution of the semidefinite program (9.15). How can we round $\boldsymbol{U}^*$ randomly to obtain a cut for G and keep it a balanced partition? In the following, we employ, besides multivariate normal rounding, an additional technique called *vertex swapping* to solve this problem (see step (5) below).

Algorithm 9.F (*Semidefinite Programming Approximation for* MAX-BISEC)

Input: A graph $G = (V, E)$, weight w_{ij} for each edge $\{i, j\} \in E$.

(1) Construct the semidefinite program (9.15).

(2) Find the optimum solution $\boldsymbol{U}^*$ of (9.15).

(3) Generate a random vector $\boldsymbol{y}$ from a multivariate normal distribution with mean $\boldsymbol{0}$ and covariance matrix $\boldsymbol{U}^*$; that is, $\boldsymbol{y} \in N(\boldsymbol{0}, \boldsymbol{U}^*)$.

(4) **If** $|\{i \mid y_i \ge 0\}| \ge n/2$ **then** $S \leftarrow \{i \mid y_i \ge 0\}$ **else** $S \leftarrow \{i \mid y_i < 0\}$.

(5) **For** each $i \in S$ **do** $\zeta(i) \leftarrow \sum_{j \notin S} w_{ij}$;
Sort S such that $S = \{i_1, i_2, \ldots, i_{|S|}\}$, with $\zeta(i_1) \ge \zeta(i_2) \ge \cdots \ge \zeta(i_{|S|})$;
Set $S_A \leftarrow \{i_1, i_2, \ldots, i_{n/2}\}$.

(6) Output the cut $(S_A, V - S_A)$. ■

To estimate the weight of the bisection cut $(S_A, V - S_A)$, let us define three random variables:

$$w = w(S) = \sum_{i \in S, j \notin S} w_{ij}, \qquad m = |S|(n - |S|),$$

and

$$z = z(S) = \frac{w}{w^*} + \frac{m}{m^*},$$

where

$$w^* = \frac{1}{4} \sum_{1 \le i,j \le n} w_{ij}(1 - u^*_{ij}) \quad \text{and} \quad m^* = \frac{n^2}{4}.$$

Lemma 9.15 *In* Algorithm 9.F, *if S satisfies $z = z(S) \ge c$, then*

$$w(S_A) = \sum_{i \in S_A, j \notin S_A} w_{ij} \ge 2(\sqrt{c} - 1)w^*.$$

Proof. Assume $w(S) = \lambda w^*$ and $|S| = \beta n$. Then $m/m^* = 4\beta(1 - \beta)$ and

$$z = \lambda + 4\beta(1 - \beta).$$

From the definition of S_A, it is easy to see that

$$w(S_A) \ge \frac{n \cdot w(S)}{2|S|}.$$

Therefore,

$$w(S_A) \ge \frac{w(S)}{2\beta} = \frac{\lambda w^*}{2\beta} = \frac{z - 4\beta(1 - \beta)}{2\beta} \cdot w^*.$$

Let us study the function

$$g(\beta) = \frac{z - 4\beta(1 - \beta)}{2\beta}.$$

Rewrite it as

$$16\beta^2 - 8\beta(2 + g(\beta)) + 4z = 0;$$

or, equivalently,

$$\big(4\beta - (2 + g(\beta))\big)^2 - \big(2 + g(\beta)\big)^2 + 4z = 0.$$

It follows that

$$\big(2 + g(\beta)\big)^2 - 4z \ge 0$$

and, hence,

$$g(\beta) \ge 2(\sqrt{z} - 1) \ge 2(\sqrt{c} - 1). \qquad \square$$

Next, we want to estimate $E(z)$. We first establish a lemma on the function $\arcsin x$.

Lemma 9.16 *For any* $0 \le x \le 1$,

$$1 - \frac{2}{\pi}\arcsin x \ge \alpha(1-x),$$

where $\alpha = 0.878567$.

Proof. From Lemma 9.8, we have

$$\frac{\pi/2-\phi}{\pi} \ge \alpha \cdot \frac{1-\cos(\pi/2-\phi)}{2},$$

for any ϕ satisfying $0 \le \pi/2 - \phi \le \pi$, or, equivalently, $-\pi/2 \le \phi \le \pi/2$. Thus, we get

$$1 - \frac{2}{\pi}\phi \ge \alpha\,(1-\sin\phi),$$

for $-\pi/2 \le \phi \le \pi/2$. □

Lemma 9.17 $E[z] \ge 2\alpha$, *where* $\alpha = 0.878567$.

Proof. Note that

$$\begin{aligned} w(S) &= \frac{1}{4}\sum_{1\le i,j\le n} w_{ij}(1 - sgn(y_i)sgn(y_j)),\\ |S|(n-|S|) &= \frac{1}{4}\sum_{1\le i,j\le n}(1 - sgn(y_i)sgn(y_j)). \end{aligned}$$

Therefore, by Lemmas 9.13 and 9.16,

$$\begin{aligned} E[w] &= \frac{1}{4}\sum_{1\le i,j\le n} w_{i,j}\left(1 - \frac{2}{\pi}\arcsin u^*_{ij}\right)\\ &\ge \frac{1}{4}\sum_{1\le i,j\le n} w_{i,j}\,\alpha\,(1-u^*_{ij}) \;=\; \alpha\, w^*. \end{aligned}$$

Also, notice that $\boldsymbol{U}^*$ satisfies $\boldsymbol{J} \bullet \boldsymbol{U}^* = 0$, or $\sum_{1\le i,j\le n} u^*_{ij} = 0$. Therefore, we have

$$\begin{aligned} E[m] &= \frac{1}{4}\sum_{1\le i,j\le n}\left(1 - \frac{2}{\pi}\arcsin u^*_{ij}\right)\\ &\ge \frac{1}{4}\sum_{1\le i,j\le n}\alpha\,(1-u^*_{ij}) \;=\; \alpha\, m^*. \end{aligned}$$

Together, we get $E[z] \ge 2\alpha$. □

When $c = 2\alpha$, $2(\sqrt{c}-1) \approx 0.651$. Therefore, we obtain the following result.

Theorem 9.18 *There is a polynomial-time randomized approximation algorithm for* MAX-BISEC, *which produces a cut with the expected total weight of the cut at least* 0.651 *times the weight of the optimal cut.*

The vector rotation technique can also be used together with multivariate normal rounding by taking $\boldsymbol{y} \in N(\boldsymbol{0}, \lambda \boldsymbol{U}^* + (1-\lambda)\boldsymbol{I})$. A further development is to replace the identity matrix $\boldsymbol{I}$ by

$$\boldsymbol{P} = \begin{pmatrix} 1 & \tau & \tau & \cdots & \tau \\ \tau & 1 & \tau^2 & \cdots & \tau^2 \\ \vdots & \vdots & \vdots & \ddots & \vdots \\ \tau & \tau^2 & \tau^2 & \cdots & 1 \end{pmatrix},$$

for some parameter τ. It has been found that sometimes this replacement may improve the performance ratio of the approximation algorithms based on multivariate normal rounding.

Exercises

9.1 Prove the following properties of positive semidefinite matrices.

(a) A matrix $\boldsymbol{A}$ is positive semidefinite if and only if $\boldsymbol{A}$ is a nonnegative linear combination of matrices of the type $\boldsymbol{v}\boldsymbol{v}^T$, where $\boldsymbol{v}$ is a vector.

(b) If $\boldsymbol{A}$ and $\boldsymbol{B}$ are positive semidefinite matrices, then $\mathrm{Tr}(\boldsymbol{A}\boldsymbol{B}) \geq 0$. Moreover, the equality sign holds if and only if $\boldsymbol{A}\boldsymbol{B} = \boldsymbol{0}$.

(c) A matrix $\boldsymbol{A}$ is positive semidefinite if and only if $\boldsymbol{A} \bullet \boldsymbol{B} \geq 0$ for every positive semidefinite matrix $\boldsymbol{B}$.

9.2 Let $\boldsymbol{Q}$ be a positive semidefinite matrix, $\boldsymbol{b}$ a vector, and c a real number. Show that the ellipsoid $\{\boldsymbol{x} \mid \boldsymbol{x}^T\boldsymbol{Q}\boldsymbol{x} + \boldsymbol{b}^T\boldsymbol{x} + c \leq 0\}$ is a spectrahedron.

9.3 Show that $\mathcal{E}_n = \{\boldsymbol{U} \in \mathcal{S}_n \mid u_{ii} = 1, \boldsymbol{U} \succeq \boldsymbol{0}\}$, called an *elliptope*, is a spectrahedron with 2^n vertices, where a vertex is a matrix in form $\boldsymbol{v}\boldsymbol{v}^T$.

9.4 A *face* of a spectrahedron is the intersection of a hyperplane and the spectrahedron.

(a) Show that the smallest face of a spectrahedron $\mathcal{G}$ containing point $\bar{\boldsymbol{x}}$ is

$$F_{\mathcal{G}}(\bar{\boldsymbol{x}}) = \{\boldsymbol{x} \in \mathcal{G} \mid \mathrm{Null}(\boldsymbol{Q}_0 - \boldsymbol{Q}(\bar{\boldsymbol{x}})) \subseteq \mathrm{Null}(\boldsymbol{Q}_0 - \boldsymbol{Q}(\boldsymbol{x}))\},$$

where $\boldsymbol{Q}(\boldsymbol{x}) = \sum_{i=1}^m x_i\boldsymbol{Q}_i$, $\mathcal{G} = \{\boldsymbol{x} \mid \boldsymbol{Q}(\boldsymbol{x}) \succeq \boldsymbol{Q}_0\}$ and, for a matrix $\boldsymbol{A}$, $\mathrm{Null}(\boldsymbol{A}) = \{\boldsymbol{y} \mid \boldsymbol{A}\boldsymbol{y} = \boldsymbol{0}\}$.

(b) Construct a spectrahedron such that the dimensions of its faces are triangular integers $k(k+1)/2$ for $k = 0, 1, \ldots, n$.

9.5 Consider a spectrahedron $\mathcal{G} = \{\boldsymbol{x} \mid \boldsymbol{Q}(\boldsymbol{x}) \succeq \boldsymbol{Q}_0\}$, where $\boldsymbol{Q}(\boldsymbol{x}) = \sum_{i=1}^m x_i\boldsymbol{Q}_i$. A *plate* of $\mathcal{G}$ of order k is defined to be the closure of a connected component of $\{\boldsymbol{x} \in \mathcal{G} \mid \mathrm{rank}(\boldsymbol{Q}_0 - \boldsymbol{Q}(\boldsymbol{x})) = k\}$.

(a) Find all plates of the following spectrahedron:

$$\left\{ \boldsymbol{x} \in \mathbb{R}^3 \;\middle|\; x_1^2 + \frac{(x_2-2)^2}{4} + \frac{x_3^2}{4} \le 1,\ \boldsymbol{x}^T\boldsymbol{x} \le 1 \right\}.$$

(b) Show that the relative interior of any face is contained in exactly one plate.

(c) Show that every spectrahedron has finitely many plates.

(d) Show that every plate of a spectrahedron is a face.

9.6 Consider the following multiquadratic program:

$$\begin{array}{lll} \text{minimize} & \boldsymbol{x}^T\boldsymbol{Q}_0\boldsymbol{x} + 2\boldsymbol{b}_0^T\boldsymbol{x} + c_0 & \\ \text{subject to} & \boldsymbol{x}^T\boldsymbol{Q}_i\boldsymbol{x} + 2\boldsymbol{b}_i^T\boldsymbol{x} + c_i = 0, & i = 1, 2, \ldots, m. \end{array}$$

First, we rewrite it as follows:

$$\begin{array}{lll} \text{minimize} & \boldsymbol{U} \bullet \boldsymbol{Q}_0 + 2\boldsymbol{b}_0^T\boldsymbol{x} + c_0 & \\ \text{subject to} & \boldsymbol{U} \bullet \boldsymbol{Q}_i + 2\boldsymbol{b}_i^T\boldsymbol{x} + c_i = 0, & i = 1, 2, \ldots, m, \\ & \boldsymbol{U} - \boldsymbol{x}\boldsymbol{x}^T = \boldsymbol{0}. & \end{array}$$

By relaxing the constraint $\boldsymbol{U} - \boldsymbol{x}\boldsymbol{x}^T = \boldsymbol{0}$ to $\boldsymbol{U} - \boldsymbol{x}\boldsymbol{x}^T \succeq \boldsymbol{0}$, we obtain

$$\begin{array}{lll} \text{minimize} & \boldsymbol{U} \bullet \boldsymbol{Q}_0 + 2\boldsymbol{b}_0^T\boldsymbol{x} + c_0 & \\ \text{subject to} & \boldsymbol{U} \bullet \boldsymbol{Q}_i + 2\boldsymbol{b}_i^T\boldsymbol{x} + c_i = 0, & i = 1, 2, \ldots, m, \\ & \boldsymbol{U} - \boldsymbol{x}\boldsymbol{x}^T \succeq \boldsymbol{0}. & \end{array}$$

This relaxation is called the *convexification relaxation* of multiquadratic programming. Prove that $\boldsymbol{U} - \boldsymbol{x}\boldsymbol{x}^T \succeq \boldsymbol{0}$ if and only if

$$\begin{pmatrix} \boldsymbol{U} & \boldsymbol{x} \\ \boldsymbol{x}^T & 1 \end{pmatrix} \succeq \boldsymbol{0}.$$

9.7 Recall that a *clique* of a graph $G = (V, E)$ is a vertex subset in which every two vertices are adjacent to each other, and an *independent set* of G is a vertex subset in which every two vertices are not adjacent to each other. Assume $G = (V, E)$ and $V = \{1, 2, \ldots, n\}$. The *characteristic vector* $\boldsymbol{x}$ of a vertex subset V' is defined by $x_i = 1$ if $i \in V'$ and $x_i = 0$ if $i \notin V'$.

(a) Prove that if $\boldsymbol{u}$ and $\boldsymbol{v}$ are characteristic vectors of a clique and an independent set, respectively, then $\boldsymbol{u}^T\boldsymbol{v} \le 1$.

(b) Let INDEP(G) be the convex hull of the characteristic vectors of all independent sets in G. Prove that INDEP(G) is a subset of the following polyhedron:

$$\text{QINDEP}(G) = \{\boldsymbol{x} \geq 0 \mid (\forall \boldsymbol{u},\ \boldsymbol{u} \text{ is a characteristic vector of a clique of } G)\ \boldsymbol{x}^T\boldsymbol{u} \leq 1\}.$$

(c) Consider the maximum independent set problem:

$$\begin{array}{lll} \text{maximize} & \sum_{i=1}^{n} x_i & \\ \text{subject to} & x_i x_j = 0, & (i,j) \in E, \\ & x_i(x_i - 1) = 0, & i \in V. \end{array}$$

Find its convexification relaxation.

9.8 Let $\mathcal{C}_n$ denote the convex hull of all matrices $\boldsymbol{v}\boldsymbol{v}^T$ for $\boldsymbol{v} \in \{-1, +1\}^n$. For a matrix $\boldsymbol{A} = (a_{ij})$, let $f_o(\boldsymbol{A}) = (f(a_{ij}))$. Show that

$$\mathcal{C}_n \subseteq \mathcal{E}_n \subseteq \left\{ \sin_o\left(\frac{\pi}{2}\boldsymbol{U}\right) \,\middle|\, \boldsymbol{U} \in \mathcal{E}_n \right\}.$$

9.9 Consider a positive semidefinite matrix $A = (a_{ij})$ of order n. Show that if $a_{ii} = 1$ for all $1 \leq i \leq n$, then $|a_{ij}| \leq 1$ for all $1 \leq i, j \leq n$.

9.10 Show that the following system of relaxed optimality conditions has a unique solution $(\boldsymbol{U}^*, \boldsymbol{x}^*, \boldsymbol{Z}^*)$:

$$\begin{array}{ll} \boldsymbol{Q}_i \bullet \boldsymbol{U} = c_i, & i = 1, 2, \ldots, m, \\ \sum_{i=1}^{m} x_i \boldsymbol{Q}_i + \boldsymbol{Z} = \boldsymbol{Q}_0, & \\ \boldsymbol{U}, \boldsymbol{Z} \succeq \boldsymbol{0}, & \\ \boldsymbol{U} \bullet \boldsymbol{Z} = 0. & \end{array}$$

9.11 Consider the Frobenius norm $\|\boldsymbol{A}\| = (\text{Tr}(\boldsymbol{A}\boldsymbol{A}^T))^{1/2}$. Show that the optimal solution of the problem of minimizing $\text{Tr}(\boldsymbol{V}^2 + \boldsymbol{V}\boldsymbol{D}_V)$ over the ellipsoid

$$\{\boldsymbol{D}_V \mid \|\boldsymbol{V}^{-1/2}\boldsymbol{D}_V\boldsymbol{V}^{-1/2}\| \leq 1\}$$

is

$$\boldsymbol{D}_V = -\boldsymbol{V}^3 / \|\boldsymbol{V}^2\|.$$

9.12 Design approximation algorithms for the following problems using the semidefinite programming relaxation with hyperplane rounding:

(a) MAX-BISEC.

(b) MAX-k-VC: Given a graph $G = (V, E)$ with nonnegative edge weights w_{ij}, find a subset $S \subseteq V$ of k vertices that maximizes the total weight of edges covered by S.

(c) MAXIMUM CUT IN A DIGRAPH (MAX-DICUT): Given a directed graph $G = (V, E)$ with nonnegative edge weights w_{ij}, find a subset $S \subseteq V$ that maximizes the total weight of the directed cut $\delta^+(S) = \{(i,j) \in E \mid i \in S, j \notin S\}$.

(d) MAX-k-UNCUT: Given a graph $G = (V, E)$ with nonnegative edge weights w_{ij} and an integer $k > 0$, find a subset $S \subseteq V$ of k vertices that maximizes the total weight of edges that do not cross S and $V - S$.

(e) DENSE-k-SUBGRAPH: Given a graph $G = (V, E)$ with nonnegative edge weights w_{ij} and an integer $k > 0$, find a subset $S \subseteq V$ of k vertices that maximizes the total weight of edges in the subgraph induced by S.

(f) MAXIMUM RESTRICTED CUT (MAX-RES-CUT): Given a graph $G = (V, E)$ with nonnegative edge weights w_{ij} and two disjoint edge subsets E_+ and E_-, find a subset $S \subseteq V$ that contains exactly one endpoint of each edge in E_- and either two endpoints or none of the endpoints of each edge in E_+, to maximize the total weight of the cut $\delta(S) = \{\{i,j\} \in E \mid i \in S, j \notin S\}$.

9.13 Study approximations to the following problems by the semidefinite programming relaxation with multivariate normal rounding:

(a) MAX-2SAT.

(b) MAX-k-VC.

9.14 Suppose there are m unit vectors $\boldsymbol{v}_1, \boldsymbol{v}_2, \ldots, \boldsymbol{v}_m$ in the unit sphere S_n. Choose a random unit vector $\boldsymbol{a}$ from S_n. Show that

$$\Pr[sgn(\boldsymbol{a}^T\boldsymbol{v}_i) = sgn(\boldsymbol{a}^T\boldsymbol{v}_j) = sgn(\boldsymbol{a}^T\boldsymbol{v}_k)] = 1 - \frac{\theta_{ij} + \theta_{ik} + \theta_{jk}}{2\pi},$$

where $\theta_{ij} = \arccos(\boldsymbol{v}_i^T\boldsymbol{v}_j)$.

9.15 Design approximation algorithms by the method of semidefinite programming for the following problems:

(a) MAX-$(n/2)$-VC: Given a graph $G = (V, E)$ with nonnegative edge weights w_{ij}, find a subset $S \subseteq V$ of $|V|/2$ vertices that maximizes the total weight of edges covered by S.

(b) MAX-$(n/2)$-DENSE-SUBGRAPH: Given a graph $G = (V, E)$ with nonnegative edge weights w_{ij}, for each edge $\{i, j\} \in E$, find a subset $S \subseteq V$ of $|V|/2$ vertices that maximizes the total weight of edges in the subgraph induced by S.

(c) MAX-$(n/2)$-UNCUT: Given a graph $G = (V, E)$ with nonnegative edge weights w_{ij}, find a subset $S \subseteq V$ of $|V|/2$ vertices that maximizes the total weight of edges that do not cross S and $V - S$.

(d) MAXIMUM BISECTION ON DIGRAPHS (MAX-DIBISEC): Given a directed graph $G = (V, E)$ with nonnegative edge weights w_{ij}, partition the vertices into two sets A and B of equal size that maximize the total weight of arcs from A to B.

9.16 Let $\boldsymbol{v}_1, \ldots, \boldsymbol{v}_5$ be five unit vectors in the n-dimensional unit sphere. Choose a random hyperplane H by uniformly choosing a random normal vector. For any set V of vectors, let $\Pr_H(V)$ denote the probability of V being separated by the random hyperplane H. Denote $\theta_{ij} = \arccos(\boldsymbol{v}_i^T \boldsymbol{v}_j)$. Prove the following facts:

(a) $\Pr_H(\boldsymbol{v}_1, \boldsymbol{v}_2, \boldsymbol{v}_3) = (\theta_{12} + \theta_{23} + \theta_{13})/(2\pi)$.

(b) $\Pr_H(\boldsymbol{v}_1, \boldsymbol{v}_2, \boldsymbol{v}_3, \boldsymbol{v}_4) = 1 - V/\pi^2$, where V is the volume of a spherical tetrahedron with dihedral angles $\lambda_{12}, \lambda_{13}, \lambda_{23}, \lambda_{14}, \lambda_{24}, \lambda_{34}$, and $\lambda_{i_1 i_2} = \pi - \theta_{i_3 i_4}$, for any permutation (i_1, i_2, i_3, i_4) of $(1, 2, 3, 4)$.

(c) $$\Pr_H(\boldsymbol{v}_1, \boldsymbol{v}_2, \boldsymbol{v}_3, \boldsymbol{v}_4, \boldsymbol{v}_5) = \tfrac{1}{2} \textstyle\sum_{1 \le i < j < k < l \le 5} \Pr_H(\boldsymbol{v}_i, \boldsymbol{v}_j, \boldsymbol{v}_k, \boldsymbol{v}_l) - \tfrac{1}{4} \textstyle\sum_{1 \le i < j \le 5} \Pr_H(\boldsymbol{v}_i, \boldsymbol{v}_j).$$

9.17 Apply the vector rotation techniques to design approximation algorithms for the following problems:

(a) SCHEDULE-PM with $m = 2$.

(b) MAXIMUM SPLITTING SET: Given a finite set $X = \{1, 2, \ldots, m\}$, a collection $\mathcal{C} = \{S_1, S_2, \ldots, S_n\}$ of subsets of X, and a nonnegative weight w_j for each subset S_j in $\mathcal{C}$, find a partition $(S, X - S)$ of X that maximizes the total weight of subsets $S_j \in \mathcal{C}$ that are split by partition $(S, X - S)$ (i.e., the total weight of sets $S_j \in \mathcal{C}$ such that $S_j \cap S \ne \emptyset \ne S_j \cap (X - S)$).

(c) MAXIMUM NOT-ALL-EQUAL SATISFIABILITY (MAX-NAE-SAT): Given n Boolean variables and m clauses over these variables, and a nonnegative weight w_j for each clause C_j, find an assignment to variables that maximizes the total weight of all clauses that contain at least one true literal and at least one false literal.

9.18 Assume $y_0, y_1, y_2, y_3, y_4 \in \{1, -1\}$. Show the following facts:

(a) $y_0 = y_1$ does not hold if and only if $(1 - y_0 y_1)/2 = 1$.

(b) $y_0 = y_1 = y_2$ does not hold if and only if $(3 - y_0 y_1 - y_0 y_2 - y_1 y_2)/4 = 1$.

(c) $y_0 = y_1 = y_2 = y_3$ does not hold if and only if

$$\frac{4 - (y_0 + y_1)(y_2 + y_3)}{4} \ge 1, \quad \frac{4 - (y_0 + y_2)(y_3 + y_1)}{4} \ge 1, \text{ and}$$
$$\frac{4 - (y_0 + y_3)(y_1 + y_2)}{4} \ge 1.$$

(d) $y_0 = y_1 = y_2 = y_3 = y_4$ does not hold if and only if

$$\frac{5 - (y_{i_0}y_{i_1} + y_{i_1}y_{i_2} + y_{i_2}y_{i_3} + y_{i_3}y_{i_4} + y_{i_4}y_{i_0})}{4} \geq 1 \text{ and}$$
$$\frac{5 - (y_{i_0} + y_{i_4})(y_{i_1} + y_{i_2} + y_{i_3}) + y_{i_0}y_{i_4}}{4} \geq 1,$$

for all permutations $(i_0, i_1, i_2, i_3, i_4)$ of $(0, 1, 2, 3, 4)$.

9.19 Consider the following generalization of the problem MAX-3SAT, where k is a constant greater than 2:

> MAX-kSAT: Given n Boolean variables and m clauses each containing at most k literals and having a nonnegative weight, find an assignment of variables such that the total weight of satisfied clauses is maximized.

Use the facts developed in Exercise 9.18 and the vector rotation technique to design approximations for MAX-3SAT and MAX-4SAT.

9.20 A function $g : \mathcal{S}_n \to \mathbb{R}$ is called a *packing* function if

(i) g is convex,

(ii) $g(\lambda \boldsymbol{M}) = \lambda g(\boldsymbol{M})$ for all $\lambda \geq 0$ and $\boldsymbol{M} \in \mathcal{S}_n$, and

(iii) $g(\boldsymbol{M}) \geq \boldsymbol{0}$ for all $\boldsymbol{M} \succeq \boldsymbol{0}$.

Show that the following functions are packing functions:

(a) $g(\boldsymbol{M}) = \boldsymbol{A} \bullet \boldsymbol{M}$, where $\boldsymbol{A} \succeq \boldsymbol{0}$.

(b) $g(\boldsymbol{M}) = \sum_{i,j} |m_{ij}| = \max\{\boldsymbol{M} \bullet \boldsymbol{Z} \mid |z_{ij}| \leq 1, 1 \leq i, j \leq n\}$, where $\boldsymbol{M} = (m_{ij})$ and $\boldsymbol{Z} = (z_{ij})$.

9.21 A semidefinite program is called a *packing* semidefinite program if it is of the following form:

$$\begin{array}{lll} \text{maximize} & \boldsymbol{C} \bullet \boldsymbol{X} & \\ \text{subject to} & g_i(\boldsymbol{X}) \leq 1, & i = 1, 2, \ldots, m, \\ & \mathrm{Tr}(\boldsymbol{X}) \leq \omega_{\boldsymbol{x}} \ (\text{or } \mathrm{Tr}(\boldsymbol{X}) = \omega_{\boldsymbol{x}}), & \\ & \boldsymbol{X} \succeq 0, & \end{array}$$

where $\boldsymbol{C} \succeq \boldsymbol{0}$ and the functions $g_i(\boldsymbol{X})$, for $i = 1, 2, \ldots, m$, are packing functions. Prove the following results on packing semidefinite programs:

(a) The semidefinite program (9.6) for MAX-CUT can be written as a packing semidefinite program.

(b) The following semidefinite program obtained from the coloring of a graph $G = (V, E)$ can be written as a packing semidefinite program:

$$\begin{array}{lll} \text{maximize} & z & \\ \text{subject to} & x_{ii} = 1, & i = 1, 2, \ldots, m, \\ & z \leq -x_{ij}, & \{i, j\} \in E, \\ & \boldsymbol{X} \succeq \boldsymbol{0}, & \end{array}$$

where $\boldsymbol{X} = (x_{ij})$.

(c) For any $\varepsilon > 0$, there exists an algorithm faster than $O(n^{3.5})$ for packing semidefinite programs, which produces a feasible solution within ε from the optimal solution.

Historical Notes

Semidefinite programming is a rapidly growing area in optimization. It first appeared in the study of graph optimization problems by Lovász [1979]. It became an active area of research starting with Alizadeh [1991], who gave the first polynomial-time algorithm for solving semidefinite programs. Later, it was found that many properties of, and algorithms for, linear programming can be extended to semidefinite programming (see Alizadeh [1991, 1995], Alizadeh et al. [1994, 1997], An et al. [1998], and de Klerk et al. [1998]). The first work on the applications of semidefinite programming to the design of approximation algorithms belongs to Goemans and Williamson [1995b]. They improved approximations for MAX-CUT and MAX 2SAT with semidefinite programming relaxation and hyperplane rounding. Feige and Goemans [1995] discovered the vector rotation technique and used it to improve the performance of hyperplane rounding. This technique is further analyzed and applied to many different problems (see Halperin et al. [2001, 2002], Alon et al. [2001], Zwick [1998, 2000, 2002], and Galbiati and Maffioli [2007]). Zwick [1999] discussed the general ideas of the outward rotation technique. Bertsimas and Ye [1998] proposed multivariate normal rounding. This rounding technique can also be used together with the vector rotation technique (see Bertsimas and Ye [1998], Han, Ye, and Zhang [2002], Han, Ye, Zhang, and Zhang [2002], Yang et al. [2003], Zhang et al. [2004], and Fu et al. [1998]). Feige and Langberg [2006] proposed a general rounding approach, which includes several well-known rounding techniques as special cases. In addition to the problems MAX-CUT and MAX-2SAT, applications of semidefinite programming in approximation have been extended to many other combinatorial optimization problems, including variations of graph-cutting and set-splitting problems [Halperin and Zwick, 2001b; Zhang et al., 2004], variations of the satisfiability problem [Halperin and Zwick, 2001a; Zhang et al., 2004], the graph coloring problem [Karger et al., 1994; Iyengar et al., 2009], and scheduling problems [Skutella, 2001; Yang et al., 2003]. See also Ye [2001], Bertsimas and Ye [1998], Frieze and Jerrum [1995], Goemans and Williamson [1995b], Nesterov

[1998], Zwick [1998, 1999, 2000, 2002], Zhao et al. [1998], and Fu et al. [1998] for other applications.

Many new directions in the research of semidefinite programming–based approximation have been explored. Arora and Kale [2007] introduced the primal-dual schema in semidefinite programming to the design of approximation algorithms. Klein and Lu [1998] and Iyengar et al. [2009] gave faster solutions for semidefinite programs arising from the study of approximations for the maximum cut and graph coloring problems. Most semidefinite programming–based approximation algorithms use random rounding. Mahajan and Ramesh [1999] gave a derandomization method for some of them. Thus, the performance ratio of some random approximation algorithms can actually be reached by deterministic algorithms. Anjos and Wolkowicz [2002] strengthened semidefinite programming relaxations and obtained a hierarchy of such relaxations. Chlamtac [2007] used this hierarchy of semidefinite programming relaxations to design new approximations. Goemans and Williamson [2004] introduced the complex semidefinite programming to the design of approximation algorithms for the problem MAX 3-CUT. For a more complete list of references, the reader is referred to Pardalos and Ramana [1997] and Pardalos and Wolkowicz [1998].

10

Inapproximability

The problems that exist in the world today cannot be solved by the level of thinking that created them.

— Albert Einstein

In this chapter, we turn our attention to a different issue about approximation algorithms. We study how to prove inapproximability results for some **NP**-hard optimization problems. We are not looking here for a lower bound for the performance ratio of a specific approximation algorithm, but, instead, we try to find a lower bound for the performance ratio of *any* approximation algorithm for a given problem. Most results in this study are based on advanced developments in computational complexity theory, which is beyond the scope of this book. Therefore, we limit ourselves to fundamental concepts and results, often with proofs omitted, which are sufficient to establish the inapproximability of many combinatorial optimization problems.

10.1 Many–One Reductions with Gap

We have seen some inapproximability results in Chapter 1. For instance, we showed that the general case of the traveling salesman problem (TSP) does not have a polynomial-time c-approximation for any $c > 1$ unless **P** = **NP**. The proof of this result is based on a simple polynomial-time reduction from the Hamiltonian circuit problem (HC) to TSP in the following form: For each instance $G = (V, E)$ of HC, the reduction maps it to an instance (H, d) of TSP, where H is the complete graph with vertex set V, and d is the cost function with the following properties (see Figure 10.1):

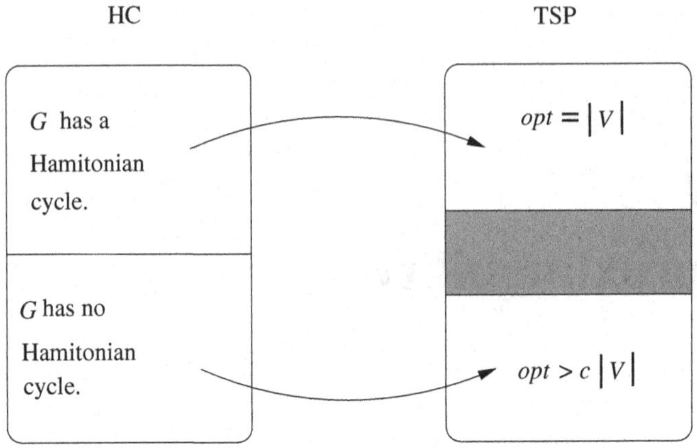

Figure 10.1: Reduction from HC to TSP.

(a) If G contains a Hamiltonian cycle, then H has a tour with cost $|V|$.

(b) If G does not have a Hamiltonian cycle, then the shortest tour of H has cost greater than $c|V|$.

Thus, there is a gap of a factor c between the shortest tours of the output graphs in the two different cases. This gap allows us to conclude that polynomial-time c-approximations do not exist for TSP unless HC can be solved in polynomial time (i.e., unless $\mathbf{P} = \mathbf{NP}$).

This proof technique can be generalized to other optimization problems. In the following, for an instance x of an optimization problem Π, we write $opt(x)$ to denote the objective function value of the optimal solution of x.

Definition 10.1 *Let* $0 < \alpha < \beta$.

(a) We say a minimization problem Π *has an* **NP**-hard gap *of* $[\alpha, \beta]$ *if there exist an* **NP**-*complete problem* Λ *and a polynomial-time many–one reduction* f *from* Λ *to* Π *with the following properties:*

(i) If $x \in \Lambda$, *then* $opt(f(x)) \le \alpha$, *and*

(ii) If $x \notin \Lambda$, *then* $opt(f(x)) > \beta$.

(b) We say a maximization problem Π *has an* **NP**-hard gap $[\alpha, \beta]$ *if there exist an* **NP**-*complete problem* Λ *and a polynomial-time many–one reduction* f *from* Λ *to* Π *with the following properties:*

(i) If $x \in \Lambda$, *then* $opt(f(x)) \ge \beta$, *and*

(ii) If $x \notin \Lambda$, *then* $opt(f(x)) < \alpha$.

Figure 10.2 shows the reduction from Λ to a minimization problem Π with a gap $[\alpha, \beta]$.

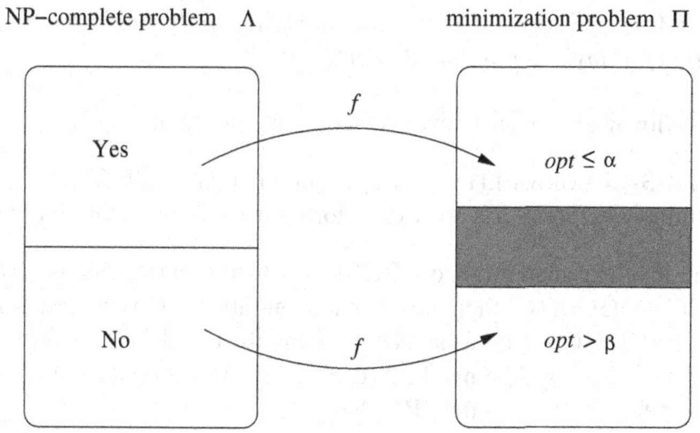

Figure 10.2: Reduction from an NP-complete problem to a minimization problem.

Lemma 10.2 *Assume that* Π *is an optimization problem with an* **NP**-*hard gap* $[\alpha, \beta]$, *with* $0 < \alpha < \beta$. *Then there is no polynomial-time* (β/α)-*approximation for problem* Π *unless* **P** = **NP**.

Proof. We prove the theorem for the case where Π is a minimization problem. The proof for maximization problems is similar.

Assume that f is a reduction from an **NP**-complete problem Λ to Π satisfying properties (i) and (ii) of Definition 10.1(a). Suppose, for the sake of contradiction, that there is a polynomial-time (β/α)-approximation $\mathcal{A}$ for problem Π. We may then construct a polynomial-time algorithm for problem Λ as follows:

(1) On input instance x of problem Λ, compute the instance $y = f(x)$ of problem Π.

(2) Run algorithm $\mathcal{A}$ on instance y to get an (β/α)-approximation s for y.

(3) Return YES if and only if the objective function value of solution s for problem Π is less than or equal to β.

It is easy to verify the correctness of the above algorithm: If $x \in \Lambda$, then $opt(y) \leq \alpha$, and hence the objective function value of any (β/α)-approximation solution for y is at most β. On the other hand, if $x \notin \Lambda$, then the objective function value of any solution for y must be greater than β. Therefore, the cutoff point β in the above algorithm solves the problem Λ at instance x correctly. □

Now, let us see some applications of this proof technique.

We first consider a simple example. A *vertex coloring* of a graph $G = (V, E)$ is a mapping $c : V \to \mathbb{Z}^+$ such that $c(u) \neq c(v)$ if $\{u, v\} \in E$.

GRAPH COLORING (GCOLOR): Given a graph $G = (V, E)$, find a vertex coloring of G using the minimum number of colors.

Theorem 10.3 *The problem* GCOLOR *does not have a polynomial-time* $((4/3)-\varepsilon)$*-approximation for any* $\varepsilon > 0$ *unless* **P** = **NP**.

Proof. The following is a well-known **NP**-complete problem:

> GRAPH-3-COLORABILITY (3GCOLOR): Given a graph $G = (V, E)$, determine whether G has a vertex coloring using at most three colors.

Let f be the identical mapping from 3GCOLOR to GCOLOR; that is, $f(G) = G$. Note that if $G \notin$ 3GCOLOR, then the chromatic number of G is at least 4. It implies that for $0 < \varepsilon < 1$, GCOLOR has an **NP**-hard gap $[3, 4 - \varepsilon]$. Note that $(4 - \varepsilon)/3 > (4/3) - \varepsilon$. Therefore, by Lemma 10.2, there is no polynomial-time $((4/3) - \varepsilon)$-approximation for GCOLOR unless **P** = **NP**. □

We now consider another problem.

> METRIC-k-CENTERS: Given n cities with a metric distance table between them, and an integer $k > 0$, select k cities to place warehouses such that the maximal distance of a city to a nearest warehouse is minimized.

It is known that METRIC-k-CENTERS has a polynomial-time 2-approximation (see Exercises 10.2 and 10.3). The following result indicates that this is the best possible.

Theorem 10.4 *There is no polynomial-time* $(2 - \varepsilon)$*-approximation for* METRIC-k-CENTERS *for any* $\varepsilon > 0$ *unless* **P** = **NP**.

Proof. In a graph $G = (V, E)$, a set $D \subseteq V$ is called a *dominating set* if every $v \in V$ either is in D or is adjacent to a vertex $u \in D$. The following problem is known to be **NP**-complete.

> DOMINATING SET (DS): Given a graph $G = (V, E)$ and an integer $k > 0$, determine whether G has a dominating set of size $\leq k$.

Define a reduction f from DS to METRIC-k-CENTERS as follows: On an instance (G, k) of DS, $f((G, k))$ consists of the graph G, a distance table d, and the same integer k, where

$$d(u, v) = \begin{cases} 1, & \text{if } \{u, v\} \in E, \\ 2, & \text{otherwise.} \end{cases}$$

We note that if G has a dominating set D of size at most k, then, for the instance (G, d, k) of problem METRIC-k-CENTERS, we can choose the cities in D to place warehouses so that every city is within distance 1 to a warehouse. On the other hand, if G does not have a dominating set of size k, then for any k choices of locations for warehouses, there must be at least one city $u \in V$ whose distance from any warehouse is at least 2. This means that METRIC-k-CENTERS has an **NP**-hard gap of $[1, 2-\varepsilon]$ for any $\varepsilon > 0$. By Lemma 10.2, there is no polynomial-time $(2-\varepsilon)$-approximation for METRIC-k-CENTERS. □

Recall the bottleneck Steiner tree problem (BNST), which asks, on a given set of terminals in the rectilinear plane, for a Steiner tree with at most k Steiner points, which minimizes the longest edge in the tree. In Section 3.4, we showed that BNST has a polynomial-time 2-approximation. The following result indicates that it is the best possible.

Theorem 10.5 *The problem* BNST *in the rectilinear plane does not have a polynomial-time* $(2-\varepsilon)$*-approximation for any* $\varepsilon > 0$ *unless* **P** = **NP**.

Proof. The following restricted version of the planar vertex cover problem is known to be **NP**-complete [Garey and Johnson, 1977, 1979]:

> PLANAR-CVC-4: Given a planar graph $G = (V, E)$ with all vertices of degree at most 4, and a positive integer $k > 0$, determine whether there is a *connected vertex cover* of G of size k.

We note that for any input instance $(G = (V, E), k)$ of PLANAR-CVC-4, we can embed G into the rectilinear plane so that all edges are horizontal or vertical segments of length at least $2k + 2$, and they do not cross each other except at the endpoints. Now, we define a set $P(G)$ of terminals for the problem BNST as follows: For each edge e of the embedded graph G of length d, we put $\lceil d \rceil - 1$ terminals on the interior of e such that the length between any two adjacent terminals is at most 1, and the first and last terminals have distance exactly 1 to the two end vertices of e. That is, the edge e of G becomes a path $p(e)$ in $P(G)$ (see Figure 10.3).

Clearly, if G has a connected vertex cover C of size k, then selecting all k vertices in C as Steiner points gives us a Steiner tree on $P(G)$ with k Steiner points such that the rectilinear length of each edge in the tree is at most 1. This means that the rectilinear length of each edge in any optimal solution of the input $P(G)$ is at most 1.

Next, assume that G has no connected vertex cover of size k. We claim that on input $P(G)$, any Steiner tree with k Steiner points must have an edge of rectilinear length ≥ 2. Suppose, for the sake of contradiction, that on input $P(G)$, there is a Steiner tree T with k Steiner points such that the rectilinear length of each edge in the tree is at most $2-\varepsilon$. Note that $P(G)$ has the following properties:

(a) Any two terminals on two different edges of the embedded G have distance at least 2.

(b) Any two terminals on two nonadjacent edges of the embedded G have distance at least $2k + 2$.

From property (b), two terminals on two nonadjacent edges cannot be connected through k Steiner points. Therefore, in any full Steiner component of T, all terminals lie on either the same edge or two adjacent edges. From property (a), we know that if a full Steiner component F of T contains two terminals lying on two different

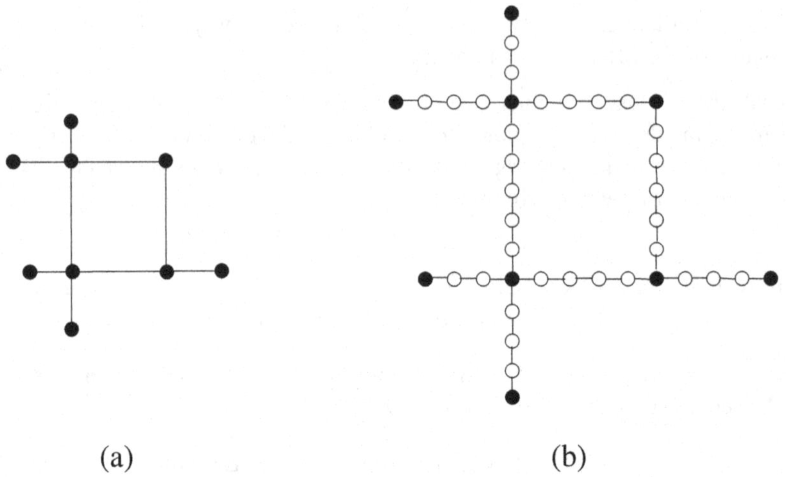

Figure 10.3: (a) A planar graph G. (b) The constructed graph $P(G)$. The dark circles $\bullet$ indicate the candidates of Steiner points, and the light circles $\circ$ indicate terminals.

edges e_1 and e_2 of G, then it must contain at least one Steiner point. Thus, we may move a Steiner point to the location of the vertex in G that covers the two edges e_1 and e_2, and remove other Steiner points in F (cf. Figure 10.3). That is, we can convert T to a new Steiner tree T' with at most k Steiner points such that all Steiner points in T' lie at the locations of the vertices in the embedded G. However, this means that the Steiner points of T' form a connected vertex cover of G of size at most k, which is a contradiction to our assumption. Thus, the claim is proven.

The above analysis showed that BNST has an **NP**-hard gap of $[1, 2-\varepsilon]$ for any $\varepsilon > 0$. The theorem now follows from Lemma 10.2. □

10.2 Gap Amplification and Preservation

In the last section, we showed how to use a reduction with a gap from an **NP**-complete problem Λ to prove an optimization problem Π having an **NP**-hard gap and establish a lower bound for the performance ratio of algorithms for Π. Sometimes, it is more convenient to reduce from an optimization problem Λ known to have an **NP**-hard gap $[\alpha, \beta]$ to another optimization problem Π to obtain an **NP**-hard gap $[\alpha', \beta']$ for Π. Such a reduction is called a *gap-preserving reduction*. If the ratio β'/α' for Π is greater than the starting ratio β/α of Λ, then we say the reduction is a *gap-amplifying reduction* (see Figure 10.4).

The following is an example of gap-amplifying reductions.

EDGE-DISJOINT PATHS (EDP): Given a graph $G = (V, E)$ and a list $L = ((s_1, t_1), (s_2, t_2), \ldots, (s_k, t_k))$ of k pairs of vertices, find edge-

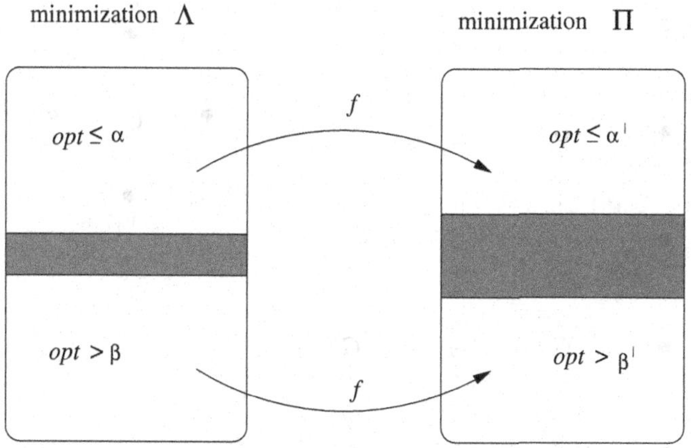

Figure 10.4: Gap amplification.

disjoint paths that maximize the number of connected pairs (s_i, t_i) in the list L.

We let EDP-c denote the problem EDP with the size of the list L equal to a constant c. The problem EDP-2 is known to be **NP**-hard; that is, it is **NP**-hard to determine whether two pairs of vertices can be connected by two edge-disjoint paths in G (see Exercise 10.7). It follows from this fact that EDP has an **NP**-hard gap $[1+\varepsilon, 2]$ for any $\varepsilon > 0$. In the following, we amplify this gap to obtain a better lower bound for approximating the problem EDP.

Theorem 10.6 *The problem* EDP *has no polynomial-time* $(m^{0.5-\varepsilon})$*-approximation for any* $0 < \varepsilon < 1/4$ *unless* **P** = **NP**, *where* m *is the number of edges in the input graph.*

Proof. We will construct a gap amplifier from EDP-2 to the general case of EDP.

Consider an instance of EDP-2 consisting of a graph $G = (V, E)$ and two pairs (u_1, v_1) and (u_2, v_2) of vertices. We construct a graph H that consists of $k(k-1)/2$ copies of G and $2k$ additional vertices $s_1, \ldots, s_k, t_1, \ldots, t_k$, which are connected as shown in Figure 10.5, where

$$k = \left\lceil \left(\frac{|E|}{2} + 1\right)^{(1-2\varepsilon)/4\varepsilon} \right\rceil.$$

That is, a copy of G is connected to other copies of G or vertices s_i, t_j through vertices u_1, u_2, v_1, and v_2. For instance, vertex u_1 of a copy of G in the main diagram of Figure 10.5 is connected to vertex v_1 of the copy of G to its left, or to a vertex s_i if it is a leftmost copy of G in the diagram. In addition, the list of pairs of vertices in H to be connected consists of (s_i, t_i), $i = 1, 2, \ldots, k$.

Clearly, if G contains two edge-disjoint paths connecting pairs (u_1, v_1) and (u_2, v_2), respectively, then H contains k edge-disjoint paths connecting all k pairs

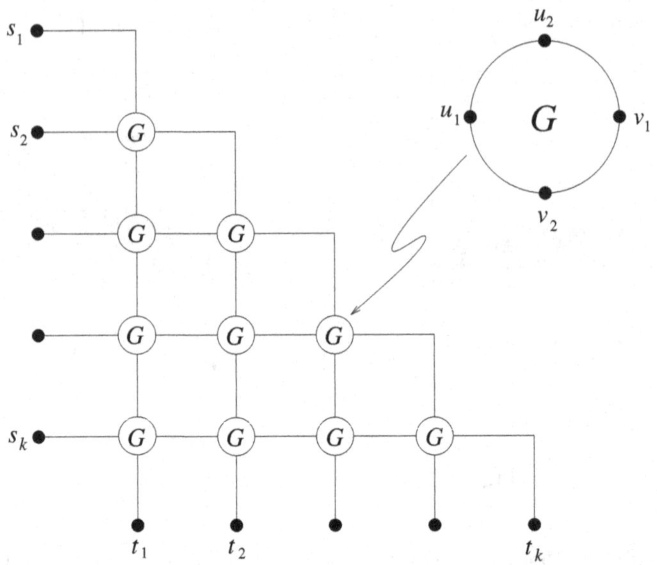

Figure 10.5: Gap-amplifying reduction from EDP-2 to EDP.

$(s_1, t_1), (s_2, t_2), \ldots, (s_k, t_k)$, respectively. On the other hand, if G does not contain two edge-disjoint paths connecting pairs (u_1, v_1) and (u_2, v_2), respectively, then H can have at most one path connecting a given pair (s_i, t_i) of vertices for some $i = 1, 2, \ldots, k$. Thus, the **NP**-hard gap $[1+\varepsilon, 2]$ of EDP-2 is amplified to a bigger **NP**-hard gap $[1+\varepsilon, k]$. Note that the number of edges in H is

$$m = \frac{k(k-1)}{2} \cdot |E| + k^2 \le k^2\Big(\frac{|E|}{2} + 1\Big) \le k^{2+(4\varepsilon)/(1-2\varepsilon)}.$$

Thus, $k \ge m^{0.5-\varepsilon}$, and the theorem follows from Lemma 10.2. □

Gap-preserving reductions are an important tool for proving the inapproximability of an optimization problem. To demonstrate its power, we borrow an inapproximability result from Section 10.4.

> MAXIMUM 3-LINEAR EQUATIONS (MAX-3LIN): Given a system of linear equations over $GF(2)$, where each equation contains exactly three variables, find an assignment to variables that satisfies the maximum number of equations.

It will be established in Section 10.4, by Håstad's three-bit PCP theorem, that MAX-3LIN has an **NP**-hard gap of $[(0.5+\varepsilon)m, (1-\varepsilon)m]$ for any $\varepsilon > 0$, where m is the number of input equations.

Theorem 10.7 *The problem* MAX-3SAT *does not have a polynomial-time* $(8/7 - \varepsilon)$*-approximation for any* $\varepsilon > 0$ *unless* **P** = **NP**.

Proof. We will construct a gap-preserving reduction from MAX-3LIN to MAX-3SAT. Consider a system $\mathcal{E}$ of m linear equations over $GF(2)$. For each equation e in $\mathcal{E}$ of the form $x_i \oplus x_j \oplus x_k = 1$, we introduce four clauses: $f_e = (x_i \vee x_j \vee x_k) \wedge (x_i \vee \bar{x}_j \vee \bar{x}_k) \wedge (\bar{x}_i \vee x_j \vee \bar{x}_k) \wedge (\bar{x}_i \vee \bar{x}_j \vee x_k)$. For each equation e' in $\mathcal{E}$ of the form $x_i \oplus x_j \oplus x_k = 0$, we also introduce four clauses: $f_{e'} = (\bar{x}_i \vee \bar{x}_j \vee \bar{x}_k) \wedge (x_i \vee x_j \vee \bar{x}_k) \wedge (x_i \vee \bar{x}_j \vee x_k) \wedge (\bar{x}_i \vee x_j \vee x_k)$. Note that the equation e (or e') and clauses in f_e (or, respectively, in $f_{e'}$) have the following relationship:

(i) If an assignment satisfies e (or, e'), then the same assignment satisfies four clauses in f_e (or, respectively, in $f_{e'}$).

(ii) If an assignment does not satisfy e (or, e'), then the same assignment satisfies exactly three clauses in f_e (or, respectively, in $f_{e'}$).

Let $f(\mathcal{E})$ be the 3CNF formula obtained from the above transformation; that is, $f(\mathcal{E})$ is the conjunct of all f_e's over all equations e in $\mathcal{E}$. We note that for any assignment, each f_e has exactly three or four satisfied clauses. Therefore, we have the following properties:

(a) If the optimal solution of MAX-3LIN on instance $\mathcal{E}$ satisfies fewer than $(0.5+\varepsilon)m$ equations in $\mathcal{E}$, then the optimal solution of MAX-3SAT on $f(\mathcal{E})$ satisfies fewer than $(3.5+\varepsilon)m$ clauses in $f(\mathcal{E})$.

(b) If there is an assignment for $\mathcal{E}$ that satisfies at least $(1-\varepsilon)m$ equations, then the same assignment satisfies at least $(4-\varepsilon)m$ clauses in $f(\mathcal{E})$.

Thus, MAX-3SAT has an **NP**-hard gap of $[(3.5+\varepsilon)m, (4-\varepsilon)m]$. By Lemma 10.2, MAX-3SAT cannot have a polynomial-time $(4-\varepsilon)/(3.5+\varepsilon)$-approximation unless **P** = **NP**. Note that

$$\frac{4-\varepsilon}{3.5+\varepsilon} \longrightarrow \frac{8}{7},$$

as $\varepsilon \to 0$. This completes the proof of this theorem. □

Theorem 10.8 *The problem* MIN-VC *does not have a polynomial-time* $(7/6-\varepsilon)$*-approximation for any* $\varepsilon > 0$ *unless* **P** = **NP**.

Proof. We construct a gap-preserving reduction from MAX-3LIN to MIN-VC. Let $\mathcal{E}$ be a system of m linear equations over $GF(2)$. For each equation of the form $x_i \oplus x_j \oplus x_k = 1$ (or of the form $x_i \oplus x_j \oplus x_k = 0$), we construct a complete graph of four vertices labeled with four satisfying assignments of the equation as shown in Figure 10.6(a) (or, respectively, in Figure 10.6(b)).

Thus, we have totally constructed m complete graphs of order 4. Next, we connect two vertices with an edge if they contain a conflicting assignment (i.e., if there exists a variable x_i such that $x_i = 0$ in the label of one vertex and $x_i = 1$ in the label of the other vertex). Now, we have obtained a graph G with $4m$ vertices with the following properties:

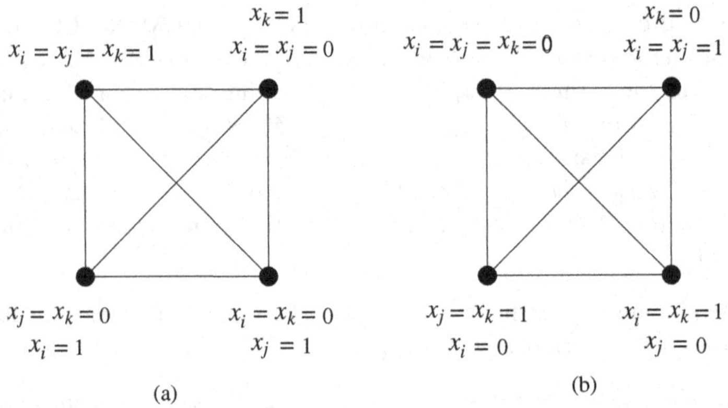

Figure 10.6: Two graphs whose four vertices are labeled with satisfying assignments of $x_i \oplus x_j \oplus x_k = 1$ [part (a)] or that of $x_i \oplus x_j \oplus x_k = 0$ [part (b)].

(a) If there is an assignment to variables that satisfies at least $(1 - \varepsilon)m$ equations in $\mathcal{E}$, then this assignment satisfies the labels of at least $(1 - \varepsilon)m$ vertices simultaneously. From our construction, these $(1 - \varepsilon)m$ vertices are independent, and so the set of remaining vertices is a vertex cover for G. Therefore, G has a vertex cover of size at most $4m - (1 - \varepsilon)m = (3 + \varepsilon)m$.

(b) If no assignment can satisfy $(0.5 + \varepsilon)m$ or more equations in $\mathcal{E}$, then no assignment can simultaneously satisfy the labels of $(0.5+\varepsilon)m$ or more vertices. As the labels of vertices in an independent set can be satisfied simultaneously, we see that every independent set of G has size less than $(0.5+\varepsilon)m$. It follows that each vertex cover has size greater than $4m - (0.5 + \varepsilon)m = (3.5 - \varepsilon)m$.

It follows that MIN-VC has an **NP**-hard gap of $[(3+\varepsilon)m, (3.5-\varepsilon)m]$ for any $\varepsilon > 0$. By Lemma 10.2, MIN-VC does not have a polynomial-time $(3.5 - \varepsilon)/(3 + \varepsilon)$-approximation unless **P** = **NP**. The proof of theorem is completed by noting that

$$\frac{3.5 - \varepsilon}{3 + \varepsilon} \longrightarrow \frac{7}{6}$$

as $\varepsilon \to 0$. □

10.3 APX-Completeness

In the last section, we used gap-preserving reductions to get strong inapproximability results. However, for problems having approximations with constant performance ratios, gap-preserving reductions are often too strong for proving their inapproximability. To study weaker inapproximability results on these problems, we introduce an approximation-preserving reduction.

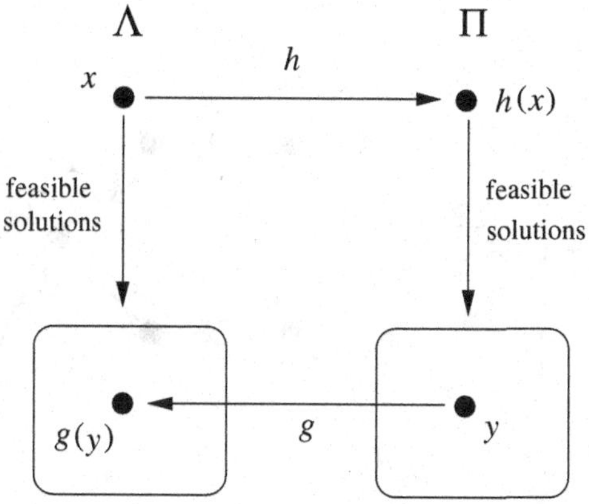

Figure 10.7: An L-reduction.

Definition 10.9 *Let Λ and Π be two optimization problems. We say Λ is L-reducible to Π, and write $\Lambda \leq_L^P \Pi$ if there are two polynomial-time mappings h and g satisfying the following conditions (see* Figure 10.7*):*

(L1) *h maps an instance x of Λ to an instance $h(x)$ of Π such that*

$$opt_\Pi(h(x)) \leq a \cdot opt_\Lambda(x)$$

for some constant a, where $opt_\Lambda(x)$ denotes the optimal objective function value of problem Λ on input x.

(L2) *g maps solutions of Π for instance $h(x)$ to solutions of Λ for instance x such that, for any solution y of $h(x)$,*

$$|obj_\Lambda(g(y)) - opt_\Lambda(x)| \leq b \cdot |obj_\Pi(y) - opt_\Pi(h(x))|$$

for some constant $b > 0$, where $obj_\Lambda(g(y))$ is the objective function value of the solution $g(y)$ for instance x.

As an example, consider the following subproblems of MIN-VC.

MIN-VC-b: Given a graph $G = (V, E)$ in which every vertex has degree at most b, find the minimum vertex cover of G.

We have the following L-reduction between these subproblems.

Theorem 10.10 *For any $b \geq 4$,* MIN-VC-b $\leq_L^P$ MIN-VC-3.

Proof. Given a graph $G = (V, E)$ in which every vertex has degree at most b, we modify graph G into a new graph G' as follows: For each vertex x of degree d in G, construct a path P_x of $2d - 1$ vertices to replace it as shown in Figure 10.8. Note that this path has a unique minimum vertex cover C_x of size $d - 1$ (the light circles

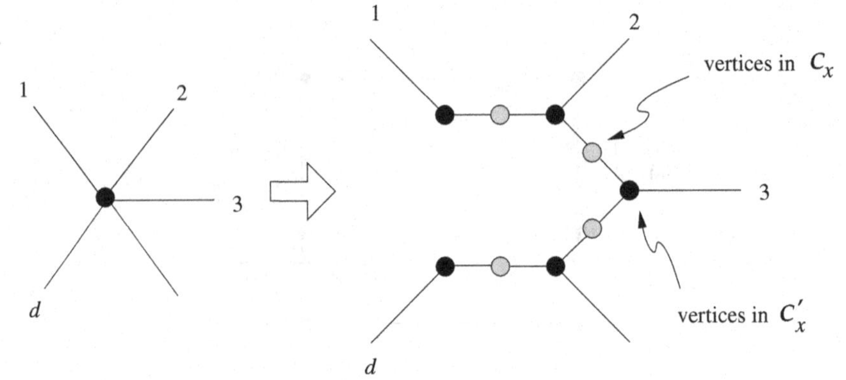

Figure 10.8: Path P_x.

in Figure 10.8). This vertex cover, however, covers only edges in path P_x. The set C'_x of vertices in P_x but not in C_x (the dark circles in Figure 10.8) is also a vertex cover of P_x. This vertex cover C'_x has size d, but it also covers all other edges that are incident on path P_x (i.e., those edges that are indicent on x in the original graph G).

Let $m = |E|$ and $n = |V|$. If G has a vertex cover S, then we can obtain a vertex cover

$$S' = \Big(\bigcup_{x\in S} C'_x\Big) \cup \Big(\bigcup_{x\notin S} C_x\Big)$$

of size $|S| + 2m - n$ for G'. Conversely, for each vertex cover S' for G', we can construct a vertex cover $S = \{x \mid C'_x \cap S' \neq \emptyset\}$ for G. Note that if, for some $x \in V$, $C'_x \cap S' \neq \emptyset$, then $P_x \cap S'$ has size at least $\deg_G(x)$. Therefore, we have $|S| \leq |S'| - (2m - n)$.

An immediate consequence of the above relationship is that

$$opt(G') = opt(G) + 2m - n,$$

where $opt(G)$ (and $opt(G')$) is the size of the minimum vertex cover in G (and, respectively, G'). Note that $m \leq b \cdot opt(G)$. Thus,

$$opt(G') \leq (2b + 1) \cdot opt(G);$$

that is, condition (L1) holds. Note that $|S| \leq |S'| - (2m - n)$ is equivalent to

$$|S| - opt(G) \leq |S'| - opt(G').$$

Therefore, condition (L2) also holds, and the proof of the theorem is complete. □

L-reductions are useful in proving problems not having PTAS, due to the following two properties.

Lemma 10.11 *If* $\Pi \leq_L^P \Gamma$ *and* $\Gamma \leq_L^P \Lambda$*, then* $\Pi \leq_L^P \Lambda$.

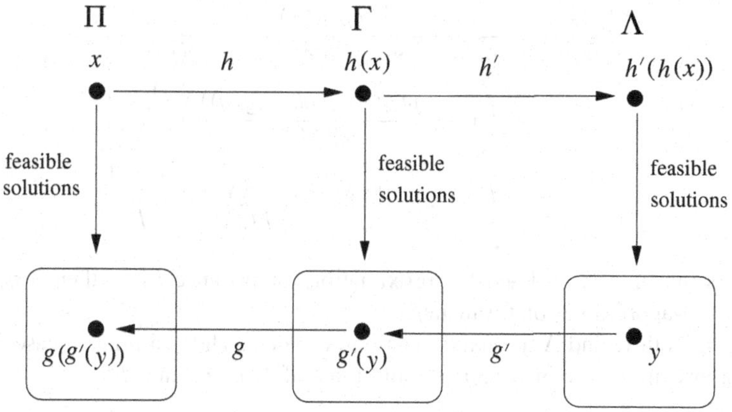

Figure 10.9: Proof of Lemma 10.11.

Proof. Suppose $\Pi \le_L^P \Gamma$ via mappings h and g and $\Gamma \le_L^P \Lambda$ via mappings h' and g'. It is easy to verify that $\Pi \le_L^P \Lambda$ via mapping $h' \circ h$ and $g \circ g'$ (see Figure 10.9). □

Lemma 10.12 *If* $\Pi \le_L^P \Lambda$ *and* Λ *has a PTAS, then* Π *has a PTAS.*

Proof. Suppose $\Pi \le_L^P \Lambda$ via mappings h and g, and let a and b be the constants satisfying conditions (L1) and (L2). Consider the following four cases. We prove that in each case, if Λ has a PTAS, then Π has a PTAS.

Case 1. Both Π and Λ are minimization problems. Then we have, for any instance x of Π and any solution y of Λ for instance $h(x)$,

$$\begin{aligned}\frac{obj_\Pi(g(y))}{opt_\Pi(x)} &= 1 + \frac{obj_\Pi(g(y)) - opt_\Pi(x)}{opt_\Pi(x)} \\ &\le 1 + ab \cdot \frac{obj_\Lambda(y) - opt_\Lambda(h(x))}{opt_\Lambda(h(x))}.\end{aligned}$$

It follows that if y is a $(1+\varepsilon)$-approximation for instance $h(x)$, then $g(y)$ is a $(1+ab\varepsilon)$-approximation for instance x.

Case 2. Π is a minimization problem and Λ is a maximization problem. Then we have, for any instance x of Π and any solution y of Λ for instance $h(x)$,

$$\begin{aligned}\frac{obj_\Pi(g(y))}{opt_\Pi(x)} &= 1 + \frac{obj_\Pi(g(y)) - opt_\Pi(x)}{opt_\Pi(x)} \\ &\le 1 + ab \cdot \frac{opt_\Lambda(h(x)) - obj_\Lambda(y)}{opt_\Lambda(h(x))} \le 1 + ab \cdot \frac{opt_\Lambda(h(x)) - obj_\Lambda(y)}{obj_\Lambda(y)}.\end{aligned}$$

It follows that if y is a $(1+\varepsilon)$-approximation for instance $h(x)$, then $g(y)$ is a $(1+ab\varepsilon)$-approximation for instance x.

Case 3. Π is a maximization problem and Λ is a minimization problem. Then we have, for any instance x of Π and any solution y of Λ on instance $h(x)$,

$$\begin{aligned}
\frac{opt_\Pi(x)}{obj_\Pi(g(y))} &= \frac{opt_\Pi(x)}{opt_\Pi(x) - opt_\Pi(x) + obj_\Pi(g(y))} \\
&= \left(1 - \frac{opt_\Pi(x) - obj_\Pi(g(y))}{opt_\Pi(x)}\right)^{-1} \\
&\leq \left(1 - ab \cdot \frac{obj_\Lambda(y) - opt_\Lambda(h(x))}{opt_\Lambda(h(x))}\right)^{-1}.
\end{aligned}$$

It follows that if y is a $(1+\varepsilon)$-approximation for instance $h(x)$, then $g(y)$ is a $1/(1-ab\varepsilon)$-approximation for instance x.

Case 4. Both Π and Λ are maximization problems. Then, similar to case 3, we have, for any instance x of Π and any solution y of Λ on instance $h(x)$,

$$\begin{aligned}
\frac{opt_\Pi(x)}{obj_\Pi(g(y))} &= \left(1 - \frac{opt_\Pi(x) - obj_\Pi(g(y))}{opt_\Pi(x)}\right)^{-1} \\
&\leq \left(1 - ab \cdot \frac{opt_\Lambda(h(x)) - obj_\Lambda(y)}{opt_\Lambda(h(x))}\right)^{-1} \\
&\leq \left(1 - ab \cdot \frac{opt_\Lambda(h(x)) - obj_\Lambda(y)}{obj_\Lambda(y)}\right)^{-1}.
\end{aligned}$$

It follows that if y is a $(1+\varepsilon)$-approximation for instance $h(x)$, then $g(y)$ is a $1/(1-ab\varepsilon)$-approximation for instance x. □

In addition to L-reductions, a weaker type of reductions, called *E-reductions*, has also been used in the study of the inapproximability of problems having constant-ratio approximations. This reduction has the following properties:

(a) If $\Pi \leq_E \Sigma$ and $\Sigma \leq_E \Lambda$, then $\Pi \leq_E \Lambda$.

(b) If $\Pi \leq_E \Lambda$ and Λ has a PTAS, then Π has a PTAS.

(c) If $\Pi \leq_L^P \Lambda$, then $\Pi \leq_E \Lambda$.

Since we will, in this section, mainly use L-reductions to establish inapproximability results, we omit the formal definition of the E-reduction and the proofs of the above properties (see Exercise 10.8).

Let **NPO** denote the class of optimization problems Π with the following properties:

(a) Its feasible solutions are polynomial-time verifiable; that is, given an instance x and a candidate y of its feasible solution, of size $|y| \leq |x|^{O(1)}$, it is decidable in time polynomial in $|x|$ whether y is a feasible solution of x.

(b) Its objective function is polynomial-time computable; that is, given an instance x and a feasible solution y of x, the objective function value $obj_\Pi(y)$ can be computed in polynomial time in $|x|$.

Let **APX** denote the class of all **NPO** problems that have polynomial-time r-approximation for some constant $r > 1$. For instance, the problems MIN-VC, EUCLIDEAN-TSP, NSMT, BNST, and METRIC-k-CENTERS all belong to **APX**. On the other hand, it is known that if **P** $\neq$ **NP**, then the problems TSP, MIN-SC, MIN-CDS, CLIQUE, and GCOLOR do not belong to **APX** (see Sections 10.5 and 10.6).

To study the inapproximability of problems in **APX**, we generalize the notion of completeness from decision problems to optimization problems. For a class $\mathcal{C}$ of optimization problems and a reduction $\leq_R$ among optimization problems, a problem Λ is called *$\mathcal{C}$-hard* if for every problem $\Pi \in \mathcal{C}$, $\Pi \leq_R \Lambda$. If Λ is already known to be in $\mathcal{C}$, then Λ is said to be $\mathcal{C}$-complete.

Papadimitriou and Yannakakis [1993] studied a subclass **MAXSNP** of **APX**, and showed **MAXSNP**-completeness, under the L-reduction, for many problems, including MIN-VC-b for $b \geq 3$. Khanna et al. [1999] showed that **APX** is the closure of **MAXSNP** under E-reduction, in the sense that every problem $\Pi \in$ **APX** is E-reducible to some problem $\Lambda \in$ **MAXSNP**. Therefore, an **MAXSNP**-complete problem under the L-reduction is also **APX**-complete under the E-reduction. (In the following, we will write **APX**-completeness to denote **APX**-completeness under the E-reduction.)

Theorem 10.13 *The problem* MIN-VC-3 *is* **APX**-*complete.*

Note that BNST and METRIC-k-CENTERS are in **APX**, but they don't have PTASs unless **P** = **NP**. Therefore, we have

Theorem 10.14 *An* **APX**-*complete problem has no PTAS unless* **P** = **NP**.

Thus, we can use L-reductions and **APX**-completeness to prove a problem in **APX** having no PTAS. The following are some examples.

> VERTEX COVER IN CUBIC GRAPHS (VC-CG): Given a cubic graph G, find a minimum vertex cover of G. (A *cubic graph* is a graph in which every vertex has degree 3.)

Theorem 10.15 *The problem* VC-CG *is* **APX**-*complete.*

Proof. Since VC-CG is clearly in **APX**, it suffices to prove that it is **APX**-hard. To do so, we construct an L-reduction from MIN-VC-3 to VC-CG.

Consider an instance of MIN-VC-3, that is, a graph $G = (V, E)$ in which each vertex has degree at most 3. Suppose that G has i vertices of degree 1 and j vertices of degree 2. Construct a new graph H as follows: H has a cycle of size $2(2i+j)$, and $2i+j$ triangles. Each triangle has two vertices connecting to two adjacent vertices in the cycle, as shown in Figure 10.10. In each triangle of H, call the vertex that is not connected to the cycle a *free vertex*. We note that we need $2i + j$ vertices to cover the cycle, and two vertices to cover each triangle. Thus, a minimum vertex cover for H has size $\geq 3(2i + j)$. In fact, it is easy to see that there exists a minimum vertex cover of H of size $3(2i + j)$ that contains all free vertices.

Next, construct a graph G' from G and H as follows:

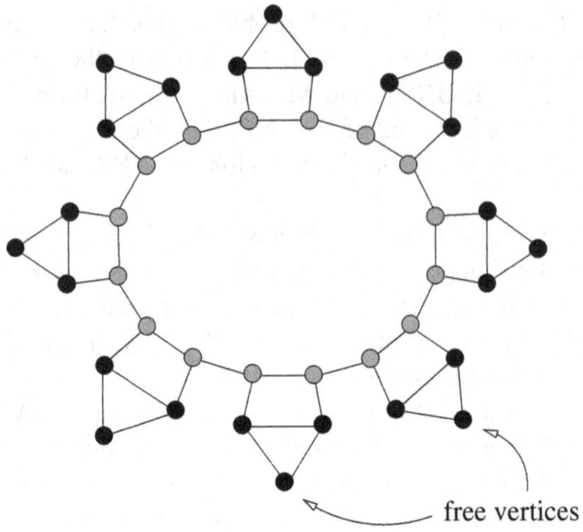

Figure 10.10: Graph H.

(a) For each vertex x of degree 1 in G, use two edges to connect x to two free vertices of H.

(b) For each vertex x of degree 2 in G, use one edge to connect x to one free vertex of H.

(c) Each free vertex of H is connected to exactly one vertex in G.

Clearly, G' is a cubic graph. In addition, G has a vertex cover of size s if and only if G' has a vertex cover of size $s' = s + 3(2i + j)$. That is,

$$opt(G') = opt(G) + 3(2i + j),$$

where $opt(G)$ (and $opt(G')$) denotes the size of the minimum vertex cover of G (and, respectively, G'). Note that G has at least $(i + 2j)/2$ edges, and each vertex in G can cover at most three edges. Therefore, $i + 2j \leq 6 \cdot opt(G)$, and so $2i + j \leq 2i + 4j \leq 12 \cdot opt(G)$. It follows that

$$opt(G') \leq 37 \cdot opt(G),$$

and condition (L1) holds for the reduction from G to G'.

Next, to prove condition (L2), we note that for each vertex cover S' of size s' in G', we can obtain a vertex cover S of size $s \leq s' - 3(2i + j)$ in G by simply removing all vertices in $S' \setminus V$. It follows that

$$s - opt(G) \leq s' - opt(G').$$

Therefore, condition (L2) also holds for this reduction. □

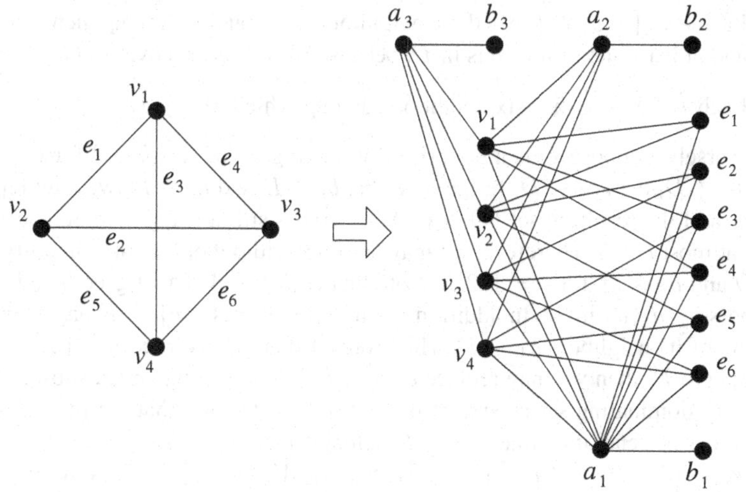

Figure 10.11: Construction from G to G'.

In the following, we consider a problem that originated from the study of social networks. We say a vertex subset $D \subseteq V$ of a graph $G = (V, E)$ is a *majority-dominating set* of G if, for every vertex v not in D, at least one half of the neighbors of v are in D.

> MAJORITY-DOMINATING SET (MAJ-DS): Given a graph $G = (V, E)$, find a majority-dominating set $D \subseteq V$ of the minimum cardinality.

Theorem 10.16 *The problem* MAJ-DS *is* **APX**-*hard.*

Proof. We will construct an L-reduction from VC-CG to MAJ-DS. For a cubic graph $G = (V, E)$, we first construct a bipartite graph $H = (V, E, F)$, where $\{v, e\} \in F$ if and only if v is an endpoint of e in G. Next, we add to H six additional vertices a_i, b_i for $i = 1, 2, 3$, and the following additional edges to form a graph G' (see Figure 10.11):

(i) $\{a_i, b_i\}$, for $i = 1, 2, 3$;

(ii) $\{a_1, e\}$, for all $e \in E$; and

(iii) $\{a_1, v\}$, $\{a_2, v\}$, $\{a_3, v\}$, for all $v \in V$.

We claim that G has a vertex cover of size at most k if and only if G' has a majority-dominating set of size at most $k + 3$.

To show our claim, we first assume G has a vertex cover C of size k. Let $D = C \cup \{a_i \mid i = 1, 2, 3\}$. In the following we verify that D is a majority-dominating set for G'.

(1) Each b_i has only one neighbor $a_i \in D$.

(2) Each $e = \{u, v\} \in E$ has three neighbors, a_1, u, and v. Among them, $a_1 \in D$ and at least one of u or v is in D, because C is a vertex cover of G.

(3) Each $v \in V - C$ has six neighbors, among which $a_1, a_2, a_3 \in D$.

Conversely, suppose D is a majority-dominating set of size $k + 3$ for G'. Note that if $b_i \notin D$, then $a_i \in D$. In the case that $b_i \in D$ and $a_i \notin D$, we may replace b_i by a_i and the resulting set $(D - \{b_i\}) \cup \{a_i\}$ is still a majority-dominating set of size at most $k + 3$. Therefore, we may assume, without of loss of generality, that $b_i \notin D$ and $a_i \in D$, for $i = 1, 2, 3$. Note that each $v \in V$ has degree 6 and it has neighbors a_1, a_2, a_3 in D. In addition, each vertex $e = \{u, v\} \in E$ has degree 3, with one of its neighbors $a_1 \in D$. Therefore, if there is a vertex $e = \{u, v\} \in E$ belonging to D, then we may replace e by u, and the resulting vertex subset is still a majority-dominating set of size at most $k + 3$. It follows that we may assume, without loss of generality, that no $e \in E$ belongs to D.

Now, let $C = D - \{a_1, a_2, a_3\}$. Then $C \subseteq V$ and $|C| \leq k$. Note that each $e = \{u, v\} \in E$ has three neighbors, a_1, u, and v. Since e has degree 3 and hence has at least two neighbors in D, we must have either $u \in C$ or $v \in C$. That is, C is a vertex cover of G. This completes the proof of our claim.

Now, suppose G has a minimum vertex cover of size opt_{VC}. Then by the claim, G' has a minimum majority-dominating set of size $opt_{\text{MDS}} = opt_{\text{VC}} + 3$. That is,

$$opt_{\text{MDS}} = opt_{\text{VC}} + 3 \leq 4 \cdot opt_{\text{VC}}.$$

Moreover, let D be a majority-dominating set of size k' for G'. Then, from the proof of our claim, we can construct a vertex cover C of size at most $k' - 3$ for G. Therefore,

$$\big||C| - opt_{\text{VC}}\big| \leq \big|k' - (opt_{\text{VC}} + 3)\big| = \big||D| - opt_{\text{MDS}}\big|.$$

Therefore, VC-CG is L-reducible to MAJ-DS. It follows that MAJ-DS is **APX**-hard. □

10.4 PCP Theorem

The following is a well-known characterization of the complexity class **NP**:

Proposition 10.17 *A language L belongs to class* **NP** *if and only if there exist a language A in class* **P**, *and a polynomial p, such that*

$$x \in L \iff (\exists y, |y| \leq p(|x|))(x, y) \in A.$$

That is, for a language $L \in \textbf{NP}$, an input x is in L if and only if there is a *proof* y of length $p(|x|)$ such that the correctness of the proof, i.e., whether $(x, y) \in A$ or not, can be verified in polynomial time. We may reformulate this characterization as a *proof system* for the language $L \in \textbf{NP}$:

(a) The proof system for L consists of a prover and a verifier.

(b) On input x, the prover presents a proof y of length $p(|x|)$ for some polynomial p.

(c) The verifier determines, from x and y, in polynomial time whether or not to accept.

(d) If $x \in L$, then there exists a proof y on which the verifier accepts.

(e) If $x \notin L$ then, for all proofs y, the verifier rejects.

The PCP theorem presents a stronger characterization of the class **NP** in terms of a new proof system. In this new proof system, the verifier can use randomness to reduce the amount of information of the proof y that he or she needs to read in order to decide whether to accept or reject. More precisely, a *probabilistically checkable proof system* $PCP_{c(n),s(n)}(r(n), q(n))$ can be described as follows:

(a) The proof system for L consists of a prover and a verifier.

(b) On input x, the prover presents a proof y of length $p(|x|)$ for some polynomial p.

(c) The verifier uses $r(n)$ random bits within polynomial time to compute $q(n)$ locations of y, and reads these $q(n)$ bits of y. Then the verifier determines, from x and the $q(n)$ bits of y, whether or not to accept x.

(d) If $x \in L$, then there exists a proof y relative to which the verifier accepts x with probability $\geq c(n)$.

(e) If $x \notin L$, then, for all proofs y, the verifier accepts x with probability $\leq s(n)$.

For example, the characterization of class **NP** given in Proposition 10.17 can be rephrased in terms of the PCP systems as follows: Every problem in **NP** has a proof system $PCP_{1,0}(0, p(n))$ for some polynomial p.

The following result is a milestone in the study of PCP systems.

Theorem 10.18 (PCP Theorem) *The problem* SAT *has a probabilistically checkable proof system* $PCP_{1,1/2}(O(\log n), O(1))$.

The result that **MAXSNP**-complete problems do not have PTAS provided **P** $\neq$ **NP** was first proved based on the PCP theorem. However, as we pointed out in Section 10.3, this conclusion can be derived without using the PCP theorem. Nevertheless, it may still provide additional information about the **NP**-hard gaps of these problems.

Theorem 10.19 *The problem* MAX-SAT *has an* **NP**-*hard gap* $[\alpha m, m]$ *for some* $0 < \alpha < 1$, *where* m *is the number of clauses in the input CNF formula.*

Skecth of Proof. We will construct a reduction from SAT to MAX-SAT with gap $[\alpha m, m]$. Let F be a Boolean formula, that is, an instance of the problem SAT. We will construct a CNF formula F' of $m = q(|F|)$ clauses, for some polynomial q, such that

(a) If $F \in$ SAT then F' is satisfiable, and

(b) If $F \notin$ SAT, then at most αm clauses in F' can be satisfied.

Let $S \in PCP_{1,1/2}(c_1 \log n, c_2)$ be a PCP system for SAT, where c_1 and c_2 are two positive constants. Assume that the prover always writes down a proof y of $p(n)$ bits, for some polynomial p, for an instance F of size n. Then the verifier of the system S works as follows:

(1) The verifier uses a random string r of $c_1 \log n$ bits to compute a set A_r of c_2 locations of the proof.

(2) The verifier reads the c_2 bits of the proof at these locations. (Call them y_r.)

(3) The verifier decides in deterministic polynomial time, from F and y_r, whether or not to accept F.

Note that the above system can be modified to execute step (3) before step (2). That is, we can use Boolean variables x_i, for $i = 1, 2, \ldots, p(n)$, to represent the ith bit of the proof y, and formulate step (3) as a Boolean formula over variables in $\{x_i \mid i \in A_r\}$. This Boolean formula has size $O(c_2)$. We can further transform this Boolean formula into a CNF formula of size $O(c_2)$ (some new variables z_j may be introduced during this transformation). Call this CNF formula F_r.

Since the verifier can use only $c_1 \log n$ random bits, there are at most $2^{c_1 \log n} = n^{O(c_1)}$ possible random strings r, and hence at most $n^{O(c_1)}$ formulas F_r. Let F' be the conjunct of all these formulas F_r. Then F' is a CNF formula of size $O(c_2) \cdot n^{O(c_1)} = n^{O(1)}$. We verify that F' satisfies the required conditions:

First, if $F \in$ SAT, then there exists a proof y relative to which the verifier accepts F with probability 1. This means that the assignment τ, with $\tau(x_i) =$ the ith bit of y, satisfies all CNF formulas F_r, and hence F' is satisfiable.

On the other hand, if $F \notin$ SAT, then the verifier accepts F with probability at most $1/2$, no matter what proof y is provided. This means that, for any assignment τ on variables x_i's, at least half of the formulas F_r are not satisfied. Assume that each CNF formula F_r contains at most c clauses, and that F' contains m clauses. Then for any assignment τ, at least $m/(2c)$ clauses of F' are not satisfied. Or, equivalently, for any assignment τ, at most $m(1 - 1/(2c))$ clauses of F' can be satisfied.

The above reduction shows that MAX-SAT has an **NP**-hard gap $[\alpha m, m]$ for $\alpha = 1 - 1/(2c)$. □

The following extension of the PCP theorem is very useful in getting better **NP**-hard gaps.

Theorem 10.20 (Håstad's 3-Bit PCP Theorem) *For any* $0 < \varepsilon < 1$, 3-SAT *has a proof system* $PCP_{1-\varepsilon,0.5+\varepsilon}(O(\log n), 3)$. *More precisely, the verifier in this system computes three locations* i, j, k *of the proof and a bit* b *from a random string of length* $O(\log n)$, *and accepts the input if and only if* $y_i \oplus y_j \oplus y_k = b$, *where* y_i *is the* i*th bit of the proof.*

Now, we apply this stronger PCP system to get the **NP**-hard gap for the problem MAX-3LIN defined in Section 10.2.

Theorem 10.21 *For any* $0 < \varepsilon < 1/4$, *the problem* MAX-3LIN *has an* **NP**-*hard gap* $[(0.5 + \varepsilon)m, (1 - \varepsilon)m]$, *where* m *is the number of equations in the input.*

Proof. We reduce 3SAT to MAX-3LIN as follows. By Håstat's 3-bit PCP theorem, 3SAT has, for any $0 < \varepsilon < 1/4$, a proof system S in $PCP_{1-\varepsilon,0.5+\varepsilon}(c \log n, 3)$, for some $c > 0$, in which the verifier produces, for any given random string r of length $c \log n$, an equation $x_i \oplus x_j \oplus x_k = b$. For each 3CNF formula F, we construct the instance E of MAX-3LIN that consists of all possible equations $x_i \oplus x_j \oplus x_k = b$ produced by the verifier of the proof system S on input F, over all possible random strings r of length $c \log n$. Since the random string r has length $c \log n$, the total number of equations in E is bounded by $2^{c \log n} = n^{O(1)}$. Therefore, this is a polynomial-time reduction.

Now we verify that this reduction preserves the **NP**-hard gap of $[(0.5+\varepsilon)m, (1-\varepsilon)m]$. First, if $F \in$ 3SAT, then there exists a proof y whose bit values satisfy the random equation $x_i \oplus x_j \oplus x_k = b$ with probability $\geq 1 - \varepsilon$. This means that there exists an assignment to variables x_i that satisfies, among m possible equations, at least $(1 - \varepsilon)m$ of them. Conversely, if $F \notin$ 3SAT, then the bit values of any given proof can satisfy a random equation $x_i \oplus x_j \oplus x_k = b$ with probability $\leq 0.5 + \varepsilon$. This means that, for any assignment to variables x_i, at most $(0.5 + \varepsilon)m$ out of m equations are satisfied.

The above reduction established the **NP**-hard gap $[(0.5 + \varepsilon)m, (1 - \varepsilon)m]$ for MAX-3SAT. □

Corollary 10.22 *The problem* MAX-3LIN *does not have a polynomial-time* $(2-\varepsilon)$-*approximation for any* $\varepsilon > 0$ *unless* **P** = **NP**.

10.5 $(\rho \ln n)$-Inapproximability

In this section, we study a class of **NPO** problems that are $(\rho \ln n)$-inapproximable for some constant $\rho > 0$ (under certain complexity-theoretic assumptions). Among such $(\rho \ln n)$-inapproximability results, the set cover problem MIN-SC plays a critical role similar to that of MAX-3LIN for the constant-ratio inapproximability results. Under the assumption that **NP** $\not\subseteq$ **DTIME**$(n^{O(\log \log n)})$,[1] many optimization problems have been proved to be $O(\rho \ln n)$-inapproximable through gap-preserving reductions from MIN-SC.

Recall that MIN-SC is the problem that, on a given set S and a collection $\mathcal{C}$ of subsets of S, asks for a subcollection $\mathcal{C}'$ of $\mathcal{C}$ of the minimum cardinality such that $\bigcup \mathcal{C} = S$. The basic $(\rho \ln n)$-inapproximability result about MIN-SC is as follows.

[1] The class **DTIME**$(n^{O(\log \log n)})$ consists of all languages that are decidable in time $n^{O(\log \log n)}$ by a deterministic Turing machine.

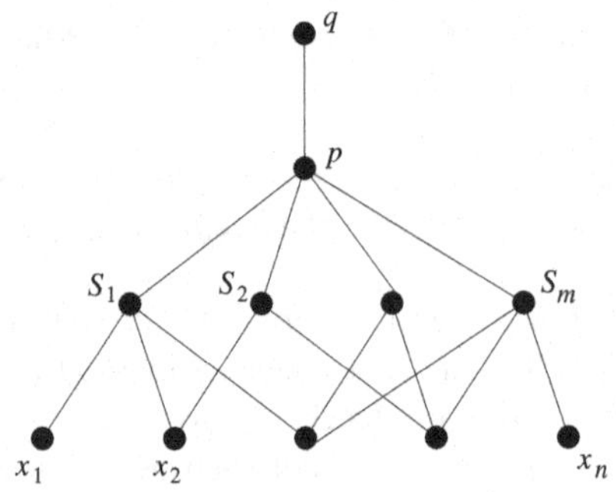

Figure 10.12: Graph G in the proof of Theorem 10.24.

Theorem 10.23 *The problem* MIN-SC *does not have a polynomial-time* $(\rho \ln n)$-*approximation for any* $0 < \rho < 1$ *unless* $\mathbf{NP} \subseteq \mathbf{DTIME}(n^{O(\log \log n)})$, *where* n *is the size of the base set* S. *Furthermore, this inapproximability result holds for the case when the size of the input collection* $\mathcal{C}$ *is no more than the size of the base set* S.

We now apply this result to establish more $(\rho \ln n)$-inapproximability results. We first look at the connected dominating set problem MIN-CDS studied in Chapter 2.

Theorem 10.24 *The problem* MIN-CDS *does not have a polynomial-time* $(\rho \ln n)$-*approximation for any* $0 < \rho < 1$ *unless* $\mathbf{NP} \subseteq \mathbf{DTIME}(n^{O(\log \log n)})$.

Proof. Suppose MIN-CDS has a polynomial-time $(\rho \ln n)$-approximation for some $0 < \rho < 1$. Choose a positive integer $k_0 > \rho/(1-\rho)$. Then $\rho(1 + 1/k_0) < 1$. Let ρ' be a positive number satisfying $\rho(1+1/k_0) < \rho' < 1$. We show that the problem MIN-SC has a polynomial-time approximation with performance ratio $\rho' \ln n$, and hence, by Theorem 10.23, $\mathbf{NP} \subseteq \mathbf{DTIME}(n^{O(\log \log n)})$.

Let $S = \{x_1, x_2, \ldots, x_n\}$ and $\mathcal{C} = \{S_1, S_2, \ldots, S_m\}$ be an input instance to MIN-SC, where each S_j, $j = 1, 2, \ldots, m$, is a subset of S. From Theorem 10.23, we may assume, without loss of generality, that $m \leq n$. We first check, for each subcollection $\mathcal{C}' \subseteq \mathcal{C}$ of size $\leq k_0$, whether it is a set cover of S or not. There are only $O(n^{k_0})$ many such subcollections, and so this step can be done in polynomial time in n.

If no set cover of cardinality $\leq k_0$ is found, then we construct a reduction from the instance $(S, \mathcal{C})$ to a graph G for problem MIN-CDS. The graph G is defined as follows: It has $m+n+2$ vertices, labeled $x_1, x_2, \ldots, x_n$, $S_1, S_2, \ldots, S_m$, p, and q. In addition, G contains the following edges: $\{p, q\}$; $\{S_j, p\}$, for all $j = 1, 2, \ldots, m$; and $\{x_i, S_j\}$ if $x_i \in S_j$ (see Figure 10.12).

Now, we observe the following relationships between $\mathcal{C}$ and G:

(1) Assume that $\mathcal{C}$ has a set cover of size k. Then graph G has a connected dominating set of size $k+1$. Indeed, if $\mathcal{C}'$ is a set cover for S, then $\mathcal{C}' \cup \{p\}$ forms a connected dominating set for G.

(2) Assume that G has a connected dominating set D of size k. Then, we can find a set cover $\mathcal{C}' \subseteq \mathcal{C}$ of size at most $k-1$. To see this, we note that if D is a connected dominating set of G, then $D' = D \cap \{S_1, S_2, \ldots, S_m, p\}$ is still a connected dominating set of G. Indeed, D must contain p in order to dominate q and to connect to any vertex S_j in D. Thus, q can be removed from set D if $q \in D$. Moreover, if $x_i \in D - D'$ for some $i = 1, \ldots, n$, then x_i must be connected to p through some vertex S_j in D. Also, every vertex S_ℓ dominated by x_i is dominated by p. Thus, x_i can be removed from D. It follows that $D' - \{p\}$ must be a set cover of S of size $k-1$.

Now, suppose the minimum set cover of $\mathcal{C}$ contains k subsets. Note that, from our preprocessing, we know that $k > k_0$. From the above two properties, we know that the minimum connected dominating set of G contains $k+1$ vertices. Applying the polynomial-time $(\rho \ln n)$-approximation for MIN-CDS on instance G, we get a connected dominating set D of G of size $\leq (\rho \ln(m+n+2))(k+1)$. From property (2), we can obtain a set cover $\mathcal{C}' \subseteq \mathcal{C}$ of S of size at most

$$\rho \ln(m+n+2)(k+1) - 1 < \rho\Big(1 + \frac{1}{k_0}\Big)\Big(1 + \frac{\ln 3}{\ln n}\Big)(\ln n)k.$$

When n is sufficiently large, $\mathcal{C}'$ is a $(\rho' \ln n)$-approximation solution for the instance $(S, \mathcal{C})$ of the problem MIN-SC. □

In Chapter 2, we showed that the weighted connected vertex cover problem (MIN-WCVC) has a polynomial-time $(1+\ln n)$-approximation. We show here that this is the best possible polynomial-time approximation for this problem.

Theorem 10.25 *There is no no polynomial-time $(\rho \ln n)$-approximation for the problem* MIN-WCVC, *for any $0 < \rho < 1$, unless* **NP** $\subseteq$ **DTIME**$(n^{O(\log \log n)})$, *where n is the number of vertices in the input graph.*

Proof. By Theorem 10.23, it suffices to show that if MIN-WCVC has a polynomial-time r-approximation, so does MIN-SC.

Let $S = \{x_1, x_2, \ldots, x_n\}$ and $\mathcal{C} = \{S_1, S_2, \ldots, S_m\}$ be an input instance to MIN-SC, where each S_j, $j = 1, 2, \ldots, m$, is a subset of S. We construct a graph G as follows: G has $n+m+1$ vertices labeled $x_1, x_2, \ldots, x_n, S_1, S_2, \ldots, S_m$, and p, and has the following edges connecting the vertices: $\{S_j, p\}$ for $j = 1, 2, \ldots, m$, and $\{x_i, S_j\}$ if $x_i \in S_j$ (see Figure 10.13). Furthermore, for each vertex in G, we assign weight to it as follows: Each vertex S_j, for $j = 1, 2, \ldots, m$, has weight $w(S_j) = 1$, and all other vertices u have weight $w(u) = 0$. We have thus obtained an instance (G, w) of MIN-WCVC.

Suppose D is an r-approximation to the problem MIN-WCVC on the instance (G, w). Let $\mathcal{C}_1 = D \cap \mathcal{C}$. Then we claim that $\mathcal{C}_1$ is a set cover of the instance $(S, \mathcal{C})$. To see this, suppose otherwise that x_i, for some $i = 1, \ldots, n$, is not covered by any subset in $\mathcal{C}_1$. Let $S_{j_1}, S_{j_2}, \ldots, S_{j_k}$ be the sets in $\mathcal{C}$ that contain x_i. Then $k \geq 1$ as $\bigcup \mathcal{C} = S$. Since $\mathcal{C}_1$ does not cover x_i, none of the sets $S_{j_1}, \ldots, S_{j_k}$ is in $\mathcal{C}_1$. It follows that $D \cap \{S_{j_1}, \ldots, S_{j_k}\} = \emptyset$. Now consider the following two cases.

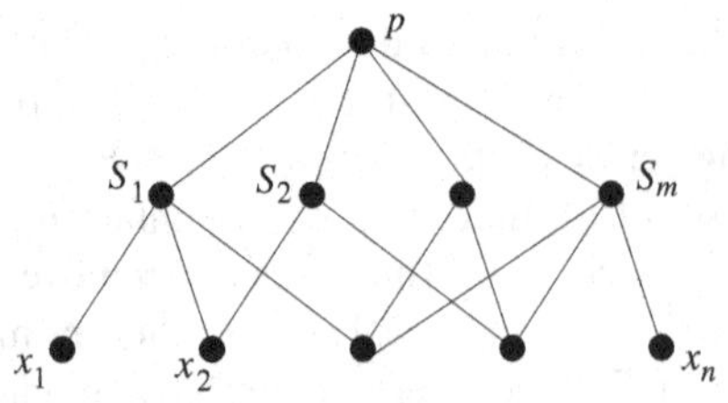

Figure 10.13: Graph G in the proof of Theorem 10.25.

Case 1. $x_i \notin D$. In this case, none of the edges between x_i and $S_{j_1}, S_{j_2}, \ldots, S_{j_k}$ in G is covered by D. This is a contradiction to the assumption that D is a vertex cover of G.

Case 2. $x_i \in D$. Since $D \cap \{S_{j_1}, \ldots, S_{j_k}\} = \emptyset$, D must contain p in order to cover edges between p and $S_{j_1}, \ldots, S_{j_k}$. However, this means that p and x_i are not connected in D, which is a contradiction to the assumption that D is connected. So, the claim is proven.

Now, from the definition of weight w, we see that $w(D) = |\mathcal{C}_1|$. We now prove that $\mathcal{C}_1$ is an r-approximation to the problem MIN-SC on the instance $(S, \mathcal{C})$. To see this, consider an optimal solution $\mathcal{C}^*$ of MIN-SC for the instance $(S, \mathcal{C})$. Let $D^* = \mathcal{C}^* \cup \{p\} \cup \{x_1, x_2, \ldots, x_n\}$. Then D^* is a connected vertex cover of G with $w(D^*) = |\mathcal{C}^*|$. Moreover, we note that D^* is a minimum connected vertex cover of G. Indeed, if there were a connected vertex cover D' of G with $w(D') < w(D^*)$, then, by the same argument above, we see that set $\mathcal{C}' = D' \cap \mathcal{C}$ would be a set cover of $(S, \mathcal{C})$ with $|\mathcal{C}'| = w(D') < w(D^*) = |\mathcal{C}^*|$, contradicting the optimality of $\mathcal{C}^*$ for the instance $(S, \mathcal{C})$. It follows that

$$\frac{|\mathcal{C}|}{|\mathcal{C}^*|} = \frac{w(D)}{w(D^*)} \leq r. \qquad \square$$

The following problem arises from the study of traffic in wireless networks:

> CONNECTED DOMINATING SET WITH SHORTEST PATHS (CDS-SP): Given a graph $G = (V, E)$, find the minimum connected dominating set C satisfying that for every pair of vertices (u, v), there is a shortest path from u to v such that all of its intermediate vertices belong to set C.

Lemma 10.26 *Let C be a connected dominating set of a graph G. Then the following two conditions about C are equivalent:*

(1) *For every pair of vertices u and v in G, there is a shortest path $(u, w_1, \ldots, w_k, v)$ such that all of its intermediate vertices $w_1, w_2, \ldots, w_k$ belong to set C.*

(2) *For every pair of vertices u and v in G of distance 2, there exists a shortest path (u, w, v) such that w belongs to set C.*

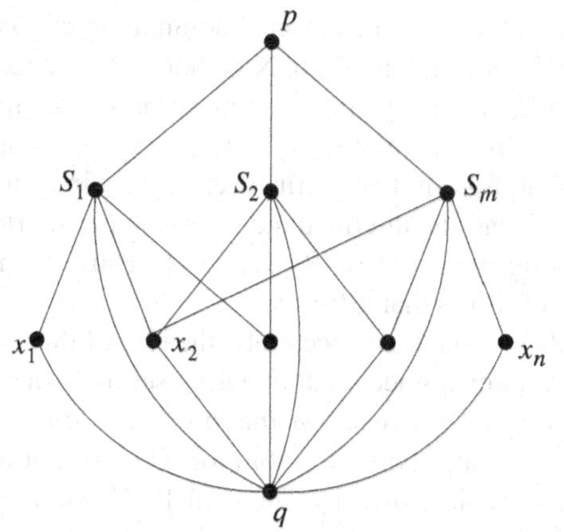

Figure 10.14: Graph G constructed in the proof of Theorem 10.27.

Proof. It is trivial to see that (1) implies (2). We now show that (2) implies (1). Consider two vertices u and v. Suppose there is a shortest path $(u, w_1, \ldots, w_k, v)$ between them. Then, by condition (2), there exist vertices $s_1, s_2, \ldots, s_k$ in C such that (u, s_1, w_2), (s_1, s_2, w_3), $(s_2, s_3, w_4), \ldots, (s_{k-1}, s_k, v)$ are all shortest paths. This implies that $(u, s_1, s_2, s_3, \ldots, s_{k-1}, s_k, v)$ is also a shortest path between u and v, with all intermediate vertices belonging in C. □

Theorem 10.27 *The problem* CDS-SP *does not have a polynomial-time $(\rho \ln \delta)$-approximation for any $0 < \rho < 1$, unless* **NP** $\subseteq$ **DTIME**$(n^{O(\log\log n)})$, *where δ is the maximum vertex degree of the input graph.*

Proof. We will construct a reduction from MIN-SC to CDS-SP. Suppose $(S, \mathcal{C})$ is an input instance of MIN-SC, where $S = \{x_1, x_2, \ldots, x_n\}$ and $\mathcal{C}$ is a collection of subsets $S_1, S_2, \ldots, S_m$ of S. We define a graph G of $m + n + 2$ vertices, labeled $S_1, \ldots, S_m, x_1, \ldots, x_n, p$, and q. In addition, it has the following edges: $\{p, S_j\}$ and $\{q, S_j\}$, for $j = 1, 2, \ldots, m$; $\{q, x_i\}$, for $i = 1, 2, \ldots, n$; and $\{x_i, S_j\}$ if $x_i \in S_j$ (see Figure 10.14).

We claim that $\mathcal{C}$ has a set cover of size at most k if and only if G has a connected dominating set of size at most $k + 1$ satisfying condition (2) of Lemma 10.26. The claim holds trivially in the case of $|\mathcal{C}| = 1$. In the following, we assume that $|\mathcal{C}| \geq 2$.

First, assume that $\mathcal{C}$ has a set cover $\mathcal{A}$ of size at most k. Then it is easy to verify that $D = \mathcal{A} \cup \{q\}$ is a connected dominating set of G satisfying condition (2) of Lemma 10.26. Indeed, for a pair of vertices u and v of distance 2 in G with $u \neq p \neq v$, (u, q, v) must be a shortest path with $q \in D$. For a pair of vertices p and v with distance 2, v must belong to $\{x_i \mid 1 \leq i \leq n\} \cup \{q\}$. If $v = x_i$ for some $i = 1, 2, \ldots, n$, then there must be a set $S_j \in \mathcal{A}$ such that $x_i \in S_j$ and, hence, (p, S_j, x_i) is a shortest path with $S_j \in D$. If $v = q$, then for any $S_j \in \mathcal{A}$, (p, S_j, q) is a required shortest path.

Conversely, assume that G has a connected dominating set D of size at most $k+1$ satisfying condition (2) of Lemma 10.26. Note that the distance from p to each x_i, for $i = 1, 2, \ldots, n$, is 2, and every shortest path from p to x_i must pass a vertex S_j for some $j = 1, 2, \ldots, m$. Therefore, $\mathcal{A} = \{S_j \mid S_j \in D\}$ is a set cover for S. Moreover, we note that, for any two distinct sets S_j, S_k in $\mathcal{C}$, the distance between vertices S_j and S_k is 2, and the intermediate vertex of any shortest path between S_j and S_k does not belong to $\mathcal{C} = \{S_1, S_2, \ldots, S_m\}$. Thus, D must contain at least one vertex not in $\mathcal{C}$. It follows that $|\mathcal{A}| \leq k$.

Let $opt_{\rm SC}$ and $opt_{\rm CDS}$ denote, respectively, the size of the minimum set cover in $\mathcal{C}$ and that of the minimum connected dominating set of G satisfying condition (2) of Lemma 10.26. The claim above shows that $opt_{\rm CDS} = opt_{\rm SC} + 1$. Now, suppose G has a polynomial-time approximation solution D of size at most $(\rho \ln \delta) opt_{\rm CDS}$ for some constant $\rho < 1$. Note that, by Theorem 10.23, we may assume that $m \leq n$. Thus, $\delta \leq 2n$. From the claim, we can find a polynomial-time approximation solution for MIN-SC of size at most

$$\rho \ln(2n)(opt_{\rm SC} + 1) < \frac{1}{2}(\rho + 1) \ln n \cdot opt_{\rm SC},$$

for sufficiently large n and sufficiently large $opt_{\rm SC}$. (Note that, for any constant α, we can check in polynomial-time whether $opt_{\rm SC} \leq \alpha$.) It follows that $\mathbf{NP} \subseteq \mathbf{DTIME}(n^{O(\log \log n)})$. □

The above results imply that the problems MIN-SC and CDS-SP are not in **APX** if $\mathbf{NP} \not\subseteq \mathbf{DTIME}(n^{O(\log \log n)})$. This result can be further improved to hold with the weaker condition of $\mathbf{P} \neq \mathbf{NP}$, using the following different lower-bound result for MIN-SC.

Theorem 10.28 *Assume that* $\mathbf{P} \neq \mathbf{NP}$. *Then there exists a constant* $c > 0$ *such that the problem* MIN-SC *does not have a polynomial-time* $(c \ln n)$*-approximation.*

Corollary 10.29 *If* $\mathbf{P} \neq \mathbf{NP}$, *then* MIN-SC *and* CDS-SP *are not in* **APX**.

10.6 n^c-Inapproximability

In this section, we study optimization problems that are not approximable with the performance ratio n^c for some constant $c > 0$, unless $\mathbf{P} = \mathbf{NP}$. We first introduce a well-known **NP**-hard optimization problem. Recall that a *clique* of a graph G is a complete subgraph of G.

> CLIQUE: Given a graph G, find a clique C of G of the maximum cardinality.

For a graph $G = (V, E)$, define its complement to be $\overline{G} = (V, \overline{E})$, where $\overline{E} = \{\{u, v\} \mid u, v \in V\} - E$. It is clear that a vertex subset $S \subseteq V$ of a graph $G = (V, E)$ is independent in G if and only if it induces a clique in $\overline{G}$. In other words, CLIQUE and MAXIMUM INDEPENDENT SET (MAX-IS) are complementary problems with the following property: An approximation algorithm for one of them can be converted to an approximation algorithm for the other one with the same performance ratio.

GRAPH COLORING (GCOLOR) and CLIQUE are the first two problems proved to be n^c-inapproximable by exploring the properties of the PCP systems.

Theorem 10.30 *The problems* CLIQUE *and* MAX-IS *do not have polynomial-time $(n^{1-\varepsilon})$-approximations for any $\varepsilon > 0$ unless* **P** = **NP**, *where n is the number of vertices in the input graph.*

Theorem 10.31 *The problem* GCOLOR *does not have a polynomial-time $(n^{1-\varepsilon})$-approximation for any $\varepsilon > 0$ unless* **P** = **NP**, *where n is the number of vertices in the input graph.*

Many n^c-inapproximability results are proved through gap-preserving reductions from these three problems. We present two examples in this section. First, we consider the following problem. Recall that for a given collection of sets, a *set packing* is a subcollection of disjoint sets.

> MAXIMUM SET PACKING (MAX-SP): Given a collection $\mathcal{C}$ of subsets of a finite set S, find a maximum set packing in $\mathcal{C}$.

Theorem 10.32 *The problem* MAX-SP *does not have a polynomial-time $(n^{1-\varepsilon})$-approximation for any $\varepsilon > 0$ unless* **P** = **NP**, *where n is the number of subsets in the input collection.*

Proof. We can reduce MAX-IS to MAX-SP. Let $G = (V, E)$ be an input instance of MAX-IS. For each $v \in V$, let E_v be the set of edges incident upon v. Consider the instance $(E, \mathcal{C})$ of MAX-SP, where $\mathcal{C} = \{E_v \mid v \in V\}$. Clearly, a vertex subset $V' \subseteq V$ is an independent set of G if and only if $\{E_v \mid v \in V'\}$ is a set packing for the collection $\mathcal{C}$. Therefore, if MAX-SP has a polynomial-time n^c-approximation for some $0 < c < 1$, so does MAX-IS, and, by Theorem 10.30, **P** = **NP**. □

The next problem is a variation of GCOLOR.

> CHROMATIC SUM (CS): Given a graph $G = (V, E)$, find a vertex coloring $\phi : V \to \mathbb{N}^+$ for G that minimizes the sum $\sum_{v \in V} \phi(v)$ of the colors.

Theorem 10.33 *The problem* CS *has no polynomial-time $(n^{1-\varepsilon})$-approximation for any $\varepsilon > 0$ unless* **P** = **NP**, *where n is the number of vertices in the input graph.*

Proof. Assume that the problem CS has a polynomial-time n^c-approximation algorithm A for some $0 < c < 1$. Let G be an input instance for the problem GCOLOR, and assume that the chromatic number of G is equal to k. Then the optimal chromatic sum of G is at most kn. Therefore, algorithm A, when run on graph G, produces a vertex coloring with the sum of colors bounded by kn^{1+c}. It follows that at least half of the vertices in G are colored by the colors in $\{1, 2, \ldots, \lfloor 2kn^c \rfloor\}$. Let us fix the coloring of these vertices. For the remaining $\lfloor n/2 \rfloor$ vertices, we apply algorithm A to these vertices again, and use up to $\lfloor 2k(n/2)^c \rfloor$ new colors to color half of these vertices. In this recursive way, we can find a vertex coloring for G using at most

$$2kn^c \sum_{i=0}^{\infty} \frac{1}{(2^c)^i} = O(kn^c)$$

colors. This means that GCOLOR has a polynomial-time $(n^{c'})$-approximation for some $c < c' < 1$. By Theorem 10.31, **P** = **NP**. □

In addition to the above three problems, the following problem also plays an important role in connecting the theory of computational complexity to the theory of inapproximability. We say a subset A of the vertex set V of a graph $G = (V, E)$ is *regular* if all vertices in A have the same degree.

> LABEL COVER (LC): Given a bipartite graph $G = (U, V, E)$, in which the set U is regular, an alphabet Σ of potential labels for vertices, and a mapping $\sigma_{(u,v)} : \Sigma \to \Sigma$, for each edge $(u, v) \in E$, find a vertex label $\tau : U \cup V \to \Sigma$ that maximizes the number of satisfied edges, where an edge (u, v) is satisfied by τ if $\sigma_{(u,v)}(\tau(u)) = \tau(v)$.

The problem LC has a polynomial-time n^c-approximation for some constant c. Indeed, the best-known performance ratio of an approximation algorithm for LC is lower than n^ε for any $\varepsilon > 0$. To further discuss the hardness of approximation for this problem, we formulate a subproblem of LC with gaps. For an input instance (G, Σ, σ) of LC, let $opt(G)$ denote the maximum number of satisfied edges by any labeling of vertices.

> LC-GAP(α, k): For an input instance $(G = (U, V, E), \Sigma, \sigma)$ of LC, with $|\Sigma| = n$, $|U| = |V| = O(n^k)$, $|E| = m$, and having the property that either $opt(G) = m$ or $opt(G) < \alpha^k m$, determine whether $opt(G) = m$ or $opt(G) < \alpha^k m$.

The following result has been proved in the theory of computational complexity.

Theorem 10.34 *There exists a constant* $0 < \alpha < 1$ *such that for every positive integer k, the problem* LC-GAP(α, k) *is not in* **P** *unless* **NP** $\subseteq$ **DTIME**(n^k).

By choosing appropriate values for k, we get the following inapproximability results for LC.

Corollary 10.35 *(a) The problem* LC *does not have a polynomial-time* $(\rho \log n)$*-approximation for any* $\rho > 0$ *unless* **NP** $\subseteq$ **DTIME**$(n^{O(\log \log n)})$.

(b) The problem LC *does not have a polynomial-time* $(2^{\log^{1-\varepsilon} n})$*-approximation for any* $\varepsilon > 0$ *unless* **NP** $\subseteq$ **DTIME**$(n^{\log^{O(1)} n})$.

More inapproximability results can be established from the above results about LC. For instance, an $O(\log n)$ lower bound for the problem MIN-SC can be proven using Corollary 10.35(a) (see Exercises 10.30 and 10.31).

It is interesting to point out that, in addition to the $(\rho \ln n)$- and n^c-inapproximability results, there are also problems of which the best performance ratio lies strictly between these two bounds. The following are two examples.

> DIRECTED STEINER TREE (DST): Given an edge-weighted directed graph $G = (V, E)$, a source node s, and a terminal set P, find a directed tree containing paths from s to every terminal in P such that the total edge-weight is minimized.

It is known that the problem DST has a polynomial-time n^c-approximation for any $c > 0$, and hence its hardness of approximation is weaker than that of CLIQUE. It is also known that DST cannot be approximated in polynomial time within a factor of $\log^{2-\varepsilon} n$ of the optimal solution unless **NP** has quasi-polynomial-time Las Vegas algorithms (i.e, unless problems in **NP** can be solved by probabilistic algorithms with zero error probability that run in time $O(n^{\log^k n})$ for some constant $k > 0$).

> GROUP STEINER TREE (GST): Given an edge-weighted graph $G = (V, E)$, a root vertex $r \in V$, and k nonempty subsets of vertices, $g_1, g_2, \ldots, g_k$, find a tree in G with the minimum total weight that contains root r and at least one vertex from each subset g_i, $i = 1, \ldots, k$.

It has been proven that the problem GST has a polynomial-time $O(\log^3 n)$-approximation, but no polynomial-time $O(\log^{2-\varepsilon} n)$-approximation for any $\varepsilon > 0$, unless **NP** has quasi-polynomial-time Las Vegas algorithms. For details of the results about these two problems, the reader is referred to Charikar et al. [1999], Garg et al. [2000], and Halperin and Krauthgamer [2003].

Exercises

10.1 Consider the problem k-CENTERS which is a generalization of the problem METRIC-k-CENTERS such that the input distance table between cities may not satisfy the triangle inequality. Prove, using the many–one reduction with gap, that there is no polynomial-time constant approximation for k-CENTER unless **P** = **NP**.

10.2 Show that the following greedy algorithm is a 2-approximation for the problem METRIC-k-CENTERS: First, pick any city to build a warehouse. In each of the subsequent $k-1$ iterations, pick a city that has the maximum distance to any existing warehouse, and place a warehouse in this city.

10.3 Let a graph G and a distance table d between vertices in G be an input instance to the problem METRIC-k-CENTERS.

(a) Sort the edges in G in nondecreasing order, and let G_i denote the graph of the same vertex set but having only the first i edges. Show that solving METRIC-k-CENTERS on instance (G, d) is equivalent to finding the minimum index i such that G_i contains a dominating set of size k.

(b) Based on part (a) above, we can design an approximation algorithm for METRIC-k-CENTERS as follows: Find the minimum index i such that G_i has a maximal independent set D of size $\leq k$, and build warehouses at each $v \in D$. Prove that this algorithm is a 2-approximation for METRIC-k-CENTERS.

10.4 Show that the bottleneck Steiner tree problem (BNST) in the Euclidean plane cannot be approximated in polynomial time with a performance ratio smaller than $\sqrt{2}$, provided $\mathbf{P} \neq \mathbf{NP}$.

10.5 Show that if $\mathbf{P} \neq \mathbf{NP}$, then the following problem has no polynomial-time $(2 - \varepsilon)$-approximation for any $\varepsilon > 0$:

> Given a set of points in the Euclidean plane and a set of disks that cover all given points, find a subset of disks covering all points such that the maximum number of disks containing a common given point is minimized.

10.6 Let $\alpha > 0$ be a constant. Show that statement (1) below implies statement (2).

(1) It is **NP**-hard to approximate CLIQUE within a factor of α.

(2) It is **NP**-hard to approximate CLIQUE within a factor of α^2.

10.7 Show the following results on the problem EDP.

(a) Given a graph G and two pairs (u_1, v_1) and (u_2, v_2) of vertices in G, it is **NP**-complete to determine whether G contains two edge-disjoint paths connecting the two given pairs, respectively.

(b) The problem EDP does not have a polynomial-time $(2 - \varepsilon)$-approximation for any $\varepsilon > 0$ unless $\mathbf{P} = \mathbf{NP}$.

(c) The problem EDP has a polynomial-time $\sqrt{m}$-approximation, where m is the number of edges in the input graph.

10.8 For a solution y to an instance x of a problem Π in **NPO**, define its error by

$$\mathcal{E}(x, y) = \max\left\{\frac{obj_\Pi(y)}{opt_\Pi(x)}, \frac{opt_\Pi(x)}{obj_\Pi(y)}\right\} - 1.$$

A problem Π is *E-reducible* to a problem Λ, denoted by $\Pi \leq_E \Lambda$, if there exist polynomial-time computable functions f, g and a constant β such that

(1) f maps an instance x of Π to an instance $f(x)$ of Λ and there exists a polynomial $p(n)$ such that $opt_\Lambda(f(x)) \leq p(|x|)opt_\Pi(x)$.

(2) g maps solutions y of $f(x)$ to solutions of x such that $\mathcal{E}(x, g(y)) \leq \beta \cdot \mathcal{E}(f(x), y)$.

Show the following:

(a) If $\Pi \leq_E \Gamma$ and $\Gamma \leq_E \Lambda$, then $\Pi \leq_E \Lambda$.

(b) If $\Pi \leq_E \Lambda$ and Λ has a PTAS, then Π has a PTAS.

(c) If $\Pi \leq_L^P \Lambda$, then $\Pi \leq_E \Lambda$.

10.9 Show that the following problems are **APX**-hard:

(a) CONNECTED-MAJ-DS: Given a connected graph $G = (V, E)$, find a connected majority-dominating set of the minimum cardinality. (A *connected majority-dominating set* is a majority-dominating set that induces a connected subgraph.)

(b) MAX-3-COLOR: Given a graph $G = (V, E)$, find a vertex coloring using three colors such that the total number of edges with two endpoints having different colors is maximized.

10.10 Show that the PCP theorem holds if and only if Theorem 10.19 holds.

10.11 Show that the problem MAX-CUT does not have a polynomial-time $(17/16 - \varepsilon)$-approximation for any $\varepsilon > 0$ unless **P** = **NP**.

10.12 Show that the problem MAX-2SAT does not have a polynomial-time $(22/21 - \varepsilon)$-approximation for any $\varepsilon > 0$ unless **P** = **NP**.

10.13 Show that the network Steiner minimum tree problem (NSMT) does not have a polynomial-time $(96/95)$-approximation unless **P** = **NP**.

10.14 Show that the problem METRIC-TSP does not have a polynomial-time $(3813/3812 - \varepsilon)$-approximation for any $\varepsilon > 0$ unless **P** = **NP**.

10.15 Design a polynomial-time $O(\ln \delta)$-approximation for the problem CDS-SP, where δ is the maximum vertex degree of the input graph.

10.16 Let $G = (V, E)$ be a connected graph, in which each edge is associated with a set of colors $c : E \to 2^{\mathbb{N}}$. A set of colors is called a *color covering* if all edges in those colors contain a spanning tree of G. Also, for each $v \in V$, we define the set of colors of v to be the set of colors associated with edges incident on v. Show that for each of the following problems, if $\mathbf{NP} \not\subseteq \mathbf{DTIME}(n^{\log^{O(1)} n})$, then it has no $(\rho \ln n)$-approximation for any $\rho < 1$:

(a) Given a graph $G = (V, E)$ and edge-color sets $c : E \to 2^{\mathbb{N}}$, find a color covering of the minimum cardinality.

(b) Given a graph $G = (V, E)$ and edge-color sets $c : E \to 2^{\mathbb{N}}$, find a subset $S \subseteq V$ of the minimum cardinality such that the set of colors of all vertices in S forms a color covering.

(c) Given a graph $G = (V, E)$ and edge-color sets $c : E \to 2^{\mathbb{N}}$, with the property that the set of edges in any fixed color forms a connected subgraph, find a color-connected subset $S \subseteq V$ of the minimum cardinality such that the set of colors of all vertices in S forms a color covering.

10.17 For each of the following problems, show that it does not have a polynomial-time approximation with performance ratio $\rho \ln n$ for any $0 < \rho < 1$ unless $\mathbf{NP} \subseteq \mathbf{DTIME}(n^{O(\log \log n)})$:

(a) WSID (defined in Section 2.5).

(b) DST.

(c) NODE WEIGHTED STEINER TREE (NWST): Given a graph with node weight and a set of terminals, find a Steiner tree interconnecting all terminals such that the total node weight is minimized.

(d) The special case of NWST in which all nodes of the input graph have weight 1.

(e) The special case of DST in which the input graph is acyclic.

10.18 Explain why the proof of Theorem 10.25 fails for the unweighted connected vertex cover problem.

10.19 Show that the problem of finding the minimum dominating set in a given graph has no polynomial-time $(\rho \ln n)$-approximation for $0 < \rho < 1$ unless $\mathbf{NP} \subseteq \mathbf{DTIME}(n^{O(\log \log n)})$.

10.20 Design a polynomial-time algorithm for the following problem:

> Given a graph $G = (V, E)$, find the minimum dominating set satisfying that for every shortest path $(u, w_1, \ldots, w_k, v)$ in G, all intermediate nodes $w_1, w_2, \ldots, w_k$ belong to the dominating set.

10.21 The *domatic number* of a graph is the maximum number of disjoint dominating sets in the graph. Show that the domatic number cannot be approximated within a factor of $\rho \ln n$ in polynomial time for any $0 < \rho < 1$ unless $\mathbf{NP} \subseteq \mathbf{DTIME}(n^{O(\log \log n)})$.

10.22 A binary matrix is *$\bar{d}$-separable* if all Boolean sums of at most d columns are distinct. Consider the following problem:

> MINIMUM $\bar{d}$-SEPARABLE SUBMATRIX (MIN-$\bar{d}$-SS): Given a binary matrix M, find a minimum $\bar{d}$-separable submatrix with the same number of columns.

Show that there is a constant $c > 0$ such that MIN-$\bar{d}$-SS has no polynomial-time $(c \ln n)$-approximation unless $\mathbf{NP} \subseteq \mathbf{DTIME}(n^{O(\log \log n)})$.

10.23 A binary matrix is *d-separable* if all Boolean sums of d columns are distinct. Consider the following problem:

> MINIMUM d-SEPARABLE SUBMATRIX (MIN-d-SS): Given a binary matrix M, find a minimum d-separable submatrix with the same number of columns.

Show that there is a constant $c > 0$ such that MIN-d-SS has no polynomial-time $(c \ln n)$-approximation unless $\mathbf{NP} \subseteq \mathbf{DTIME}(n^{O(\log \log n)})$.

10.24 A binary matrix is *d-disjunct* if for every $d+1$ columns $C_0, C_1, \ldots, C_d$, there is a row at which C_0 has entry 1 but all of $C_1, \ldots, C_d$ have entry 0. Consider the following problem:

> MINIMUM d-DISJUNCT SUBMATRIX (MIN-d-DS): Given a binary matrix M, find a minimum d-disjunct submatrix with the same number of columns.

(a) Show that there is a constant $c > 0$ such that MIN-d-DS has no polynomial-time $(c \ln n)$-approximation unless $\mathbf{NP} \subseteq \mathbf{DTIME}(n^{O(\log \log n)})$.

(b) Show that the special case of MIN-d-DS in which each row of the binary matrix contains at most two 1s is **APX**-complete.

10.25 Consider the following problem:

> BUDGETED MAXIMUM COVERAGE: Given a finite set S, a weight function $w : S \to \mathbb{N}$ on elements of S, a collection $\mathcal{C}$ of subsets of set S, a cost function $c : \mathcal{C} \to \mathbb{N}$ on sets in $\mathcal{C}$, and a budget L, find a subcollection $\mathcal{C}' \subseteq \mathcal{C}$ with its total cost no more than the budget L such that the total weight of the covered elements is maximized.

Show that this problem does not have a polynomial-time $(\frac{e}{e-1} - \varepsilon)$-approximation for any $\varepsilon > 0$ unless $\mathbf{NP} \subseteq \mathbf{DTIME}(n^{O(\log \log n)})$.

10.26 Show that for any $\varepsilon > 0$, it is **NP**-hard to approximate the following problem within a factor of $n^{1-\varepsilon}$:

> Given a graph G, find a maximal independent set in G of the minimum cardinality.

10.27 Study the hardness of approximation for the following problems:

> CONNECTED SET COVER: Given a collection $\mathcal{C}$ of a finite set S and a graph G with vertex set $\mathcal{C}$, find a minimum set cover $\mathcal{C}' \subseteq \mathcal{C}$ such that the subgraph induced by $\mathcal{C}'$ is connected.

> MAXIMUM DISJOINT SET COVER: Given a collection $\mathcal{C}$ of a finite set S, find a partition of $\mathcal{C}$ into the maximum number of parts such that each part is a set cover.

10.28 Consider the following problem:

> MAXIMUM CONSTRAINT GRAPH (MAX-CG): Given an alphabet Σ and a directed graph $G = (V, E)$ with each edge $(u, v) \in E$ labeled with a mapping $\sigma_{(u,v)} : \Sigma \to \Sigma$, find a mapping $\tau : V \to \Sigma$ that maximizes the number of satisfied edges, where an edge (u, v) is satisfied if $\sigma_{(u,v)}(\tau(u)) = \tau(v)$.

Answer the following questions and prove your answers:

(a) Is MAX-CG in **APX**?

(b) Is MAX-CG **APX**-hard for the alphabet Σ with $|\Sigma| \geq 2$?

10.29 Show that every **APX**-complete problem has an **NP**-hard gap $[\alpha, \beta]$ with ratio β/α greater than 1.

10.30 Let B be a ground set, and $\mathcal{C} = \{C_1, \ldots, C_m\}$ a collection of subsets of B. We say $(B, \mathcal{C})$ is an (m, ℓ)-system if any subcollection of ℓ subsets chosen from $\{C_1, \ldots, C_m, \overline{C}_1, \ldots, \overline{C}_m\}$ that covers B must contain both C_i and $\overline{C}_i$ for some $i = 1, 2, \ldots, m$. Prove by the probabilistic method that, for any $0 < \ell < m$, there exists an (m, ℓ)-system with a ground set B of size $O(2^\ell \ell \log m)$.

10.31 In this exercise, we construct a reduction from LC to MIN-SC to establish an $O(\log n)$ lower bound for the performance ratio for any approximation of MIN-SC. Let $(G = (U, V, E), \Sigma, \sigma)$ be an input instance of LC, with $|\Sigma| = n$, $|U| = |V| = O(n^k)$, and $|E| = m$. Choose $\ell = O(\log n)$ and $k = O(\log \log n)$ so that $\alpha^k \ell^2 < 2$. Let $\mathcal{C} = \{C_1, \ldots, C_m\}$ be an (m, ℓ)-system with a ground set B, as constructed from Exercise 10.30. Let $S = E \times B$, and define a collection $\mathcal{F}$ of subsets of S as follows: For each vertex $v \in V$ and $x \in \Sigma$, construct a subset $S_{v,x}$ of $E \times B$ as

$$S_{v,x} = \bigcup_{u:(u,v)\in E} \{(u, v)\} \times C_x.$$

For each vertex $u \in U$ and $x \in \Sigma$, construct a subset $S_{u,x}$ of $E \times B$ as

$$S_{u,x} = \bigcup_{v:(u,v)\in E} \{(u, v)\} \times \overline{C}_{\sigma_{(u,v)}(x)}.$$

Prove that this reduction has the following two properties:

(1) If the instance (G, Σ, σ) of LC has a labeling τ that satisfies all edges, then the instance $(S, \mathcal{F})$ of MIN-SC has a set cover of size $2n$.

(2) If every labeling for the instance (G, Σ, σ) of LC can satisfy at most $\alpha^k m$ edges, then every set cover for the instance $(B, \mathcal{F})$ of MIN-SC has size at least $\ell n/4$.

10.32 Show that the problem LC with the gap $[m/\log^3 m, m]$ is not in **P** unless **NP** $\subseteq$ **DTIME**$(n^{O(\log \log n)})$, where m is the number of edges in the input graph.

Historical Notes

Inapproximability results and the concept of approximation-preserving reductions have been studied since the 1970s (see, e.g., Garey and Johnson [1976], Sahni and Gonzalez [1976], Ko [1979], and Ausiello et al. [1980]). However, the development of the theory of inapproximability flourished only in the 1990s through the study of PCP systems, which was inspired by the study of interactive proof systems [Feige et al., 1991]. The notion of L-reductions was introduced by Papadimitriou and Yannakakis [1988]. They also introduced the class **MAXSNP** and showed many **MAXSNP**-complete problems. Khanna et al. [1999] generalized it to **APX**-completeness. **APX**-hardness of VC-CG (Theorem 10.15) and MAJ-DS (Theorem 10.16) are from Du, Gao, and Wu [1997] and Zhu et al. [2010], respectively.

The PCP theorem, with its application to the inapproximability of MAX-SAT was established in Arora et al. [1992, 1998] and Arora and Safra [1992, 1998], and received a lot of attention. Nowadays, due to the work of Khanna et al. [1999], the PCP theorem is no longer required to get the inapproximability of MAX-SAT or many other optimization problems. However, the PCP system remains an important tool to study inapproximability. Håstad's 3-Bit PCP theorem [Håstad, 2001] is an important version. Many constant lower bounds for performance ratios were established from this theorem, including MAX-3SAT, MIN-VC, METRIC TSP [Böckenhauser et al., 2000], and NSMT [Chlebik and Chlebikoca, 2002]. Another important result is the proof for the lower bound of the performance ratio of MIN-SC. Lund and Yannakakis [1993] obtained the first lower bound that MIN-SC does not have a polynomial-time $(\rho \ln n)$-approximation for any $0 < \rho < 1/4$ unless $\mathbf{NP} \subseteq \mathbf{DTIME}(n^{poly(\log n)})$. The current best bounds (Theorems 10.23 and 10.28) are given by Feige [1996] and Raz and Safra [1997], respectively. The $(\rho \ln n)$-inapproximability of MIN-CDS (Theorem 10.24), MIN-WCVC (Theorem 10.25), and CDS-SP (Theorem 10.27) are from Guha and Khuller [1998], Fujito [2001], and Ding et al. [2010], respectively. For the problem CLIQUE, Håstad [1999] established the lower bound $n^{1-\varepsilon}$ for its performance ratio, using a stronger complexity-theoretic assumption of $\mathbf{NP} \neq \mathbf{ZPP}$. Zuckerman [2006, 2007] derandomized his construction and weakened the assumption to $\mathbf{P} \neq \mathbf{NP}$. The best-known approximation algorithm for GCOLOR generates a coloring of size within a factor $O(n(\log n)^{-3}(\log\log n)^2)$ of the chromatic number [Halldórsson, 1993]. The $(n^{1-\varepsilon})$-inapproximability for GCOLOR was proved by Zuckerman [2006, 2007] under the assumption $\mathbf{P} \neq \mathbf{NP}$. The inapproximability of CHROMATIC SUM (Theorem 10.33) is due to Bar-Noy et al. [1998]. The problem LABEL COVER and its inapproximability (Theorem 10.34) are studied in Arora et al. [1993].

Exercise 10.5 is from Erlebach and van Leeuwen [2008]. The notion of E-reductions and its basic properties (Exercise 10.8) are due to Khanna et al. [1999]. The lower bound of $96/95$ for the performance ratio of NSMT (Exercise 10.13) is from Chlebik and Chlebikoca [2002]. The lower bound of $3813/3812$ for the performance ratio of METRIC-TSP (Exercise 10.14) is from Böckenhauer et al. [2000]. The inapproximability of domatic numbers (Exercise 10.21) is due to Feige et al. [2002]. Exercises 10.22, 10.23, and 10.24(a) are from Du and Hwang [2006], and Exercise 10.24(b) is from Wang et al. [2007]. The inapproximability of BUDGETED MAXIMUM COVERAGE (Exercise 10.25) is due to Khuller et al. [1999]. Exercise 10.26 is from Halldórsson [1993]. The problem CONNECTED SET COVER is studied in Zhang, Gao, and Wu [2009].

Bibliography

Agarwal, P.K., van Kreveld, M. and Suri, S. [1998], Label placement by maximum independent set in rectangles, *Comput. Geom. Theory Appl.* **11**, 209–218.

Ageev, A.A. and Sviridenko, M. [2004], Pipage rounding: A new method of constructing algorithms with proven performance guarantee, *J. Comb. Optim.* **8**, 307–328.

Agrawal, A., Klein, P. and Ravi, R. [1995], When trees collide: An approximation algorithm for the generalized Steiner problem on networks, *SIAM J. Comput.* **24**, 440–456.

Alizadeh, F. [1991], Combinatorial Optimization with Interior Point Methods and Semidefinite Matrices, *Ph.D. Thesis*, Computer Science Department, University of Minnesota, Minneapolis, Minnesota.

Alizadeh, F. [1995], Interior point methods in semidefinite programming with applications to combinatorial optimization, *SIAM J. Optim.* **5**, 13–51.

Alizadeh, F., Haeberly, J.-P. A. and Overton, M. [1994], Primal-dual interior point methods for semidefinite programming, Technical Report 659, Computer Science Department, Courant Institute of Mathematical Sciences, New York University, New York.

Alizadeh, F., Haeberly, J.-P. A. and Overton, M. [1997], Complementarity and nondegeneracy in semidefinite programming, *Math. Program.* **77**, 111–128.

Alon, N., Goldreich, O., Hastad, J. and Peralta, R. [1992], Simple constructions of almost k-wise independent random variables, *Random Struc. Algorithms* **3**, 289–304.

Alon, N., Sudakov, B. and Zwick, U. [2001], Constructing worst case instances for semidefinite programming based approximation algorithms, *SIAM J. Disc. Math.* **15**, 58–72.

Alzoubi, K.M., Wan, P. and Frieder, O. [2002], Message-optimal connected dominating sets in mobile ad hoc networks, *Proceedings, 3rd ACM International Symposium on Mobile ad hoc Networking and Computing*, pp. 157–164.

Ambühl, C. [2005], An optimal bound for the MST algorithm to compute energy efficient broadcast trees in wireless networkds, *Proceedings, 32nd International Colloquium on Automata, Languages and Programming*, Lecture Notes in Computer Science **3580**, Springer, pp. 1139–1150.

Ambühl, C., Erlebach, T., Mihalák, M. and Nunkesser, M. [2006], Constant-approximation for minimum-weight (connected) dominating sets in unit disk graphs, *Proceedings, 9th International Workshop on Approximation Algorithms for Combinatorial Optimization Problems* and *International Workshop on Randomization and Computation*, Lecture Notes in Computer Science **4110**, Springer, pp. 3–14.

An, L.T.H., Tao, P.D. and Muu, L.D. [1998], A combined d.c. optimization-ellipsoidal branch-and-bound algorithm for solving nonconvex quadratic programming problems, *J. Comb. Optim.* **2**, 9–28.

Anjos, M.F. and Wolkowicz, H. [2002], Strengthened semidefinite relaxations via a second lifting for the Max-Cut problem, *Disc. Appl. Math.* **119**, 79–106.

Arkin, E.M., Mitchell, J.S.B. and Narasimhan, G. [1998], Resource-constructed geometric network optimization, *Proceedings, 14th Symposium on Computational Geometry*, pp. 307–316.

Armen, C. and Stein, C. [1996], A $2\frac{2}{3}$-approximation algorithm for the shortest superstring problem, *Proceedings, 7th Symposium on Combinatorial Pattern Matching*, Lecture Notes on Computer Science **1075**, Springer, pp. 87–101.

Arora, S. [1996], Polynomial-time approximation schemes for Euclidean TSP and other geometric problems, *Proceedings, 37th IEEE Symposium on Foundations of Computer Science*, pp. 2–12.

Arora, S. [1997], Nearly linear time approximation schemes for Euclidean TSP and other geometric problems, I, *Proceedings, 38th IEEE Symposium on Foundations of Computer Science*, pp. 554–563.

Arora, S. [1998], Polynomial-time approximation schemes for Euclidean traveling salesman and other geometric problems, *J. Assoc. Comput. Mach.* **45**, 753–782.

Arora, S., Babai, L., Stern, J. and Sweedyk, Z. [1993], The hardness of approximate optima in lattices, codes, and systems of linear equations, *Proceedings, 34th IEEE Symposium on Foundations of Computer Science*, pp. 727–733.

Arora, S., Grigni, M., Karger, D., Klein, P. and Woloszyn, A. [1998], A polynomial time approximation scheme for weighted planar graph TSP, *Proceedings, 9th ACM-SIAM Symposium on Discrete Algorithms*, pp. 33–41.

Arora, S. and Kale, S. [2007], A combinatorial, primal-dual approach to semidefinite programs, *Proceedings, 39th ACM Symposium on Theory of Computing*, pp. 227–236.

Arora, S., Lund, C., Motwani, R., Sudan, M. and Szegedy, M. [1992], Proof verification and hardness of approximation problems, *Proceedings, 33rd IEEE Symposium on Foundations of Computer Science*, pp. 14–23.

Arora, S., Lund, C., Motwani, R., Sudan, M. and Szegedy, M. [1998], Proof verification and hardness of approximation problems, *J. Assoc. Comput. Mach.* **45**, 753–782.

Arora, S., Raghavan, P. and Rao, S. [1998], Polynomial time approximation schemes for Euclidean k-medians and related problems, *Proceedings, 30th ACM Symposium on Theory of Computing*, pp. 106–113.

Arora, S., Rao, S. and Vazirani, U. [2004], Expender flows, geometric embeddings, and graph partitionings, *Proceedings, 36th ACM Symposium on Theory of Computing*, pp. 222–231.

Arora, S. and Safra, S. [1992], Probabilistic checking of proofs: A new characterization of NP, *Proceedings, 33rd IEEE Symposium on Foundations of Computer Science*, pp. 2–13.

Arora, S. and Safra, S. [1998], Probabilistic checking of proofs: A new characterization of NP, *J. Assoc. Comput. Mach.* **45**, 70–122.

Ausiello, G., D'Atri, A. and Protasi, M. [1980], Structural preserving reductions among convex optimization problems, *J. Comput. Systems Sci.* **21**, 136–153.

Bafna, V., Berman, P. and Fujito, T. [1999], A 2-approximation algorithm for the undirected feedback vertex set problem, *SIAM J. Disc. Math.* **12**, 289–297.

Baker, B.S. [1983], Approximation algorithms for NP-complete problems on planar graphs, *Proceedings, 24th IEEE Symposium on Foundations of Computer Science*, pp. 265–273.

Baker, B.S. [1994], Approximation algorithms for NP-complete problems on planar graphs, *J. Assoc. Comput. Mach.* **41**, 153–180.

Bar-Ilan, J., Kortsarz, G. and Peleg, D. [2001], Generalized submodular cover problem and applications, *Theoret. Comput. Sci.* **250**, 179–200.

Bar-Noy, A., Bar-Yehuda, R., Freund, A., Naor, J. and Shieber, B. [2001], A unified approach to approximating resource allocation and scheduling, *J. Assoc. Comput. Mach.* **48**, 1069–1090.

Bar-Noy, A., Bellare, M., Halldórsson, M.M., Shachnai, H. and Tamir, T. [1998], On chromatic sums and distributed resource allocation, *Inform. Comput.* **140**, 183–202.

Bar-Yehuda, R., Bendel, K., Freund, A. and Rawitz, D. [2004], Local ratio: A unified framework for approximation algorithms. In memoriam: Shimon Even 1935–2004, *ACM Comput. Surv.* **36**, 422–463.

Bar-Yehuda R. and Even, S. [1981], A linear time approximation algorithm for the weighted vertex cover problem, *J. Algorithms* **2**, 198–203.

Bar-Yehuda, R. and Even, S. [1985], A local-ratio theorem for approximating the weighted vertex cover problem, *Annals Disc. Math.* **25**, 27–46.

Bar-Yehuda, R. and Rawitz, D. [2004], Local ratio with negative weights, *Oper. Res. Lett.* **32**, 540–546.

Bar-Yehuda, R. and Rawttz, D. [2005a], On the equivalence between the primal-dual schema and the local ratio technique, *SIAM J. Disc. Math.* **19**, 762–797.

Bar-Yehuda, R. and Rawitz, D. [2005b], Using fractional primal-dual to schedule split intervals with demands, *Proceedings, 13th European Symposium on Algorithms*, Lecture Notes in Computer Science **3669**, Springer, pp. 714–725.

Bellare, M., Goldreich, O. and Sudan, M. [1995], Free bits, PCPs and non-approximability—towards tight results, *Proceedings, 36th IEEE Symposium on Foundations of Computer Science*, pp. 422–431.

Berman, P., DasGupta, B., Muthukrishnan, S. and Ramaswami, S. [2001], Efficient approximation algorithms for tiling and packing problem with rectangles, *J. Algorithms* **41**, 443–470.

Bertsimas, D. and Teo, C.-P. [1998], From valid inequalities to heuristics: A unified view of primal-dual approximation algorithms in covering problems, *Oper. Res.* **46**, 503–514.

Bertsimas, D., Teo, C.-P. and Vohra, R. [1999], On dependent randomized rounding algorithms, *Oper. Res. Lett.* **24**, 105–114.

Bertsimas, D. and Ye, Y. [1998], Semidefinite relaxations, multivariate normal distributions, and order statistics, in *Handbook of Combinatorial Optimization*, Vol. 3, D.-Z. Du and P.M. Pardalos (eds.), Kluwer, pp. 1–19.

Bland, R.G. [1977], New finite pivoting rules of the simplex method, *Math. Oper. Res.* **2**, 103–107.

Blum, A., Jiang, T., Li, M., Tromp, J. and Yannakakis, M. [1991], Linear approximation of shortest superstrings, *Proceedings, 23rd ACM Symposium on Theory of Computing*, pp. 328–336.

Blum, A., Jiang, T., Li, M., Tromp, J. and Yannakakis, M. [1994], Linear approximation of shortest superstrings, *J. Assoc. Comput. Mach.* **41**, 630–647.

Böckenhauer, H.-J., Hromkovic, J., Klasing, R., Seibert, S. and Unger, W. [2000], An improved lower bound on the approximability algorithms of metric TSP with sharpened triangle inequality, *Proceedings, 17th Symposium on Theoretical Aspects of Computer Science*, Lecture Notes on Computer Science **1770**, Springer, pp. 382–394.

Borchers, A. and Du, D.-Z. [1995], The k-Steiner ratio in graphs, *Proceedings, 27th ACM Symposium on Theory of Computing*, pp. 641–649.

Butenko, S. and Ursulenko, O. [2007], On minimum connected dominating set problem in unit-ball graphs, preprint.

Byrka, J. [2007], An optimal bifactor approximation algorithm for the metric uncapacitated facility location problem, *Proceedings, 10th International Workshop on Approximation Algorithms for Combinatorial Optimization Problems*, pp. 29–43.

Cadei, M., Cheng, M.X., Cheng, X. and Du, D.-Z. [2002], Connected domination in multihop ad hoc wireless networks, *Proceedings, 6th Joint Conference on Information Science*, pp. 251–255.

Calinescu, G., Chekuri, C., Pál, M. and Vondrk, J. [2007], Maximizing a submodular set function subject to a matroid constraint, *Proceedings, 12th International Integer Programming and Combinatorial Optimization Conference*, Lecture Notes in Computer Science **4513**, Springer, pp. 182–196.

Chan, T.M. [2003], Polynomial-time approximation schemes for picking and piercing fat objects, *J. Algorithms* **46**, 178–189.

Chan, T.M. [2004], A note on maximum independent sets in rectangle intersection graphs, *Inform. Process. Lett.* **89**, 19–23.

Charikar, M., Chekuri, C., Cheung, T.-Y., Dai, Z., Goel, A., Guha, S. and Li, M. [1999], Approximation algorithms for directed Steiner problems, *J. Algorithms* **33**, 73–91.

Charnes, A. [1952], Optimality and degeneracy in linear programming, *Econometrica* **20**, 160–170.

Chen, J.-C. [2007], Iterative rounding for the closest string problem, *Computing Research Repository*, arXiv:0705.0561.

Chen, Y.P. and Liestman, A.L. [2002], Approximating minimum size weakly-connected dominating sets for clustering mobile ad hoc networks, *Proceedings, 3rd ACM International Symposium on Mobile ad hoc Networking and Computing*, pp. 165–172.

Cheng, X., DasGupta, B. and Lu, B. [2001], A polynomial time approximation scheme for the symmetric rectilinear Steiner arborescence problem, *J. Global Optim.* **21**, 385–396.

Cheng, X., Huang, X., Li, D., Wu, W. and Du, D.-Z. [2003], Polynomial-time approximation scheme for minimum connected dominating set in ad hoc wireless networks, *Networks* **42**, 202–208.

Cheng, X., Kim, J.-M. and Lu, B. [2001], A polynomial time approximation scheme for the problem of interconnecting highways, *J. Comb. Optim.* **5**, 327–343.

Cheriyan, J., Vempala, S. and Vetta, A. [2006], Network design via iterative rounding of setpair relaxations, *Combinatorica* **26**, 255–275.

Chlamtac, E. [2007], Approximation algorithms using hierarchies of semidefinite programming relaxations, *Proceedings, 48th IEEE Symposium on Foundations of Computer Science*, pp. 691–701.

Chlebik, M. and Chlebikoca, J. [2002], Approximation hardness of the Steiner tree problem on graphs, *Proceedings, 8th Scandinavia Workshop on Algorithm Theory*, Lecture Notes on Computer Science **2368**, Springer, pp. 170–179.

Christofides, N. [1976], Worst-case analysis of a new heuristic for the travelling salesman problem, Technical Report, Graduate School of Industrial Administration, Carnegie-Mellon University, Pittsburgh, PA.

Chudak, F.A., Goemans, M.X., Hochbaum, D.S. and Williamson, D.P. [1998], A primal-dual interpretation of two 2-approximation algorithms for the feedback vertex set problem in undirected graphs, *Oper. Res. Lett.* **22**, 111–118.

Chung, F.R.K. and Gilbert, E.N. [1976], Steiner trees for the regular simplex, *Bull. Inst. Math. Acad. Sinica* **4**, 313–325.

Chung, F.R.K. and Graham, R.L. [1985], A new bound for Euclidean Steiner minimum trees, *Ann. N.Y. Acad. Sci.* **440**, 328–346.

Chung, F.R.K. and Hwang, F.K. [1978], A lower bound for the Steiner tree problem, *SIAM J. Appl. Math.* **34**, 27–36.

Chvátal, V. [1979], A greedy heuristic for the set-covering problem, *Math. Oper. Res.* **4**, 233–235.

Cook, S.A. [1971], The complexity of theorem-proving procedures, *Proceedings, 3rd ACM Symposium on Theory of Computing*, pp. 151–158.

Cormem, T.H., Leiserson, C.E. and Rivest, R.L. [1990], *Introduction to Algorithms*, McGraw-Hill, New York.

Courant, R. and Robbins, H. [1941], *What Is Mathematics?*, Oxford University Press, New York.

Czumaj, A., Gasieniec, L., Piotrow, M. and Rytter, W. [1994], Parallel and sequential approximation of shortest superstrings, *Proceedings, 4th Scandinavian Workshop on Algorithm Theory*, pp. 95–106.

Dahlhaus, E., Johnson, D.S., Papadimitriou, C.H., Seymour, P.D. and Yannakakis, M. [1994], The complexity of multiterminal cuts, *SIAM J. Comput.* **23**, 864–894.

Dai, D. and Yu, C. [2009], A $5+\epsilon$-approximation algorithm for minimum weighted dominating set in unit disk graph, *Theoret. Comput. Sci.* **410**, 756–765.

Dantzig, G.B. [1951], Maximization of a linear function of variables subject to linear inequalities, in *Activity Analysis of Production and Allocation* (Cowles Commission Monograph 13), T.C. Koopmans (ed.), John Wiley, New York, pp. 339–347.

Dantzig, G.B. [1963], *Linear Programming and Extensions*, Princeton University Press, Princeton, NJ.

Dantzig, G.B., Ford, L.R. and Fulkerson, D.R. [1956], A primal-dual algorithm for linear programs, in *Linear Inequalities and Related Systems*, H.W. Kuhn and A.W. Tucker (eds.), Princeton University Press, Princeton, NJ, pp. 171–181.

Das, B. and Bharghavan, V. [1997], Routing in ad hoc networks using minimum connected dominating sets, *Proceedings, IEEE International Conference on Communications*, Vol. 1, pp. 376–380.

Deering, S., Estrin, D., Farinacci, D., Jacobson, V., Lui, C. and Wei, L. [1994], An architecture for wide area multicast routing, *Proceedings, ACM SIGCOMM 1994*, pp. 126–135.

Ding, L., Gao, X., Wu, W., Lee, W., Zhu, X. and Du, D.-Z [2010], Distributed construction of connected dominating sets with minimum routing cost in wireless networks, *Proceedings, 30th International Conference on Distributed Computing Systems*, pp. 448–457.

Drake, D.E. and Hougardy, S. [2004], On approximation algorithms for the terminal Steiner tree problem, *Inform. Process. Letters* **89**, 15–18.

Du, D.-Z. [1986], *On heuristics for minimum length rectangular partitions*, Technical Report, Mathematical Sciences Research Institute, University of California, Berkeley.

Du, D.-Z., Gao, B. and Wu, W. [1997], A special case for subset interconnection designs, *Disc. Applied Math.* **78**, 51–60.

Du, D.-Z., Graham, R.L, Pardalos, P.M., Wan, P.-J., Wu, W. and Zhao, W. [2008], Analysis of greedy approximation with nonsubmodular potential functions, *Proceedings, 19th ACM-SIAM Symposium on Discrete Algorithms*, pp. 167–175.

Du, D.-Z., Hsu, D.F. and Xu, K.-J. [1987], Bounds on guillotine ratio, *Congressus Numerantium* **58**, 313–318.

Du, D.-Z. and Hwang, F.K. [1990], The Steiner ratio conjecture of Gilbert–Pollak is true, *Proc. National Acad. Sci.* **87**, 9464–9466.

Du, D.-Z. and Hwang, F.K. [2006], *Pooling Designs and Nonadaptive Group Testing*, World Scientific, Singapore.

Du, D.Z., Hwang, F.K., Shing, M.T. and Witbold, T. [1988], Optimal routing trees, *IEEE Trans. Circuits* **35**, 1335–1337.

Du, D.-Z. and Ko, K.-I. [2000], *Theory of Computational Complexity*, Wiley Interscience, New York.

Du, D.-Z. and Ko, K.-I. [2001], *Problem Solving in Automata, Languages, and Complexity*, John Wiley & Sons, New York.

Du, D.-Z. and Miller, Z. [1988], Matroids and subset interconnection design, *SIAM J. Disc. Math.* **1**, 416–424.

Du, D.-Z., Pan, L.Q., and Shing, M.-T. [1986], *Minimum edge length guillotine rectangular partition*, Technical Report 02418-86, Mathematical Sciences Research Institute, University of California, Berkeley.

Du, D.-Z. and Zhang, Y. [1990], On heuristics for minimum length rectilinear partitions, *Algorithmica* **5**, 111–128.

Du, D.-Z., Zhang, Y. and Feng, Q. [1991], On better heuristic for Euclidean Steiner minimum trees, *Proceedings, 32nd IEEE Symposium on Foundations of Computer Science*, pp. 431–439.

Du, H., Jia, X., Wang, F., Thai, M. and Li, Y. [2005], A note on optical network with non-splitting nodes, *J. Comb. Optim.* **10**, 199–202.

Du, X., Wu, W. and Kelley, D.F. [1998], Approximations for subset interconnection designs, *Theoret. Comput. Sci.* **207**, 171–180.

Eriksson, H. [1994], MBONE: The multicast backbone, *Comm. Assoc. Comput. Mach.* **37**, 54–60.

Erlebach, T., Jansen, K. and Seidel, E. [2001], Polynomial-time approximation schemes for geometric graphs, *Proceedings, 12th ACM-SIAM Symposium on Discrete Algorithms*, pp. 671–679.

Erlebach, T. and van Leeuwen, E.J. [2008], Approximating geometric coverage problems, *Proceedings, 19th ACM-SIAM Symposium on Discrete Algorithms*, pp. 1267–1276.

Feige, U. [1996], A threshold of $\ln n$ for approximating set cover (preliminary version), *Proceedings, 28th ACM Symposium on Theory of Computing*, pp. 314–318.

Feige, U. [1998], A threshold of $\ln n$ for approximating set cover, *J. Assoc. Comput. Mach.* **45**, 634–652.

Feige, U. and Goemans, M.X. [1995], Approximating the value of two prover proof systems, with applications to MAX 2SAT and MAX DICUT, *Proceedings, 3rd Israel Symposium on Theory of Computing and Systems*, pp. 182–189.

Feige, U., Goldwasser, S., Lovsz, L., Safra, S. and Szegedy, M. [1991], Approximating clique is almost NP-complete, *Proceedings, 32nd IEEE Symposium on Foundations of Computer Science*, pp. 2–12.

Feige, U., Halldórsson, M., Kortsarz, G. and Srinivasan, A. [2002], Approximating the domatic number, *SIAM J. Comput.* **32**, 172–195.

Feige, U. and Langberg, M. [2001], Approximation algorithms for maximization problems arising in graph partition, *J. Algorithms* **41**, 174–211.

Feige, U. and Langberg, M. [2006], The RPR^2 rounding technique for semidefinite programs, *J. Algorithms* **60**, pp. 1–23.

Fleischer, L., Jain, K. and Williamson, D.P. [2001], An iterative rounding 2-approximation algorithm for the element connectivity problem, *Proceedings, 42nd IEEE Symposium on Foundations of Computer Science*, pp. 339–347.

Foulds, L.R. and Graham, R.L. [1982], The Steiner problem in phylogeny is NP-complete, *Adv. Appl. Math.* **3**, 43–49.

Freund, A. and D. Rawitz, D. [2003], Combinatorial interpretations of dual fitting and primal fitting, *Proceedings, 1st Workshop on Approximation and Online Algorithms*, Lecture Notes in Computer Science **2909**, Springer, pp. 137–150.

Frieze, A. and Jerrum, M. [1995], Improved approximation algorithms for MAX k-CUT and MAX BISECTION, *Proceedings, 4th International Integer Programming and Combinatorial Optimization Conference*, Lecture Notes in Computer Science **920**, Springer, pp. 1–13.

Fu, M., Luo, Z.-Q. and Ye, Y. [1998], Approximation algorithms for quadratic programming, *J. Comb. Optim.* **2**, 29–50.

Fujito, T. [1998], A unified approximation algorithm for node-deletion problems, *Disc. Appl. Math.* **86**, 213–231.

Fujito, T. [1999], On approximation of the submodular set cover problem, *Oper. Res. Lett.* **25**, 169–174.

Fujito, T. [2001], On approximability of the independent/connected edge dominating set problems, *Infom. Process. Lett.* **79**, 261–266.

Fujito, T. and Yabuta, T. [2004], Submodular integer cover and its application to production planning, *Proceedings, 2nd International Workshop on Approximation and Online Algorithms*, Lecture Notes in Computer Science **3351**, Springer, pp. 154–166.

Funke, S., Kesselman, A., Meyer, U. and Segal, M. [2006], A simple improved distributed algorithm for minimum CDS in unit disk graphs, *ACM Trans. Sensor Networks* **2**, 444–453.

Gabow, H.N. and Gallagher, S. [2008], Iterated rounding algorithms for the smallest k-edge connected spanning subgraph, *Proceedings, 19th ACM-SIAM Symposium on Discrete Algorithms*, pp. 550–559.

Gabow, H.N., Goemans, M.X., Tardos, E. and Williamson, D.P. [2009], Approximating the smallest k-edge connected spanning subgraph by LP-rounding, *Networks* **53**, pp. 345–357.

Galbiati, G. and Maffioli, F. [2007], Approximation algorithms for maximum cut with limited unbalance, *Theoret. Comput. Sci.* **385**, 78–87.

Gandhi, R., Khuller, S., Parthasarathy, S. and Srinivasan, A. [2006], Dependent rounding and its applications to approximation algorithms, *J. Assoc. Comput. Mach.* **53**, 324–360.

Gao, X., Huang, Y., Zhang, Z. and Wu, W. [2008], $(6 + \varepsilon)$-approximation for minimum weight dominating set in unit disk graphs, *Proceedings, 14th International Conference on Computing and Combinatorics*, pp. 551–557.

Garey, M.R., Graham, R.L. and Johnson, D.S. [1977], The complexity of computing Steiner minimal trees, *SIAM J. Appl. Math.* **32**, 835–859.

Garey, M.R. and Johnson, D.S. [1976], The complexity of near-optimal graph coloring, *J. Assoc. Comput. Mach.* **23**, 43–49.

Garey, M.R. and Johnson, D.S. [1977], The rectilinear Steiner tree is NP-complete, *SIAM J. Appl. Math.* **32**, 826–834.

Garey, M. R. and Johnson, D. S. [1979], *Computers and Intractability: A Guide to the Theory of NP-Completeness*, W. H. Freeman and Company, New York.

Garg, N., Konjevod, G. and Ravi, R. [2000], A polylogarithmic approximation algorithm for the group Steiner tree problem, *J. Algorithms* **37**, 66–84.

Ge, D., He, S., Ye, Y. and Zhang, S. [2010], Geometric rounding: A dependent rounding scheme, *J. Comb. Optim.* (to appear).

Ge, D., Ye, Y. and Zhang, J. [2010], Linear programming-based algorithms for the fixed-hub single allocation problem, preprint.

Gilbert, E.N. and Pollak, H.O. [1968], Steiner minimal trees, *SIAM J. Appl. Math.*, **16**, 1–29.

Goemans, M.X., Goldberg, A., Plotkin, S., Shmoys, D., Tardos, E. and Williamson, D.P. [1994], Approximation algorithms for network design problems, *Proceedings, 5th ACM-SIAM Symposium on Discrete Algorithms*, pp. 223–232.

Goemans, M.X. and Williamson, D.P. [1994], New $\frac{3}{4}$-approximation algorithms for the maximum satisfiability problem, *SIAM J. Disc. Math.* **7**, 656–666.

Goemans, M.X. and Williamson, D.P. [1995a], A general approximation technique for constrained forest problems, *SIAM J. Comput.* **24**, 296–317.

Goemans, M.X. and Williamson, D.P. [1995b], Improved approximation algorithms for maximum cut and satisfiability problems using semidefinite programming, *J. Assoc. Comput. Mach.* **42**, 1115–1145.

Goemans, M.X. and Williamson, D.P. [1997], The primal-dual method for approximation algorithms and its application to network design problems. in *Approximation Algorithms for NP-Hard Problems*, D. Hochbaum (ed.), PWS Publishing Company, Boston, MA, Chapter 4.

Goemans, M.X. and Williamson, D.P. [2004], Approximation algorithms for Max-3-Cut and other problems via complex semidefinite programming, *J. Comput. System Sci.* **68**, 442–470.

Gonzalez, T. and Zheng, S.Q. [1985], Bounds for partitioning rectilinear polygons, *Proceedings, 1st Symposium on Computational Geometry*, pp. 281–287.

Gonzalez, T. and Zheng, S.Q. [1989], Improved bounds for rectangular and guillotine partitions, *J. Symbolic Comput.* **7**, 591–610.

Graham, R.L. [1966], Bounds on multiprocessing timing anomalies, *Bell System Tech. J.* **45**, 1563–1581.

Graham, R.L. and Hwang, F.K. [1976], Remarks on Steiner minimal trees, *Bull. Inst. Math. Acad. Sinica* **4**, 177–182.

Guha, S. and Khuller, S. [1998a], Approximation algorithms for connected dominating sets, *Algorithmica* **20**, 374–387.

Guha, S. and Khuller, S. [1998b], Improved methods for approximating node weighted Steiner trees and connected dominating sets, Lecture Notes on Computer Science **1530**, Springer, pp. 54–66.

Guha, S. and Khuller, S. [1998c], Greedy strikes back: Improved facility location algorithms, *Proceedings, 9th ACM-SIAM Symposium on Discrete Algorithms*, pp. 228–248.

Guo, L., Wu, W., Wang, F. and Thai, M. [2005], Approximation for minimum multicast route in optical network with nonsplitting nodes, *J. Comb. Optim.* **10**, 391–394.

Gusfield, D. and Pitt, L. [1992], A bounded approximation for the minimum cost 2-SAT problem, *Algorithmica* **8**, 103–117.

Halldórsson, M.M. [1993a], A still better performance guarantee for approximate graph coloring, *Inform. Process. Lett.* **45**, 19–23.

Halldórsson, M.M. [1993b], Approximating the minimum maximal independence number, *Inform. Process. Lett.* **46**, 169–172.

Halldórsson, M.M. and Wattenhofer, R. [2009], Wireless communication is in APX, *Proceedings, 36th International Colloquium on Automata, Languages and Programming, Part I*, pp. 525–536.

Halperin, E. and Krauthgamer, R. [2003], Polylogarithmic inapproximability, *Proceedings, 35th ACM Symposium on Theory of Computing*, pp. 585–594.

Halperin, E., Livnat, D. and Zwick, U. [2002], MAX CUT in cubic graphs, *Proceedings, 13th ACM-SIAM Symposium on Discrete Algorithms*, pp. 506–513.

Halperin, E., Nathaniel, R. and Zwick, U. [2001], Coloring k-colorable graphs using smaller palettes, *Proceedings, 12th ACM-SIAM Symposium on Discrete Algorithms*, pp. 319–326.

Halperin, E. and Zwick, U. [2001a], Approximation algorithms for MAX 4-SAT and rounding procedures for semidefinite programs, *J. Algorithms* **40**, 184–211.

Halperin, E. and Zwick, U. [2001b], A unified framework for obtaining improved approximation algorithms for maximum graph bisection problems, *Proceedings, 8th International Integer Programming and Combinatorial Optimization Conference*, Lecture Notes in Computer Science **2081**, Springer, pp. 210–225.

Han, Q., Ye, Y., Zhang, H. and Zhang, J. [2002], On approximation of max-vertex-cover, *Eur. J. Operat. Res.* **143**, 342–355.

Han, Q., Ye, Y. and Zhang, J. [2002], An improved rounding method and semidefinite programming relaxation for graph partition, *Math. Program.* **92**, 509–535.

Håstad, J. [1999], Clique is hard to approximate within n to the power $1 - \epsilon$, *Acta Math.* **182**, 105–142.

Håstad, J. [2001], Some optimal inapproximability results, *J. Assoc. Comput. Mach.* **48**, 798–859.

Hausmann, D., Korte, B. and Jenkyns, T.A. [1980], Worst case analysis of greedy type algorithms for independence systems, *Math. Program. Study* **12**, 120–131.

Hochbaum, D.S. [1997a], Approximating covering and packing problems: Set cover, vertex cover, independent set, and related problems, in *Approximation Algorithms for NP-Hard Problems*, D.S. Hochbaum (ed.), PWS Publishing Company, Boston, pp. 94–143.

Hochbaum, D.S. [1997b], Various notions of approximations: good, better, best, and more, in *Approximation Algorithms for NP-hard Problems*, D.S. Hochbaum (ed.) PWS Publishing Company, Boston, pp. 346–398.

Hochbaum, D.S. and Maass, W. [1985], Approximation schemes for covering and packing problems in image processing and VLSI, *J. Assoc. Comput. Mach.* **32**, 130–136.

Hsieh, S.Y. and Yang, S.-C. [2007], Approximating the selected-internal Steiner tree, *Theoret. Comput. Sci.* **38**, 288–291.

Hunt III, H.B., Marathe, M.V., Radhakrishnan, V., Ravi, S.S., Rosenkrantz, D.J. and Stearns, R.E. [1998], Efficient approximations and approximation schemes for geometric problems, *J. Algorithms* **26**, 238–274.

Hwang, F.K. [1972], On Steiner minimal trees with rectilinear distance, *SIAM J. Appl. Math.* **30**, 104–114.

Ibarra, O.H. and Kim, C.E. [1975], Fast approximation algorithms for the knapsack and sum of subset proble, *J. Assoc. Comput. Mach.* **22**, 463–468.

Iyengar, G., Phillips, D.J. and Stein, C. [2009], Approximating semidefinite packing program, *Optimization Online*, http://www.optimization-online.org/DB_HTML/2009/06/2322.html.

Jain, K. [2001], A factor 2 approximation algorithm for the generalized Steiner network problem, *Combinatorica* **21**, 39–60.

Jain, K., Mahdian, M., Markakis, E., Saberi, A. and Vazirani, V. [2003], Greedy facility location algorithms analyzed using dual-fitting with factor-revealing LP, *J. Assoc. Comput. Mach.* **50**, 795–824.

Jain, K. and Vazirani, V. [2001], Approximation algorithms for metric facility location and k-median problems, using the primal-dual schema and Lagrangian relaxation, *J. Assoc. Comput. Mach.* **48**, 274–296.

Jenkyns, T.A. [1976], The efficacy of the "greedy" algorithm, *Congressus Numerantium* **17**, 341–350.

Jiang, T., Lawler, E.B. and Wang, L. [1994], Aligning sequences via an evolutionary tree: Complexity and algorithms, *Proceedings, 26th ACM Symposium on Theory of Computing*, pp. 760–769.

Jiang, T. and Wang, L. [1994], An approximation scheme for some Steiner tree problems in the plane, *Proceedings, 5th International Symposium on Algorithms and Computation*, Lecture Notes in Computer Science **834**, Springer, pp. 414–422.

Johnson, D.S. [1974], Approximation algorithms for combinatorial problems, *J. Comput. Systems Sci.* **9**, 256–278.

Johnson, N. and Kotz, S. [1972], *Distributed in Statistics: Continuous Multivariate Distribution*, John Wiley & Sons, New York.

Karger, D., Motwani, R. and Sudan, M. [1994], Approximate graph coloring by semidefinite programming, *Proceedings, 35th IEEE Symposium on Foundations of Computer Science*, pp. 1–10.

Karloff, H. and Zwick, U. [1997], A 7/8-approximation algorithm for MAX 3SAT?, *Proceedings, 38th IEEE Symposium on Foundations of Computer Science*, pp. 406–415.

Karmarkar, N. [1984], A new polynomial-time algorithm for linear programming, *Proceedings, 16th ACM Symposium on Theory of Computing*, pp. 302–311.

Karp, R.M. [1972], Reducibility among combinatorial problems, in *Complexity of Computer Computations*, E.E. Miller and J.W. Thatcher (eds.), Plenum Press, New York, pp. 85–103.

Karp, R.M. [1977], Probabilistic analysis of partitioning algorithms for the traveling salesman problem in the plane, *Math. Operat. Res.* **2**, 209–224.

Khachiyan, L.G. [1979], A polynomial algorithm for linear programming, *Doklad. Akad. Nauk., USSR Sec.* **244**, 1093–1096.

Khanna, S., Motwani, R., Sudan, M. and Vazirani, U. [1999], On syntactic versus computational views of approximability, *SIAM J. Comput.* **28**, 164–191.

Khanna, S., Muthukrishnan, S. and Paterson, M. [1998], On approximating rectangle tiling and packing, *Proceedings, 9th ACM-SIAM Symposium on Discrete Algorithms*, pp. 384–393.

Khot, S. [2001], Improved inapproximability results for MaxClique, chromatic number and approximate graph coloring, *Proceedings, 42nd IEEE Symposium on Foundations of Computer Science*, pp. 600–609.

Khot, S. [2002], On the power of unique 2-prover 1-round games, *Proceedings, 34th ACM Symposium on Theory of Computing*, pp. 767–775.

Khot, S., Kindler, G., Mossel, E. and O'Donnell, R. [2007], Optimal inapproximability results for MAX-CUT and other 2-variable CSPs?, *SIAM J. Comput.* **37**, 319–357.

Khuller, S., Moss, A. and Naor, J. [1999], The budgeted maximum coverage problem, *Inform. Process. Lett.* **70**, 39–45.

Kim, D., Wu, Y., Li, Y., Zou, F. and Du, D.-Z. [2009], Constructing minimum connected dominating sets with bounded diameters in wireless networks, *IEEE Trans. Parallel Distributed Systems* **20**, pp. 147–157.

Klee, V.L. and Minty, G.J. [1972], How good is the simplex algorithm?, in *Inequalities III*, O. Shisha (ed.), Academic Press, New York, pp. 159–175

Klein, P. and Lu, H.-I. [1998], Space-efficient approximation algorithms for MAXCUT and COLORING semidefinite programs, *Proceedings, 9th International Symposium on Algorithmd and Computation*, Lecture Notes in Computer Science **1533**, Springer, pp. 387–396.

Klein, P. and Ravi, R. [1995], A nearly best-possible approximation for node-weighted Steiner trees, *J. Algorithms* **19**, 104–115.

Klerk, E. de [2002], *Aspects of Semidefinite Programming: Interior Point Algorithms and Selected Applications (Applied Optimization)*, Kluwer Academic Publishers, Dordreht, The Netherlands.

Klerk, E. de, Roos, C. and Terlaky, T. [1998], Polynomial primal-dual affine scaling algorithms for semidefinite programming, *J. Comb. Optim.* **2**, 51–70.

Ko, K. [1979], *Computational Complexity of Real Functions and Polynomial Time Approximation*, Ph.D. Thesis, Ohio State University, Columbus, Ohio.

Komolos, J. and Shing, M.T. [1985], Probabilistic partitioning algorithms for the rectilinear Steiner tree problem, *Networks* **15**, 413–423.

Korte, B. and Hausmann, D. [1978], An analysis of the greedy heuristic for independence systems, *Ann. Disc. Math.* **2**, 65–74.

Korte, B. and Vygen, J. [2002], *Combinatorial Optimization: Theory and Algorithms*, 2nd Ed., Springer, Berlin.

Kosaraju, S.R., Park, J.K. and Stein, C. [1994], Long tour and shortest superstring, *Proceedings, 35th IEEE Symposium on Foundations of Computer Science*, pp. 166–177.

Kumar, A., Manokaran, R., Tulsiani, M. and Vishnoi, N. [2010], On the optimality of a class of LP-based algorithms, manuscript.

Lenstra, J.K., Shmoys, D.B. and Tardos, E. [1990], Approximation algorithms for scheduling unrelated parallel machines, *Math. Program.* **46**, 259–271.

Levcopoulos, C. [1986], Fast heuristics for minimum length rectangular partitions of polygons, *Proceedings, 2nd Symposium on Computational Geometry*, pp. 100–108.

Lewin, M., Livnat, D. and Zwick, U. [2002], Improved rounding techniques for the MAX 2-SAT and MAX DI-CUT problems, *Proceedings, 9th International Conference on Integer Programming and Combinatorial Optimization*, pp. 67–82.

Li, D., Du, H., Wan, P., Gao, X., Zhang, Z. and Wu, W. [2008], Minimum power strongly connected dominating sets in wireless networks, *Proceedings, 2008 International Conference on Wireless Networks*, pp. 447–451.

Li, D., Du, H., Wan, P., Gao, X., Zhang, Z. and Wu, W. [2009], Construction of strongly connected dominating sets in asymmetric multihop wireless networks, *Theoret. Comput. Sci.* **410**, 661–669.

Li, X., Gao, X. and Wu, W. [2008], A better theoretical bound to approximate connected dominating set in unit disk graph, *Proceedings, 3rd International Conference on Wireless Algorithms, Systems and Applications*, Lecture Notes in Computer Science **5258**, Springer, pp. 162–175.

Li, Y., Thai, M.T., Wang, F., Yi, C.W., Wan, P.-J. and Du, D.-Z. [2005], Greedy construction of connected dominating sets in wireless networks, *Europe J. Wireless Comm. Mobile Comput.* **5**, 927–932.

Lin, G.H. and Xue, G. [1999], Steiner tree problem with minimum number of Steiner points and bounded edge-length, *Inform. Process. Lett.* **69**, 53–57.

Lin, G.H. and Xue, G. [2002], On the terminal Steiner tree problem, *Inform. Process. Lett.* **84**, 103–107.

Ling, A., Tang, L. and Xu, C. [2010], Approximation algorithms for MAX RES CUT with limited unbalanced constraints, *J. Appl. Math. Comput.* **33**, 357–374.

Lingas, A. [1983], Heuristics for minimum edge length rectangular partitions of rectilinear figures, *Proceedings, 6th GI-Conference*, pp. 199–210.

Lingas, A., Pinter, R.Y., Rivest, R.L. and Shamir, A. [1982], Minimum edge length partitioning of rectilinear polygons, *Proceedings, 20th Allerton Conference on Communication, Control and Computing*, pp. 53–63.

Lovász, L. [1975], On the ratio of optimal integral and fractional covers, *Disc. Math.* **13**, 383–390.

Lovász, L. [1979], On the Shannon capacity of a graph, *IEEE Trans. Inform. Theory* **IT-25**, 1–7.

Lu, B. and Ruan, L. [2000], Polynomial time approximation scheme for the rectilinear Steiner arborescence problem, *J. Comb. Optim.* **4**, 357–363.

Lund, C., and Yanakakis, M. [1994], On the hardness of approximating minimization problems, *J. Assoc. Comput. Mach.* **41**, 960–981.

Mahajan, S. and Ramesh, H. [1999], Derandomizing approximation algorithms based on semidefinite programming, *SIAM J. Comput.* **28**, 1641–1663.

Mahdian, M., Ye, Y. and Zhang, J. [2002], Improved approximation algorithms for metric facility location problems, *Proceedings, 5th International Workshop on Approximation Algorithms for Combinatorial Optimization Problems*, pp. 229–242.

Mandoiu, I. and Zelikovsky, A. [2000], A note on the MST heuristic for bounded edge-length Steiner trees with minimum number of Steiner points, *Inform. Process. Lett.* **75**, 165–167.

Manki, M., Du, H., Jia, X., Huang, C.X., Huang, C.-H. and Wu, W. [2006], Improving construction for connected dominating set with Steiner tree in wireless sensor networks, *J. Global Optim.* **35**, 111–119.

Melkonian, V. and Tardos, É. [2004], Algorithms for a network design problem with crossing supermodular demands, *Networks* **43**, 256–265.

Min, M., Huang, S.C.-H., Liu, J., Shragowitz, E., Wu, W., Zhao, Y. and Zhao, Y. [2003], An approximation scheme for the rectilinear Steiner minimum tree in presence of obstructions, in *Novel Approaches to Hard Discrete Optimization*, Fields Institute Communications Series, American Mathematical Society, **37**, pp. 155–163.

Min, M., Du, H., Jia, X., Huang, C.X., Huang, S. C-H. and Wu, W. [2006], Improving construction for connected dominating set with Steiner tree in wireless sensor networks, *J. Global Optim.* **35**, 111–119.

Mitchell, J.S.B. [1996a], Guillotine subdivisions approximate polygonal subdivisions: A simple new method for the geometric k-MST problem, *Proceedings, 7th ACM-SIAM Symposium on Discrete Algorithms*, pp. 402–408.

Mitchell, J.S.B. [1996b], Guillotine subdivisions approximate polygonal subdivisions: A simple polynomial-time approximation scheme for geometric k-MST, TSP, and related problem, manuscript.

Mitchell, J.S.B. [1997], Guillotine subdivisions approximate polygonal subdivisions: Part III — Faster polynomial-time approximation scheme for geometric network optimization, *Proceedings, 9th Canadian Conference on Computational Geometry*, pp. 229–232.

Mitchell, J.S.B. [1999], Guillotine subdivisions approximate polygonal subdivisions: Part II — A simple polynomial-time approximation scheme for geometric k-MST, TSP, and related problem, *SIAM J. Comput.* **29**, 515–544.

Mitchell, J.S.B., Blum, A., Chalasani, P. and Vempala, S. [1999], A constant-factor approximation algorithm for the geometric k-MST problem in the plane, *SIAM J. Comput.* **28**, 771–781.

Navarra, A. [2005], Tight bounds for the minimum energy broadcasting problem, *Proceedings, 3rd International Symposium on Modeling and Optimization in Mobile, ad hoc, and Wireless Networks*, pp. 313–322.

Nemhauser, G.L. and Wolsey, L.A. [1999], *Integer and Combinatorial Optimization*, John Wiley & Sons, New York.

Nesterov, Y.E. [1998], Semidefinite relaxation and nonconvex quadratic optimization, *Optim. Method. Software* **9**, 141–160.

Nielsen, F. [2000], Fast stabbing of boxes in high dimensions, *Theoret. Comput. Sci.* **246**, 53–72.

Papadimitriou, C. and Yannakakis, M. [1988], Optimization, approximations, and complexity classes, *Proceedings, 20th ACM Symposium on Theory of Computing*, pp. 229–234.

Pardalos, P.M. and Ramana, M. [1997], Semidefinite programming, in *Interior Point Methods of Mathematical Programming*, Kluwer, Docdreht, The Netherlands, pp. 369–398.

Pardalos, P.M. and Wolkowicz, H. [1998], *Topics in Semidefinite and Interior-Point Methods*, American Mathematical Society, Providence, RI.

Prisner, E. [1992], Two algorithms for the subset interconnection design problem, *Networks* **22**, 385–395.

Ramamurthy, B., Iness, J. and Mukherjee, B. [1997], Minimizing the number of optical amplifiers needed to support a multi-wavelength optical LAN/MAN, *Proceedings, 16th IEEE Conference on Computer Communications*, pp. 261–268.

Rao, S.B. and Smith, W.D. [1998], Approximating geometrical graphs via "spanners" and "banyan," *Proceedings, 30th ACM Symposium on Theory of Computing*, pp. 540–550.

Ravi, R. and Kececioglu, J.D. [1995], Approximation methods for sequence alignment under a fixed evolutionary tree, *Proceedings, 6th Symposium on Combinatorial Pattern Matching*, Lecture Notes on Computer Science **937**, Springer, pp. 330–339.

Ravi, R. and Klein, P. [1993], When cycles collapse: A general approximation technique for constrained two-connectivity problems, *Proceedings, 3rd MPS Conference on Integer Programming and Combinatorial Optimization*, pp. 39–56.

Raz, R. and Safra, S. [1997], A sub-constant error-probability low-degree test, and a sub-constant error-probability PCP characterization of NP, *Proceedings, 28th ACM Symposium on Theory of Computing*, pp. 474–484.

Robin, G. and Zelikovsky, A. [2000], Improved Steiner trees approximation in graphs, *Proceedings, 11th ACM-SIAM Symposium on Discrete Algorithms*, pp. 770–779.

Ruan, L., Du, H., Jia, X., Wu., W., Li, Y. and Ko, K. [2004], A greedy approximation for minimum connected dominating sets, *Theoret. Comput. Sci.* **329**, 325–330.

Rubinstein, J.H. and Thomas, D.A. [1991], The Steiner ratio conjecture for six points, *J. Combinatorial Theory, Ser. A*, **58**, 54–77.

Sahni, S. [1975], Approximate algorithms for the 0/1 knapsack problem, *J. Assoc. Comput. Mach.* **22**, 115–124.

Sahni, S. and Gonzalez, T. [1976], P-complete approximation algorithms, *J. Assoc. Comput. Mach.* **23**, 555–565.

Salhieh, A., Weinmann, J., Kochha, M. and Schwiebert, L. [2001], Power efficient topologies for wireless sensor networks, *Proceedings, 30th International Workshop on Parallel Processing*, pp. 156–163.

Sankoff, D. [1975], Minimal mutation trees of sequences, *SIAM J. Appl. Math.* **28**, 35–42.

Schreiber, P. [1986], On the history of the so-called Steiner Weber problem, *Wiss. Z. Ernst-Moritz-Arndt-Univ. Greifswald, Math.-nat.wiss. Reihe* **35**.

Sivakumar, R., Das, B. and Bharghavan, V. [1998], An improved spine-based infrastructure for routing in ad hoc networks, *Proceedings, 3rd IEEE Symposium on Computers and Communications*.

Skutella, M. [2001], Convex quadratic and semidefinite programming relaxations in scheduling, *J. Assoc. Comput. Mach.* **48**, 206–242.

Slavik, P. [1997], A tight analysis of the greedy algorithm for set cover, *J. Algorithms* **25**, 237–254.

Stojmenovic, I., Seddigh, M. and Zunic, J. [2002], Dominating sets and neighbor elimination based broadcasting algorithms in wireless networks, *IEEE Trans. Parallel Distr. Systems* **13**, 14–25.

Tarhio, J. and Ukkonen, E. [1988], A greedy approximation algorithm for constructing shortest common superstrings, *Theoret. Comput. Sci.* **57**, 131–145.

Teng, S.-H. and Yao, F.F. [1997], Approximating shortest superstrings, *SIAM J. Comput.* **26**, 410–417.

Thai, M.T., Wang, F., Liu, D., Zhu, S. and Du, D.-Z. [2007], Connected dominating sets in wireless networks with different transmission ranges, *IEEE Trans. Mobile Comput.* **6**, 1–9.

Turner, J.S. [1989], Approximation algorithms for the shortest common superstring problem, *Inform. Comput.* **83**, 1–20.

Vavasis, S.A. [1991], Automatic domain partitioning in tree dimesions, *SIAM J. Sci. Stat. Comput.* **12**, 950–970.

Wan, P.-J., Alzoubi, K.M. and Frieder, O. [2002], Distributed construction of connected dominating set in wireless ad hoc networks, *Proceedings, 21st Joint Conference of IEEE Computer and Communications Societies*.

Wan, P.-J., Wang, L. and Yao, F.F. [2008], Two phased approximation algorithms for minimum CDS in wireless ad hoc networks, *Proceedings, 28th IEEE International Conference on Distributed Computing Systems*, pp. 337–344.

Wang, F., Du, H., Jia, X., Deng, P., Wu, W. and MacCallum, D. [2007], Non-unique probe selection and group testing, *Theoret. Comput. Sci.* **381**, 29–32.

Wang, L. and Du, D.-Z. [2002], Approximations for bottleneck Steiner trees, *Algorithmica* **32**, 554–561.

Wang, L. and Gusfield, D. [1996], Improved approximation algorithms for tree alignment, *Proceedings, 7th Symposium on Combinatorial Pattern Matching*, Lecture Notes on Computer Science **1075**, Springer, pp. 220–233.

Wang, L. and Jiang, T. [1996], An approximation scheme for some Steiner tree problems in the plane, *Networks* **28**, 187–193.

Wang, L., Jiang, T. and Gusfield, D. [1997], A more efficient approximation scheme for tree alignment, *Proceedings, 1st International Conference on Computational Biology*, pp. 310–319.

Wang, L., Jiang, T., and Lawler, E.L. [1996], Approximation algorithms for tree alignment with a given phylogeny, *Algorithmica* **16**, 302–315.

Wang, W., Zhang, Z., Zhang, W. and Du, D.-Z. [2009], An approximation algorithm for the t-latency bounded information propagation problem in social networks, preprint.

Wesolowsky, G. [1993], The Weber problem: History and perspective. *Location Science* **1**, 5–23.

Williamson, D.P. [2002], The primal dual method for approximation algorithms, *Math. Program.* **91**, 447–478.

Williamson, D.P., Goemans, M.X., Mihail, M. and Vazirani, V.V. [1995], A primal-dual approximation algorithm for generalized Steiner network problems. *Combinatorica* **15**, 435–454.

Willson, J., Gao, X., Qu, Z., Zhu, Y., Li, Y. and Wu, W. [2009], Efficient distributed algorithms for topology control problem with shortest path constraints, *Disc. Math., Algorithms and Applications* **1**, 437–461.

Wolsey, L.A. [1980], Heuristic analysis, linear programming and branch and bound, *Math. Program. Study* **13**, 121–134.

Wolsey, L.A. [1982a], An analysis of the greedy algorithm for submodular set covering problem, *Combinatorica* **2**, 385–393.

Wolsey, L.A. [1982b], Maximizing real-valued submodular function: Primal and dual heuristics for location problems, *Math. Operat. Res.* **7**, 410–425.

Wu, J. and Li, H.L. [1999], On calculating connected dominating set for efficient routing in ad hoc wireless networks, *Proceedings, 3rd ACM International Workshop on Discrete Algorithms and Methods for Mobile Computing and Communications*, pp. 7–14.

Wu, W., Du, H., Jia, X., Li, Y. and Huang, C.H. [2006], Minimum connected dominating sets and maximal independent sets in unit disk graphs, *Theoret. Comput. Sci.* **352**, 1–7.

Yan, S., Deogun, J.S. and Ali, M. [2003], Routing in sparse splitting optical networks with multicast traffic, *Comput. Networks* **41**, 89–113.

Yang, H., Ye, Y. and Zhang, J. [2003], An approximation algorithm for scheduling two parallel machines with capacity constraints, *Disc. Appl. Math.* **130**, 449–467.

Yannakakis, M. [1994], On the approximation of maximum satisfiability, *J. Algorithms* **3**, 475–502.

Ye, Y. [2001], A .699-approximation algorithm for Max-Bisection, *Math. Program.* **90**, 101–111.

Zelikovsky, A. [1993], The 11/6-approximation algorithm for the Steiner problem on networks, *Algorithmica* **9**, 463–470.

Zelikovsky, A. [1997], A series of approximation algorithms for the acyclic directed Steiner tree problem, *Algorithmica* **18**, 99–110.

Zhang, J., Ye, Y. and Han, Q. [2004], Improved approximations for max set splitting and max NAE SAT, *Disc. Appl. Math.* **142**, 133–149.

Zhang, Z., Gao, X. and Wu, W. [2009], Algorithms for connected set cover problem and fault-tolerant connected set cover problem, *Theoret. Comput. Sci.* **410**, 812–817.

Zhang, Z., Gao, X., Wu, W. and Du, D. [2009], A PTAS for minimum connected dominating set in 3-dimensional wireless sensor networks, *J. Global Optimiz.* **45**, 451–458.

Zhao, Q., Karisch, S.E., Rendl, F. and Wolkowicz, H. [1998], Semidefinite programming relaxations for the quadratic assignment problem, *J. Comb. Optim.* **2**, 71–109.

Zhu, X., Yu, J., Lee, W., Kim, D., Shan, S. and Du, D.-Z. [2010], New dominating sets in social networks, *J. Global Optim.* (published online in 2010).

Zong, C. [1999], *Sphere Packing*, Springer-Verlag, New York.

Zou, F., Li, X., Kim, D. and Wu, W. [2008a], Two constant approximation algorithms for node-weighted Steiner tree in unit disk graphs, *Proceedings, 2nd International Conference on Combinatorial Optimizationa and Applications*, pp. 21–24.

Zou, F., Li, X., Kim, D. and Wu, W. [2008b], Construction of minimum connected dominating set in 3-dimensional wireless network, *Proceedings, 3rd International Conference on Wireless Algorithms, Systems, and Applications,*, Lecture Notes in Computer Science **5258**, Springer, pp. 134–140.

Zou, F., Wang, Y., Xu, X., Li, X., Du, H., Wan, P.-J. and Wu, W. [2011], New approximations for minimum-weighted dominating sets and minimum-weighted connected dominating sets on unit disk graphs, *Theoret. Comput. Sci.* **412**, 198–208.

Zuckerman, D. [2006], Linear degree extractors and the inapproximability of max clique and chromatic number, *Proceedings, 38th ACM Symposium on Theory of Computing*, pp. 681–690.

Zuckerman, D. [2007], Linear degree extractors and the inapproximability of Max Clique and Chromatic Number, *Theory Comput.* **3**, 103–128

Zwick, U. [1998], Approximation algorithms for constraint satisfaction problems involving at most three variables per constraint, *Proceedings, 9th ACM-SIAM Symposium on Discrete Algorithms*, pp. 201–210.

Zwick, U. [1999], Outward rotations: A tool for rounding solutions of semidefinite programming relaxations, with applications to Max Cut and other problems, *Proceedings, 10th ACM-SIAM Symposium on Discrete Algorithms*, pp. 679–687.

Zwick, U. [2000], Analyzing the MAX 2-SAT and MAX DI-CUT approximation algorithms of Feige and Goemans, manuscript.

Zwick, U. [2002], Computer assisted proof of optimal approximability results, *Proceedings, 13th ACM-SIAM Symposium on Discrete Algorithms*, pp. 496–505.

Index